PHYSICS OF SOLIDS AT HIGH PRESSURES

PHYSICS OF SOLIDS AT HIGH PRESSURES

EDITED BY

C. T. Tomizuka

UNIVERSITY OF ARIZONA
TUCSON, ARIZONA

R. M. Emrick

UNIVERSITY OF ARIZONA
TUCSON, ARIZONA

Proceedings of the First International Conference on the Physics of Solids at High Pressures held in Tucson, Arizona April 20-23, 1965. Jointly sponsored by the University of Arizona and by the Air Force Office of Scientific Research, Office of Aerospace Research, United States Air Force, under AFOSR Grant Number 740-65.

1965

Academic Press New York and London

ACADEMIC PRESS INC.
111 FIFTH AVENUE
NEW YORK, NEW YORK 10003

United Kingdom Edition
Published by
ACADEMIC PRESS INC. (LONDON) LTD.
BERKELEY SQUARE HOUSE, LONDON W. 1

Library of Congress Catalog Card Number: 65-27552

PRINTED IN THE UNITED STATES OF AMERICA

PREFACE

In recent years the use of high pressure as a parameter for the investigation of solid state phenomena has become widespread. The diversity of the techniques employed and the effects studied are manifested in a very extensive literature. Despite such a broad interest, many theoreticians and even many experimentalists were unaware of the great insights to be gained from high pressure studies. It therefore seemed appropriate for an international conference to bring together workers active in this field with theoreticians, active or not. Their purpose would be to define the problems to be solved and to outline the most fruitful course of approach to these problems.

In order to achieve this purpose, the USAFOSR elected to sponsor the First International Conference on the Physics of Solids at High Pressures. Host institution and co-sponsor of the Conference was the University of Arizona, Tucson, Arizona. The use of the term "First International" was chosen in recognition of the fact that for the first time there was a clearly definable need to give direction in a particular area of research. The development of excellent high-pressure facilities in many countries required that the conference be international. Indeed, nine countries were represented. The use of the term First International Conference in no way implies plans for a Second. The need for such a conference has not yet arisen and may never arise.

The spontaneity and enthusiasm of the discussion, both formal and informal, indicated that the goal of the Conference had been achieved. Paramount in contributing to the success was the excellence of the papers which were delivered. The distinguished audience responded by keeping the discussion lively and to the point. Also contributing to the informal discussion was the elegant but relaxed setting of the Arizona Inn, where the Conference was held.

This Conference Proceedings contains the formal papers and formally presented discussion. No attempt was made to record the spontaneous discussion after each paper so as not to hinder the free exchange of ideas. It is felt that the sense of the informal discussions will soon be appearing in the literature.

Six formal sessions were held. The general topics of each session and the chairmen are outlined in the Table of Contents. In addition to the Welcome by William J. Price. Executive Director USAFOSR, which is reproduced here, the conferees were welcomed by Richard A. Harvill, President of the University of Arizona. Governor of Arizona, Sam Goddard, sent a telegram of welcome which was read to the participants. Since this Proceedings serves also as the Conference Report, the papers are presented in the order in which they were given. This also preserves the logical order of their presentation. Professor Ziman was unfortunately unable to attend the Conference. Copies of his manuscript were distributed to the participants rather than having it read by the Session Chairman. Professor Vereschagin's travel schedule prevented him from submitting a formal manuscript. In order that his contribution be recorded, we have taken the liberty of editing the tape recordings which were made. We accept full responsibility for any errors which we may have introduced in our paraphrasing of Professor Vereschagin's talk.

We have used the term Prepared Discussion to indicate contributed remarks which were made formally after the appropriate invited paper. The term Discussion indicates spontaneous remarks which were made during the discussion period and later written-up and submitted by their originator.

The Conference Committee wishes to thank the authors of the papers and all those who contributed to the discussions. Thanks are due also those graduate students who took care of the many details which are necessary to keep a program running smoothly. Mrs. Anne Olson and the entire staff of the Arizona Inn did an excellent job of providing superb accommodations and personal service to such a large group. The Conference secretaries, Mrs. Dorothy Conniff and Miss Betty King provided invaluable assistance. For the seemingly endless task of correcting and retyping the conference manuscripts, special thanks are due Mrs. Dorothy Conniff.

The Conference Committee wishes to express their appreciation for the sponsorship by the Office of Scientific Research, United Air Force and by the University of Arizona.

July 1965

Roy M. Emrick
Carl T. Tomizuka

LIST OF CONTRIBUTORS

First International Conference on Physics of Solids at High Pressures
April 20 - 22, 1965, Tucson, Arizona

CONFERENCE COMMITTEE

C. T. Tomizuka, Chairman
University of Arizona

R. M. Emrick
University of Arizona

H. G. Drickamer
University of Illinois

A. W. Lawson
University of California
Riverside

David Lazarus
University of Illinois

William Paul
Harvard University

Max Swerdlow
AF Office of
Scientific Research

LIST OF CONFEREES

H. D. Adams	U. S. Naval Weapons Laboratory, Dahlgren, Va.
Y. Adda	Atomic Energy Commission, Saclay, France
P. N. Adler	General Telephone & Electronics Laboratory, Bayside, Long Island, N. Y.
E. D. Albrecht	Lawrence Radiation Laboratory, Livermore, Calif.
P. B. Alers	U. S. Naval Research Labs., Washington, D. C.
O. L. Anderson	Bell Telephone Laboratories, Murray Hill, N. J.
T. N. Anderson	University of Utah, Salt Lake City, Utah
J. F. Andrew	Los Alamos Scientific Laboratory, Los Alamos, N. M.
R. Aumont	High Pressure Laboratory, C. N. R. S., Bellevue, France
J. D. Barnett	U. S. Army Electronics, Ft. Monmouth, N. J.
Jane K. Bassett	University of Rochester, Rochester, N. Y.
W. A. Bassett	University of Rochester, Rochester, N. Y.
G. Beardsley	University of North Carolina, Chapel Hill, N. C.
R. E. Behringer	Office of Naval Research, Pasadena, Calif.
M. Beyeler	Atomic Energy Commission, Saclay, France
A. Bienenstock	Harvard University, Cambridge, Mass.

LIST OF CONFEREES

J. Blackledge	University of Denver, Denver, Colorado
A. P. Blandin	University of Paris, Orsay, France
H. Blum	Brookhaven National Lab., Upton, Long Island, N. Y.
R. J. Borg	Lawrence Radiation Lab., Livermore, Calif.
H. Brooks	Harvard University, Cambridge, Mass.
R. G. Brophy	Michelson Lab., U. S. Naval Ordnance Test Station, China Lake, Calif.
E. Brown	Rensselaer Polytechnic Institute, Troy, N. Y.
B. Buescher	University of Arizona, Tucson, Arizona
B. S. Chandrasekhar	Western Reserve University, Cleveland, Ohio
J. C. Clegg	Brigham Young University, Provo, Utah
R. V. Coleman	University of Virginia, Charlottesville, Va.
E. M. Compy	U. S. Naval Research Lab., Washington, D. C.
C. Coston	University of Illinois, Urbana, Ill.
L. T. Crane	National Science Foundation, Washington, D. C.
W. B. Daniels	Princeton University, Princeton, N. J.
B. C. Deaton	General Dynamics, Ft. Worth, Texas
P. S. De Carli	Stanford Research Institute, Menlo Park, Calif.
K. L. De Vries	University of Utah, Salt Lake City, Utah
R. H. Dickerson	University of Illinois, Urbana, Ill.
S. F. Dickinson	Air Force Cambridge Research Laboratory, Bedford, Mass.
D. G. Doran	Stanford Research Institute, Menlo Park, Calif.
H. Drickamer	University of Illinois, Urbana, Ill.
H. C. Duecker	National Bureau of Standards, Washington, D. C.
J. S. Dugdale	National Research Council, Ottawa, Canada
G. E. Duvall	Washington State University, Pullman, Wash.
J. W. Edington	Battelle Memorial Institute, Columbus, Ohio
H. Ehrenreich	Harvard University, Cambridge, Mass.
N. G. Einspruch	Texas Instruments, Inc., Dallas, Texas
R. M. Emrick	University of Arizona, Tucson, Arizona
J. C. Erdmann	Boeing Scientific Research Labs., Seattle, Wash.
C. Eubig	University of Arizona, Tucson, Arizona
R. N. Euwema	Wright-Patterson AFB, Ohio

LIST OF CONFEREES

G. Fair	NASA, Cleveland, Ohio
L. M. Falicov	University of Chicago, Chicago, Ill.
E. Fawcett	Bell Telephone Laboratories, Murray Hill, N. J.
J. Feinleib	Lincoln Laboratory, M. I. T., Lexington, Mass.
D. Finnemore	Iowa State University, Ames, Iowa
D. B. Fitchen	Cornell University, Ithaca, N. Y.
D. Francois	Atomic Energy Commission, Paris, France
P. Freud	Battelle Memorial Institute, Columbus, Ohio
J. N. Fritz	Los Alamos Scientific Lab., Los Alamos, N. M.
H. Fritzsche	University of Chicago, Chicago, Ill.
H. Fujiwara	Hiroshima University, Hiroshima City, Japan
H. C. Gatos	Lincoln Lab., Massachusetts Inst. of Tech., Lexington, Mass.
T. H. Geballe	Bell Telephone Labs., Murray Hill, N. J.
S. Geller	North American Aviation, Thousand Oaks, Calif.
A. A. Giardini	University of Georgia, Athens, Georgia
A. Gilbert	Battelle Memorial Institute, Columbus, Ohio
L. A. Girifalco	University of Pennsylvania, Philadelphia, Pa.
E. Gregory	Air Reduction Company, Inc., Murray Hill, N. J.
H. H. Grimes	NASA, Cleveland, Ohio
R. E. Hanneman	G. E. Research Lab., Schenectady, N. Y.
W. A. Harrison	G. E. Research Lab., Schenectady, N. Y.
C. Helmick	University of Arizona, Tucson, Arizona
F. Herman	Lockheed Research Labs., Palo Alto, Calif.
Y. Hiki	University of Illinois, Urbana, Ill.
T. Hirone	Tohoku University, Sendai, Japan
C. G. Homan	Watervliet Arsenal, Watervliet, N. Y.
R. P. Huebener	Argonne National Laboratory, Argonne, Ill.
H. B. Huntington	Rensselaer Polytechnic Institute, Troy, N. Y.
R. Jaggi	I. B. M. Research Lab., Zurich, Switzerland
J. C. Jamieson	University of Chicago, Chicago, Ill.
A. Jayaraman	Bell Telephone Laboratories, Murray Hill, N. J.
C. Johansen	Brigham Young University, Provo, Utah

Q. Johnson	Lawrence Radiation Lab., Livermore, Calif.
J. A. Kafalas	Lincoln Laboratory, M. I. T., Lexington, Mass.
H. Kedesdy	U. S. Army Electronics, Fort Monmouth, N. J.
C. J. Kevane	Arizona State University, Tempe, Arizona
R. W. Keyes	I. B. M. Research, Yorktown Heights, N. Y.
A. F. Kip	University of California, Berkeley, Calif.
W. Klement	University of California, Berkeley, Calif.
W. Kohn	University of California, La Jolla, Calif.
R. Kossowsky	Curtiss Wright Corp., Wood-Ridge, N. J.
J. A. Krumhansl	Cornell University, Ithaca, N. Y.
A. Kyrala	Arizona State University, Tempe, Arizona
D. B. Larson	Lawrence Radiation Lab., Livermore, Calif.
J. Laurent	C. N. R. S., Bellevue, France
A. W. Lawson	University of California, Riverside, Calif.
D. Lazarus	University of Illinois, Urbana, Ill.
W. J. Leahy	USAF, Edwards, Calif.
A. D. Le Claire	UK Atomic Energy Research Establishment, Harwell, England
E. L. Lee	Lawrence Radiation Lab., Livermore, Calif.
D. Liberman	Los Alamos Scientific Lab., Los Alamos, N. M.
R. Liptai	Argonne National Laboratory, Argonne, Ill.
E. C. Lloyd	National Bureau of Standards, Washington, D. C.
R. C. Lowell	University of Arizona, Tucson, Arizona
D. W. Lynch	Iowa State University, Ames, Iowa
M. Marezio	Bell Telephone Laboratories, Murray Hill, N. J.
H. Margolin	New York University, University Heights, Bronx, N. Y.
J. Matricon	Atomic Energy Commission, Saclay, France
R. Matters	Allis Chalmers Mfg. Co., Milwaukee, Wisc.
B. T. Matthias	University of California, La Jolla, Calif.
P. P. McArdle	University of Arizona, Tucson, Arizona
D. E. McDonald	University of Arizona, Tucson, Arizona
D. B. McWhan	Bell Telephone Laboratories, Murray Hill, N. J.
E. Michigan	University of Arizona, Tucson, Arizona

LIST OF CONFEREES

F. Milstein	University of California, Los Angeles, Calif.
S. Minomura	Institute for Solid State Physics, University of Tokyo, Minato-ku, Tokyo, Japan
I. Mirtskhuleva	State University, Tiflis Georgian SSR, USSR
T. Mitsui	University of Arizona, Tucson, Arizona (Hokkaido University, Sapporo, Japan)
N. H. Nachtreib	University of Chicago, Chicago, Ill.
H. M. Nelson	Brigham Young University, Provo, Utah
M. F. Nicol	University of California, Los Angeles, Calif.
R. L. Orbach	University of California, Los Angeles, Calif.
W. J. O'Sullivan	Sandia Laboratory, Albuquerque, N. M.
M. Pagannone	University of Rome, Rome, Italy
W. Paul	Harvard University, Cambridge, Mass.
C. B. Pierce	Sandia Laboratory, Albuquerque, N. M.
D. H. Pipkorn	Lawrence Radiation Lab., Livermore, Calif.
A. B. Pippard	Cambridge University, Cambridge, England
W. Z. Plachy	University of California, Van Nuys, Calif.
E. G. Ponyatovski	Institute for Ferrous Metals, Moscow, USSR
W. J. Price	AF Office of Scientific Research, Washington, D. C.
E. Rapoport	University of California, Los Angeles, Calif.
J. T. Ratchford	AF Office of Scientific Research, Washington, D. C.
J. P. Remeika	Bell Telephone Laboratories, Murray Hill, N. J.
S. A. Rice	University of Chicago, Chicago, Ill.
L. B. Robinson	University of California, Los Angeles, Calif.
Laura Roth	Tufts University, Belmont, Mass.
S. J. Rothman	Argonne National Laboratory, Argonne, Ill.
R. R. Roy	Arizona State University, Tempe, Arizona
E. B. Royce	Lawrence Radiation Lab., Livermore, Calif.
A. Ruoff	Cornell University, Ithaca, N. Y.
E. I. Salkovitz	Advanced Research Projects Agency, Washington, D. C.
G. A. Samara	Sandia Laboratory, Albuquerque, N. M.
J. E. Schirber	Sandia Laboratory, Albuquerque, N. M.

LIST OF CONFEREES

D. E. Schuele	Case Institute, Cleveland, Ohio
C. M. Schwartz	Battelle Memorial Institute, Columbus, Ohio
G. Shirane	Brookhaven National Laboratory, Upton, Long Island, N. Y.
D. B. Singer	Aerospace Corporation, San Bernardino, Calif.
M. J. Skove	Clemson University, Clemson, S. C.
L. M. Slifkin	University of North Carolina, Chapel Hill, N. C.
C. S. Smith	Case Institute, Cleveland, Ohio
H. I. Smith	Boston College, Chestnut Hill, Mass.
A. Sosin	Atomics International, Canoga Park, Calif.
K. F. Sterrett	Northrop Space Labs., Hawthorne, Calif.
D. K. Stevens	Atomic Energy Commission, Washington, D. C.
A. Stewart	University of North Carolina, Chapel Hill, N. C.
E. P. Stillwell	Clemson University, Clemson, S. C.
R. G. Stoner	Arizona State University, Tempe, Arizona
H. M. Strong	G. E. Research Lab., Schenectady, N. Y.
D. L. Styris	University of Arizona, Tucson, Arizona
C. Swenson	C. S. I. R. O., Sydney, Australia (Iowa State University)
M. Swerdlow	AF Office of Scientific Research, Washington, D. C.
A. Taylor	Westinghouse Electric Corp., Monroeville, Pa.
S. M. Taylor	Terminal Ballistics Lab., Aberdeen Proving Ground, Md.
R. Thomson	University of Illinois, Urbana, Ill.
C. T. Tomizuka	University of Arizona, Tucson, Arizona
Y. Tomono	University of Arizona, Tucson, Arizona (Inst. for Solid State Physics, Univ. of Tokyo)
M. Tosi	Argonne National Laboratory, Argonne, Ill.
L. C. Towle	Allis Chalmers Mfg. Co., Milwaukee, Wisc.
H. Van Fleet	Brigham Young University, Provo, Utah
A. Van Valkenburg	National Science Foundation, Washington, D. C.
R. Venuti	University of Denver, Denver, Colo.
L. F. Vereschagin	Academy of Science, Moscow, USSR
R. E. Weigle	Watervliet Arsenal, Watervliet, N. Y.
G. K. White	Bell Telephone Laboratories, Murray Hill, N. J.

LIST OF CONFEREES

R. F. Wood	Oak Ridge National Lab., Oak Ridge, Tenn.
E. N. Yakovlev	Institute of High Pressure Physics, Academy of Science, Moscow, USSR
Z. Zeitlin	Barogenics, White Plains, N. Y.
R. Zeto	Rocketdyne, Canoga Park, Calif.

CONTENTS

SESSION B
ELECTRONIC PROPERTIES II
Chairman: H. BROOKS

SESSION C
MAGNETISM
Chairman: L. F. VERESCHAGIN

SESSION D

LATTICE DEFECTS

Chairman: A. D. LE CLAIRE

SESSION E

PHASE TRANSFORMATIONS

Chairman: H. B. HUNTINGTON

SESSION F

PHASE TRANSFORMATIONS AND LATTICE PROPERTIES

Chairman: A. B. PIPPARD

WELCOMING INTRODUCTION

W. J. Price, Executive Director
Air Force Office of Scientific Research
Washington, D. C.

I am delighted to be able to be with you here at the opening session of this conference. This is truly an international gathering. I note that a significantly high portion of the participants are from outside the United States. I want to thank all of you for coming, and wish to extend a particularly warm welcome to our visitors from abroad.

The organizers of the conference have done a remarkable job. We are fortunate indeed to be able to meet here jointly with the University of Arizona, whose hospitality means so much to the success of our purpose.

That purpose is to further the knowledge and understanding of the application of high pressure techniques to research in solid state physics. The importance of the subject is evident by the high professional repute of you who are attending. The exchange of information and free discussion here will accelerate our basic understanding of this active research area. I am sure the publication of the papers, a very impressive list, will have a strong impact throughout the scientific community.

In the course of the past ten years there has been a significant surge of research interest in the subject of high pressure solid state research, and the number of research workers in the field has multiplied fourfold. During this time the need became clearly evident for a full-fledged international conference in order to clarify the status of the field and to restate principal problem areas.

Some of these major problems lie in electronic properties of metals, semiconductors and even molecular crystals. We would hope to shed light on one of the modern mysteries of solid state physics--ferromagnetism in metals and alloys. We will learn more detailed information about lattice defects and phase transformations and transitions in crystalline solids.

It is evident then that this conference will be of considerable benefit to workers in high pressure techniques and to solid state physicists in general, because we have with us not only some of the true pioneers but also a very large percentage of those who are active in this research. Moreover there is an outstanding group of theorists who we hope will be inspired by the proceedings.

I take special pride in the fact that the Air Force Office of the Scientific Research has helped to support these sessions. My organization has rather broad responsibilities for the support of fundamental research in the physical, mathematical, life, and information sciences.

Further, AFOSR, together with other Federal Government agencies sponsoring research, has a responsibility to stimulate research in areas that are hindered by lack of fundamental knowledge. AFOSR also has a clear-cut responsibility to bring new research knowledge to the attention of those who are working both on related research and on technological problems. To meet this need we carry out an extensive program of support via topical conferences, symposia, and other working sessions.

AFOSR is currently supporting fundamental research in the sciences at more than 200 colleges, universities and research organizations in the United States and perhaps 50 abroad. We are especially pleased to know that a number of you here today are supported in your research by AFOSR.

This research support is a direct effort to create the new knowledge technology must have in order to progress, and to provide products, industries, and economic expansion that the growing numbers of people on this planet so ardently desire. I need not remind you how closely related your own particular interests are to industrial development. It is this expanding technological capability that will enable, for example, developing nations the world over to realize their full potentialities.

This group today comprises an excellent example of what has come to be called the "invisible colleges," groups of scientists working on the frontiers of a given specialized area of interest. They are not grouped geographically, but rather are widely scattered--but their effectiveness is in their communication with each other. There is direct correlation between how rapidly research progresses and how well these members of invisible colleges communicate.

I might warn the speakers that the organizing committee has threatened not to allow any speaker to leave the ranch until his manuscript is submitted.

The real success of the conference will be in the face-to-face discussions and informal conversations that a conference in such beautiful surroundings as these can make possible.

SESSION A

ELECTRONIC PROPERTIES I

Chairman: H. EHRENREICH

INFLUENCE OF PRESSURE ON THE FERMI SURFACE IN METALS

Walter A. Harrison
General Electric Research Laboratory
Schenectady, New York

INTRODUCTION

Perhaps it is appropriate to begin this conference with the discussion of problems which are reasonably well in hand theoretically. This is not to say that the influence of pressure on Fermi surfaces is entirely understood - it is to say that I will restrict my discussion to problems upon which we have a reasonable grip. In particular, I will talk about simple metals - the polyvalent and alkali metals - rather than transition metals and noble metals. I would like to present a simple model pseudopotential which seems to give a very good account of a wide range of properties of these simple metals. I will then use this model to explore the effect of pressure on the Fermi surface.

The pseudopotential method itself is simply a reformulation of the Schroedinger Equation and does not constitute an approximation. It is applicable to semiconductors and semimetals as well as metals, and such applications will be described later in this session by Professor Falicov. The great power of the pseudopotential formulation is that it provides a very convenient starting point for approximate treatments of properties. In metals the important approximation is made in treating the pseudopotential as a perturbation and computing energies to second order. In addition, for the purposes of our discussion, we will replace the pseudopotential operator by a simple local potential, that is, a scalar function of position.

The resulting model has much in common with the nearly-free-electron theory described by Mott and Jones (1), however, it differs somewhat in emphasis and in the properties we choose to treat. In addition, we incorporate at the outset the usual approximations of diffraction theory. For that reason we call it the diffraction model.

THE DIFFRACTION MODEL

We consider a metal as a collection of ions located at the position $r_1, r_2, \ldots r_N$. Associated with these ions are NZ electrons (Z being the valence) which do not interact with each other, but which interact with the ions through a local pseudopotential W(r).

We assume that this local pseudopotential can be written as the superposition of individual ionic pseudopotentials w(r) centered at the N ion sites. These individual pseudopotentials may overlap each other; all we require is that the total pseudopotential may be written in the form

$$W(r) = \Sigma_j \, w(|\vec{r}-\vec{r}_j|). \qquad (1)$$

Finally, we include three contributions to each individual ionic pseudopotential. The first is the potential due to the ionic charge; this may be written -Ze /r. The second is a repulsive term which is localized to the core and which arises ultimately from the Pauli principle or from the requirement that the conduction-electron wave functions be orthogonal to the core wave functions. We take this term to be of the form e^{-r/r_c}; this particular form is motivated by the form taken in the full pseudopotential theory. The final contribution is the potential due to screening by the conduction electrons. This contribution comes ultimately from the interactions between electrons; our approach here is simply to replace those interactions by a screening of the pseudopotential. This screening will be accomplished by dividing each Fourier component of the unscreened pseudopotential by the corresponding static Hartree dielectric function for free electrons. For a Fourier component of wave number q this dielectric function is given by

$$\epsilon(q) = 1 + \frac{me^2}{2\pi k_F \hbar^2 \eta^2}\left[\frac{1-\eta^2}{2\eta}\ln\left|\frac{1+\eta}{1-\eta}\right| + 1\right] \tag{2}$$

where $\eta = q/2k_F$.

The repulsive term from the core tends to cancel the attractive coulomb potential of the valence charge so the resulting pseudopotential is small. This will make it possible to treat the pseudopotential as a perturbation.

Having specified the pseudopotential we may write the wave equation (time-independent in this case) for the pseudo wave function describing the electrons

$$-\frac{\hbar^2 \nabla^2}{2m}\phi + W(\vec{r})\phi = E\phi \tag{3}$$

In calculating any property we will treat the pseudopotential as a first-order term. The zero-order states are plane waves, normalized in the volume of the metal Ω. The pseudopotential will enter always through plane-wave matrix elements of the pseudopotential. These matrix elements may readily be factored as in diffraction theory.

$$\langle \vec{k}+\vec{q}\,|\,W\,\,|\vec{k}\rangle = \frac{1}{\Omega}\sum_j \int e^{-i(\vec{k}+\vec{q})\cdot\vec{r}}\,w(\vec{r}-\vec{r}_j)\,e^{i\vec{k}\cdot\vec{r}}\,d\tau$$

$$= \frac{1}{N}\sum_j e^{-i\vec{q}\cdot\vec{r}_j}\,\frac{1}{\Omega_0}\int e^{-i\vec{q}\cdot\vec{r}}\,w(r)\,d\tau$$

where Ω_0 is the atomic volume. The first factor in the final form is called the <u>structure factor</u>;

$$S(\vec{q}) = \frac{1}{N} \Sigma_j e^{-i\vec{q}\cdot\vec{r}_j} \tag{4}$$

It depends only upon the individual ion positions. For perfect lattices it will vanish unless $\vec{q}$ is a lattice wave number (2π times a reciprocal lattice vector).Then it will be of order unity. The second factor is called the form factor;

$$w_q = \frac{1}{\Omega_o} \int e^{-i\vec{q}\cdot\vec{r}} w(r) d\tau \tag{5}$$

It is independent of the positions of the ions and depends only upon the individual ionic pseudopotentials. For the model given above it may be readily evaluated and is given by

$$w_q = \frac{-(4\pi Ze^2/q^2) + \beta/(1+(qr_c)^2)^2}{\Omega_o \epsilon(q)}$$

where β represents the strength of the core repulsion.

If we know the two parameters β and r_c for a given metal we may evaluate w_q directly and have all the information required to compute the energy eigenstates for any arrangement of the ions of interest. r_c is of the order of the radius of the core. It is sufficiently small that for the q's entering most electronic properties it is a reasonable approximation to take qr_c equal to zero; this is equivalent to replacing the core by a point. We will make this simplification in the remainder of our discussion and use the form

$$w_q = (-4\pi Ze^2/q^2 + \beta)/(\Omega_o \epsilon(q)) \tag{6}$$

I have computed form factors for a series of metals (2) using the full pseudopotential method. These are displayed in Figure 1.

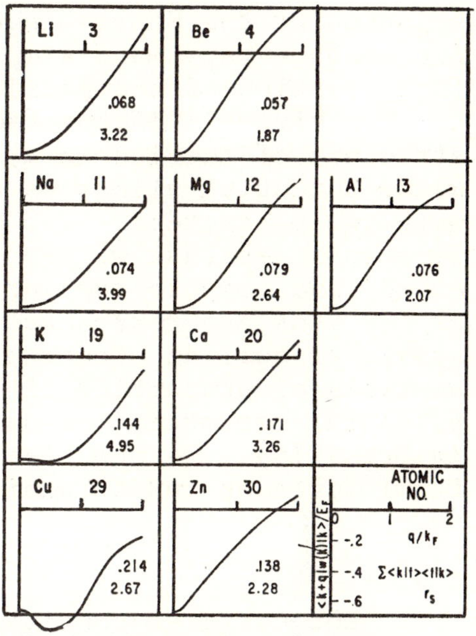

Fig. 1. Form factors w_q for a series of metals. The key appears to the lower right; $\Sigma\langle k|t\rangle\langle t|k\rangle$ is a measure of the overlap between core wave functions and plane waves or the core size; $(4\pi r_s^3/3)^{-1}$ is the electron density

These may be fit rather well by Equation 6. The values of β obtained by fitting at $q = 2k_F$ are given in Table 1.

Table 1. Strength of Core Repulsion β for Several Metals in Ry. - Atomic Units of Volume

Li	29	Be	31		
Na	27	Mg	42	Al	37
K	32	Ca	51		
Cu	2	Zn	27		

Estimates of β for a number of other metals were obtained by studying the resistivity of alloys involving these metals and are listed elsewhere (3).

Before discussing Fermi surfaces in some detail, we might summarize what we have accomplished. We have replaced the very complicated problem of solving the Schroedinger Equation in a metal by a simple model problem. The treatment of any property of a metal which can be done by a suitable solution of the Schroedinger Equation is now replaced by a very much simpler model problem. The factorization of the matrix elements has enabled us to summarize all of the required knowledge of the structure of the ion in a single curve, and the model potential allows us to characterize that curve with two parameters. The problem has been so greatly simplified that we can even sum the electronic energy and compute the variation of the total energy as the ions are rearranged; that is, we can treat atomic as well as electronic properties. This analysis of properties of metals using the pseudopotential method is the subject of a book currently in press (4). We now restrict our attention to Fermi surfaces and to their variation with volume.

FERMI SURFACES

Let us first consider the limiting case as the pseudopotential approaches zero. The Hamiltonian approaches simply the kinetic energy and the eigenstates approach plane waves. The ground state of the system will have all states occupied within a Fermi sphere; the radius k_F of that sphere depends upon the electron density:

$$\frac{2}{(2\pi)^3} \Omega \frac{4\pi}{3} k_F^3 = NZ$$

or

$$k_F = \left(\frac{3\pi^2 Z}{\Omega_o}\right)^{1/3} \tag{7}$$

This is the Fermi surface when the pseudopotential vanishes. Note that as the atomic volume decreases under pressure the Fermi surface grows but remains spherical.

If the pseudopotential is not quite zero, we may think of it as giving rise to diffraction. The electrons do not see the potential unless they satisfy the Bragg condition; that is, unless their wave numbers lie on a Bragg reflection plane. The

Bragg condition is simply

$$2\vec{k}.\vec{q} = -q^2$$

where $\vec{q}$ is a lattice wave number. This is simply the condition that $\vec{k}$ lie on the bisector of a lattice wave number. Under such circumstances the electron will be diffracted to the opposite Bragg reflection plane.

Of course the number of electrons which happen to lie on a Bragg reflection plane is of measure zero and this diffraction tends to be unimportant. However, when a magnetic field is applied, the electron wave numbers move in circles on the Fermi sphere (as the electron orbits in the field). Thus electrons on much of the Fermi surface intersect a Bragg reflection at some time in their orbit and the orbits of most electrons are modified by the Bragg reflections. Electrons with wave numbers near the intersection of the Fermi surface and a Bragg-reflection plane diffract to a corresponding piece of the Fermi surface at the opposite Bragg reflection plane. Thus it is reasonable to cut up the sphere and reassemble it such that pieces connected by a Bragg reflection are placed together. This leads to a set of Fermi surfaces, each made up of spherical surfaces. Electrons with wave numbers on one section of Fermi surface move always on that section. The construction of such surfaces is straightforward (5) and leads to Fermi surfaces such as that for aluminum given in Figure 2.

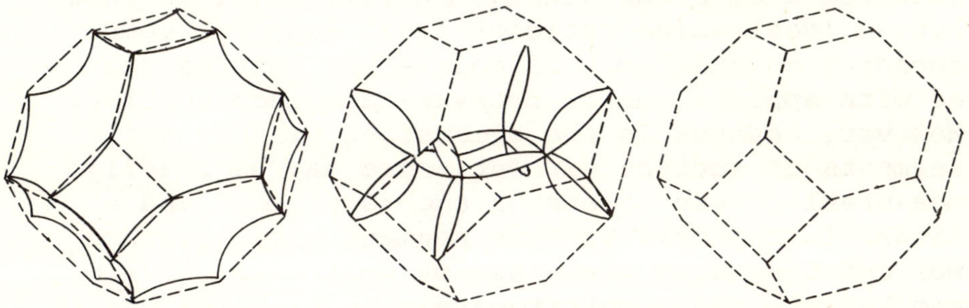

Fig. 2. The Fermi surface of aluminum obtained by reassembling spheres.

Such Fermi surfaces have been obtained and published for most simple metals (6). It is not difficult to show that the shape of the orbits which each electron takes in the crystal in the presence of a magnetic field is the same as the shape of some section of the Fermi surface constructed in this way. As a consequence, these Fermi surface shapes play a fundamental role in determining the electronic properties of metals in a magnetic field.

It is interesting to note now what happens when pressure is applied and the volume changes isotropically. We know first that the Fermi sphere grows as the reciprocal of the lattice distance; this follows from Equation 7. The lattice wave numbers also grow inversely as the reciprocal of the lattice distance since the corresponding wavelengths are related to the lattice distance by a constant geometrical factor. Thus the shape of the Fermi surfaces we have constructed does not change as metal is isotropically compressed. They are simply scaled up. Since the scaling will generally be small, this change is not easy to measure and is not particularly interesting.

In spite of this fact, these nearly-free-electron Fermi surfaces can change in shape under hydrostatic pressure if the resulting distortion is not isotropic. Such anisotropic distortions occur in hexagonal close-packed metals, where the axial ratio changes with pressure. The type of modification which this makes in the Fermi surface can be seen in Figure 3 where a segment of the Fermi surface for a divalent hex. close-packed metal is shown for an ideal axial ratio of 1.633 and that corresponding to cadmium. Of course the changes expected with application of pressure are much smaller. However, changes in the dimensions of very small segments of surface with pressure may be readily measurable. Zinc contains such a segment and a change in the area of this segment with pressure was detected some years ago by Dmitrenko, Verkin, and Lazarev (7). Unfortunately that change was in a direction opposite to what we would expect using

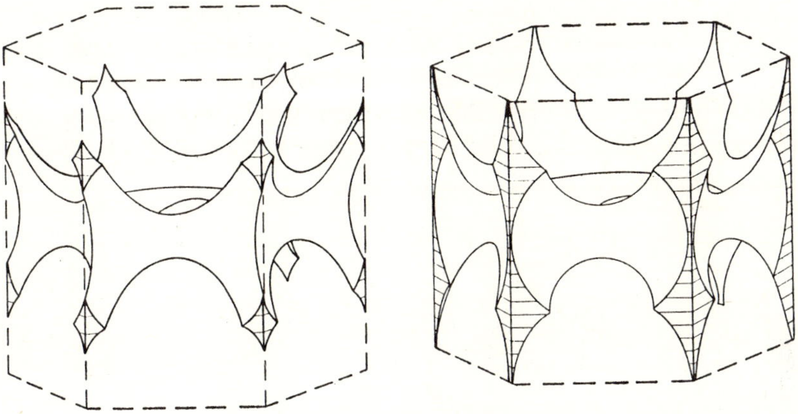

Fig. 3. A segment of the Fermi surface for divalent hexagonal close-packed metals in the nearly free-electron approximation. The figure on the left corresponds to an axial ratio of 1.633; that on the right, to 1.862.

the model surface (6), but more recent work has cast some doubt on the validity of the earlier result (8). Hopefully this question will be cleared up by further de Haas-van Alphen work on zinc under pressure.

Let us next consider the influence of still larger pseudopotentials; that is, the influence of the pseudopotential to higher order. For finite strength of the pseudopotential, it is not correct to assert as we did earlier that the electron does not see the periodic potential unless its wave number lies on a Bragg reflection plane. An electron will "begin to diffract" as it nears the Bragg plane, and will continue to diffract as it leaves it. More precisely, the Fermi surface becomes distorted in the neighborhood of the Bragg plane. In terms of the Fermi surfaces such as those shown in Figures 2 and 3, the sharp corners on the surfaces are rounded off. The shapes of the electron orbits in a magnetic field are still directly related to sections of the Fermi surface and these orbits in real space are rounded off in the same way. In cases where the Fermi surface sections are particularly thin, this rounding off may even change the connectivity of the orbits. Such a change of the Fermi surface in calcium is shown in Figure 4. In some cases the small seg-

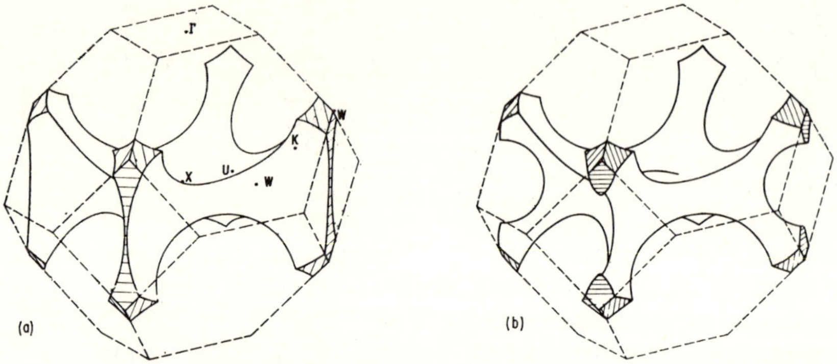

Fig. 4. A segment of the Fermi surface of calcium computed (a) in the limit of vanishing pseudopotential, and (b) assuming a pseudopotential of a strength obtained from Figure 1.

ments are eliminated entirely; this is the case with the smallest fragments in aluminum shown in Figure 2. The calculation of such modifications of the Fermi surface is quite straightforward (9) once the form factors of Equation 5 are known.

The question of interest here is how these Fermi surfaces change under pressure which of course depends upon how the form factors change with the volume. These changes can be computed directly from Equation 6 assuming that β does not change with volume. This assumption turns out to be a very good one - probably as good as the model itself. This may be seen from Figure 5 where we fit the model form, Equation 6, to the form factor for aluminum computed from the full pseudopotential theory. We then recompute the form factor from the

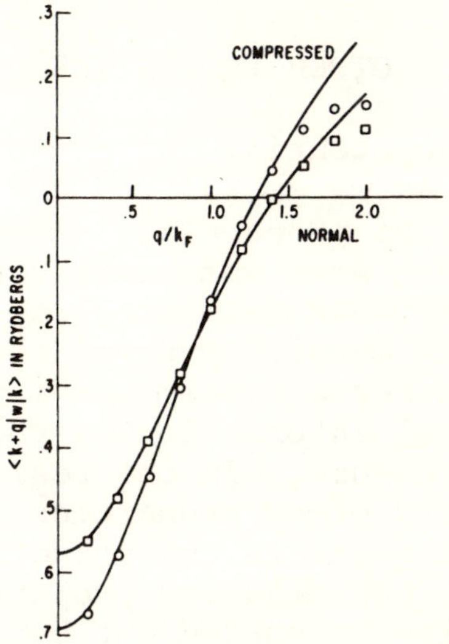

Fig. 5. Form factors computed from the full pseudopotential theory for aluminum at normal volume (square points) and for a lattice distance reduced by 10% (circular points). Curves represent the corresponding results for the model form with β = 43.8 ry-atomic units of volume. This value was chosen to give the correct abscissa intercept for the normal volume.

---

full theory for a contracted volume and compare with the model form using the same value of β.

Again the change in volume shown is large compared to that which is obtainable with reasonable pressures and the changes in form factors may be expected to be much smaller. Nonetheless, in a Fermi surface where the form factor corresponding to a particular Bragg plane is very small the fractional change under pressure may be appreciable. Such accidentally small form factors are not uncommon. They apparently occur in zinc (10), cadmium, and calcium (2). (Small form factors are also found which arise from spin-orbit splitting in some metals, but these should not change rapidly with volume). Hopefully among such metals we might find a form factor which goes through zero as pressure is increased with a conspicuous change in the connectivity of the Fermi surface.

OTHER PROPERTIES

The pseudopotential method which we have been describing is, as we emphasized earlier, applicable to a wide range of properties. Among the electronic properties are some which are very sensitive to the form factors; one such property is the resistivity of a liquid metal. Perhaps the behavior of such a property for a system under pressure would give experimental information on the change in form factors with pressure. In the resistivity of the metal, as in many other properties, there are at the same time changes in the structure factors which influence the measured property. However, these may usually be sorted out by other measurements; in the case of the liquid metal these would be measurements of the structure by x-rays or neutrons.

Perhaps the example of liquid metal resistivity is not a practical one. Nevertheless it illustrates the point that we have a sufficient understanding of the simple metals that we may consider almost any property of interest with enough accuracy to tell us whether measurable effects should be expected and to enable us to relate the results of the measurements to the most fundamental properties of the metals.

References

1. N. F. Mott and H. Jones, Theory of the Properties of Metals and Alloys, (Dover Publications, New York, 1958).
2. W. A. Harrison, Phys. Rev. 131, 2433(1963).
3. W. A. Harrison, Rev. Mod. Phys. 36, 256(1964).
4. W. A. Harrison, Pseudopotentials in the Theory of Metals, (W. A. Benjamin, New York), in press.
5. W. A. Harrison, Phys. Rev. 116, 555(1959).
6. W. A. Harrison, Phys. Rev. 118, 1190(1960).
7. I. M. Dmitrenko, B. I. Verkin, and B. G. Lazarev, Soviet Physics, JETP 6, 223(1958).
8. D. Lazarus, private communication.
9. See for example, W. A. Harrison, Phys. Rev. 118, 1182(1960).
10. W. A. Harrison, Phys. Rev. 126, 497(1962).

SOME PRESSURE EFFECTS IN THE MONOVALENT METALS

J. S. Dugdale
National Research Council
Division of Pure Physics
Ottawa, Canada

In this paper I would like to review some of the experimental information that we now have about the effect of pressure on electrical resistivity and related parameters in the monovalent metals. I have in mind two main problems: (1) to understand the effect of pressure on the electrical resistivity due to thermal vibrations (the "ideal" resistivity) at high temperatures, and (2) to understand the effect of pressure on the electrical resistivity due to small amounts of impurity or physical imperfections (the residual resistivity).

In general, the residual resistivity can only be studied at low temperatures where the lattice vibrations have effectively died out. This does, however, introduce practical difficulties because at the lowest temperatures it is only possible to produce hydrostatic pressures up to about 30 atmospheres (with liquid helium); above this all materials are solid. In studying the effect of pressure on electrical resistivity at low temperatures our technique (1), (2) has been to use fluid helium at, say, the boiling point of hydrogen, 20.4°K, where the available pressure range is 1800 atm (the solidification pressure of helium at this temperature), or, if measurements are required in the helium range of temperatures, to use solid helium as the pressure transmitting medium. Templeton (3), however, has shown that even within the restricted pressure range afforded by liquid He^4 at about 1°K (25 atm) valuable information can be obtained about the distortion under pressure of the Fermi surfaces of the noble metals (see below).

1. Electrical resistivity due to thermal vibrations

1.1 The alkali metals

The electrical resistivity of a metal that arises from the lattice vibrations depends on (a) the Fermi surface, (b) the dispersion relations of the normal modes of the lattice (in brief the phonons), and (c) the electron-phonon interaction.

Theoretical calculations (4), (5) have shown that in the alkali metals the electrical resistivity depends very largely on the number of Umklapp processes that are possible; this is particularly important at low temperatures where normal processes can cause only small-angle scattering whereas Umklapp processes can produce large-angle scattering, but even at high temperatures Umklapp processes are of over-riding importance. Since in the alkali metals the Fermi surfaces are all approximately spherical (the maximum radial deviation from a sphere is found in lithium and caesium where it amounts to about 5% (6)), the Fermi surfaces do not touch the Brillouin zone boundaries and the distinction between normal and Umklapp processes is clear. The probability of an Umklapp process depends not only on the geometry of the Fermi surface in relation to the Brillouin zone boundary, but also on the velocity of the appropriate lattice waves. These two effects are so intimately related that they cannot *a priori* be separated (7). This means that the shape of the Fermi surface and the elastic anisotropy of the metal are equally important in determining electrical resistivity in the alkali metals. To determine the effect of pressure on resistivity we need to know, therefore, both how the Fermi surface changes with pressure and how the elastic constants change with pressure. Indeed, to get the full story we really need to know (by means of slow neutron spectrometry, for example,) how the complete dispersion relations of the lattice depend on pressure. It does seem, however, that the elastic constants by themselves provide a fairly good basis from which to calculate the full dispersion curves (8).

When a metal is subjected to pressure the amplitude of the lattice vibrations in general decreases. This, therefore, tends to decrease the electrical resistivity at a given temperature. We can make an approximate allowance for this effect at high temperatures in the following way. We may write for the ideal electrical resistivity at high temperatures ($T > \Theta$)

$$\rho_i = \frac{KT}{M\Theta^2} \qquad (1)$$

In this expression T is the absolute temperature, M the mass of the ions, Θ the characteristic temperature of the lattice, and K represents all the effects of the electron-phonon interaction. This relationship simply asserts that the high temperature resistivity is proportional to the mean square amplitude of the lattice vibrations. For the volume derivative we therefore deduce

$$\frac{\partial \ln \rho_i}{\partial \ln V} = \frac{d \ln K}{d \ln V} + 2\gamma , \qquad (2)$$

where $\gamma = - d \ln \Theta / d \ln V$. As a first approximation we can take γ as equal to the Grüneisen parameter; its value can then be found from the equation of state and the specific heat of the solid.

In Table 1 values of $\partial \ln \rho_i / \partial \ln V$ and of $d \ln K / d \ln V$ are listed for the monovalent metals; the main theoretical problem is to account for these values of $d \ln K / d \ln V$. Some calculations of this quantity have been made by Hasegawa (8) and they are also shown in Table 1. In the alkali metals we have as yet no direct measurements of the change of shape of the Fermi surface by pressure, but theoretical calculations by Ham (9) show that in sodium and potassium the distortions of the Fermi surface with pressure should be small at low pressures. For these two metals experiment (10), (11) shows that the elastic anisotropy does not change with pressure, at least up to pressures of 10,000 atm. On the basis of this information Hasegawa has calculated the pressure coefficient of the ideal electrical resistivity in sodium and potassium and has found quite good agreement with experiment. Since the elastic anisotropy and the properties of the Fermi surface do not change with pressure the geometry of the scattering processes in k-space is unchanged and the decrease in the value of K with increasing pressure is mainly due to the decrease in the electron-phonon matrix element. Basically it appears that the scattering (for a given amplitude of lattice vibrations) diminishes with increased pressure because the Fermi energy increases.

Table 1. Change of Ideal Resistivity with Volume and Related Parameters in the Monovalent Metals

Metal	$\frac{\partial \ln \rho_i}{\partial \ln V}$ $0^{\circ}C$	$\frac{d \ln K}{d \ln V}$ (Experimental)	Change in elastic anisotropy with pressure $\frac{d \ln A}{d \ln V}$	Distortion of Fermi surface with pressure	$\frac{d \ln K}{d \ln V}$ (Theoretical: Hasegawa (8))
Li	-0.49	-2.3	-0.4		-3.7
Na	4.6	2.0	0	(Probably	1.8
K	5.6	3.0	0	(small	1.9
Rb	$4._3$	$2._3$			
Cs	$3._1$	$1._1$			
				$\left[\frac{d \ln r_N}{d \ln V}\right]^{*}_{distortion}$	
Cu	3.0	-1.0	-0.87	-1.1±0.2	
Ag	3.9	-0.9	-0.84		
Au	5.5	-0.7	-2.1	-1.4±0.2	

*This measures the relative change of neck radius due to distortion of the Fermi surface with change of volume. These results are due to Templeton (3).

Rubidium under pressure tends to behave rather like sodium and potassium, at least at fairly low pressures, and it may be that its behaviour can be explained in the same way as that of sodium and potassium although no comparable detailed calculations have yet been made on rubidium.

In lithium Ham's calculations (9) show that the Fermi surface should become more distorted under pressure. It is therefore possible that the increased distortion of the Fermi surface, which would tend to enhance the Umklapp contribution to the scattering, might account for the anomalous positive pressure coefficient of the ideal resistivity of lithium at high temperatures (12). Indeed, Hasegawa (8) has shown that on the basis of Ham's estimates of the Fermi surface distortion a strongly positive pressure coefficient would result. In addition, experiments (13) have shown that the elastic anisotropy of lithium (as measured by the anisotropy parameter, $A = \frac{2c_{44}}{c_{11}-c_{12}}$) increases with pressure (see Table 1). This effect too may be important in determining the pressure coefficient of the ideal resistivity.

In caesium the ideal resistivity first decreases and then increases under pressure (14). This is probably due to a combination of distortion of the Fermi surface and a change in elastic anisotropy, but we have so far no experimental information which enables us to decide which, if either, of these effects is the dominant one.

1.2 The noble metals

In the noble metals the experimental information is more complete but the theoretical problem is more difficult and there seem to be no detailed calculations of the electrical resistivity and its pressure dependence comparable to those on the lighter alkali metals. A summary of the information that relates to pressure effects is included in Table 1.

1.2.1 Fermi surface

The Fermi surfaces of the noble metals are rather distorted and touch the (111) faces of the

Brillouin zone (15) (cf. Figure 1).

Very recently Templeton (3) has shown that it is possible to measure the changes in the Fermi surfaces of the noble metals under very small pressures. So far, indeed, he has confined himself to truly hydrostatic pressures which near 1°K means about 25 atm. His method has been to observe how the phase of the de Haas-van Alphen oscillations changes with pressure. He has applied the method directly to observe the change in area of the extremal belly orbits in all three metals. However, to measure the distortion of the Fermi surface (and not just the change of size brought about by the volume change) he has measured the relative phase change of belly and neck oscillations as the pressure is changed (see Figure 1).

To give some idea of the sensitivity of this method it may be noted that at about 50 kgauss the extremal belly orbits enclose roughly 10,000 quantized electron orbits of smaller radius. Thus one cycle of the corresponding de Haas-van Alphen oscillations corresponds to a change in extremal cross-section of one part in 10,000 and since it is possible to detect changes of less than 1/500 of a cycle the sensitivity approaches one part in 10^7. A sensitivity of 1 part in 10^6 is possible under favorable conditions in comparing belly and neck oscillations. It may also be noted that the advantages of hydrostatic pressure are two-fold. In the first place, any damage done to the single crystals due to compression should be negligible. In the second place, it is easy to vary the pressure and observe continuously the changes in de Haas-van Alphen signals that are produced.

Some results from Templeton's experiments are summarized in Table 1; they show, for copper and gold, the relative change in neck radii due to distortion of the Fermi surface under compression. It is seen that the area of contact increases in both cases.

1.2.2 The elastic constants

The change of the elastic constants of the noble metals under pressure has also been measured (16) and it is found that the elastic anisotropy parameter, A, increases with pressure in all of them.

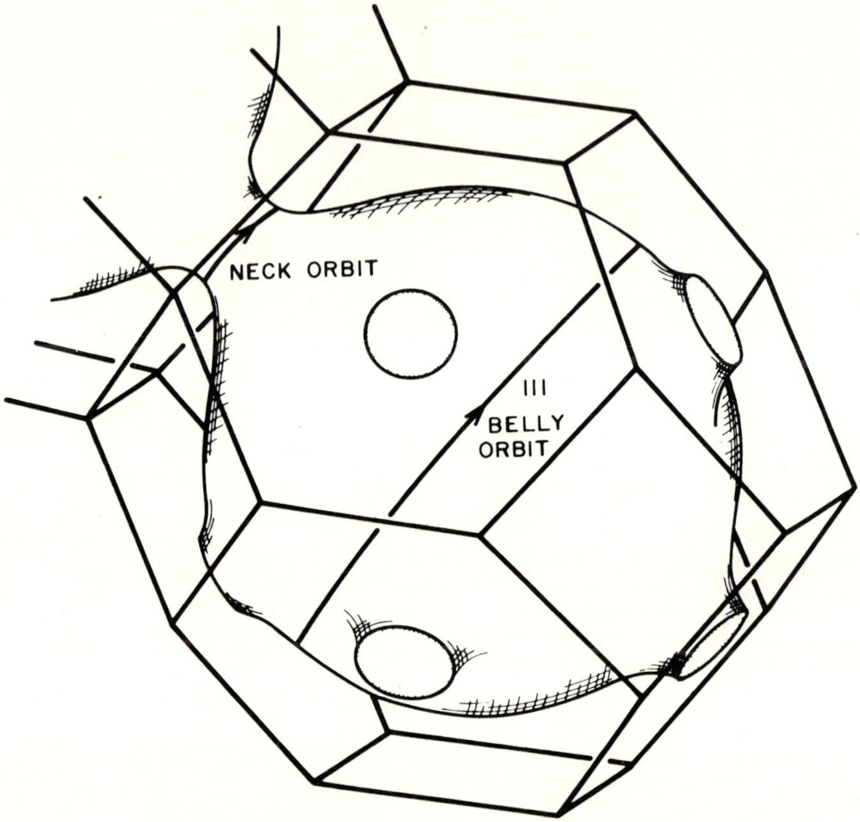

Fig. 1. The diagram illustrates schematically the Fermi surface and first Brillouin zone of Cu, Ag or Au. By applying a magnetic field in the (111) direction the belly and neck orbits shown in the figure can be directly compared by means of the de Haas-van Alphen effect.

From these results, it appears, therefore, that the distortion of the Fermi surface and the change in elastic constants may contribute significantly to the increase in the value of K with pressure in the noble metals. However, the fact that the large value of d ln A/d ln V in gold does not seem to be reflected in the pressure dependence of ρ_i suggests that, at least at high temperatures, the elastic anisotropy is probably not a major factor in determining the resistivity of these metals.

2. Residual resistivity

2.1 The alkali metals

2.1.1 Impurity scattering

So far there appears to have been no systematic study of the effect of pressure on the residual resistance of the alkali metals containing impurities. Isolated measurements have, however, been reported (17).

2.1.2 Dislocations

If the explanation of the pressure dependence of the ideal resistivity of sodium and potassium at high temperatures given by Hasegawa is correct, a similar argument should also apply to the resistivity due to dislocations. The perturbing potential which is important for the scattering of the electrons arises in both cases from the displacement of the ionic positions. Since, as we have seen, the pressure dependence of the ideal resistivity is not due to any change in the geometry of the scattering processes but to a change in the relevant matrix elements, this should alter the probability of phonon scattering and dislocation scattering in the same way. In the case of dislocation scattering there will, of course, be no term 2γ corresponding to the change in amplitude of the lattice vibrations under pressure (Eq. (2)).

One might, therefore, expect that in sodium the volume coefficient of the resistivity due to dislocations would have a value of about +1.8 (see Table 1). In the pure specimens measured by Dugdale and Gugan (18) (with resistance ratios of about 3,000) a negative pressure coefficient of residual resistivity was indeed found corresponding to a value of $\partial \ln \rho/\partial \ln V = 1.4 \pm 0.4$. Most of the residual resistivity in such specimens, which are heavily deformed because of the martensitic transformation, is probably due to dislocations and this value therefore lends support to the idea that their volume coefficient of resistivity is, apart from the amplitude term, similar to that of phonon scattering.

2.2 The noble metals

Linde (19) has made extensive studies of the pressure dependence of the residual resistivity due to impurities in the noble metals and Bridgman (20) has also made some measurements of this kind (see also Crisp et al. (21)). Recently in this laboratory we have been extending this work to survey fairly systematically the effect of pressure on the residual resistivity due to a wide range of non-transition metals dissolved in the noble metals. Linde's pressure measurements were made at about 0°C and Bridgman's at room temperature; this method is thus not very suitable for determining the pressure coefficient of residual resistivity due to homovalent impurities which cause comparatively little scattering; for these alloys low temperature measurements are needed.

The results of measurements from several sources are included in Table 2. It is seen that for the most part the more strongly heterovalent impurities tend to give consistently positive volume coefficients of resistivity with magnitude about 1, but even here there is considerable variation among the values. Among the homovalent and divalent impurities the variation is even more marked and quite a number of impurities have coefficients with a negative sign.

Let us compare these results with those predicted by simple free electron theory. Mott and Jones (23) give an expression for the residual resistivity of a dilute alloy in which the effect of the impurity can be represented by means of a screened Coulomb potential. If the screening is calculated by the Thomas-Fermi method this model predicts that $d \ln \rho_0/d \ln V$ should have a value somewhere between +1 and +1/3 depending on the

Table 2. Volume Derivatives of Resistivity due to Impurities in the Noble Metals

$d \ln \rho_o / d \ln V$		Parent metal		
Impurity	Valency	Cu	Ag	Au
Cu	1		+0.4	-0.3
Ag	1	-0.6		+1.2
Au	1	-0.7	+1.0	
Zn	2	0.3	0.9	-0.9*
Cd	2	-0.3	1.0	
Hg	2		1.7*	
Ga	3	0.9	1.1*	1.1*
In	3	0.7	1.6	0.1
Tl	3		1.4	
Ge	4	0.6**	0.8	0.8
Sn	4	0.9**	1.0*	1.4*
Pb	4			
As	5	0.6*	0.4*	
Sb	5	0.9*	0.9	0.8
Bi	5			

*Measured by Linde (19).

**Measured by Linde (19) and by Dugdale and Phillips (22).

Remainder measured by Dugdale and Phillips (22).

Crisp et al. (21) have measured Zn, Ga, Ge and As in copper; their values (deduced from measurements at 50°C) are slightly higher than those quoted here.

screening radius. For a given matrix the screening radius is fixed (by the electron density) so that in this model there would be no variation from one impurity to another. Moreover, even if one took account of the complicated Fermi surface involved this could not by itself explain the diversity of the results for different impurities in the same matrix. The variety of values observed experimentally and the fact that some impurities yield negative values suggest that the details of the potentials associated with the particular impurities are important. This, as Ziman (24) has stressed, is true of the actual values of ρ_0 so that it is perhaps even more to be expected in the volume derivatives of ρ_0. (25)

ACKNOWLEDGMENTS

I would like to thank Dr. A. M. Guénault for many valuable discussions and Dr. I. M. Templeton for access to his results before publication.

REFERENCES

1. J.S. Dugdale and J.A. Hulbert, Canad.J.Phys. 35, 720 (1957).
2. J.S. Dugdale and D. Gugan, Proc.Roy.Soc. A241, 397 (1957).
3. I.M. Templeton, Bull.Amer.Phys.Soc. 10, 349 (1965). Some of the data in this abstract are incorrect: the corrected data for copper are given in the text. Full details will be published.
4. J.M. Ziman, Proc.Roy.Soc. A226, 436 (1954).
5. M. Bailyn, Phys.Rev. 120, 381 (1960).
6. A.T. Stewart, J.J. Donaghy, J.H. Kussmiss and D.M. Rockmore, Bull.Amer.Phys.Soc. 9, 238 (1964) (Li); D. Schoenberg and P.J. Stiles, Proc.Roy. Soc. (London) A281, 62 (1964) (Na, K, Rb); K. Okumura and I.M. Templeton, Proc.Roy.Soc. (London) to appear (Cs).
7. Cf. M. Bailyn, Phil.Mag. 5, 1059 (1960).
8. A. Hasegawa, J.Phys.Soc. Japan 19, 504 (1964).
9. F.S. Ham, Phys.Rev. 128, 82 (1962).
10. W.B. Daniels, Phys.Rev. 119, 1246 (1960).
11. P.A. Smith and C.S. Smith, J.Chem.Phys.Solids 26, 279 (1965).

12. This explanation was originally proposed by M.H. Cohen and V. Heine, Advanc.Phys. 7, 395 (1958).
13. A.L. Jain, Phys.Rev. 123, 1234 (1961).
14. P.W. Bridgman, Proc.Amer.Acad.Arts Sci. 60, 385 (1925).
15. See for example D. Schoenberg, Phil.Trans.Roy. Soc. 255, 85 (1962).
16. W.B. Daniels and C.S. Smith, Phys.Rev. 111, 713 (1958).
17. e.g. P.W. Bridgman, Amer.Acad.Arts Sci. 83, 151 (1954); D.Gugan, Canad.J.Phys. 41, 1381 (1963).
18. J.S. Dugdale and D. Gugan, Proc.Roy.Soc. A270, 186 (1962).
19. J.O. Linde, Thesis (Lond.) 1939, as reported by A.N. Gerritsen, Encyclopedia of Physics XIX, 214 (1956).
20. P.W. Bridgman, Proc.Amer.Arts Sci. 84, 131, 179 (1955).
21. R.S. Crisp, W.G.Henry, E. Whalley and R.W.Wilson (to be published).
22. J.S. Dugdale and D. Phillips (to be published); these results are at present only preliminary.
23. N.F. Mott and H. Jones, Properties of Metals and Alloys (Oxford University Press, London, 1936), p. 294.
24. J.M. Ziman, Advanc.Phys. 13, 89 (1964).
25. The striking correlation in the monovalent metals between $d \ln K/d \ln V$ and $\partial \ln \sigma/\partial \ln E$ derived from the high temperature thermoelectric power has already been emphasized (18). A comparison between volume coefficients of residual resistivity and corresponding energy derivatives (derived from the thermoelectric power characteristic of that impurity) may also prove valuable.

SOME REMARKS ON MAGNETOSTRICTION IN NON-FERROMAGNETIC SOLIDS, WITH REFERENCE TO CERTAIN PRESSURE EFFECTS

B. S. Chandrasekhar: Department of Physics
Western Reserve University
Cleveland, U.S.A.

Consider a crystalline solid in a uniform field H. Then

$$e = -(\partial G/\partial \sigma)_H \tag{1}$$

$$M = -(\partial G/\partial H)_\sigma \tag{2}$$

where e and σ are the strain and stress in the crystal (taken to be uniaxial and uniform for simplicity), M is the magnetization, and G is the Gibbs potential. From (1) and (2) we get

$$C \equiv (\partial e/\partial H) = (\partial M/\partial \sigma)_H \tag{3}$$

which is the fundamental equation to magnetostriction. The following applications to pressure problems are possible from equations (1)-(3):

(a) In a normal metal at low temperatures, where G shows the characteristic Landau oscillations periodic in reciprocal field, the magnetostrictive strain e shows similar oscillations with the same periodicity.[1]

---

[1] B. S. Chandrasekhar, Phys. Letters 6, 27 (1963)
B. A. Green, Jr. and B. S. Chandrasekhar, Phys. Rev. Letters 11, 331 (1963)
D. M. Sparlin, B. S. Chandrasekhar, E. Fawcett, Bull. Am. Phys. Soc. 10, 350 (1965).

More relevantly to the present discussion, the amplitude of the magnetostriction oscillations is proportional to ($\partial A/\partial \sigma$), where A is the appropriate extremal cross-section of the Fermi surface. The value of ($\partial A/\partial \sigma$) can in fact be extracted from a comparison of the amplitudes of the magnetostriction and the de Haas-van Alphen torque on the same sample. This technique provides an alternative (which is sometimes more convenient) to the direct measurement of the pressure dependence of the de Haas-van Alphen effect.

(b) In a many-valley semiconductor or semimetal, a stress would in general cause relative shifts of the band edges and a consequent redistribution of carriers. The magnetization would therefore be stress-dependent, and from equation (3) one would expect a finite magnetostriction which, *inter alia*, should be related to deformation potentials. A preliminary elaboration of this idea has already been given.[2]

---

[2] J. J. Hall and R. W. Keyes, Bull. Am. Phys. Soc. 9, 96 (1964).

(c) In superconductors, the magnetostriction takes the special form of a change in dimensions on going from the normal to the superconducting state in a magnetic field, and is related to the volume dependence of the condensation energy for the transition.

(d) The magnetostriction in magnetically ordered systems is of course well known, and is discussed in other contributions to this Conference.

The point being made here is that magnetostriction can be used to obtain the pressure dependence of certain physical parameters which characterize a solid without actually doing the pressure measurements.

THE INFLUENCE OF PRESSURE ON THE BAND STRUCTURE OF SOLIDS

L. M. Falicov*

Department of Physics
and Institute for the Study of Metals
University of Chicago, Chicago, Illinois**

1. INTRODUCTION

In the first paper of this conference Dr. Harrison discusses how the influence of pressure on the band structure and Fermi surface of metals can be studied and interpreted in terms of a simple pseudopotential approach. In this paper a very similar approach is used to study the changes under pressure in the band structure of semiconductors and semimetals. This method gives in principle a quantitative way of interpreting the experiments and justifies the simple qualitative ideas obtained from crystal-structure and crystal-chemistry considerations. In addition it makes it possible to understand some of the phase transitions which take place under pressure, as for instance in the case of the Group V semimetals. In the remainder of this section a short summary of the pseudopotential method is given. The rest of the paper will be restricted to specific examples. Section 2 reviews the simple arguments obtained from crystal chemistry. Detailed results of calculations are given in section 3 for the semiconductors, and in section 4 for the semimetals.

It has been repeatedly proved (1) that the one-electron Schrödinger equation for the perfect crystal,

$$[\frac{p^2}{2m} + V(\vec{r})]\, \psi_{n\vec{k}}(\vec{r}) = E_n(\vec{k})\, \psi_{n\vec{k}}(\vec{r}) \qquad (1)$$

* Alfred P. Sloan Research Fellow.
** Work supported in part by the National Science Foundation and the Office of Naval Research.

can be rewritten

$$\left[\frac{p^2}{2m} + W(\vec{r})\right] \varphi_{n\vec{k}}(\vec{r}) = E_n(\vec{k})\, \varphi_{n\vec{k}}(\vec{r}) \,, \qquad (2)$$

where Eqs. (1) and (2) have the same eigenvalues $E_n(\vec{k})$. $W(\vec{r})$ is in general a non-local operator, and φ_{nk} is a smooth function which is free from the typical atomic-like oscillations of the real wave function $\psi_{n\vec{k}}$ close to the ion cores. Consequently an expasion of $\varphi_{n\vec{k}}$ in plane waves will converge rather rapidly and $W(\vec{r})$, the pseudopotential, may then be considered as a perturbation. The band structure is then determined by finding the roots E of

$$\det \left| W_{\vec{k},\vec{k}'} + \left(\frac{\hbar^2 k^2}{2m} - E\right) \delta_{\vec{k},\vec{k}'} \right| = 0 \qquad (3)$$

where $W_{\vec{k},\vec{k}'}$ is the matrix element of $W(\vec{r})$ between two plane waves.

The real simplification in the calculation appears when W(r) is assumed to be a superposition of local, spherically-symmetric, atomic-like pseudopotentials centered about each ion site, i.e.

$$W(\vec{r}) = \sum_{\substack{j \\ \text{ion} \\ \text{sites}}} w(|\vec{r}-\vec{r}_j|). \qquad (4)$$

In this case the matrix elements of the pseudopotential between two plane waves of wave-vectors $\vec{k}+\vec{q}$ and $\vec{k}$ can be written as

$$W_{\vec{k}+\vec{q},\vec{k}} = W_{\vec{q}} = S(\vec{q}) \cdot w(q) \qquad (5)$$

where $S(\vec{q})$ is the structure factor

$$S(\vec{q}) = \frac{1}{N} \sum_{\substack{j \\ \text{ion sites}}} \exp\, [-i\vec{q} \cdot \vec{r}_j) \qquad (6)$$

and w(q) is the form factor

$$w(q) = \frac{1}{\Omega_0} \int \exp\, [-i\vec{q} \cdot \vec{r}]\, w(|\vec{r}|)\, d^3\vec{r} \,. \qquad (7)$$

It is important to note that:

i) Eq. (6) is zero unless $\vec{q}$ is a reciprocal lattice vector $\vec{G}$;

ii) for crystal lattices with more than one atom per unit cell symmetry may require some $S(\vec{G})$ to be

zero;
iii) Eq. (7) shows that w(q) is independent of structure, depends only on the length of $\vec{q}$ and not on its direction, and, for constant $w(|\vec{r}|)$, is inversely proportional to the atomic volume Ω_o. It is of interest to emphasize at this point that the application of external pressure may modify the matrix elements $W_{\vec{q}}$ (and through them the electronic band structure $E_n(\vec{k})$) in various ways:
a) A change in the symmetry of the lattice, which radically changes the structure factors and the band structure.
b) A change in the relative position $\vec{r}_j$ of the atoms in the unit cell, which modifies the numerical values of $S(\vec{G})$.
c) A change in the lattice parameters, which varies the atomic volume Ω_o and scales w(q) accordingly.
d) The change in lattice parameters also alters the length $|\vec{G}|$ of the reciprocal lattice vectors and with it the values $q = |\vec{G}|$ where w(q) should be evaluated.
e) A change in atomic volume also affects the screening properties of the electrons and consequently modifies the functional dependence of w(q); this effect however is only important for small values of q and in most cases is of no consequence for $w(|\vec{G}|)$, $\vec{G} \neq 0$.

2. CRYSTAL-STRUCTURE AND CRYSTAL-CHEMISTRY CONSIDERATIONS

In what follows we restrict our attention to elements of the groups IV (C, Si, Ge, Sn) and V (As, Sb, Bi). However, before going into the details of the calculations, a few symmetry considerations together with the known chemical properties of these elements shed enough light into the physical properties of the solids to permit a qualitative understanding of the experiments.

Detailed calculations put these ideas on a more quantitative basis.

The group IV elements, with four valence electrons, show in the atom a ground state configuration s^2p^2. It is also well known that the sp^3 configuration lies, in energy, quite close to the ground state; from the point of view of the chemical bonding, the sp^3 configuration is energetically more favorable and it gives rise to four equivalent bonds tetrahedrally coördinated. This geometrical configuration favors, for the group IV elements, a crystallization in a diamond type lattice, i.e., a face-centered cubic lattice with two atoms per primitive cell placed at the points (0,0,0) and $\frac{1}{4}(\underline{a}_o, \underline{a}_o, \underline{a}_o)$, where $\underline{a}_o$ is the side of the cube. The diamond structure is known to be very stable and, due to its high symmetry, it is completely determined with the specification of only one parameter, i.e. $\underline{a}_o$.

The application of hydrostatic pressure is thus expected to produce only a change in $\underline{a}_o$, which in turn affects the band structure through the mechanisms (c), (d) and (e) of the last section. Changes in symmetry, with a corresponding more drastic change in electronic properties, would only occur through the application of shears or uniaxial stresses; Professor Fritzsche discusses these cases in more detail in a subsequent paper.

The Group V elements, with five valence electrons, have a ground-state atomic configuration s^2p^3. Chemically, however, they show a strong tendency to tri-valent behavior and this, added to a somewhat increased separation of the s and p terms seems to indicate that the s-electrons participate less in the chemical bonds of the Group V than in the Group IV elements. The electronic configuration responsible for the bonding is then p^3, which gives rise to three p-bonds at right angles and a resulting octahedral coordination. This coordination would in turn favor a crystal with simple-cubic structure (2). The simple cubic structure is however not stable because of the two following reasons:

i) A pentavalent simple-cubic solid with one atom per unit cell should have two and a half Brillouin zones occupied by the electrons. This gives complete metallic character and almost complete loss of directional covalent bonding.

ii) Born and his collaborators (3) have proved that under the assumption of central, pairwise interactions between atoms, the simple-cubic lattice is unlikely to be stable; it is interesting to note that in their arguments the tendency to instability increases with the size of the unit cell.

The final, stable structure of the Group V, the A7 structure, is nontheless not too different from the simple-cubic structure. It is obtained from the simple cube by two independent distortions: a shear and an internal displacement. The shear is along a body diagonal and transforms the unit cube into a rhombohedron. This diagonal retains its 3-fold symmetry and becomes the trigonal axis of the A7 structure. The second distortion can be easily visualized if the simple rhombohedral lattice is thought of as being composed of two interpenetrating face-centered rhombohedral lattices: each point of either lattice is in the center of the cell of the other lattice. The A7 structure is obtained then by shifting one of these lattices toward the other along the trigonal axis. This results in a face-centered rhombohedral structure with two atoms per unit cell. The complete structure is specified by three parameters: $\underline{a}$, the side of the rhombohedron; α, the rhombohedral angle; $\underline{u}$, half of the distance between the atoms in a unit cell, in units of the rhombohedron diagonal ($\underline{u} \leq 0.25$). It is worth pointing out that for special values of α and $\underline{u}$ higher symmetries are obtained; e.g. $\alpha = 60^o$ $\underline{u} = 0.25$ is the simple cubic lattice; $\alpha = 60^o$ $\underline{u} = 0.125$ is the diamond structure; $\alpha \neq 60^o$ $\underline{u} = 0.25$ is a simple rhombohedral lattice. In the first two cases $\underline{a}_o$, the side of the cube is $\underline{a}_o = \sqrt{2}\ \underline{a}$.

The application of hydrostatic pressure to a semimetal would change in general the three parameters $\underline{a}$, α and $\underline{u}$. Changes in $\underline{a}$ and α modify the band structure only through mechanisms (c), (d) and (e) of section 1; changes in $\underline{u}$ on the other hand, do not alter the atomic volume Ω_o or the reciprocal lattice vectors $\vec{G}$, but modify the structure factor $S(\vec{G})$ (mechanism(b)). Finally, if α and $\underline{u}$ take some of the special values mentioned above, a phase transition takes place and the electronic properties are drastically changed (mechanism (a)). The simple stability arguments given before (3) seem to indicate that as the pressure is increased and the volume decreased, there should be a tendency towards a simple-cubic structure, i.e. $\alpha \to 60^o$, $\underline{u} \to 0.25$. This is confirmed by experiment (4, 5).

3. THE GROUP IV SEMICONDUCTORS

Silicon and Germanium are probably the two solids for which the electronic band structure is best known, both from the theoretical and the experimental points of view. In addition, experimental data on Si and Ge under pressure (6-9) are readily available and therefore they seem to be the ideal choice for a theoretical study.

The top of the fourth (valence) band in both Ge and Si lies at $\vec{k} = 0$, the point Γ at the center of the Brillouin zone, and corresponds to the so-called Γ'_{25} level (10). All other energies will be measured with respect to this level. In the fifth (conduction) band there are three local minima of importance, the so-called L_1, Γ_2' and Δ_1 band edges. Their energies are

	Si	Ge
L_1	2.0 eV	0.65 eV
Δ_1	1.1 eV	0.80-0.85 eV
Γ_2'	~ 4.0 eV	0.80 eV

The indirect energy gaps E_g are thus 0.65 eV ($\Gamma'_{25} \rightarrow L_1$) in Ge and 1.1 eV ($\Gamma'_{25} \rightarrow \Delta_1$) in Si. The direct energy gap E_d is 0.80 eV in Ge ($\Gamma'_{25} \rightarrow \Gamma'_2$) and has a very large value (~3eV) in Si. A schematic plot of the energy band structure in Ge is shown in Fig. 1.

Bassani and Brust (12) and Brust (13) were able to obtain an accurate description of the band structure of Ge and Si by means of pseudopotential approach in which only three $w(|\vec{G}|)$ were non-zero. The values used in the calculation were

	$w(\lvert\vec{G}\rvert)$ Si	$w(\lvert\vec{G}\rvert)$ Ge
$\vec{G} = (111)$	-2.992 eV	-3.128 eV
$\vec{G} = (220)$	+0.544 eV	0 eV
$\vec{G} = (311)$	+1.088 eV	+0.816 eV
$\vec{G} = (400)$ or higher	0 eV	0 eV

In order to study the change in band structure with lattice parameter, it is necessary to know the complete function $w(q)$ as well as $w(|\vec{G}|)$. In the first paper of this conference Dr. Harrison uses a model in which $w(q)$ depends on four parameters: the charge of the ion, the radius of the ion core, the density of valence electrons and the strength of the core-electron repulsion. Bassani and Brust (12) use a similar approach and obtain (14)

$$w(q) = \frac{1}{\Omega_0}\left[-\frac{A}{q^2+\kappa^2} + B\,\frac{\epsilon^2-q^2}{[\epsilon^2+q^2]^4}\right] \qquad (8)$$

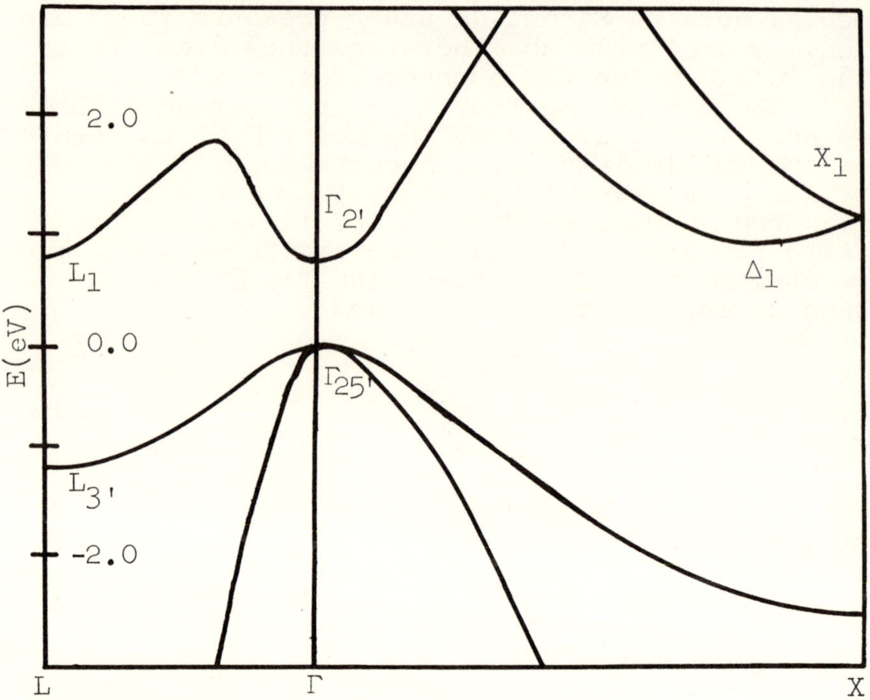

Fig. 1. The band structure of Ge.

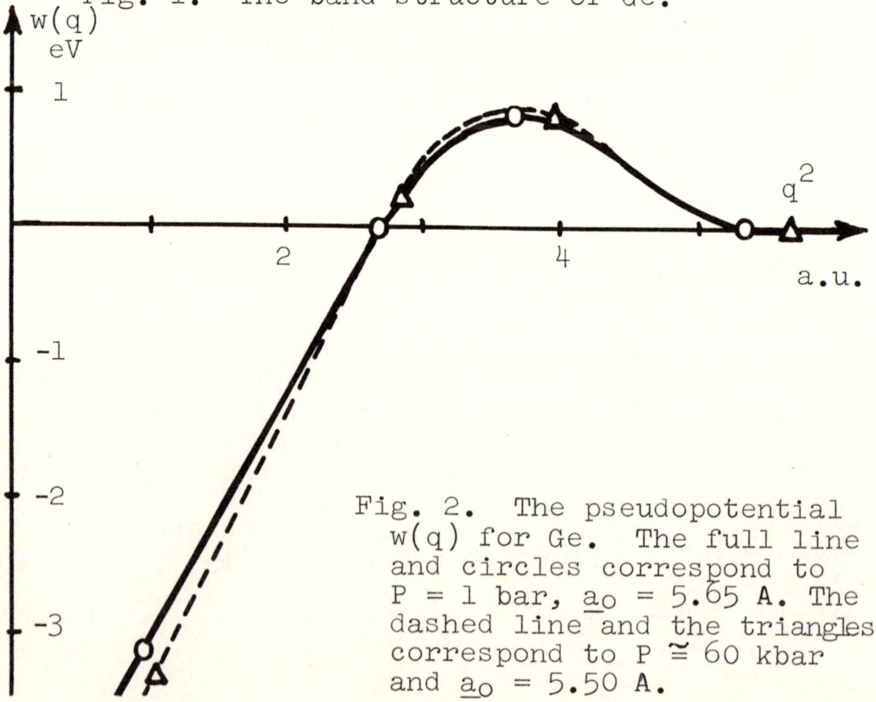

Fig. 2. The pseudopotential w(q) for Ge. The full line and circles correspond to P = 1 bar, a_o = 5.65 A. The dashed line and the triangles correspond to P ≅ 60 kbar and a_o = 5.50 A.

where $\varkappa$, ϵ, A and B are chosen to give the known values at $q = |\vec{G}|$.

In Fig. 2 we have plotted in a w(q) versus q^2 graph: (i) the four values of $w(|\vec{G}|)$ which give the observed band structure for Ge [circles]; (ii) a graphical interpolation which represents w(q) for a_o = 5.65 A, P = 1 bar [full line]; (iii) a scaling of the previous curve by a factor of 1.084, corresponding to an identical change in volume, i.e. a_o= 5.50 A, P $\simeq$ 60 kbars [dotted line]; (iv) the new four values of $w(|\vec{G}|)$ for a_o = 5.50 A [triangles]. These turn out to be in good agreement with the values 3.210 eV for (111), 0.163 eV for (220) and 0.979 for (311) used by Bassani and Brust (12) in their calculations. They have chosen a very high pressure in order to make the alterations in the energy bands more evident.

Figure 3 shows the results of the calculation together with the known experimental results for Ge and Si. It is seen that the agreement is not more than fair, but the theory is good enough to distinguish between sensitive and insensitive levels and explains beyond doubt the reason for the initial increase and subsequent decrease of the energy gap with pressure (8) as a change in the conduction band minimum from the L_1 edge to the Δ_1 edge.

4. THE GROUP V SEMIMETALS

The band structures of As, Sb and Bi are not yet as well understood as those of Si and Ge. This is due to the low symmetry of the crystal structure which causes some complications in the theoretical approach. However, the experimental information is quite abundant (15) and this, together with some preliminary band structure calculations (16-18) already in press or under way make it possible to form a fairly clear picture of the electronic properties of the semimetals.

The semimetallic character implies that a small portion at the top of the fifth (valence) band lies higher in energy than the bottom of the sixth (conduction) band. The Fermi level should be, by necessity, somewhere in between these two values. The (negative) difference in energy between the bottom of the sixth band and the top of the fifth is called the overlap energy and can be interpreted as a negative energy gap. General considerations

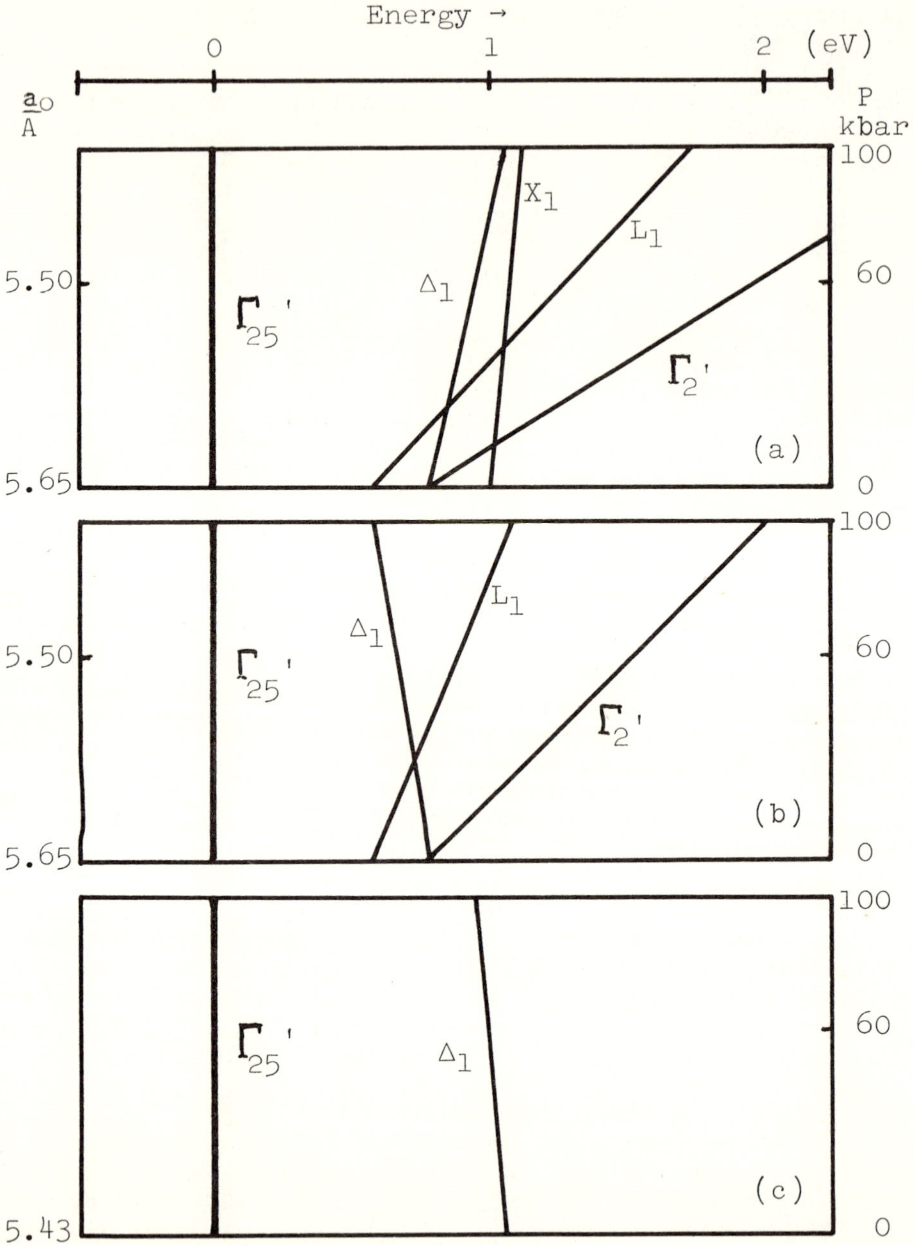

Fig. 3. Variation of the band edges as a function of pressure in Ge and Si. (a) Ge from the theoretical calculation of Bassani and Brust (12). (b) Ge, extrapolated from the experimental results (6-9,11). (c) Si, extrapolated from experimental data (8,11).

(2, 16, 17, 18) show that the bottom of the conduction band, i.e. the so-called electron carriers, should be at or close to the point L of the Brillouin zone (19); the top of the valence band on the other hand, i.e. the hole carriers, should be located at or close to the point T. The actual location and quantity of these maxima and minima are not definitely known yet, but there is strong evidence that:
(i) in Bi (20) there are three (electron) minima at the three equivalent L points and one (hole) minimum at T;
(ii) in Sb (18, 15, 21) there are three electron minima at L and six equivalent (hole) maxima along binary axes off the T point;
(iii) in As (22, 23, 16) it seems to be definite that there are six hole maxima similar to those of Sb, but it is not certain whether the electron minima are three (at L) or six (off L). These possibilities can all be obtained with rather small variations in the pseudopotential and, what is more important and difficult to calculate, by varying the strength of the spin-orbit interaction in the crystal. Fig. 4 shows a schematic representation of the band structure in Sb, in which the spin-orbit term is considered as an adjustable parameter.

As was mentioned in section 2, the application of pressure may change the three parameters of the crystal structure, namely a the lattice spacing, α the rhombohedral angle and u the internal displace-parameter. In general it can be proved from symmetry considerations and detailed calculations that for $u = 0.25$ the system will be metallic; for $\alpha = 60^\circ$ $u \neq 0.25$ the system will be semiconducting (i.e. the energy gap will be positive) and angles α very different from 60° and $u \neq 0.25$ will favor a semimetallic structure (i.e. a negative energy gap). In order to perform an accurate calculation for any of the semimetals, it is necessary to know, in addition to the pseudopotential at atmospheric pressure, the variation of a, u and α with pressure. This is not known for any of the semimetals; however: (i) the variation of a and α with pressure is known for Sb (4), and although the variation of u has not been measured (it would involve measurements of the x-ray line intensities), it is known that u does not take the value 0.25 up to P = 70 kbar. These measurements indicate that Sb becomes "more cubic"

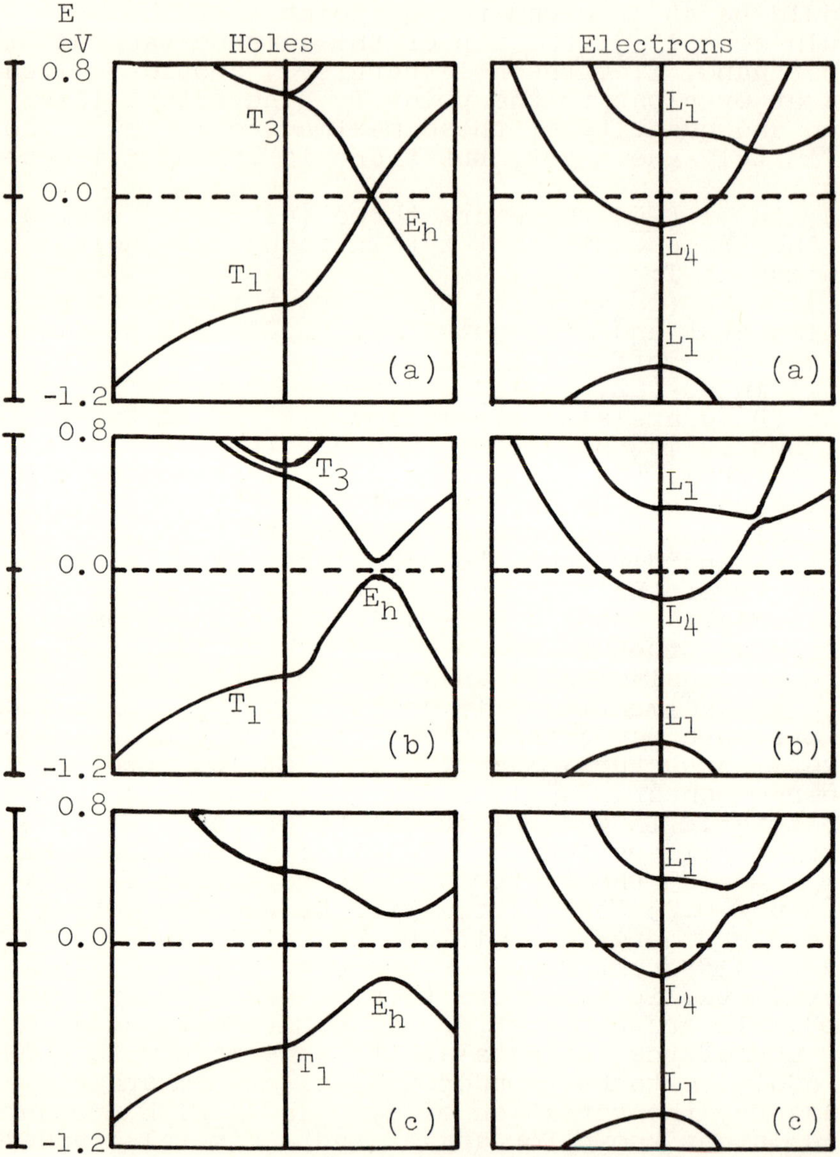

Fig. 4. A schematic representation of the band structure of Sb. (a) without spin-orbit interaction; (b) with a small value of the spin-orbit splitting; (c) with a large value of the spin-orbit splitting.

under pressure, and has a phase transition to simple cubic structure at P = 70 kbar, in good agreement with the simple ideas of section 2.

(ii) Bi (5) has a smiilar behavior and the well-known first-order phase transition (24) at room temperature and P = 24.8 kbars for BiI to BiII in which the resistance drops abruptly by a factor of 6, corresponds to the transition between the face-centered rhombohedral structure and the simple cubic BiII.

The experiments (5, 25-28) also indicate that Bi at smaller pressures ($P \cong 19$ kbars) has a higher order phase transition from a semimetallic to a semiconducting phase, i.e. the energy gap goes through zero from a negative to a positive value. In all probability, although no such evidence has been found experimentally as yet, Sb undergoes a similar transformation.

In order to put these ideas on a more quantitative basis we have performed a model calculation applicable to the crystal parameters of Sb. We have assumed a pseudopotential w(q) similar to that of Ge [Fig.2] which has been approximated (16) by the formula

$$w(q) = A_1(q^2-A_2)[\exp A_3\ (q^2-A_4)\ +1]^{-1} \qquad (9)$$

For the crystal parameters $\underline{a} = 4.301$ A, $\alpha = 60^o$, $\underline{u} = 0.233$, the values chosen for the constant were

$A_1 = 1.615$ eV $\qquad A_2 = 2.78$ atomic units
$A_3 = 2.38$ atomic units $\qquad A_4 = 3.70$ atomic units.

Energy band structures were computed:
(i) as a function of $\underline{a}$, for constant α and $\underline{u}$,
(ii) as a function of α, for constant $\underline{a}$ and $\underline{u}$.

In all calculations A_1 was scaled so as to be inversely proportional to the atomic volume Ω_o,

$$\Omega_o = \tfrac{1}{2}\ (1-\cos\alpha)\ \sqrt{1+2\cos\alpha}\ \ \underline{a}^3\ . \qquad (10)$$

The results are shown in Figure 5. The values at $\underline{a} = 4.301$ A, $\alpha = 57^o 14'$, $\underline{u} = 0.233$ correspond to the band structure of Sb at low temperatures and atmospheric pressures in this model.

In the whole region of interest the bottom of the sixth band corresponds to the L_4 level, while the top of the fifth band lies in between the T_1 and T_3 levels, at approximately an energy

$$E_h = \tfrac{1}{2}T_1 + \tfrac{1}{2}T_3 - \tfrac{1}{2}\,\delta_{so}, \qquad (11)$$

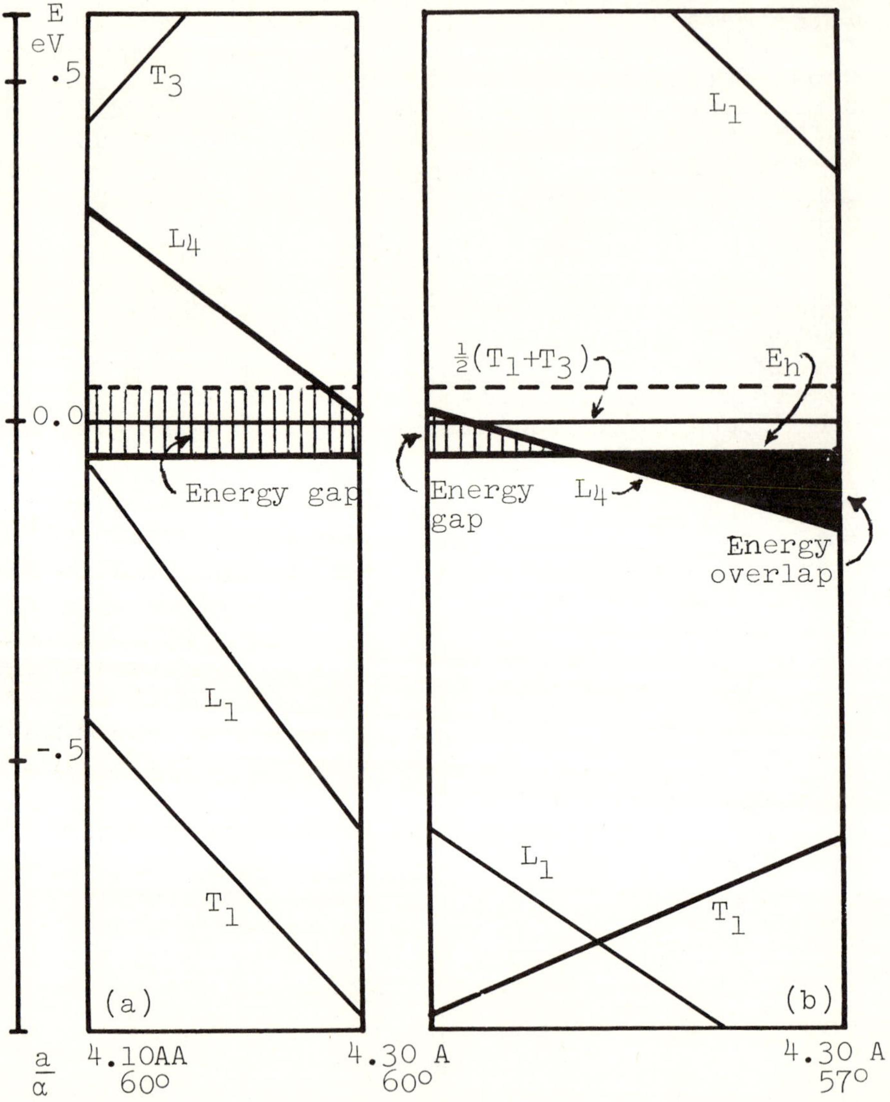

Fig. 5. A model calculation for the change in energy levels in Sb as a function of (a) lattice parameter a, and (b) rhombohedral angle α.

where δ_{so} is the value of the spin-orbit splitting. We have arbitrarily chosen in Fig. 5 a value of δ_{so} = 0.05 eV, but it can be considered an adjustable parameter which displaces the E_h line parallel to itself. It is seen that as α goes from the value of 57°14' toward 60° and as a decreases in magnitude, L_4 increases its energy with respect to E_h, i.e. the energy gap

$$E_g = L_4 - \tfrac{1}{2}T_1 - \tfrac{1}{2}T_3 + \tfrac{1}{2}\delta_{so} \qquad (12)$$

changes from negative (i.e. semimetallic overlap) to positive (i.e. semiconducting energy gap), giving rise to a high order phase transition as found experimentally in Bi (5, 25-28).

In order to make more realistic calculations or more accurate determinations of the variation of the energy levels with pressure, it is necessary to determine the pseudopotential more accurately (18), to know more about spin-orbit effects (18) and to measure from x-ray line intensities the variation of u with pressure. As far as the author is aware, no information is available at present on As, and the semimetallic-semiconductor transition in Sb has not been studied experimentally. However, all indications seem to predict a similar type of behavior for the three Group V semimetals.

ACKNOWLEDGEMENTS

The author is indebted to Dr. P. J. Lin for several stimulating conversations and for her help in the calculations. He is also grateful to Dr. R. Jaggi, Mr. J. Vanderkooy, Mr. L. Windmiller and Prof. M. G. Priestley for communicating their results prior to publication.

We are grateful to the National Science Foundation and the Office of Naval Research for direct financial support of this work. In addition the research benefited from partial support of related solid state theory by NASA and general support of the Institute for the Study of Metals by the NSF and ARPA.

REFERENCES

1. J. C. Phillips and L. Kleinman, Phys. Rev. 116, 287 (1959); M. H. Cohen and V. Heine, Phys. Rev. 122, 1821 (1961); B. J. Austin, V. Heine

and L. J. Sham, Phys. Rev. 127, 276 (1962).
2. M. H. Cohen, L. M. Falicov and S. Golin, IBM J. Res. Develop. 8, 215 (1964).
3. M. Born and K. Huang, Dynamical Theory of Crystal Lattices (Oxford at Clarendon Press, London, 1954) p. 143.
4. L. F. Vereshchagin and S. S. Kabalkina, J. Extpl. Theoret. Phys. (USSR) 47, 414 (1964) [Translation: Soviet Physics JETP 20, 274 (1965).]
5. R. Jaggi, private communication and to be published.
6. W. Paul and H. Brooks, Phys. Rev. 94, 1128 (1954).
7. W. Paul and D. M. Warschauer, J. Phys. Chem. Solids 5, 89 (1958); ibid 5, 102 (1958); ibid 6, 6 (1958).
8. T. E. Slykhouse and H. G. Drickamer, J. Phys. Chem. Solids 7, 210 (1958).
9. M. Cardona and W. Paul, J. Phys. Chem. Solids 17, 138 (1960).
10. The labeling of states corresponds to the notation of L. B. Bouckaert, R. Smoluchowski and E. Wigner, Phys. Rev. 50, 58 (1936).
11. For a very complete review of the experimental information see W. Paul and D. M. Warschauer, The Role of Pressure in Semiconductor Research, in Solids under Pressure, edited by W. Paul and D. M. Warschauer (McGraw-Hill, New York, 1963) p. 179.
12. F. Bassani and D. Brust, Phys. Rev. 131, 1524 (1963).
13. D. Brust, Phys. Rev. 134, A1337 (1964).
14. Equation (8) corresponds to an atomic-like potential screened by a volume-independent dielectric constant and a repulsive orthogonalization term due to a 3s Slater core orbital.
15. For a summary of the present state of the field and its many facets see the proceedings of the Topical Conference on Semimetals, New York, IBM J. Res. Develop. 8, 215 ff (1964), and the many references quoted there.
16. L. M. Falicov and S. Golin, Phys. Rev. 137, A871 (1965).
17. S. Golin, to be published (Arsenic).
18. L. M. Falicov and P. J. Lin, to be published (Antimony).
19. The notation for the energy levels is that of

M. H. Cohen, Phys. Rev. 121, 387 (1961) and of reference (16).

20. J. J. Hall and S. H. Koenig, IBM J. Res. Develop. 8, 241 (1964).
21. L. Windmiller, private communcation and to be published.
22. Y. Schapira and J. Williamson, Phys. Letters 14, 73 (1965).
23. J. Vanderkooy, private communication and to be published; M. G. Priestley, private communication.
24. F. P. Bundy, Phys. Rev. 110, 314 (1958).
25. R. Jaggi, A. L. Jain and H. Weibel, Phys. Letters 7, 181 (1963).
26. R. Jaggi, Physics of Semiconductors, Proceedings of the 7th International Conference, (Dunod, Paris, 1964) p. 413.
27. N. B. Brandt, Yu. P. Gaidukov, E. S. Itskevich and N. Ya. Minina, J. Exptl. Theoret. Phys. (USSR) 47, 455 (1964) [Translation: Soviet Physics JETP 20, 301 (1965)].
28. D. Balla and N. B. Brandt, J. Exptl. Theoret. Phys. (USSR) 47, 1653 (1964).

Effect of Pressure on the Connectivity of the Fermi Surface

J. E. SCHIRBER
Sandia Laboratory
Albuquerque, New Mexico

I. INTRODUCTION

The effect of pressure upon the detailed topology of the Fermi surface of metals is of increasing interest. With the free electron model, one can make an unambiguous prediction of the sign as well as a semiquantitative estimate of the magnitude of the pressure effect. Therefore measurements of the behavior of the Fermi surface topology under pressure provide sensitive tests of the validity of the nearly free electron model, particularly in the case of anisotropic metals. Further interest in the pressure work stems from the predictions made by Lifshitz (1) of striking changes in the thermodynamic and transport properties of a metal at "electron transitions" occurring when the connectivity of the surface is broken.

In this discussion we will limit ourselves to consideration of the galvomagnetic measurements under pressure as the preponderance of work to date is of this type and because these measurements are most easily related to the connectivity of the Fermi surface. Available results of such measurements on Cu, Ag, In, Cd, Pb and Zn are reviewed. The work on Zn is discussed in some detail as it is the most complete and best demonstrates both the advantages and limitations of this type of measurement under pressure.

II. BRIEF THEORETICAL BACKGROUND

The theory of the galvomagnetic behavior of metals in the high field region has been developed in detail by Lifshitz and Peschanskii (2) and Alekseevskii (3), et al. For the purposes of this discussion we need consider only the case of magnetoresistance. We define n_e and n_h as the total number of electrons and holes per unit cell summed over the respective electron and hole sheets. A metal is said to be compensated if $n_e=n_h$. Compensation is observed to occur for all non ferromagnetic metals when the product of the valence times the number of atoms in the unit cell is even. Following Fawcett (4), five cases for the qualitative field dependence of the magnetoresistance $\rho(H)$ in the high field limit are listed in Table 1.

Table 1. Field Dependence of $\rho(H)$

	Orbit Type (state of compensation)	$\rho(H)$
I	all closed (non compensated)	saturates
II	all closed (compensated)	$\sim H^2$
III	open in a single direction	$\sim H^2 \cos^2 \alpha^+$
IV	open in two directions	saturates
V	singular field direction‡	saturates

$^+\alpha$ is the angle between the open orbit direction and the current.

‡intersection of two or more open orbits resulting in closed trajectories

Case III can be further subdivided depending upon whether the orbits permitted in a given field direction are periodic in the reciprocal lattice or not. Periodic open orbits are found in planes of applied magnetic field and so give rise to lines on a stereogram. These great circle lines may be of any length. Aperiodic open orbits are generated for solid angles of applied field centered about symmetry axes and are represented on the stereogram by two dimensional areas.

The boundaries of both two-dimensional regions and one-dimensional regions less than 360° in length (whiskers) usually can be related to dimensions of the Fermi surface with the help of the nearly-free electron model (5). Priestley first pointed this out for the case of Cu (6). For non-compensated metals the boundaries of these regions are marked by peaks in the magnetoresistance given by the $H^2\cos^2\alpha$ rule so that the dimensions of the regions are readily determined. For compensated metals, the magnetoresistance varies as H^2 over most of the stereogram and the boundaries of these regions must be determined by more elaborate field scanning techniques (7)(8). Briefly, use is made of the $\cos^2\alpha$ term of Case III to mark the edge of a region supporting open orbits as the field sweep scans the region at right angles to the direction of sweep. The edge of the region is distinguished by the disappearance of a characteristic cusp-like trough in the plot of $\rho(H)$ vs sweep angle. (See Figure 2, reference 7 or Figure 1, reference 8.) In dealing with the hexagonal metals, the free electron theory must be modified by the inclusion of spin-orbit interactions (9)(10) and magnetic breakdown effects (11) in order to agree with the experimental observations. For our purposes it suffices to state that magnetic breakdown can occur across small energy gaps of width $\Delta\epsilon$ between sheets of the Fermi surface if $(\epsilon_F \hbar\omega_c)^{\frac{1}{2}} > \Delta\epsilon$ (12)(13), where ϵ_F is the Fermi energy and ω_c is the cyclotron frequency eH/m^*c. The dimensions of regions on the various sheets of the Fermi surface for which this inequality is satisfied enter into some of the equations relating the size of features on the magnetoresistance stereogram to details of the Fermi surface. This will be elaborated upon in the discussion of Zn in Section IV.

III. PRESSURE TECHNIQUE

Except for the work of Gaidukov and Itskevich (14), all the data discussed were taken using solid He pressure generation techniques. The method consists of the isobaric freezing of He about a sample and gives all indications of yielding hydrostatic pressures at liquid He temperatures (15)(16). A schematic of the system is shown in Figure 1. The high pressure side of a 14 kbar Harwood intensifier is charged with 2-4 kbar of He gas. Final pressure is generated in a single stroke of the intensifier with the Be-Cu sample chamber (1/8-in. ID x 7/16-in. OD) at a temperature just above the freezing temperature of He at the desired pressure. The sample chamber is then cooled very slowly from below by lowering to a colder level in the dewar while the pressure is maintained at a constant value.

This nearly isobaric freezing has two distinct advantages: 1.) the pressure drop from the melting curve to 4°K can be quite accurately calculated from the known isochores for solid He (17), 2.) the final pressures are closer (~90%) to the highest pressure the system will hold than for cooling a fixed quantity of gas. The difficulty in keeping the system leak tight to He gas sets a practical upper limit with this system at ~9 kbars at 4°K. Since a leak usually results in the loss of a carefully oriented high purity single crystal sample, most of the data are taken below 6 kbar.

The two current and two potential leads exit through a frozen oil (Octal S) seal such as described by Dugdale and Hulbert (18). The pressure on the low side of the intensifier is read with a 0-20,000 psi Heise gauge. The intensifier ratio (nominally 15:1) was determined to 7 kbar by putting a 0-100,000 psi Heise gauge on the high side of the intensifier. The estimated uncertainty in the pressures at 4.0°K is less than 5%.

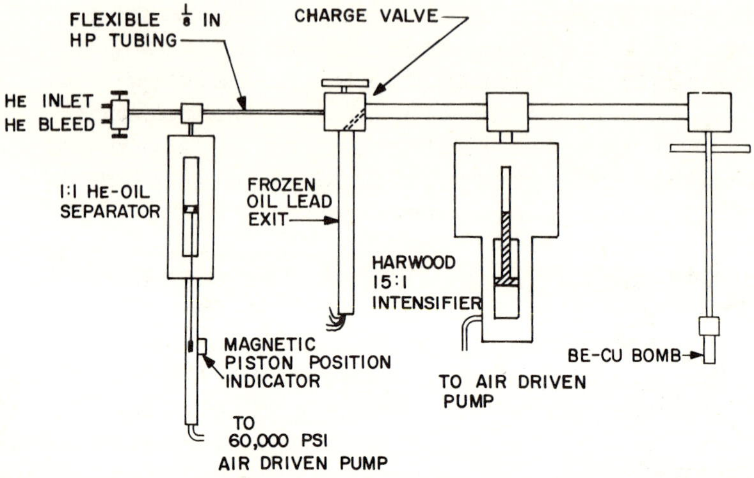

Figure 1. A Schematic of the 9 kbar solid He pressure apparatus.

IV. RESULTS

The ρ(H,P) experiments can be separated into two groups: 1.) those experiments where the possibility of changes in connectivity is investigated (as would be manifested by the appearance of new open orbits or the quenching of existing ones), 2.) those experiments in which the change in actual dimensions of the Fermi surface are investigated. The work on Cd and In are in the former group, while that on Cu, Ag, Pb and Zn are in the latter.

A. Cadmium. Within the framework of the free electron model, the hexagonal series Mg-Zn-Cd have very similar Fermi surfaces differing in detail because of the monotonic increase in c/a in the series. For example, the horizontal waists of the monster hole surface which are necessary for the existence of open trajectories in the basal plane are observed experimentally (19) to decrease in size as c/a increases. The waists are finite in size in Zn (see Figure 2) and are pinched off in Cd and consequently open orbits are observed in the basal plane of the former metal but not of the latter. Such orbits give rise to "whiskers" centered at the c axis on the magnetoresistance stereogram whose lengths are related to the size of the waists (see Sect. IV-D and Figure 3). Because of the large anisotropy of the compressibility of Cd, its c/a becomes equal to that of Zn at hydrostatic pressures of about 13 kbar.

The magnetoresistance of Cd was investigated at pressures to nearly 9 kbar (20). There was no indication that any new orbits were opened. The sample axis was progressively tipped (±7 deg) relative to the axis of sweep to insure that the very small features expected if trajectories just opened were not missed due to slight sample misorientation. This result agrees with the earlier work of Gaidukov and Itskevich to 7 kbar (14).

B. Indium. The free electron second zone hole surface of In intersects the tetragonally distorted fcc Brillouin zone at the corners of the square faces whose normals are in the [100] and

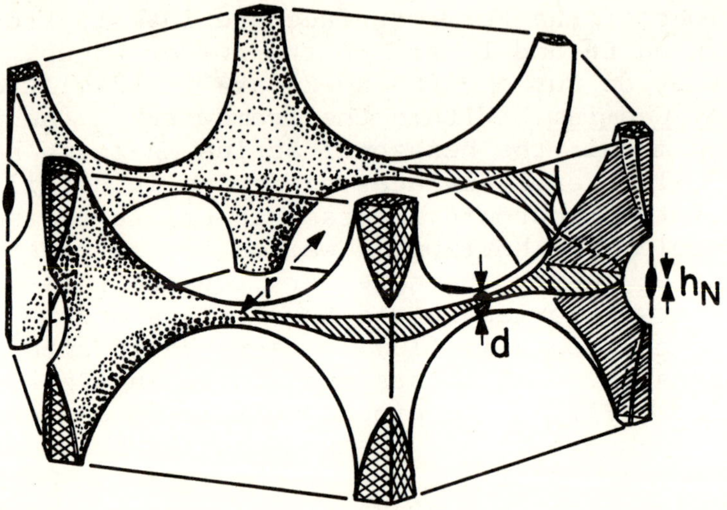

Figure 2. "Monster" and "needle" sheets of the Fermi surface of Zn showing dimensions related to the connectivity in the basal plane.

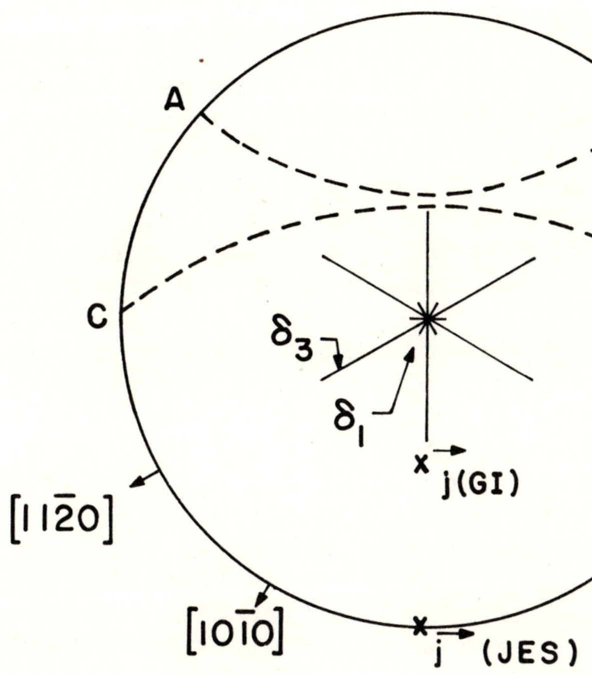

Figure 3. [0001] projection showing regions of applied magnetic field supporting open orbits in Zn. AB is the field sweep taken with our crossed field geometry. CD is the sweep used by GI.

[010] directions. Open trajectories are then expected particularly in the plane perpendicular to the tetrad axis. Both galvomagnetic (21) and magnetoacoustic (22) measurements indicate that no open trajectories exist in any direction in In.

Hydrostatic pressure increases the tetragonal distortion of the lattice and so increases the likelihood of open orbits. We have investigated the magnetoresistance of In at pressures up to ~7.5 kbar and see no indication of open trajectories. As in the work on Cd, the region of the stereogram where the new orbits would be expected was scanned by progressively tipping the axis of the crystal relative to the axis of sweep of the applied field.

C. Copper and Silver. The first attempt to make quantitative measurements of the effect of pressure on sizes of features of the Fermi surface using magnetoresistance techniques was the work of Caroline and Schirber (16) on Cu and Ag. The Fermi surfaces of the noble metals consist of spheres connected in the extended zone scheme by necks in [111] directions. As pointed out by Priestley (5), the size of these necks can be related to the dimensions of areas on the magnetoresistance stereogram which support open orbits. The edges of these areas are marked by sharp peaks in the transverse magnetoresistance. Measurement of the transverse magnetoresistance of Cu and Ag to pressures of 2.7 kbar showed that the diameter of the necks relative to the size of the Brillouin zone was unchanged with pressure. This corresponds to a linear scaling of the dimensions of the Fermi surface with those of the Brillouin zone. The precision of the measurement was such that a change of 0.2%/kbar for Cu and 0.3% for Ag in the ratio of the neck size to the zone dimension could have been detected. Very recently Templeton has reported de Haas van Alphen measurements on Cu to pressures of 25 bar which show this ratio constant to within ±0.06%/kbar (23). The Fermi surface of Pb has also been shown to scale in this manner (24).

This scaling of the Fermi surface of cubic

materials with pressure (which at these low pressures is equivalent to scaling with the volume in reciprocal space) is what might be expected from free electron considerations since the fractional occupation of the zone remains fixed. On the other hand, the features of the surface supporting connectivity are often just those features which are least well predicted by theory (e.g., the free electron model gives a closed Fermi surface for the noble metals) so that it is dangerous to generalize this scaling behavior.

D. Zinc. The sheets of the Fermi surface of interest to us in Zn are the monster hole surface and the needles (see Figure 2). The basal plane open orbits which can be related to dimensions of the surface require 1.) that the height, d, of the horizontal waists be finite (as in the section on Cd above) and 2.) that there be a finite height h_N on the needle in which an electron can break down the gap between the needle and the monster. Open orbits are then possible in the basal plane in both the $[10\bar{1}0]$ and $[11\bar{2}0]$ directions and give rise to the δ_1 and δ_3 whiskers respectively as shown on the stereogram in (Figure 3)(8)(25). The angular length of these whiskers can be related to d, h_N, r (the radius of the monster), r_n (the radius of the needle) and k_a (one half the reciprocal lattice vector in the basal plane) as follows:

$$\tan \delta_1 = \tfrac{1}{2}(h_N+d)/r$$

$$\tan \delta_3 = (h_N+d)/[\sqrt{3}r-(2/\sqrt{3})k_a+r_N].$$

From the form of these two equations it is apparent that it is impossible to separate h_N and d. Therefore δ_3 has received most of the attention in the pressure work since any change in the mean dimension $(h_N+d)/2$ should be most easily observed in δ_3 (for Zn, $\delta_1=5.63°$, $\delta_3=42.5°$ (8)). The experimental uncertainty in the determination of both δ_1 and δ_3 is of the order of 0.1 deg which amounts to a detectable change of about 0.23% in δ_3 compared to about 2% in δ_1. Furthermore Gaidukov and

Itskevich (GI) reported that the δ_3 whiskers shortened with pressure (14). From their results GI inferred that the connectivity of the Fermi surface of Zn in the basal plane would be broken by pressures of about 30 kbar. This is a surprising result as the free electron model would predict an increase in d and a net decrease in the denominator of the equation for tan δ_3. While this model says nothing about h_N, the empirical data for the series Cd-Zn-Mg show a monotonic increase in the length of the whiskers from Cd(δ_1=δ_3=0, no open trajectories in the basal plane) to Mg (δ_1=15 deg, δ_3=60 deg). (19)(26)

We have made a quantitative measurement of the effect of pressure on δ_3 which, contrary to the results of GI, shows that δ_3 increases at the rate of 0.066 $\pm$0.010 deg/kbar (27). (See Figure 4).

The tipping techniques mentioned in Section II are relatively simple to use at zero pressure, but rigid high pressure hardware coupled with the constraints of dewars and electromagnet pole-piece geometries make these techniques difficult in pressure applications. The δ_3(P) experiment requires a tip of $\pm$43 deg in order to sample the extremities of the region (see Figure 3). This problem was solved by adding a variable vertical field generated in a superconducting solenoid to the horizontal field of a rotatable electromagnet (27). The dewar and solenoid system is fixed to the electromagnet so that there is no relative motion of the two magnets. The direction of the resultant field is reproducible to better than 0.1 deg. The field sweeps are "latitude" lines on a stereogram (Figure 3) rather than the typical great circle sweeps observed in transverse magnetoresistance experiments and in tipping geometries where the axis of the crystal is tipped relative to the axis of rotation of the electromagnet (7).

The crossed field geometry allows us to eliminate error due to sample shifts in the temperature/pressure cycling. The angular distance between the end of the whiskers and A and B in

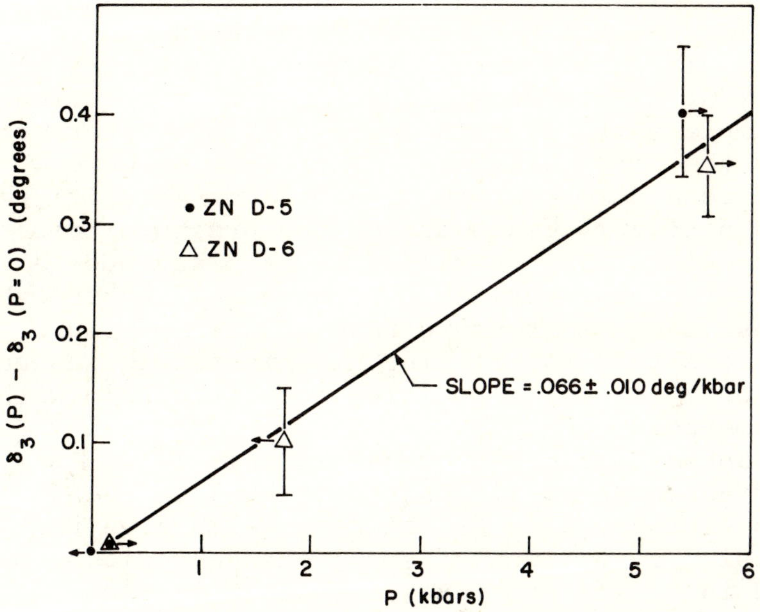

Figure 4. Change in the δ_3 whisker of Zn with pressure. Arrows denote the direction the pressure was changed to arrive at the point.

Figure 3 allow correction for any rotation of the crystal about its c axis. A shift of the c axis from the horizontal plane contributes no error as both ends of the whisker are effectively measured by rotating the horizontal field 180 deg and measuring the whisker centered on the negative c axis. The great circle sweep (CD in Figure 3) used by GI allows no similar internal correction for sample shift as this sweep does not intersect any of the other whiskers. Therefore GI were forced to assume that their sample orientation remained perfectly fixed during the pressure/temperature cycling. The result of GI is based upon a qualitative observation that the cusp-like trough was less deep at pressure than at zero pressure. We believe that the filling of this trough was due to a shift in sample position not detectable with their method.

Subsequent measurements in this laboratory of the effect of high fields (to 60 kOe) on the length of the δ_1 and δ_3 whiskers in Zn gave the surprising result that a small change could be detected in δ_1 but not in δ_3. Just as in the pressure work it had been assumed from the form of the equations for tan δ_1 and tan δ_3 that any change in h_N with field would result in a much greater angle change in δ_3 than in δ_1. This field dependence can be understood by adjustment of the angle of divergence of the gap between the needle and the monster (see Figure 21, reference 25). Another way of saying this is that h_N varies differently with field in the equations for δ_1 and δ_3.

Our field results suggest that h_N might also vary differently with pressure in the equations for tan δ_1 and tan δ_3 thus negating the above arguements about the relative change in δ_1 and δ_3. We have measured the effect of pressure to 5 kbar on the short whisker δ_1 and find that this is indeed the case. The relative change in δ_1 with pressure is nearly an order of magnitude larger than in δ_3. The results of a representative run are shown in Figure 5. A completely independent determination

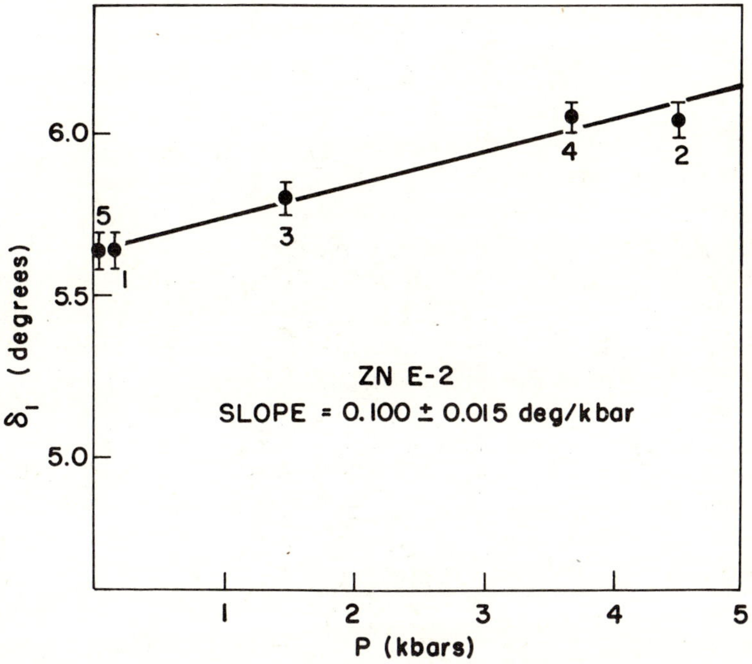

Figure 5. Change in the δ_1 whisker of Zn with pressure. The points are numbered in the order in which they were taken.

of the angle at which the δ_1 whisker disappears is possible in this case since the sweep intersects the δ_3 whiskers on either side of the feature of interest. Tip angles determined from the angular spacing of these flanking δ_3 troughs were in good agreement with the angles given by the calibrations of the crossed fields.

While a reasonable model can be envisioned for the δ dependence of $\Delta h_N(H)$, it is not apparent what form the δ dependence $\Delta h_N(P)$ should take. If no δ dependence is included, the pressure dependences of δ_1 and δ_3 give a relation between dr/dP and dr_N/dP. The free electron model predicts that dr/dP is negative so that the largest reported value of $(1/r_N)dr_N/dP$ (the measured values vary from -0.15/kbar (28) to 0.18/kbar (29)) is too small by about a factor of two. Determinations of $r_N(P)$ and $d(P)$ by de Haas-van Alphen measurements are needed to clarify this situation.

V. SUMMARY

To date no information has been obtained concerning the "electron transition" discussed by Lifshitz. It may be necessary to investigate this effect by alloying so that the limited pressures available at present will be sufficient to cause the effect. Alloying increases the fields necessary to achieve the high field limit but advances in superconducting solenoid technology soon may be able to make up the difference.

The quantitative measurements of Fermi surface features as a check on predictions of the free electron theory appear most interesting in those materials having anisotropic compressibilities. Here, pressure can change relative dimensions of the Brillouin zone. The cubic metals investigated to date appear to scale in a linear fashion with the reciprocal lattice. The heavy reliance on a model in order to relate the magnetoresistance features to dimensions of the surface makes it highly desirable to complement these measurements with extremal area measurements as a function of pressure. The d.c. field de Haas-van Alphen measurements in high homogeneity superconducting solenoids promise to fill this need in the very near future (23).

VI. ACKNOWLEDGMENTS

The author thanks Professor R. G. Chambers for suggesting the original experiments on the noble metals and D. Caroline for communicating his results prior to publication. Helpful conversations with Dr. A. Narath, Dr. W. J. O'Sullivan, and Dr. D. C. Wallace and the technical assistance of D. D. Sand are gratefully acknowledged.

REFERENCES

1. I. M. Lifshitz, Soviet Phys. JETP 11, 1130 (1960).
2. I. M. Lifshitz and V. G. Peschanskii, Soviet Phys. JETP 8, 875 (1959) and 11, 137 (1960).
3. N. E. Alekseevskii, Yu. P. Gaidukov, I. M. Lifshitz and V.G. Peschanskii, Soviet Phys. JETP 12, 837 (1961).
4. E. Fawcett, Advan. Phys. 13, 139 (1964).
5. W. A. Harrison, Phys. Rev. 118, 1190 (1960).
6. M. G. Priestley, Phil. Mag. 5, 111 (1960).
7. J. E. Schirber, Phys. Rev. 131, 2459 (1963).
8. J. E. Schirber, Proceedings of the Ninth International Conference on Low Temperature Physics, Columbus, Ohio, 1964 (to be published).
9. M. H. Cohen and L. M. Falicov, Phys. Rev. Letters 5, 544 (1960).
10. L. M. Falicov and M. H. Cohen, Phys. Rev. 130, 92 (1963).
11. M. H. Cohen and L. M. Falicov, Phys. Rev. Letters 7, 231 (1961).
12. E. I. Blount, Phys. Rev. 126, 1636 (1962).
13. A. B. Pippard, Proc. Roy. Soc. (London) A270, 1 (1962).

14. Yu. P. Gaidukov and E. S. Itskevich, Soviet Phys. JETP 18, 51 (1964).
15. J. E. Schirber and C. A. Swenson, Phys. Rev. 123, 1115 (1961).
16. D. Caroline and J. E. Schirber, Phil. Mag. 8, 71 (1963).
17. J. S. Dugdale, Nuovo Cimento 9, Suppl. 27 (1958).
18. J. S. Dugdale and J. A. Hulbert, Canad. J. Phys. 35, 720 (1957).
19. See Ref. 4 for magnetoresistance references and Alexander D. C. Grassie, Phil. Mag. 9, 847 (1964) for dHvA reference.
20. J. E. Schirber, Bull. Am. Phys. Soc. 9, 98 (1964).
21. N. E. Alekseevskii and Yu. P. Gaidukov, Soviet Phys. JETP 10, 481 (1960).
22. J. A. Rayne and B. S. Chandrasekhar, Phys. Rev. 125, 1952 (1962).
23. I. M. Templeton, Bull. Am. Phys. Soc. 10, 349 (1965). See J. S. Dugdale, this volume.
24. D. Caroline, unpublished results.
25. R. W. Stark, Phys. Rev. 135, A1698 (1964).
26. R. W. Stark, T. G. Eck and W. L. Gordon, Phys. Rev. 133, A443 (1964).
27. J. E. Schirber, Phys. Rev. Letters 14, 66 (1965).
28. I. M. Dmitrenko, B. I. Verkin and B. G. Lazarev, Soviet Phys. JETP 8, 229 (1959).
29. K. S. Balain, C. G. Grenier and J. M. Reynolds, Phys. Rev. 119, 939 (1960).

Possible Uses of High Pressure Techniques for the Study of the Electronic States of Molecular Crystals

Stuart A. Rice and Joshua Jortner

Department of Chemistry and Institute for the Study of Metals
University of Chicago, Chicago, Illinois 60637

I. INTRODUCTION

In this paper we wish to survey the current theory of the electronic states of molecular crystals of aromatic molecules, and to examine how experimental studies at high pressure may be used to solve some of the problems that arise in the interpretation of observations.

The theory of light absorption by molecular crystals was first studied by Frenkel (1) and Peierls (2) and has since been extended by many workers (3). The conventional theory is based on the observation that, at low pressure, molecular crystals are held together by weak dispersion forces so that, to a first approximation, the manifold of electronic states of each molecule can be considered to be unaltered by the crystal field. Because of the small but finite intermolecular forces, it is impossible to construct a stationary state for a perfect crystal in which some one selected molecule is excited. The correct zero order states for the crystal with one quantum of electronic excitation correspond to a delocalization of the electronic energy over all molecules. It is convenient to describe the quantum of excitation as an exciton wave of momentum $\hbar \underline{k}$ which propagates through the crystal with a velocity determined by the magnitude of the intermolecular interaction.

Now, aromatic compounds have a number of electronic transitions for which the spatial extent of the electron density is almost the same in both the ground state and the excited state. In this case, for the zero pressure crystal geometry, the overlap integrals between molecular orbitals on adjacent molecules are of the order of 10^{-2} to 10^{-3}, and considerable simplification of the analysis is feasible. For, in the case of small intermolecular overlap, the Heitler-London formalism may be used to construct proper zero order wave functions. Following the pioneering work of Davydov (3a), the Heitler-London scheme has often been used to construct both the ground state and excited state wave functions, and much effort has gone into calculations of energy shifts and splittings associated with the interaction of a molecule and the weak crystal field (3).

An alternate description of the exciton was given by Wannier (4), who constructed exciton states from the states of the conduction band and the valence band. Using these, a wave packet was built describing the localization of the electron due to the electron-hole (Coulomb) interaction. When the energy eigenvalues corresponding to the relative motion of the electron and hole are close to the ionization continuum , the effective wave equation becomes identical to the Schrodinger equation for the hydrogen atom. This is an adequate description only for large radius excitons, and the effect of the medium enters explicitly by dielectric screening. We shall later discuss charge transfer excitons in aromatic crystals, which are closely related to the Wannier model.

Consider first the properties of the crystal at zero pressure. It is convenient to start our discussion by assuming that the electronic states of the individual molecules are known, and to consider the following effects of the crystalline medium on the manifold of the electronic states:

(1) Spectral Shifts: Environmental medium effects lead to spectral shifts representing the end effects of very subtle balances between intermolecular interactions. These typical solvent effects are not of special interest for our present discussion.

(2) Removal of electronic degeneracy: The crystal field leads to a splitting of degenerate electronic states. A typical example is the doubly degenerate E_u state of benzene, which is expected to split in the crystal.

(3) Violation of selection rules: Symmetry forbidden electronic transitions which are vibronically induced in the isolated molecule become formally allowed in the crystal. Typical examples are transitions to the $^1B_{2u}$ and $^1B_{1u}$ excited states of benzene.

(4) The Davydov splitting: In crystals containing more than one molecule per unit cell a splitting of the excited state may be observed, due to interactions between translationally inequivalent molecules.

(5) Crystal field mixing: Configuration interaction between exciton states arising from crystal field perturbation is of considerable importance in determining the intensities and the splitting of the Davydov components.

(6) Appearance of excited states of the crystal which are not observed in the isolated molecule: Wannier exciton states have been observed experimentally in inorganic semiconductors and in rare gas crystals (5). These states have not yet been observed in organic

crystals.

(7) Effects of exciton trapping on the emission spectrum: When the rate of intermolecular electronic excitation transfer is less than the rate of formation of an excited deformed region in the crystal, excitation trapping will occur. The formation of a localized exciton may lead to a marked change in the emission spectrum. Red shifts of the order of 0.5-2.0 eV in the emission spectrum (relative to the absorption spectrum) have been observed in some aromatic crystals (Pyrene and Perylene) and in rare gas crystals (7).

What changes may be expected to occur in the states of the crystal when the crystal is compressed? Clearly, the magnitude of the spectral shift, mentioned in (1), will change markedly as the density of the crystal is increased. Provided that the compression of the crystal does not lead to a change of phase (and hence crystal geometry), and that the overlap in the compressed phase is still small, the effects (2), (3), (4), and (5) will be altered in magnitude, but otherwise basically the same in character. If the compression is sufficient to make the overlap between molecules large, the Heitler-London description of the excited states becomes a poor approximation, and the character of the excited crystal states begins to lose all resemblance to the excited states of the parent free molecule. Thus, at high pressure, it is anticipated that the effects (6) and (7) will be enhanced. Finally, if the compression is large enough, the crystal ceases to be "molecular" and becomes essentially metallic.

In the following we consider, in order, the theory of singlet exciton states, charge transfer exciton states, triplet exciton states, electron and hole mobility, exciton-exciton annihilation, and related topics. Finally, we shall also consider the changes brought about by large scale compression of the molecular solid, and examine both the experiments already reported and those not yet performed that are most likely to lead to greater understanding of the properties of molecular solids.

II. SINGLET EXCITON STATES

A. General Comments

The Hamiltonian for a crystal containing N unit cells and h molecules per unit cell may be written in the form

$$H = \sum_{\mu=1}^{h} \sum_{k=1}^{N} H_{k\mu} + \sum_{k\mu < \ell\nu} V_{k\mu,\ell\nu} \tag{1}$$

where the double indices $k\mu$ and $\ell\nu$ label different molecules, $H_{k\mu}$ is the Hamiltonian for an isolated molecule and $V_{k\mu,\ell\nu}$ is the intermolecular pair interaction. Since in the low pressure limit the coupling between molecules is small compared with intramolecular interactions, the Heitler-London formalism may be used. The ground state of the crystal is, then, represented in the form

$$\Psi^{\circ} = \mathcal{A} \prod_{k\mu=1}^{Nh} \varphi_{k\mu}^{\circ} \tag{2}$$

where the $\varphi_{k\mu}^{\circ}$ are the ground state (antisymmetrized) wave functions of the individual molecules and $\mathcal{A}$ is the antisymmetrization operator permuting electrons between the molecules.

The symmetry adapted wave functions corresponding to the fth excited state of the crystal are, in the same formalism,

$$\Psi_i^f = \frac{1}{\sqrt{h}} \sum_{\alpha=1}^{h} B_\alpha^i \, \Phi_\alpha^f \tag{3}$$

$$\Phi_\alpha^f = \left(\frac{h}{N}\right)^{1/2} \sum_{m=1}^{N} e^{i\mathbf{k}\cdot\mathbf{r}_{m\alpha}} \phi_{m\alpha}^f \tag{4}$$

$$\phi_{m\alpha}^f = \mathcal{A}\, \varphi_{m\alpha}^f \prod_{\ell\nu \neq m\alpha}' \varphi_{\ell\nu}^{\circ} \tag{5}$$

The state (5), representing a single excitation, is not an eigenfunction of the tight binding Hamiltaonian. This Hamiltonian is characterized by the full symmetry of the crystal lattice, so that the wave functions (3) should belong to an irreducible representation of the crystal space group. The entire set of symmetry operations includes proper and improper rotations, glide planes and screw axes, and primitive translations. The translations form an invariant subgroup. The cosets of the translation subgroup from the factor group, which is isomorphic to a point group.

The functions Φ_α^f are known as one site excitons, $\mathbf{r}_{m\alpha}$ is the vector from the origin to the center of the molecule $m\alpha$, and the coefficients B_α^i can be found by diagonalizing the h x h secular determinant or by using the symmetry properties of the group of the wave vector.

For the case where there exists one state for each representation of the factor group, the energies of the Davydov components of this excited state, relative to the ground state, are given by

$$E_f^i = E_f^o + D_f + M_f^i(\underset{\sim}{k})\,; \quad i = 1, \ldots, h. \tag{6}$$

In Eq. (6) E_f^o is the excitation energy of the gree molecule and D_f represents the energy shift of the center of gravity of the band:

$$D_f = \sum_m \sum_\mu \{\langle a\varphi^f_{n\nu}\varphi^f_{n\nu} | V_{n\nu,m\mu} | a\varphi^o_{m\mu}\varphi^o_{m\mu}\rangle - \langle a\varphi^o_{n\nu}\varphi^o_{n\nu} | V_{n\nu,m\mu} | a\varphi^o_{m\mu}\varphi^o_{m\mu}\rangle\}. \tag{7}$$

Since D_f leads to a uniform shift of all the energy levels, it may be thought of as an environmental effect. On the other hand, $M_f^i(\underset{\sim}{k})$ is different for each representation of the factor group and thereby leads to the splitting of the crystal level into h components (Davydov splitting):

$$M_f^i = \sum_{m \neq n} \sum_\mu \{(B_\nu^i)^* B_\mu^i e^{i\underset{\sim}{k}\cdot(\underset{\sim}{r}_{n\nu} - \underset{\sim}{r}_{m\mu})} \times \langle a\varphi^f_{n\nu}\varphi^o_{m\mu} | V_{n\nu,m\mu} | \varphi^o_{n\nu}\varphi^f_{m\mu}\rangle\}. \tag{8}$$

In Eq. (8), $n\nu$ is an arbitrarily chosen reference molecule.

As specific examples, we consider the compounds naphthalene and anthracene, which crystallize in the monoclinic system with space group D_{2h}^5 and have two molecules per unit cell. Corresponding to every excited state it is possible to form a one site exciton for each site in the unit cell. We denote these one site excitons by $\Phi_1^f(k)$ and $\Phi_2^f(k)$. An elementary argument leads to the linear combinations (for the case where $\underset{\sim}{k}$ is parallel to or perpendicular to the monoclinic axis)

$$\Psi_-^f(\underset{\sim}{k}) = \frac{1}{\sqrt{2}}\{\Phi_1^f(\underset{\sim}{k}) - \Phi_2^f(\underset{\sim}{k})\}$$

$$\Psi_+^f(\underset{\sim}{k}) = \frac{1}{\sqrt{2}}\{\Phi_1^f(\underset{\sim}{k}) + \Phi_2^f(\underset{\sim}{k})\} \tag{9}$$

The vectors defining the unit cell are $\underset{\sim}{a}$, $\underset{\sim}{b}$ and $\underset{\sim}{c}$. (See Figure 1). In this crystal lattice a type 1 molecule may be transformed into a type 2 molecule by reflection in the ac plane followed by a glide of (1/2) $(\underset{\sim}{a} + \underset{\sim}{b})$. At $\underset{\sim}{k} = \underset{\sim}{0}$ the group of the wave vector is isomorphous with the point group C_{2h}.

We have considered thus far only the formal outline of the analysis. Using a more detailed representation of the molecular wave functions, and when only one excited state is considered, the excited state wave functions are

$$\Psi_{\pm}^{f}(\underset{\sim}{k}) = \sum_{n} e^{i\underset{\sim}{k}\cdot\underset{\sim}{r}_{n1}}\Big[a\,\varphi_{n1}^{f}\prod_{mj\neq n1}\varphi_{mj}^{o} \pm a\,e^{i\underset{\sim}{k}\cdot\underset{\sim}{\tau}}\varphi_{n2}^{f}\prod_{mj\neq n2}\varphi_{mj}^{o}\Big], \tag{10}$$

where $\underset{\sim}{\tau} = (1/2)(\underset{\sim}{a} + \underset{\sim}{b})$.

For $\underset{\sim}{k} = \underset{\sim}{0}$, Ψ_{+}^{f} and Ψ_{-}^{f} correspond to the B_u and A_u representations of the factor group C_{2h}[5]. The exciton band structure is determined by

$$M_f^{\pm} = E_f^{\pm}(\underset{\sim}{k}) = \sum_{\substack{m1 \\ (m\neq n)}}(J_{n1,m1}^{f} + K_{n1,m1}^{f})\,e^{i\underset{\sim}{k}\cdot(\underset{\sim}{r}_{m1}-\underset{\sim}{r}_{n1})} \pm \sum_{m\gamma}(J_{n1,m\gamma}^{f} + K_{n1,m\gamma}^{f})\,e^{i\underset{\sim}{k}\cdot(\underset{\sim}{r}_{m2}-\underset{\sim}{r}_{n1})} \tag{11}$$

where the excitation transfer matrix element is

$$J_{n1,m\mu}^{f} = \langle \varphi_{n1}^{f}\varphi_{m\mu}^{o} | V_{n1,m\mu} | \varphi_{n1}^{o}\varphi_{m\mu}^{f}\rangle \tag{12}$$

while the electron and excitation exchange matrix element is

$$K_{n1,m\mu}^{f} = \langle (\sum_{P}(-)^{P}P)\varphi_{n1}^{f}\varphi_{m\mu}^{o} | V_{n1,m\mu} | \varphi_{n1}^{o}\varphi_{m\mu}^{f}\rangle. \tag{13}$$

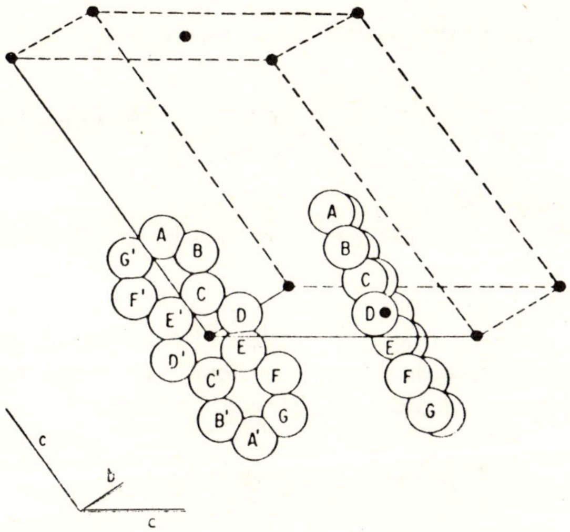

Fig. 1. The crystal structure of anthracene.

Only terms involving excitation transfer and electron exchange between translationally nonequivalent molecules lead to a Davydov splitting, while interactions between translationally equivalent molecules contribute to the band shift. The Davydov splitting, in first order, is given by

$$\Delta E_f = E_f^+(\underset{\sim}{k}) - E_f^-(\underset{\sim}{k}) \tag{14}$$

and, bearing in mind the necessity for momentum conservation, $\underset{\sim}{k} = \underset{\sim}{q}$, we obtain

$$\Delta E_f = 2 \sum_{\substack{m2 \\ (\underset{\sim}{k}=\underset{\sim}{q})}} \left(J^f_{n1,m2} + K^f_{n1,m2} \right) e^{i\underset{\sim}{k}\cdot(\underset{\sim}{r}_{m2} - \underset{\sim}{r}_{n1})} \tag{15}$$

The exciton theory just sketched must be extended by taking into account crystal field induced mixing of excited molecular states into the state f. The tight binding Hamiltonian (1) must be diagonalized in the basis set $\Psi^f(\underset{\sim}{k})$, where the indices f refer to all the molecular excited states. The diagonal matrix elements are obtained from Eq. (11). The off diagonal elements assume the form:

$$H^{\pm}_{fg} = \langle \Psi^f_{\pm}(\underset{\sim}{k}) | H | \Psi^g_{\pm}(\underset{\sim}{k}) \rangle, \tag{16}$$

$$\begin{aligned} H^{\pm}_{fg} = & \sum_{m1 \neq n1} \Gamma^{fg}_{m1,n1} \cos[(\underset{\sim}{r}_{n1} - \underset{\sim}{r}_{m1}) \cdot \underset{\sim}{k}] \\ & \pm \sum_{m2} \Gamma^{fg}_{m2,n1} \cos[(\underset{\sim}{r}_{n1} - \underset{\sim}{r}_{m2}) \cdot \underset{\sim}{k}] \\ & + \sum_{mi \neq n1} \Delta^{fg}_{mi,n1}, \end{aligned} \tag{17}$$

where

$$\Gamma^{fg}_{mi,nl} = \frac{1}{2}\left\{\langle \mathcal{A}\varphi^f_{nl}\varphi^0_{mi}|V_{nl,mi}|\mathcal{A}\varphi^0_{nl}\varphi^g_{mi}\rangle + \langle \mathcal{A}\varphi^g_{nl}\varphi^0_{mi}|V_{nl,mi}|\mathcal{A}\varphi^0_{nl}\varphi^f_{mi}\rangle\right\} \tag{18}$$

$$\Delta^{fg}_{mi,nl} = \langle \mathcal{A}\varphi^f_{nl}\varphi^0_{mi}|V_{nl,mi}|\mathcal{A}\varphi^g_{nl}\varphi^0_{mi}\rangle . \tag{19}$$

To reduce Eqs. (11)-(19) to numerical form both the intermolecular potential and the molecular wave functions must be specified. The intermolecular pair potential has the form

$$V_{k,\ell} = \sum_{I,J}{}' \frac{Z_I Z_J}{R_{IJ}} e^2 - \sum_{I,j}{}' \frac{Z_I}{R_{Ij}} e^2 - \sum_{J,j} \frac{Z_J}{R_{Jj}} e^2 + \sum_{i>j}{}' \frac{e^2}{r_{ij}} , \tag{20}$$

where I and J refer to the nuclei in the different molecules k and ℓ, i refers to the electrons on k, ...

In many applications of exciton theory to the singlet excited states of aromatic crystals the following procedure has been adopted:

(a) A multipole expansion for the matrix elements of $V_{k\ell}$ is introduced and the series truncated after only a few (usually the dipole) terms.

(b) The polarization of the transition is chosen on the basis of experimental or theoretical arguments.

(c) The dipole terms are evaluated using the experimental oscillator strength (obtained from solution spectra).

(d) For allowed optical transitions the higher multipoles are not calculated. Some calculations have been reported (7) in which the octupole transition moment is used as a variable parameter to obtain agreement with experiment. It is not apparent without detailed calculation how many terms have to be included in the multipole expansion. Similar problems are encountered in the computation of dispersion forces between aromatic molecules.

(e) The dipole sum is often truncated at a relatively small distance (20–30 Å).

An alternative procedure is to use the best available molecular wavefunctions and compute the matrix elements of $V_{k\ell}$ directly, thus bypassing the use of the multipole expansion. In recent work (8) on the singlet exciton states $^1B_{2u}$(p) and $^1B_{3u}{}^+$ (β) of the anthracene crystal, we have computed the intermolecular interactions directly out to relatively large distances (50 Å). These calculations make possible the evaluation of short range interactions due to higher multipoles. For the region from 50 Å to the bounding surface dipole sums were used. In previous work, Craig and Walsh (9) had made use of the Ewald-Kornfeld procedure for evaluating dipole sums (characterized by $\underset{\sim}{k} = 0$). It is well known that dipole sums are only conditionally convergent (10), since an interaction of the form $(\cos \Theta_{ij} - 3 \cos \Theta_i \cos \Theta_j) R_{ij}^{-3}$ decreases with increasing R_{ij} only as fast as the volume of space (number of interacting molecules) increases. The dipole sums for $\underset{\sim}{k} = 0$, therefore depend on the shape of the region considered. While the neglect of the momentum of a photon ($\hbar \underset{\sim}{q}$) leads to a satisfactory treatment of the selection rules for photon absorption, this approximation is equivalent to the assumption that all the transition dipoles in the crystal are in phase. Momentum conservation requires (when the interaction between the radiation field and the phonon field is neglected) that $\underset{\sim}{k} = \underset{\sim}{q}$. The ratio of k to the magnitude of the primitive translation vector of the reciprocal lattice of the crystal is of the order of 10^{-3}, and the assumption $\underset{\sim}{k} = \underset{\sim}{0}$ is justified only for short range interactions, i.e., octupole-octupole or electron exchange interactions.

Hall (11) and Amos (12) have shown that when only dipole-dipole interactions are included in description of the lowest excited states of molecular crystals, the classical theory of long waves developed by Born and Huang can be employed. The effect of higher excited states may be included in terms of a frequency independent polarizability. But if direct dipole-dipole sums are made, including the modulation introduced by momentum conservation, it is found that the convergence length is of the order of the wavelength of light. Under these conditions, retardation of the intermolecular interaction due to the finite velocity of light cannot be neglected. Such effects have recently been studied by Simpson (13).

B. Evaluation of the Matrix Elements

In order to make clear the pressure dependence of the electronic interaction energy of the molecular crystal, it is necessary to examine some of the details of the calculation of the matrix elements.

Consider first the direct evaluation of the intermolecular interactions within the inner zone. The matrix elements of $V_{k\ell}$ were calculated to distances of the order of 50 Å using the best available π electron wave functions. For larger distances the dipole summation was used. Because the interactions are sensitive to the magnitude of the electronic overlap and the molecules are far enough apart that the asymptotic behavior of the wavefunction is important, we must examine the choice of molecular wave functions. The molecular wave functions used were antisymmetrized products of Huckel molecular orbitals, and semi-empirical configuration interaction wave functions in the π electron approximation computed from the Pariser-Parr theory.

The J and K integrals required for the calculation of the first order Davydov splitting can now be represented in terms of the molecular wave functions. Using Huckel wave functions these integrals for the $^1B_{2u}$ (p) excited state assume the form:

$$J^{p}_{n1,m\mu} = 2\langle u^{7}_{m\mu}(1)\, u^{8}_{m\mu}(1) | r_{12}^{-1} | u^{7}_{n1}(2)\, u^{8}_{n1}(2)\rangle$$

$$\equiv 2\langle u^{7}_{m\mu} u^{8}_{m\mu} | u^{7}_{n1} u^{8}_{n1}\rangle \tag{21}$$

$$K^{p}_{n1,m\mu} = -\langle u^{7}_{m\mu}(1)\, u^{7}_{n1}(1) | r_{12}^{-1} | u^{8}_{m\mu}(2)\, u^{8}_{n1}(2)\rangle$$

$$\equiv -\langle u^{7}_{m\mu} u^{7}_{n1} | u^{8}_{m\mu} u^{8}_{n1}\rangle \tag{22}$$

where $u_a^{\,i}$ represents the molecular orbital i located on molecule a.

Somewhat more complex expressions are obtained using the configuration interaction (Pariser-Parr) wave functions.

The molecular orbitals u^m are now represented in terms of a linear combination of carbon atom 2 pz wave functions, ω^i,

$$u^{m} = \sum_i \omega^{i} c_i^{m} \tag{23}$$

where the C_i^m are Huckel coefficients without overlap. Using these functions, the matrix elements are reduced to coulomb integrals:

$$\langle u^7_{m\mu} u^8_{m\mu} | u^7_{n1} u^8_{n1} \rangle = \sum_{i,j,k,l} C_i^7 C_j^8 C_k^7 C_l^8 \langle \omega^i_{m\mu}(1) \omega^j_{m\mu}(1) | r_{12}^{-1} | \omega^k_{n1}(2) \omega^l_{n1}(2) \rangle \tag{24}$$

In the calculations reported herein, three and four center integrals were neglected. Order of magnitude estimates based on the Mulliken approximation indicate that the contribution of these high order terms will not exceed 10%. The calculation of the coulomb integrals requires careful choice of the carbon atom wave functions. Unlike the case of electron exchange and charge transfer interactions, these coulomb integrals are not sensitive to the behavior of the molecular wave functions at large distances. However, in order to obtain maximum accuracy, SCF 2p carbon atom wave functions were used for the computation of the coulomb integrals. The same wave functions were used for the computation of the exchange integrals $K_{n\nu, m\mu}{}^f$.

The matrix elements necessary for the computation of the energy shifts and splittings were computed directly using an IBM 7094. The interactions calculated for the inner zone were extended to about 60 Å in all directions, thereby including approximately 2000 molecules. The interactions summed over translationally equivalent and translationally nonequivalent molecules for the anthracene p band are displayed in Table I. Here HU and PA represent the interactions calculated using the Huckel and Pariser wave functions. It is important to note at this point that both wave functions overestimate the transition dipole moment to the $^1B_{2u}$ (p) state and therefore a scaling procedure must be devised to correct the calculated matrix elements. In Table I we compare the total interaction matrix element sums with the first term of the multipole expansion, i.e., the dipole-dipole sums over the same region:

Table I. Singlet First Order Exciton Interaction Terms for Anthracene Crystal

(All Energies in cm^{-1} Units)

Ia

Anthracene - p band

Translationally Inequivalent Molecules

Spherical Convergence

R	I_{dd}^{p}	HU	δHU	PA	δPA
20	35	-267	-442	- 99	-235
25	36	-266	-446	- 97	-237
30	26	-313	-443	-131	-242
35	30	-294	-444	-118	-235
40	57	-162	-447	- 24	-246
45	81	- 46	-451	+ 55	-259
50	90	44	-451	120	-266
55	100	89	-456	153	-272
60	110	95	-455	157	-272
65	110	91	-459	154	-275

Slab Convergence

No. Cells					
2	-169	-1257	-405	-796	-137
3	-134	-1080	-410	-674	-152
4	- 71	- 777	-422	-460	-190
5	- 19	- 527	-432	-282	-204
6	+ 27	- 302	-437	-124	-229
7	+ 72	- 85	-445	+ 30	-250
8	+110	88	-462	+152	-277

Table I. Singlet First Order Exciton Interaction Terms for Anthracene Crystal

(All Energies in cm^{-1} Units)

Ib

Anthracene - p band

Translationally Equivalent Molecules

Spherical Convergence

R	$I_{d-d}{}^{p}$	HU	δHU	PA	δPA
20	-334	-1114	556	-1025	280
25	-369	-1278	567	-1140	299
30	-362	-1248	570	-1116	295
35	-377	-1317	568	-1167	303
40	-399	-1424	570	-1242	314
45	-427	-1561	574	-1339	326
50	-449	-1603	582	-1410	341
55	-457	-1694	591	-1433	349
60	-457	-1704	583	-1439	343
65	-457	-1702	583	-1439	343

Slab Convergence

No. Cells					
2	-287	- 899	546	- 861	258
3	-300	- 948	552	- 905	265
4	-351	-1195	560	-1080	289
5	-336	-1120	560	-1027	283
6	-428	-1566	574	-1342	325
7	-426	-1556	574	-1340	321
8	-457	-1702	583	-1438	343

$$I_{dd} = \sum_{m\mu} e^{i\mathbf{q}\cdot(\mathbf{r}_{m\mu}-\mathbf{r}_{n1})} \times \left[\frac{\boldsymbol{\mu}_{n1}\cdot\boldsymbol{\mu}_{m\mu}}{|\mathbf{r}_{m\mu}-\mathbf{r}_{n1}|^3} - \frac{3(\boldsymbol{\mu}_{n1}\cdot(\mathbf{r}_{m\mu}-\mathbf{r}_{n1}))(\boldsymbol{\mu}_{m\mu}\cdot(\mathbf{r}_{m\mu}-\mathbf{r}_{n1}))}{|\mathbf{r}_{m\mu}-\mathbf{r}_{n1}|^5}\right] \tag{25}$$

In Eq. (25), $\boldsymbol{\mu}$ represents the transition dipole moment. For short range interactions $\mathbf{q}\cdot(\mathbf{r}_{m\mu} - \mathbf{r}_{n1}) \approx 0$, and the modulation of the dipole-dipole sums is irrelevant.

Recent work (14) has shown that dipole sums within a sphere tend to oscillate about a limit, while sums within a slab the shape of the unit cell tend to a limit more regularly. For this reason we have carried out the dipole sums needed to estimate the multipole interaction in both geometries.

The contribution I_m of higher order multipoles (greater than dipoles) to the interaction energy may be computed by subtracting from the inner zone total interaction energy, the interaction energy due to just the transition dipoles. The latter must be scaled by a numerical factor C which is the ratio between the squares of the calculated and experimental transition moments of the free molecule. For the p state of anthracene we take $C_H = 5$ for the Huckel and $C_p = 3.9$ for the Pariser wave function. Thus, the contribution of the higher multipoles is given by $\delta_{HU} = HU-5I_{dd}$ and $\delta_{PA} = PA-3.9I_{dd}$ for Huckel and Pariser wave functions, respectively. It is found that the higher multipole interaction contribution in the anthracene p band is of short range, as expected. To obtain the correct energy it is necessary to scale the magnitude of the higher multipoles. As seen from Table I the higher multipole contributions computed on the Huckel basis and on the Pariser basis show that the entire multipole expansion for the $^1B_{2u}$ (p) state scales approximately just as do the dipole-dipole sums, so that the contribution of the higher multipoles in the p band is just $I_m = PA/3.9-I_{dd}$, i.e., $I_m = -92\ \text{cm}^{-1}$.

We now consider the interactions at intermolecular pair separations greater than 50 Å. As shown, these long range interactions are due only to dipole-dipole terms and it has been demonstrated[15] that the contribution of dipole-dipole interactions between distant neighbors cannot be neglected for small values of q.

In our treatment of long range interactions a continuum model introduced by Heller and Marcus was extended to the case of more than one molecule per unit cell. We shall consider again the + and − states of the crystal as linear combinations of unit cell wave functions $(1/\sqrt{2})(\phi_{m1} \pm \exp(i\underset{\sim}{k}\cdot\underset{\sim}{\tau})\,\phi_{m2})$. An application of the multipole expansion to the interaction matrix element in the basis of the unit cell functions leads to the interaction of unit cell transition dipoles, ... It is seen that the − state dipole is oriented along the $\underset{\sim}{b}$ axis and the + state dipole is oriented in the plane perpendicular to the $\underset{\sim}{b}$ axis. We shall consider only the dipole-dipole interaction in this scheme. Since the minimum distance between molecules is of the order of 50 Å it is assumed that the contribution of higher multipoles has converged within the inner excluded region.

Because the wave functions belong to different irreducible representations, the + and − states do not interact with light whose k vector is parallel to $\underset{\sim}{b}$ or perpendicular to $\underset{\sim}{b}$. Therefore we need consider only the case of interaction between parallel dipoles,

$$I_{dd} = |\underset{\sim}{\mu}|^2 \sum_{R>R_0}^{\infty} \left[\frac{1 - 3\cos^2(\underset{\sim}{\mu}\cdot\underset{\sim}{R})}{R^3} \right] e^{i\underset{\sim}{k}\cdot\underset{\sim}{R}} \tag{26}$$

with $\underset{\sim}{\mu}$ the dipole moment of the unit cell. Now the unit cell is small relative to the volume excluded from (26) (~ 125 A^3 compared with ~ 125 x 10^{3}A)^3 so that the sum may be replaced by an integral over a density of dipoles. Of course, the direct summation for the inner volume must be added to complete the evaluation of the total interaction energy. Indeed, as shown, within this volume we have included all interactions. For the region outside R_0, Eq. (26) becomes

$$I_{dd} = \frac{|\underset{\sim}{\mu}|^2}{V} \int_{R_0}^{\infty} \frac{e^{i\underset{\sim}{k}\cdot\underset{\sim}{R}}\,(1 - 3\cos^2(\underset{\sim}{\mu}\cdot\underset{\sim}{R}))}{R^3}\, d^3R$$

$$= \frac{8\pi}{3V} |\underset{\sim}{\mu}|^2 P_2(\cos(\underset{\sim}{\mu}\cdot\underset{\sim}{k}))[j_0(|\underset{\sim}{k}|R_0) + j_2(|\underset{\sim}{k}|R_0)] \tag{27}$$

where V is the volume of a unit cell, P_2 is the second Legendre polynomial and j_0 and j_2 are spherical Bessel functions of order 0, 2 respectively. Since the excluded volume is characterized by a radius of 50 Å, $kR_o \sim 0.2$ and $j_0(kR_o) + j_2(kR_o) \approx 1$.

We have thus shown that introduction of the momentum conservation restriction implies that the dipole sums in the outer zone become absolutely convergent for an infinite crystal and independent of the boundary conditions. It should be noted at this point that we have thus far been concerned only with the electrostatic intermolecular interactions. Since the convergence length for these interactions is of the order of magnitude of 1/k, it is necessary to include the effects of retardation of the interaction potential. Following the treatment of Born and Wolf (16) and of Anex and Simpson (17) we take for the field of a small oscillating dipole

$$\underset{\sim}{E}_i = e^{ikR_i}\left[\frac{3(\underset{\sim}{\mu}\cdot\underset{\sim}{R}_i)\underset{\sim}{R}_i - R_i^2\underset{\sim}{\mu}}{R_i^5}(1 - ikR_i) - k^2\left(\frac{(\underset{\sim}{\mu}\cdot\underset{\sim}{R}_i)\underset{\sim}{R}_i - R_i^2\underset{\sim}{\mu}}{R_i^3}\right)\right] \tag{28}$$

The interaction energy between such dipoles is $\underset{\sim}{\mu}\cdot\underset{\sim}{E}_i$ summed over all the dipoles of the crystal. In this calculation only the real part of (28) (corresponding to the energy) need be considered. Using the continuum approximation Eq. (26) is replaced by

$$\begin{aligned} I &= \mathcal{R}e\left[\sum_i \underset{\sim}{\mu}\cdot\underset{\sim}{E}_i\right] \\ &= \frac{1}{V}\int e^{i\underset{\sim}{k}\cdot\underset{\sim}{R}}\left\{\frac{\mu^2R^2 - 3(\underset{\sim}{\mu}\cdot\underset{\sim}{R})}{R^5}(\cos(kR) + kR\sin(kR)) + k^2\left(\frac{(\underset{\sim}{\mu}\cdot\underset{\sim}{R})^2 - \mu^2R^2}{R^3}\cos(kR)\right)\right\} d^3R \end{aligned} \tag{29}$$

Application of this analysis to a spherical crystal (characterized by a radius R_f), excluding the inner zone (of radius R_o, with $kR_o \ll 1$) leads to the result

$$I = \frac{8\pi}{3V} |\underset{\sim}{\mu}|^2 P_2(\cos(\underset{\sim}{\mu}\cdot\underset{\sim}{k})) + \frac{4\pi}{3V} |\underset{\sim}{\mu}|^2 \sin^2(kR_f) \times \times \left[P_2(\cos(\underset{\sim}{\mu}\cdot\underset{\sim}{k})) + 1 \right]. \tag{30}$$

The first term in Eq. (30) is the interaction energy in the static dipole approximation (no retardation included), while the second term, which is modulated by a sinusoidal variation with R_f, is the retardation correction.

For a macroscopic sample the interaction energy cannot depend on the exact position of the surface, and it seems legitimate to replace $\sin^2(kR_f)$ in Eq. (30) by its average value. Under these conditions the contribution of the retardation term to the interaction energy is 20%. Further refinement of the theory is necessary to take into account the difference between the wave vector corresponding to the propagation of the interaction and the incident wave vector which defines the modulation of the wave function. When reasonable values for the effective refractive index are employed (corresponding to a self-consistent calculation of the fields) the corrections arising from the retardation effect become small. Although the theoretical situation is not entirely satisfactory, it has been demonstrated that to sufficient accuracy the retardation effects are not important, and Eq. (27) can be used for the calculation of long-range interactions.

C. Numerical Results for Singlet Exciton States of the Anthracene Crystal

The computation of the J and K integrals within the inner zone followed by the calculation of long range dipole-dipole interactions leads to the first order Davydov splitting. It is now well known that first order theory is not sufficient to account for the Davydov splitting and the intensity ratios in aromatic crystals; i.e., the off diagonal matrix elements of the crystal Hamiltonian between singlet exciton states must be included. Examination of the effect of crystal field mixing between the p and β states of the anthracene requires the evaluation of the integrals $\Gamma_{m2,n1}^{p\beta}$, $\Gamma_{m1,n1}^{p\beta}$ and $\Delta_{n1,mi}^{p\beta}$, Eq. (18). Previous work has employed the multipole expansion for the pair interaction term, keeping only the dipole-dipole interaction. In this approximation the $\Delta_{n1,mi}$ (i = 1,2) matrix elements vanish (9). Using the m.o. π electron scheme, we write, in the Huckel approximation,

$$\Delta_{nm} = \sum_{r=1}^{7} \left[\langle u_n^8 u_n^9 | u_m^r u_m^r \rangle - \langle u_n^6 u_n^7 | u_m^r u_m^r \rangle \right]. \tag{31}$$

Neglecting the contribution of three and four center integrals and assuming the validity of the pairing property for the molecular orbitals of an alternant hydrocarbon we find that $\Delta_{nm} = 0$.

The $\Gamma_{nm}^{p\beta}$ integrals were evaluated numerically for molecules within a sphere of 50 Å. The matrix elements take the form

$$\Gamma_{nm}^{p\beta} = \frac{1}{\sqrt{2}} \left[\langle u_n^8 u_n^6 | u_m^7 u_m^8 \rangle + \langle u_n^9 u_n^6 | u_m^7 u_m^8 \rangle + \langle u_n^7 u_n^8 | u_m^6 u_m^8 \rangle + \langle u_n^7 u_n^8 | u_m^6 u_m^9 \rangle \right]. \tag{32}$$

Only two center integrals were included in this calculation, using Huckel and Pariser wave functions. As in the case of the first order calculation, the contribution of the higher multipoles was estimated by evaluating the contribution of the dipole-dipole terms only, scaling this dipole term by the ratios between the calculated and experimental transition moments for the band (the transition moment for the β band is assumed to be correctly given by the theory), and subtracting the scaled dipole-dipole contribution from the total interaction energy. The contribution of the higher multipoles is of the order of 25% of the total interaction energy within the 50 Å sphere. For the region outside the 50 Å sphere it was assumed that convergence of the higher multipole interactions had been achieved, and only dipole-dipole terms remained to be summed from 50 Å to infinity. This summation was carried out using the continuum model.

A crystal field mixing scheme involving only the p and β states is not complete. We have extended the treatment taking into account all one electron excited states arising from $\pi \rightarrow \pi^*$ transitions and characterized by a nonzero oscillator strength.

Throughout the previous discussion we have assumed that the Davydov splitting is large relative to the vibrational spacing. In this strong coupling limit the wave function is represented by the product of the excited state electronic wave function and the ground state vibrational wave function. There will be, then, two well-separated zero-zero lines. In the weak coupling limit, the Davydov splitting is small compared to the vibrational spacing and instead of starting with the total electronic wave function a different formalism must be used. In the weak coupling limit the molecular wave functions are constructed from vibronic functions, i.e.,

$$\varphi_n^{f(i)} = \varphi_n^{f} \chi_n^{f(i)} \tag{33}$$

where $\chi_n^{f(i)}$ is a vibrational wave function for the nth excited molecule in the ith vibrational state. When the functions defined by Eq. (33) are used as the basis for calculating the matrix elements of the intermolecular potential, the integrals are all modified as follows:

$$\langle \varphi_n^{f(i)} \varphi_m^{o} | V_{nm} | \varphi_n^{o} \varphi_m^{f(j)} \rangle = J_{nm} \langle \chi_n^{o(o)} | \chi_n^{f(i)} \rangle \langle \chi_m^{o(o)} | \chi_m^{f(j)} \rangle . \tag{34}$$

Therefore, in the weak coupling limit the total electronic matrix element is modified by vibrational overlap factors. It should be noted that, in this case, off diagonal matrix elements between vibronic states corresponding to the same electronic transition must be included. If these off diagonal terms are neglected the vibrational sum rule, $\sum_i |\langle \chi^{o(o)} | \chi^{f(i)} \rangle|^2 = 1$, implies that the total electronic contribution to the Davydov splitting can be obtained by summation over all the experimental splittings corresponding to the individual vibronic components.

Since the total calculated splitting in the p state of anthracene is of the order of 500 cm^{-1} compared with a (symmetric) vibrational spacing of the order of 1400 cm^{-1}, we shall examine the theoretical predictions in the weak coupling limit. On the other hand, for the β band the application of the strong coupling scheme seems most appropriate. For the computation of the Davydov splittings in the vibronic components of the band we have considered four vibrational states with the diagonal and off diagonal matrix elements given by (33).

The configuration interaction crystal field mixing matrix elements between the vibronic components of the p band and the β band are given by $\Gamma_{nm} \langle \chi^{o(o)} | \chi^{p(i)} \rangle$.

The vibrational overlap integrals $\langle \chi^{o(o)} | \chi^{p(i)} \rangle$ can be obtained from the squares of the relative intensities (normalized to unity) of the vibronic components in the solution spectra. Alternatively, a semi-empirical procedure (18) for the evaluation of these integrals can be employed.

In Tables II and III we compare the results of the theoretical analysis for the splitting and the intensity ratios of the vibronic components of the anthracene p band with the available experimental data (7b, 1g). The second order theory (including crystal field mixing of the p and β states) leads to good agreement with experimental splittings. The polarization ratios can be properly accounted for when the effect of configuration interaction with higher excited states is taken into account.

As a result of this analysis it is concluded that:

(a) The problem of calculating energies in molecular crystals cannot be reduced to the summation of dipole-dipole interactions. From π electron theory it has been demonstrated that short range higher multipole interactions yield important contributions to the diagonal and off diagonal matrix elements. The most striking case is that of the diagonal terms for the p state, where the higher multipoles reduce the contribution within the sphere of 50 Å from 100 cm^{-1} for dipole-dipole terms to 16 cm^{-1}.

(b) Electron exchange interactions in the allowed singlet state ($^1B_{2u}$ in anthracene) are found to be small.

(c) The long range dipole-dipole interactions are of great importance, and any meaningful theory of singlet exciton states arising from allowed transitions must take the long range of the interaction into account. Long range interactions involving modulated dipole sums are absolutely convergent, and under the experimental conditions employed to date, the Davydov splitting is predicted to be independent of the crystal thickness.

(d) In the analysis presented herein the effects of configuration interaction between singlet exciton states has been carefully examined. It is found that the mixing between p and β singlet states has a large effect on the Davydov splitting, while inclusion of higher excited π states has only a minor effect on the splittings. However, to account for the polarization ratios in the vibronic components of the p band, higher excited states must be included.

Table II. Davydov Splittings in Crystalline Anthracene

p band	Theory cm^{-1}	Experiment cm^{-1}
Dipole terms to 55 Å	+200	
Higher multipoles	-180	
Long range dipole terms (from 55 Å to infinity)	+632	
Electron exchange	-44	
First order splitting	608	
Second order splitting (β state included)	350	450[a], 400[b]
β band		
First order splitting	33800	
Second order splitting	34495	(15000)

[a]Reference (19)

[b]Reference (7b)

Table III. Davydov Splittings in Vibronic Components of p State (Assuming Strong Coupling in All Other States)

	Calculated								Exp.	
	Only p State in Conf. Int.				All π States in Conf. Int.					
Band	A_u (∥b)	B_u (⊥b)	ΔE cm^{-1}	P(b/a)	A_u (∥b)	B_u (⊥b)	ΔE cm^{-1}	P(b/a)	ΔE [(a)] cm^{-1}	P(b/a) [(b)]
0-0	- 404	- 219	185	2.5	- 437	- 230	207	3.5	220 (191)	5
0-1	1143	1228	85	1.6	1119	1221	102	2.5	141 (150)	4.5
0-2	2640	2690	50	0.5	2632	2686	54	1.9	58 (80)	3
0-3	4133	4153	20	---	4130	4152	22	---	----	---
0-4	5563	5573	10	---	5560	5573	13	---	----	---

- Davydov splitting, P(b/a) polarization ratio

[(a)] First set of data from Reference (19); data in parentheses from Reference (7b).

[(b)] Reference (19).

III. CHARGE-TRANSFER EXCITONS

A. General Remarks

In the preceding section we have considered the theory of neutral (Frenkel) exciton states, i.e., states of the crystal in which the hole and the electron are localized on the same molecule. Frenkel's theory was extended by Slater and Shockley (22) and by Wannier (4) who considered the excitation waves

$$E(\underset{\sim}{k},\underset{\sim}{R}) = \frac{A}{\sqrt{N}}\sum_{m} e^{i\underset{\sim}{k}\cdot\underset{\sim}{R}_m} \,|\underset{\sim}{R}_m - \tfrac{1}{2}\underset{\sim}{R}, \underset{\sim}{R}_m + \tfrac{1}{2}\underset{\sim}{R}\rangle \qquad (35)$$

where $|\underset{\sim}{R}_m, \underset{\sim}{R}_n\rangle$ represents a crystal state in which an electron is removed from the molecule at $\underset{\sim}{R}_m$ and placed on the molecule at $\underset{\sim}{R}_n$. Such an excitation wave may be considered to describe the coherent motion of the center of mass of the electron and the associated hole with relative position vector, $\underset{\sim}{R}$, fixed. By expanding the wave function in terms of the excitation waves $E(\underset{\sim}{k}, \underset{\sim}{R})$, an effective wave equation for the relative motion of the electron and hole was derived by Wannier (4). Under certain conditions, the effective wave equation becomes identical to the Schroedinger equation for the hydrogen atom. The hydrogen-like limit is only accurate if the energy eigenvalue corresponding to the relative motion of the electron and hole is close to the ionization continuum. This circumstance is not realized in crystals of aromatic molecules at low pressures, but may be approached at high pressures.

As described in the last section, because of the great similarities between the vapor and crystal spectra of aromatic compounds, the states $E(\underset{\sim}{k}, \underset{\sim}{0})$ (i.e., Frenkel states) have been used to describe the exciton states of such crystals. Clearly, a better description (and one which will be needed at high pressures) must include the effect of excitation waves with $\underset{\sim}{R} \neq \underset{\sim}{0}$. These states we denote charge-transfer or ion-pair exciton states. In general, because of the very small electronic overlap in crystals of aromatic compounds at low pressure, it is to be expected that configuration mixing of neutral and ion-pair exciton states will be of importance primarily when the bandwidth corresponding to the neutral exciton state is very small. Such cases arise for the triplet exciton bands of most organic crystals, and for the bands which arise from weak singlet-singlet transitions, as in naphthalene and benzene. At high pressures, when the electronic overlap is greatly increased, the excited states of the crystal undoubtedly depart markedly from the Frenkel description, whereupon the effect of ion-pair states becomes very important in defining the structure of the exciton band.

Consider first the low pressure limit. Then the lowest lying charge transfer exciton states are those with an electron and a hole on adjacent molecules. Simple arguments based on the magnitude of overlap integrals suggest that, of all possible charge-transfer states, these particular states will interact most strongly with the neutral exciton states. Since the case $\underset{\sim}{k} = \underset{\sim}{0}$, i.e., vanishing crystal momentum, is most important if attention is focused on optical excitation, we shall consider only this case. The procedure employed for the construction of the ion-pair exciton states is essentially that suggested by Slater and Shockley (22): localized, antisymmetrized charge transfer states are constructed from the positive and negative ion wave functions. As in the case of neutral excitons, the states so defined are not eigenstates of the tight binding Hamiltonian, and the localized functions must be transformed to construct charge-transfer exciton states which belong to the representation of the crystal lattice space group.

Consider the class of crystals described by the symmetry $C_{2h}^5 \equiv P_{2_1/a}$. Such crystals have a monoclinic cell containing two molecules: examples include naphthalene and anthracene.

Our first step is to construct the several exciton wave functions belonging to $\underset{\sim}{k} = \underset{\sim}{0}$. Symmetry-adapted exciton states can be obtained with the use of the Wigner projection operator (23)

$$\underset{\sim}{W}_{ts}^{i} = (\ell_i/h) \sum_{\underset{\sim}{Q}} D^i(\underset{\sim}{Q})_{ts}^{*} \underset{\sim}{Q}, \tag{36}$$

where $\underset{\sim}{Q}$ is a symmetry operation, ℓ_i is the dimension of the ith irreducible representation, h is the order of the group, and $D^i(\underset{\sim}{Q})_{ts}$ is the ts element of the matrix which represents $\underset{\sim}{Q}$ in the ith representation.

We take antisymmetrized products of molecular and ionic wave functions for the localized states from which symmetry-adapted functions are formed. Such localized states are denoted as follows:

$$|\underset{\sim}{R}_m\rangle = \mathcal{Q}\, \varphi_{m1}^{f} \prod_{\ell\alpha \neq m1}' \varphi_{\ell\alpha}^{0}\,; \quad \ell = 1, \ldots, N; \quad \alpha = 1, 2; \tag{37}$$

$$|\underset{\sim}{R}_m, \underset{\sim}{R}_n\rangle = \mathcal{Q}_I\, \varphi_{m1}^{+} \varphi_{n1}^{-} \prod_{\ell\gamma \neq m1, n1}'' \varphi_{\ell\gamma}^{0}\,. \tag{38}$$

The superscripts 0, f, +, and - refer to the ground state, excited state, positive ion, and negative ion, respectively, and $\mathcal{A}$ and $\mathcal{A}_I$ are the antisymmetrizing operators

$$\mathcal{A} = [(a!)^{2N}/(2Na)!]^{1/2} \sum_P (-)^P P,$$

$$\mathcal{A}_I = [(a!)^{2N-2}(a+1)!(a-1)!/(2Na)!]^{1/2} \sum_P (-)^P P. \tag{39}$$

The operator P interchanges electrons only between molecules, since the molecular and ionic wave functions are already antisymmetrized with respect to their electron coordinates. As usual, N is the total number of unit cells, and a is the number of electrons on each neutral molecule. We use $\underset{\sim}{R}_m$ for the vector with components equal to the coordinates of the center of Molecule m.

On application of the Wigner projection operator (36) to the localized states defined in Eqs. (37) and (38), there are obtained linear combinations of all the localized states which form the basis sets of the irreducible representations. From Eq. (37) we obtain the well-known neutral-exciton functions

$$\Psi_{A_u} = \frac{1}{\sqrt{2N}} \sum_m^N [|\underset{\sim}{R}_m\rangle - |\underset{\sim}{R}_m + \underset{\sim}{\tau}\rangle], \tag{40}$$

$$\Psi_{B_u} = \frac{1}{\sqrt{2N}} \sum_m^N [|\underset{\sim}{R}_m\rangle + |\underset{\sim}{R}_m + \underset{\sim}{\tau}\rangle] \tag{41}$$

where $\underset{\sim}{\tau}$ is the vector extending from the corner nearest to the origin to the center of the second molecule in the unit cell, i.e., $\underset{\sim}{\tau} = \frac{1}{2}(\underset{\sim}{a} + \underset{\sim}{b})$. The functions Ψ_{A_u} and Ψ_{B_u} are those used earlier for the calculation of the Davydov splitting. Now, the coupling between the neutral and ion-pair excitons decreases exponentially with the distance between the centers of the positive ion and the negative ion. For this reason, we consider only those ion-pair excitons where the electron and the hole are located on adjacent molecules, corresponding to the limitation to nearest-neighbor interactions. It may be shown that the coupling between the neutral and ion-pair excitons is determined by overlap, hybrid, and exchange integrals, all of which fall off exponentially with increasing distance. Because of this, the approximation of considering only the mixing of near-neighbor ion-pair excitons and neutral excitons may be seen to be valid.

With the above simplification, we need consider only the following localized states:

$$
\begin{aligned}
&|\mathbf{R}_m, \mathbf{R}_m+\mathbf{c}\rangle, \quad &&|\mathbf{R}_m+\mathbf{c}, \mathbf{R}_m\rangle;\\
&|\mathbf{R}_m, \mathbf{R}_m+\boldsymbol{\tau}\rangle, \quad &&|\mathbf{R}_m+\boldsymbol{\tau}, \mathbf{R}_m\rangle;\\
&|\mathbf{R}_m, \mathbf{R}_m+\mathbf{b}\rangle, \quad &&|\mathbf{R}_m+\mathbf{b}, \mathbf{R}_m\rangle;\\
&|\mathbf{R}_m, \mathbf{R}_m+\mathbf{a}+\mathbf{c}\rangle, \quad &&|\mathbf{R}_m+\mathbf{a}+\mathbf{c}, \mathbf{R}_m\rangle;\\
&|\mathbf{R}_m, \mathbf{R}_m+\mathbf{c}+\boldsymbol{\tau}\rangle, \quad &&|\mathbf{R}_m+\mathbf{c}+\boldsymbol{\tau}, \mathbf{R}_m\rangle.
\end{aligned}
\tag{42}
$$

From these states are obtained seven independent, symmetry-adapted states for the A_u and B_u representations. These are

$$
\begin{aligned}
|\mathbf{0}, \mathbf{c}; \mp\rangle = \frac{1}{\sqrt{4N}} \sum_m^N \{ &|\mathbf{R}_m, \mathbf{R}_m+\mathbf{c}\rangle \\
&\mp |\mathbf{R}_m-\mathbf{b}+\boldsymbol{\tau}, \mathbf{R}_m-\mathbf{b}-\mathbf{c}+\boldsymbol{\tau}\rangle \\
&\mp |\mathbf{R}_m-\mathbf{a}+\boldsymbol{\tau}, \mathbf{R}_m-\mathbf{a}+\mathbf{c}+\boldsymbol{\tau}\rangle \\
&+ |\mathbf{R}_m, \mathbf{R}_m-\mathbf{c}\rangle \},
\end{aligned}
\tag{43}
$$

$$
\begin{aligned}
|\mathbf{0}, \mathbf{b}; \mp\rangle = \frac{1}{\sqrt{4N}} \sum_m^N \{ &|\mathbf{R}_m, \mathbf{R}_m+\mathbf{b}\rangle \\
&\mp |\mathbf{R}_m-\mathbf{b}+\boldsymbol{\tau}, \mathbf{R}_m+\boldsymbol{\tau}\rangle \\
&\mp |\mathbf{R}_m-\mathbf{a}+\boldsymbol{\tau}, \mathbf{R}_m-\mathbf{a}-\mathbf{b}+\boldsymbol{\tau}\rangle \\
&+ |\mathbf{R}_m, \mathbf{R}_m-\mathbf{b}\rangle \},
\end{aligned}
\tag{44}
$$

$$
\begin{aligned}
|\mathbf{0},\mathbf{a}+\mathbf{c};\mp\rangle = \frac{1}{\sqrt{4N}}\sum_{m}^{N}\{ & |\mathbf{R}_m,\mathbf{R}_m+\mathbf{a}+\mathbf{c}\rangle \\
& \mp|\mathbf{R}_m-\mathbf{b}+\boldsymbol{\tau},\mathbf{R}_m-\mathbf{a}-\mathbf{b}-\mathbf{c}+\boldsymbol{\tau}\rangle \\
& \mp|\mathbf{R}_m-\mathbf{a}+\boldsymbol{\tau},\mathbf{R}_m+\mathbf{c}+\boldsymbol{\tau}\rangle \\
& +|\mathbf{R}_m,\mathbf{R}_m-\mathbf{a}-\mathbf{c}\rangle\},
\end{aligned}
\tag{45}
$$

$$
\begin{aligned}
|\mathbf{0},\boldsymbol{\tau};\mp\rangle = \frac{1}{\sqrt{4N}}\sum_{m}^{N}\{ & |\mathbf{R}_m,\mathbf{R}_m+\boldsymbol{\tau}\rangle \\
& \mp|\mathbf{R}_m-\mathbf{b}+\boldsymbol{\tau},\mathbf{R}_m\rangle \\
& \mp|\mathbf{R}_m-\mathbf{a}+\boldsymbol{\tau},\mathbf{R}_m\rangle \\
& +|\mathbf{R}_m,\mathbf{R}_m-\mathbf{a}-\mathbf{b}+\boldsymbol{\tau}\rangle\},
\end{aligned}
\tag{46}
$$

$$
\begin{aligned}
|\mathbf{0},\mathbf{c}+\boldsymbol{\tau};\mp\rangle = \frac{1}{\sqrt{4N}}\sum_{m}^{N}\{ & |\mathbf{R}_m,\mathbf{R}_m+\mathbf{c}+\boldsymbol{\tau}\rangle \\
& \mp|\mathbf{R}_m-\mathbf{b}+\boldsymbol{\tau},\mathbf{R}_m-\mathbf{c}\rangle \\
& \mp|\mathbf{R}_m-\mathbf{a}+\boldsymbol{\tau},\mathbf{R}_m+\mathbf{c}\rangle \\
& +|\mathbf{R}_m,\mathbf{R}_m-\mathbf{c}-\boldsymbol{\tau}\rangle\},
\end{aligned}
\tag{47}
$$

$$
\begin{aligned}
|\boldsymbol{\tau},\mathbf{0};\mp\rangle = \frac{1}{\sqrt{4N}}\sum_{m}^{N}\{ & |\mathbf{R}_m+\boldsymbol{\tau},\mathbf{R}_m\rangle \\
& \mp|\mathbf{R}_m,\mathbf{R}_m-\mathbf{b}+\boldsymbol{\tau}\rangle \\
& \mp|\mathbf{R}_m,\mathbf{R}_m-\mathbf{a}+\boldsymbol{\tau}\rangle \\
& +|\mathbf{R}_m-\mathbf{a}-\mathbf{b}+\boldsymbol{\tau},\mathbf{R}_m\rangle\},
\end{aligned}
\tag{48}
$$

$$
\begin{aligned}
|\mathbf{c}+\boldsymbol{\tau}, \mathbf{0}; \mp\rangle = \frac{1}{\sqrt{4N}} \sum_{m}^{N} \{ & |\mathbf{R}_m+\mathbf{c}+\boldsymbol{\tau}, \mathbf{R}_m\rangle \\
& \mp |\mathbf{R}_m-\mathbf{c}, \mathbf{R}_m-\mathbf{b}+\boldsymbol{\tau}\rangle \\
& \mp |\mathbf{R}_m+\mathbf{c}, \mathbf{R}_m-\mathbf{a}+\boldsymbol{\tau}\rangle \\
& + |\mathbf{R}_m-\mathbf{c}-\boldsymbol{\tau}, \mathbf{R}_m\rangle \}.
\end{aligned}
\tag{49}
$$

The negative and positive signs on the right-hand side of Eqs. (43) through (49) refer to the A_u and B_u states, respectively. In the above functions we have neglected the overlap integral in the normalization factors.

We consider now the configuration interaction between the neutral and charge-transfer states (in the oriented gas scheme), and its effect on the Davydov splitting. As long as the mixing matrix elements are small compared with the separation between the two exciton states, the application of perturbation theory is legitimate. Now, our basis functions are not orthogonal functions, but rather than employ an orthogonalization procedure (i.e., the Schmidt or the symmetric orthogonalization method) we prefer to apply second-order perturbation theory for a nonorthogonal set of unperturbed basis functions. Our calculations show that the effects of the nonorthogonality corrections are relatively small. A further error is introduced into the second-order perturbation term by the use of only a limited number of ion-pair exciton wave functions. However, this error is negligibly small. The energy, to the second order, is given by

$$
E\langle \mathbf{0},\mathbf{0};\mp|\mathbf{1}|\mathbf{0},\mathbf{0};\mp\rangle = \langle \mathbf{0},\mathbf{0};\mp|H|\mathbf{0},\mathbf{0};\mp\rangle + \sum_{i,j} \frac{|L_1(i,j) - L_2 L_3|^2}{L_2 L_4 - L_5}
$$

$$
L_1(i,j) = \langle \mathbf{0},\mathbf{0};\mp|H|\mathbf{R}_i, \mathbf{R}_j;\mp\rangle,
$$

$$L_2 = \langle \ell, \mathbf{0}; \mp | H | \ell, \mathbf{0}; \mp \rangle ,$$

$$L_3 = \langle \ell, \mathbf{0}; \mp | \mathbf{1} | \mathbf{R}_i, \mathbf{R}_j; \mp \rangle ,$$

$$L_4 = \langle \mathbf{R}_i, \mathbf{R}_j; \mp | \mathbf{1} | \mathbf{R}_i, \mathbf{R}_j; \mp \rangle ,$$

$$L_5 = \langle \mathbf{R}_i, \mathbf{R}_j; \mp | H | \mathbf{R}_i, \mathbf{R}_j; \mp \rangle ,$$

where H is the Hamiltonian for the system, $\mathbf{1}$ is the unit operator, and $| \ell, \mathbf{0}; \mp \rangle$ denotes the neutral exciton state. The summation over the charge-transfer states $| \mathbf{R}_i, \mathbf{R}_j; \mp \rangle$ implies, in our approximation, a sum over the states listed in Eqs. (43) through (49). The problem reduces now to the computation of the appropriate matrix elements.

We now proceed to evaluate the contribution of the ion-pair exciton states to the Davydov splitting. The matrix elements connecting the various exciton states can be represented in terms of matrix elements between localized charge transfer and neutral excited states. For example, the nondiagonal matrix element between the states 5, 6, and 8 is given by

$$\langle \ell, \mathbf{0}; \mp | H | \ell, \boldsymbol{\tau}; \mp \rangle$$

$$= \frac{1}{2\sqrt{2}\,N} \sum_{m,n} \Big[\langle \mathbf{R}_n | H | \mathbf{R}_m, \mathbf{R}_m + \boldsymbol{\tau} \rangle$$

$$+ \langle \mathbf{R}_n | H | \mathbf{R}_m, \mathbf{R}_m - \boldsymbol{\tau} \rangle$$

$$+ \langle \mathbf{R}_n + \boldsymbol{\tau} | H | \mathbf{R}_m - \mathbf{b} + \boldsymbol{\tau}, \mathbf{R}_m \rangle$$

$$+ \langle \mathbf{R}_n + \boldsymbol{\tau} | H | \mathbf{R}_m - \mathbf{a} + \boldsymbol{\tau}, \mathbf{R}_m \rangle$$

$+$ Continued next page

$$\mp\langle \mathbf{R}_n | H | \mathbf{R}_m - \mathbf{b} + \boldsymbol{\tau}, \mathbf{R}_m \rangle$$
$$\mp\langle \mathbf{R}_n | H | \mathbf{R}_m - \mathbf{a} + \boldsymbol{\tau}, \mathbf{R}_m \rangle$$
$$\mp\langle \mathbf{R}_n + \boldsymbol{\tau} | H | \mathbf{R}_m, \mathbf{R}_m + \boldsymbol{\tau} \rangle$$
$$\mp\langle \mathbf{R}_n + \boldsymbol{\tau} | H | \mathbf{R}_m, \mathbf{R}_m - \boldsymbol{\tau} \rangle \Big] . \qquad (51)$$

Similar expressions can be readily derived for the other matrix elements.

The largest contribution to these matrix elements arises from terms of the form $\langle \mathbf{R}_p | H | \mathbf{R}_q, \mathbf{R}_s \rangle$ when p = q or p = s. Terms which do not satisfy one of these conditions yield three-center integrals with the centers located on different molecules. These are small terms, and they are neglected herein. The same consideration applies, of course, to the calculation of the overlap integrals. We therefore consider only those matrix elements where the molecular excitation and the negative (or positive) ion are located on the same molecule.

For the purpose of estimating the nonorthogonality corrections we may set

$$\langle \mathbf{0} | H | \mathbf{0} \rangle \equiv \langle \mathbf{0}, \mathbf{0}; \mp | H | \mathbf{0}, \mathbf{0}; \mp \rangle \approx \langle \mathbf{R}_m | H | \mathbf{R}_m \rangle$$

with negligible error. With these rules, it is apparent that only the following four ion-pair exciton wave functions $|\mathbf{0}, \boldsymbol{\tau}; \mp\rangle$, $|\boldsymbol{\tau}, \mathbf{0}; \mp\rangle$, $|\mathbf{0}, \mathbf{c}+\boldsymbol{\tau}; \mp\rangle$, and $|\mathbf{c}+\boldsymbol{\tau}, \mathbf{0}; \mp\rangle$ have different mixing coefficients for the states of A_u and B_u symmetry, and only these four states will contribute to the Davydov splittings. We consider, then, the terms

$$K_1^{\mp} = \langle \mathbf{0}, \mathbf{0}; \mp | H | \mathbf{0}, \boldsymbol{\tau}; \mp \rangle - \langle \mathbf{0} | H | \mathbf{0} \rangle \langle \mathbf{0}, \mathbf{0}; \mp | \mathbf{0}, \boldsymbol{\tau}; \mp \rangle,$$
$$K_2^{\mp} = \langle \mathbf{0}, \mathbf{0}; \mp | H | \boldsymbol{\tau}, \mathbf{0}; \mp \rangle - \langle \mathbf{0} | H | \mathbf{0} \rangle \langle \mathbf{0}, \mathbf{0}; \mp | \boldsymbol{\tau}, \mathbf{0}; \mp \rangle,$$
$$K_3^{\mp} = \langle \mathbf{0}, \mathbf{0}; \mp | H | \mathbf{0}, \mathbf{c}+\boldsymbol{\tau}; \mp \rangle - \langle \mathbf{0} | H | \mathbf{0} \rangle \langle \mathbf{0}, \mathbf{0}; \mp | \mathbf{0}, \mathbf{c}+\boldsymbol{\tau}; \mp \rangle,$$
$$K_4^{\mp} = \langle \mathbf{0}, \mathbf{0}; \mp | H | \mathbf{c}+\boldsymbol{\tau}, \mathbf{0}; \mp \rangle - \langle \mathbf{0} | H | \mathbf{0} \rangle \langle \mathbf{0}, \mathbf{0}; \mp | \mathbf{c}+\boldsymbol{\tau}, \mathbf{0}; \mp \rangle. \qquad (52)$$

Off-diagonal matrix elements between a localized excitation and a charge-transfer state may be represented in the form

$$\begin{aligned} B(i,j) &= \langle \mathbf{R}_i | H | \mathbf{R}_i, \mathbf{R}_j \rangle \\ &\qquad - \langle \mathbf{R}_i | \mathbf{R}_i, \mathbf{R}_j \rangle \langle \mathbf{R}_i | H | \mathbf{R}_i \rangle, \\ C(i,j) &= \langle \mathbf{R}_i | H | \mathbf{R}_j, \mathbf{R}_i \rangle \\ &\qquad - \langle \mathbf{R}_i | \mathbf{R}_j, \mathbf{R}_i \rangle \langle \mathbf{R}_i | H | \mathbf{R}_i \rangle, \end{aligned} \tag{53}$$

whereupon the matrix elements (52) become

$$\begin{aligned} 2K_1^{\mp} &= \sqrt{2}\{[B(\boldsymbol{\rho},\boldsymbol{\tau}) + B(\boldsymbol{\tau},\boldsymbol{\rho})] \mp [C(\boldsymbol{\rho},\boldsymbol{\tau}) + C(\boldsymbol{\tau},\boldsymbol{\rho})]\}, \\ 2K_2^{\mp} &= \sqrt{2}\{[C(\boldsymbol{\rho},\boldsymbol{\tau}) + C(\boldsymbol{\tau},\boldsymbol{\rho})] \mp [B(\boldsymbol{\rho},\boldsymbol{\tau}) + B(\boldsymbol{\tau},\boldsymbol{\rho})]\}, \\ 2K_3^{\mp} &= \sqrt{2}\{[B(\boldsymbol{\rho},\boldsymbol{\sigma}+\boldsymbol{\tau}) + B(\boldsymbol{\sigma}+\boldsymbol{\tau},\boldsymbol{\rho})] \\ &\qquad \mp [C(\boldsymbol{\rho},\boldsymbol{\sigma}+\boldsymbol{\tau}) + C(\boldsymbol{\sigma}+\boldsymbol{\tau},\boldsymbol{\rho})]\}, \\ 2K_4^{\mp} &= \sqrt{2}\{[C(\boldsymbol{\rho},\boldsymbol{\sigma}+\boldsymbol{\tau}) + C(\boldsymbol{\sigma}+\boldsymbol{\tau},\boldsymbol{\rho})] \\ &\qquad \mp [B(\boldsymbol{\rho},\boldsymbol{\sigma}+\boldsymbol{\tau}) + B(\boldsymbol{\sigma}+\boldsymbol{\tau},\boldsymbol{\rho})]\}. \end{aligned} \tag{54}$$

By definition, the Davydov splitting is

$$\Delta E = E(B_u) - E(A_u)$$

and the contribution of the ion-pair exciton states to this splitting is readily obtained from Eq. (50):

$$\begin{aligned} \Delta E &= \sum_{i=1}^{4} \frac{[(K_i^{+})^2 - (K_i^{-})^2]}{\Delta} \\ &= \frac{16}{\Delta}[\bar{C}(\boldsymbol{\rho},\boldsymbol{\tau})\bar{B}(\boldsymbol{\rho},\boldsymbol{\tau}) + \bar{C}(\boldsymbol{\rho},\boldsymbol{\sigma}+\boldsymbol{\tau})\bar{B}(\boldsymbol{\rho},\boldsymbol{\sigma}+\boldsymbol{\tau})], \\ \bar{C}(\boldsymbol{\rho},\boldsymbol{\tau}) &= \tfrac{1}{2}[C(\boldsymbol{\rho},\boldsymbol{\tau}) + C(\boldsymbol{\tau},\boldsymbol{\rho})]. \end{aligned} \tag{55}$$

In Eq. (55), Δ is the energy difference between the charge-transfer and the neutral exciton states.

The configuration-interaction scheme just presented considered the total electronic interaction. We have, however, to consider the common case of electronic bands where the intensity is spread over several vibronic components. We are interested in the configuration interaction between charge-transfer excitons and some triplet and singlet neutral exciton states, both characterized by relatively small intermolecular electronic-coupling matrix elements. Under these circumstances, the application of the weak coupling scheme is legitimate, and in the following we consider configuration interaction between the individual vibronic components of the charge-transfer and the neutral exciton states.

In the weak coupling limit, the wave functions for the neutral excited molecule and for the positive and negative ions are represented as a product of electronic and vibrational wave functions. The localized excitation and charge-transfer states are taken to be the vibronic components

$$|\underset{\sim}{R}_m\rangle \chi_{m1}^{f(p)} \prod_{\ell\alpha \neq m1}' \chi_{\ell\alpha}^{0(q)}$$

and

$$|\underset{\sim}{R}_m, \underset{\sim}{R}_n\rangle \chi_{m1}^{+(k)} \chi_{n2}^{-(\ell)} \prod_{\ell\gamma \neq m1, n2}'' \chi_{\ell\gamma}^{0(q)} \tag{56}$$

The indices p, q, k, and ℓ refer to the quantum states of the molecular vibrations. We consider the system to be at a relatively low temperature so that the ground state molecules are in the zeroth vibrational state, and $q = 0$ for all $\ell\alpha$ and $\ell\gamma$. The electronic integrals (54) are now replaced by integrals over both the vibrational and the electronic coordinates. The off diagonal matrix elements B and C (Eq. (53)) take the form

$$\begin{aligned} C(i,j)V^{p\ell k} = [&\langle \underset{\sim}{R}_i|H|\underset{\sim}{R}_j, \underset{\sim}{R}_i\rangle \\ &- \langle \underset{\sim}{R}_i|\underset{\sim}{R}_j, \underset{\sim}{R}_i\rangle\langle \underset{\sim}{R}_i|H|\underset{\sim}{R}_i\rangle] \times \\ &\times \langle \chi_i^{f(p)}|\chi_i^{-(\ell)}\rangle\langle \chi_\ell^{0(0)}|\chi_j^{+(k)}\rangle, \end{aligned} \tag{57}$$

$$B(i,j)V^{pk\ell} = [\langle \underset{\sim}{R}_i|H|\underset{\sim}{R}_i, \underset{\sim}{R}_j\rangle - \langle \underset{\sim}{R}_i|\underset{\sim}{R}_i, \underset{\sim}{R}_j\rangle\langle \underset{\sim}{R}_i|H|\underset{\sim}{R}_i\rangle] \times \langle \chi_i^{f(p)}|\chi_i^{+(k)}\rangle\langle \chi_j^{o(o)}|\chi_j^{-(\ell)}\rangle \tag{58}$$

where the coefficients $V^{pk\ell}$ and $V^{p\ell k}$ are the vibrational overlap factors. It is necessary to examine the effect of the k and ℓ positive and negative ion vibronic states on the fp neutral exciton vibronic state. For the sake of simplicity, we consider only one progression in the fp neutral exciton state, the totally symmetric vibrational mode. Furthermore, only the totally symmetric progressions will be considered in the vibronic states k, ℓ of the ions. This is a serious approximation. However, the vibrational overlap integrals $V^{pk\ell}$ and $V^{p\ell k}$ cannot be obtained from experimental spectroscopic data, and our approximate treatment should lead to a reasonable estimate of the vibronic coupling.

The contribution to the Davydov splitting of a given vibronic band of a neutral exciton by configuration interaction will be

$$\Delta E^{oo\to fp} = \sum_{k,\ell} \frac{16[\bar{B}(\underset{\sim}{\varrho},\underset{\sim}{\tau})\bar{C}(\underset{\sim}{\varrho},\underset{\sim}{\tau}) + \bar{B}(\underset{\sim}{\varrho},\underset{\sim}{\varsigma}+\underset{\sim}{\tau})\bar{C}(\underset{\sim}{\varrho},\underset{\sim}{\varsigma}+\underset{\sim}{\tau})]}{[E(fi)-E(+k,-\ell)][V_C^{p\ell k}V_B^{pk\ell}]^{-1}}, \tag{59}$$

where E (fi) is the energy of the neutral exciton vibronic state, while E($+k$, $-\ell$) is the energy of the vibronic charge-transfer state with the positive and negative ions in the vibrational states k and ℓ

If the energy denominator is large enough not to be strongly dependent on the vibronic charge-transfer state, it may be replaced by a mean-energy difference Δ , so that

$$\Delta E^{oo\to fp} = \Delta E \sum_{k,\ell} V^{p\ell k} V^{pk\ell}; \tag{60}$$

whereupon the total electronic contribution ΔE is just modified by multiplying by the vibrational overlap integrals.

The summation appearing in Eq. (60), taken over all the vibronic charge-transfer states, can be considerably simplified by writing

$$\sum_{k,l}' V^{pkl} V^{plk} = \sum_{k}' \langle \chi_i^{f(p)} | \chi_i^{+(k)} \rangle \langle \chi_j^{0(0)} | \chi_j^{+(k)} \rangle \times \sum_{l}' \langle \chi_i^{f(p)} | \chi_i^{-(l)} \rangle \langle \chi_j^{0(0)} | \chi_j^{-(l)} \rangle . \tag{61}$$

The intramolecular vibrational overlap integrals are, of course, independent of the labeling of the particular molecule, so that the subscripts i and j can be omitted. Hence the sums in Eq. (61) reduce to

$$\sum_{k}' \langle \chi^{f(p)} | \chi^{+(k)} \rangle \langle \chi^{+(k)} | \chi^{0(0)} \rangle = \langle \chi^{f(p)} | \chi^{0(0)} \rangle \tag{62}$$

so that

$$\sum_{l,k}' V^{pkl} V^{plk} = \langle \chi^{0(0)} | \chi^{f(p)} \rangle^2 , \tag{63}$$

and the contribution of the charge-transfer states to the Davydov splitting in any vibronic band is just given by

$$\Delta E^{00 \to fp} = \Delta E \langle \chi^{0(0)} | \chi^{f(p)} \rangle^2 . \tag{64}$$

We have therefore demonstrated that, when the separation between the charge-transfer and the neutral exciton states is large compared to the vibronic spread of the individual electronic states, the total configuration-interaction matrix element is multiplied by a vibrational overlap factor specifying the transition probability of the isolated molecule to the particular excited vibronic state. This is the same result as obtained for the weak coupling case for neutral exciton states.

B. Location of the Charge-Transfer States

In order to proceed with the configuration-interaction scheme outlined in the previous section, the position of the charge-transfer exciton band must be determined. To date, charge-transfer states in aromatic crystals have not been observed. It should be noted that the intensity of the optical transition to the charge-transfer exciton state should be very low (24, 25), and overlap between charge-transfer and neutral exciton absorption bands will make impossible the observation of the former states. The approximate location of the charge-transfer exciton state relative to the ground state of the crystal can be estimated from the simple expression

$$E(C.T.) = I + EA + G + P, \tag{65}$$

where I and EA are the ionization potential and the electron affinity of the free molecule, G is the Coulomb interaction between the positive and negative ion, and P represents the polarization energy of the lattice due to the ion pair.

The ionization potentials of aromatic molecules are well known, but the electron affinities are not known and must be taken from theoretical calculations. The electron affinities used were calculated by Hoyland and Goodman in the π-electron SCF scheme, including orbital deformation effects.

The Coulomb interaction energy was estimated using two different approximations: (a) Following Lyons (26), the ion-pair interaction energy is computed as that between two charges located at the centers of the molecules. (b) The charges on the positive and negative ions were assumed to be distributed on the ring carbons with densities determined by Huckel-type wave functions. It is surprising that the difference between the point charge and the distributed charge approximation to G is only 10%. The polarization energy P was estimated by Lyons, assuming that the ion-pair constitutes a dipole with magnitude determined by the distance between neighboring sites. This leads to P = −0.8 eV for anthracene and P = −0.6 eV for naphthalene. A more recent evaluation of these polarization energies, again using a classical ion-pair model but taking into account the anisotropic polarizability of the aromatic molecule (27), leads to a value of P = −1.7 eV for anthracene. The cited classical calculation of the polarization energy raises questions similar to those previously encountered in the study of the exciton states in alkali halides on the basis of the electron-transfer model (28). The ion-pair does not represent a stationary state, and the charge distribution in the true excited state is characterized by a multipole moment higher than a dipole, thus leading to a decrease of the polarization energy arising from long-range interactions (29). In view of these uncertainties, we take for the calculation of Δ, P $\approx$ −1 eV for both anthracene and naphthalene. Since the value of Δ for the configuration interaction between triplet and charge-transfer states is large, this is not expected to cause any serious error.

The calculated positions of the charge-transfer states for naphthalene and anthracene crystals are displayed in Table IV. From our estimates, these states are located in the vicinity of the p band of the aromatic crystal. Since exchange interactions between the component ions and the ion-pair are expected to be quite small (probably of the order of 10 cm^{-1}), the energy difference between the singlet and triplet charge-transfer exciton states may be taken to be negligible for our purposes.

From the known locations of the first naphthalene and anthracene triplet states, we estimate the energy denominators as Δ = -1.7 eV for naphthalene and Δ = -1.5 eV for anthracene.

Table IV. The Location of Charge-Transfer States in Some Aromatic Crystals (All Energies in Electron Volts)

	Naphthalene	Anthracene
I	8.3	7.4
EA	-0.2	-0.5
$G(\boldsymbol{\rho}, \boldsymbol{\tau})$[a]	-2.65(-2.83)	-2.50(-2.75)
$G(\boldsymbol{\rho}, \boldsymbol{\sigma}+\boldsymbol{\tau})$	-1.90(-1.83)	-1.56(-1.46)
P	(-1.0)	(-1.0)
E(CT)	4.4 ± 0.5	3.4 ± 0.5

[a](G($\boldsymbol{\rho}, \boldsymbol{\tau}$) and G($\boldsymbol{\rho}, \boldsymbol{\sigma}+\boldsymbol{\tau}$) refer to the Coulomb interactions in the exciton states $|\boldsymbol{\rho}, \boldsymbol{\tau}\rangle$ and $|\boldsymbol{\rho}, \boldsymbol{\sigma}+\boldsymbol{\tau}\rangle$, respectively, calculated by the distributed charge approximation. Numbers in parentheses are calculated from the point-charge approximation.

C. Evaluation of the Electronic Matrix Elements

The calculation of the electronic matrix elements $\langle \mathbf{R}_m | H | \mathbf{R}_m \mathbf{R}_n \rangle$ and $\langle \mathbf{R}_m | H | \mathbf{R}_n \mathbf{R}_m \rangle$ is straightforward although somewhat lengthy. Unlike some more conventional treatments within the framework of perturbation theory, the matrix elements contributing to charge-transfer interactions cannot be expressed in powers of some perturbation parameter. Instead, they consist of molecular integrals involving the excited molecule and the positive- and negative-ion wave functions. The wave functions of the isolated molecule, and of the ion-pair, are again assumed to be represented in the SCF π-electron approximation. Our calculation involves long-range intermolecular interaction, where the behavior of the wave functions at large distances is crucial. Approximating the carbon atomic orbitals by a single Slater-type orbital characterized by an orbital exponent ξ = 1.625 a.u. is clearly inadequate, as it seriously underestimates the magnitude of the tails of the wave functions. In the present case, we have followed our previous procedure and represented the π-type molecular orbital in the form of a linear combination of Slater-type orbitals (LCSTO), with the best available carbon-atom wave functions also represented by a LCSTO.

The molecular wave function for a pair of identical neutral aromatic molecules, one of them excited to the first triplet state, can be represented in the form

$$|\underset{\sim}{R}_m^t, \underset{\sim}{R}_n^0\rangle = \mathcal{Q}[u_m^1(1)\,\bar{u}_m^1(2)\ldots\bar{u}_m^{N/2}(N-1)\times \bar{u}_m^{(N/2)+1}(N)\,u_n^1(N+1)\,\bar{u}_n^1(N+2)\times \cdots \times u_n^{N/2}(2N-1)\,\bar{u}_n^{N/2}(2N)]. \tag{66}$$

The subscript refers to the molecule, while the superscript labels the particular MO. The bar refers to spin β, while the absence of a bar refers to spin α. N is the number of atoms in the molecule. The lowest-lying triplet state of the aromatic molecules of interest involves the excitation $u^{N/2} \rightarrow u^{(N/2)+1}$. The triplet state represented in Eq. (66) corresponds to the spin components S = 1, S_z = -1. $\mathcal{Q}$ represents the antisymmetrization operator $\Sigma_P(-)^P P$ permuting all the electrons of the system.

The triplet ion-pair wave function for the same spin components, S = 1, S_z = -1, is given by

$$|\underset{\sim}{R}_m, \underset{\sim}{R}_n\rangle = \mathcal{Q}[u_m^1(1)\,\bar{u}_m^1(2)\ldots\bar{u}_m^{N/2}(N-1)\times \bar{u}_n^{(N/2)+1}(N)\,u_n^1(N+1)\,\bar{u}_n^1(N+2)\times \cdots \times u_n^{N/2}(2N-1)\,\bar{u}_n^{N/2}(2N)]. \tag{67}$$

The total Hamiltonian for the two molecules is taken in the form $H = H_m + H_n + H_{mn}$, where H_m and H_n are the Hamiltonian operators for the isolated molecules m and n, while H_{mn} is the intermolecular interaction potential.

We have assumed that the molecular orbitals are known, and that the molecular wave functions are not distorted by the interaction between the excited- and ground-state molecules, or by the interaction between the positive and the negative ions. The calculation of the matrix elements can then be considerably simplified by the introduction of core potentials. In the configurations of the neutral excited molecular pair and the ion-pair, the occupation numbers of orbitals on the two molecules are different, and care must be used in the definition of the core potentials and the interaction potential. The simplest wave functions which will allow us to consider all the interactions pertinent for the present discussion are those involving four electrons, which in the state $|\underset{\sim}{R}^f_m \underset{\sim}{R}^\circ_n\rangle$ occupy the highest filled orbital on n and the orbital $u_m^{N/2}$ and $u_m^{(N/2)+1}$ so that

$$|\underset{\sim}{R}^f_m \underset{\sim}{R}^\circ_n\rangle = Q\left[\bar{u}_m^{N/2}(1)\,\bar{u}_m^{(N/2)+1}(2)\,u_n^{N/2}(3)\,\bar{u}_n^{N/2}(4)\right],$$

$$|\underset{\sim}{R}_m \underset{\sim}{R}_n\rangle = Q\left[\bar{u}_m^{N/2}(1)\,\bar{u}_n^{(N/2)+1}(2)\,u_n^{N/2}(3)\,\bar{u}_n^{N/2}(4)\right].$$

The core Hamiltonian H^c for the molecules m and n will be defined in the form (30)

$$H_m^c = V_m^{Nuc} + \sum_{s=2}^{N/2-1}(2J_m^s - K_m^s),$$

$$H_n^c = V_n^{Nuc} + \sum_{s=1}^{N/2-1}(2J_n^s - K_n^s),$$

(68)

where V_m^{Nuc} and V_n^{Nuc} are the bare nuclear potentials of Molecules m and n, and J^s and K^s are the Coulomb and exchange operators defined by the well-known relations

$$J_m^s u_r^i(1) = \int \frac{|u_m^s(2)|^2}{r_{12}}\, d\tau_2\, u_r^i(1),$$

$$K_m^s u_r^i(1) = \int \frac{u_m^s(2)\, u_r^i(2)}{r_{12}}\, d\tau_2\, u_m^s(1),$$

(69)

where r = m, n. The Hamiltonian for the four-electron system may now be recast in the form

$$H = \sum_{i=1}^{4} [T_i + H_m^c(i) + H_n^c(i) + \sum_{j<i}' \frac{1}{r_{ij}}], \tag{70}$$

where the T_i are the kinetic-energy operators. This scheme can be easily extended including all the nonorthogonality corrections for all the π electrons.

Finally, the Coulomb interaction is described in terms of the Goeppert-Mayer-Sklar potential

$$V_A^{GMS} = V_A^{Nuc} + \sum_{s=1}^{N/2} 2 J_A^s, \tag{71}$$

which involves the screening of the nuclear charge by the Hartree field.

Table V. Configuration-Interaction Matrix Elements for Triplet States with Charge-Transfer States

	$\bar{B}(\boldsymbol{\rho}, \boldsymbol{\xi})$ (10^{-2} eV)	$\bar{B}(\boldsymbol{\rho}, \boldsymbol{\sigma}+\boldsymbol{\xi})$ (10^{-2} eV)	$\bar{C}(\boldsymbol{\rho}, \boldsymbol{\xi})$ (10^{-2} eV)	$\bar{C}(\boldsymbol{\rho}, \boldsymbol{\sigma}+\boldsymbol{\xi})$ (10^{-2} eV)
Anthracene	-2.30	-0.39	-0.78	+0.85
Naphthalene	-0.76	+1.52	-1.30	+1.20

D. Some Numerical Results

Our calculations lead to the following contribution by the ion-pair states to the Davydov splittings of the first triplet state (see Table V for the configuration interaction matrix elements):

$$\text{naphthalene} = 34/\Delta \quad (\text{eV})\ \text{cm}^{-1},$$

$$\text{anthracene} = 24/\Delta \quad (\text{eV})\ \text{cm}^{-1}.$$

Setting $\Delta = -1.7$ eV for naphthalene and $\Delta = -1.5$ eV for anthracene, the contributions of the ion-pair states to the first triplet exciton state are finally obtained. As will be seen in Section IV, the ion-pair states increase the triplet band splitting due to exchange interactions by about 50%. It should also be noted that both the (neutral exciton) exchange (interaction) and the charge-transfer contributions to the triplet band Davydov splitting are found to be of the same sign, and awkward cancellations do not occur in the cases investigated by us.

The contribution of the ion-pair states to the Davydov splitting is of considerable importance for those singlet exciton states where the energy denominator Δ is small. In these cases, contributions to of the order of 100-200 cm^{-1} are quire plausible. This is, indeed, the case for the first absorption band of naphthalene which has recently been studied by us (31).

The study of the naphthalene crystal is of particular interest because the lowest electronic transition (α) in the naphthalene crystal shows a Davydov splitting of about 150 cm^{-1} in the 0-0 vibrational component (32), while the parent free molecule transition is so weak that the observed splitting cannot be accounted for in terms of transition dipole interactions (7). This failure of the usual formulation of exciton theory led Craig and co-workers to consider higher transition multipole interactions as the source of the splitting (7). Indeed, the transition octopole moments required to fit the observed spectrum are not large, and it is conceivable that they are good approximations to the true moments. On the other hand, the octopole moments estimated using either Huckel or Pariser wave functions (33) vanish in the approximation of zero differential overlap (34). Wheland orbitals (including nearest neighbor overlap) lead to transition octupole moments in fair agreement with those suggested by Craig. However, it must be remembered that the use of Wheland orbitals leads to a transition dipole moment a factor of 10-15 larger than is observed (35), and therefore the estimated octopole moment may be seriously in error. In view of this uncertainty in the magnitude of the transition moments, we prefer to interpret the splitting in terms of the configuration mixing of neutral and ion-pair exciton states.

The results of the calculation of the effect of mixing of charge transfer states into the α and p exciton states, assuming the lowest charge-transfer states to be 500 cm^{-1} and 1000 cm^{-1} above the α band, are given in Table VI. The experimental results are also given in Table VI.

We now calculate the polarization ratio of the α band from our results. The oscillator strength of the α band in the free molecule is approximately 0.001, but about 90% is vibronically induced, so that the true oscillator strength is approximately 0.0001. The polarization ratio is easily obtained from

Table VI. Calculated Splittings and Observed Splittings for Naphthalene

Electronic State	Energy of A_u State ($\parallel b$) Relative to Gas Value	Energy of B_u State ($\perp b$) Relative to Vapor Value	$\Delta E^*_{(calc)}$	$\Delta E_{(exp)}$
	Charge Transfer State 500 cm^{-1} above α Band			
α	-178 cm^{-1}	-310 cm^{-1}	-132 cm^{-1}	-186 cm^{-1} (a)
p	-1470 cm^{-1}	-630 cm^{-1}	840 cm^{-1}	320 cm^{-1} (b)
β	-4400 cm^{-1}	32000 cm^{-1}	36400 cm^{-1}	--
	Charge Transfer State 1000 cm^{-1} above α Band			
α	-100 cm^{-1}	-193 cm^{-1}	-93 cm^{-1}	
p	-1460 cm^{-1}	-620 cm^{-1}	840 cm^{-1}	
β	-4400 cm^{-1}	32000 cm^{-1}	36400 cm^{-1}	

$\Delta E = E_{B_u} - E_{A_u}$

(a) D. P. Craig and H. C. Wolf, J. Chem. Phys. 41, 2057 (1964). This is the total splitting in the entire vibronic band. The splitting in the 0-0 band is 150 cm^{-1} and in the 0-1 band 36 cm^{-1} for the progression that is entirely built on a_g vibrations.

(b) D. S. McClure and O. Schnepp, J. Chem. Phys. 23, 1575 (1955). The number quoted is the splitting in the 0-0 band.

$$P^f(\underset{\sim}{b}/\underset{\sim}{a}) = \left[\left(\sum_i c_{fi}^-(p_1^f - p_2^f)\cdot\underset{\sim}{b}\right)\Big/\left(\sum_i c_{fi}^+(p_1^f + p_2^f)\cdot\underset{\sim}{a}\right)\right]^2 \tag{72}$$

where the $c_{fi}^{\pm}$ are the coefficients of the ith state in the wave function of the fth state, and p_i^f is the transition moment of the ith state on one set of molecules. The results of this calculation are given in Table VII. The polarization ratio is extremely sensitive to the actual transition moment of the α band and, in fact, agreement between theory and experiment can be obtained by assuming that the oscillator strength represents 80% vibronically induced character and 20% "true" electronic character. As before (8), we have also carried through the analysis including the higher π -electron states. The results are not much different from those obtained with these states omitted (see Table VII).

In the vapor phase, the first singlet state (α) of naphthalene has an interesting spectrum (36). There are three vibrational progressions observed: one is entirely an a_g vibration, while the other two are a_g progressions built on a b_{1g} vibration. Since the lowest state of naphthalene is a B_{3u} state the addition of a b_{1g} vibration transforms the total wave function to have symmetry b_{2u}, which symmetry is the same as that of the second excited state, thereby permitting vibronic mixing of the p (B_{2u}) state into the α (B_{3u}) state. We write the ground vibronic function of the α state (to zero order) as

$$\Phi^\alpha = \phi^\alpha_{elec} \prod_\sigma \chi_\sigma^{(0)} \tag{73}$$

where the $\chi_\sigma^{(i)}$ are vibrational wave functions of the σ th symmetry (normal) coordinate with i quanta. Thus the progression built solely on a_g vibrations may be derived from

$$\Phi^\alpha_{a_g} = \phi^\alpha_{elec}\, \chi_{a_g}^{(i)} \prod_\sigma{}' \chi_\sigma^{(0)} \tag{74}$$

where only one normal coordinate is assumed to have changed from the ground state vibronic wave function. Actually, of course, almost all the normal coordinates of the excited state are slightly changed from the ground state. This will reduce the calculated splittings, but we

Table VII. Calculated and Observed Polarization Ratios for Naphthalene Transitions

State	Charge Transfer 500 cm^{-1} above α	Charge Transfer 1000 cm^{-1} above α	Oriented Gas	Exp.
α	100.	0.1	0.25	>10*
p	4.0	4.0	7.7	3.0
β	0.6	0.6	0.25	--

α state is 80% vibronically induced, all π states included.

* In the 0–0 band; in the 0–1 band the polarization ratio is ~ 4.

assume that only one normal mode changes substantially and that the effect of this will mask the effects of the other vibrations. The progression built on a b_{1g} vibration may be obtained from the wave function

$$\Phi^{\alpha}_{b_{1g}} = \phi^{\alpha}_{elec}\,\chi^{(1)}_{b_{1g}}\,\chi^{(i)}_{a_g}\prod_{\sigma}{}''\chi^{(0)}_{\sigma} + c\,\phi^{\beta}_{elec}\,\chi^{(i)}_{a_g}\prod_{\sigma}{}'\chi^{(0)}_{\sigma}, \tag{75}$$

where c is the mixing coefficient and may be estimated from the vapor spectrum. The reader should note that the charge-transfer state will mix into both wave functions, but that the extent of mixing into the second wave function will be reduced by an extra vibrational overlap from the mixing into the first wave function. If we consider the mixing of the charge transfer states into the $\Phi^{\alpha}(b_{3g})$ state we find for the B and C matrix elements (now including vibrational effects and denoted B_{vibr}, etc.)

$$B_{vibr} = B_{elec}\left\langle \chi^{(1)}_{b_{1g}}\,\chi^{(i)}_{a_g}\prod_{\sigma}{}''\chi^{(0)}_{\sigma} \,\middle|\, \chi^{+}_{Total}\right\rangle \times \left\langle \prod_{\sigma}\chi^{(0)}_{\sigma} \,\middle|\, \chi^{-}_{Total}\right\rangle, \tag{76}$$

$$C_{vibr} = C_{elec}\left\langle \chi^{(1)}_{b_{1g}}\,\chi^{(i)}_{a_g}\prod_{\sigma}{}''\chi^{(0)}_{\sigma} \,\middle|\, \chi^{-}_{Total}\right\rangle \times \left\langle \prod_{\sigma}\chi^{(0)}_{\sigma} \,\middle|\, \chi^{+}_{Total}\right\rangle. \tag{77}$$

Now if the b_{1g} vibration is on the positive ion, the $C_{vibr} = 0$ (since χ^{ground}_{total} is an a_g function) similarly if the b_{1g} vibration is on the negative ion then $B_{vibr} = 0$. Thus the mixing of the crystal charge transfer states of A_u and B_u symmetry into the neutral exciton states of corresponding symmetry will be identical (or at most differ in sign) and thus the Davydov splitting in the transitions built on a b_{3g} vibration will <u>not</u> be affected by presence of charge transfer states. The magnitude of the splitting in these bands (built on a false origin) can be estimated in the following manner: the coefficient of vibronic mixing may be estimated from the ratio of oscillator strengths of the α and p bands, i.e., $c \approx (f_\alpha / f_p)^{1/2}$; since the oscillator strength of the transition is 80 to 90% vibronically induced, we find c $\approx$ 1/10. Therefore, the entire Davydov splitting in the bands built on a b_{1g} vibration will be given by c^2 x p band splitting which is 8 cm^{-1}. In the different vibrational components, we should then have a splitting of $\sim$ 2 cm^{-1}.

We conclude that the progression built on the b_{1g} vibrations should have a much smaller Davydov splitting than the progression built on the a_g vibration, in agreement with the experimental data.

The above argument and the agreement it gives with experiment rest on the assumption that the charge transfer state lies very close to the α state and above it. There is no evidence supporting this assumption. Crude estimates of the energy of the charge transfer state put it in the vicinity of the α band in naphthalene, but it is impossible to go further at present.

IV. TRIPLET EXCITONS

A. General Remarks

The lowest excited state of all aromatic molecules is a triplet, to which direct optical transition from the ground state is extremely weak because of spin selection rules (37-39). Although direct observation of the Davydov splitting of a triplet level has only been reported for one compound (bromnaphthalene) (40) there have been many recent investigations of the dynamical properties of triplet excitons. It has now been unambiguously established that triplet excitons are mobile in aromatic crystals and various phenomena involving triplet energy transfer have been demonstrated. The following observations are pertinent to the present discussion:

(a) From electron spin resonance studies it has been shown that triplet energy is transferred between widely separated guest molecules in a host crystal (20). For the case of the guests phenanthrene and naphthalene in the host biphenyl, with the guest molecules oriented with their principal magnetic axes approximately parallel to that of the biphenyl, it has been shown that an important pathway for energy transfer involves a vibrationally excited state of the lowest triplet state of phenanthrene (with energy matching the triplet band energy of the host crystal) (41).

(b) Optical emission studies of isotopically mixed benzene (21) and naphthalene (42) crystals have shown that trap-to-trap migration of triplet excitation energy occurs in these systems (43). A recent study of the phosphorescence spectra of isotopically mixed benzene crystals leads to a value of 12 cm^{-1} for the nearest neighbor pair triplet interaction matrix element (44). The total triplet electronic matrix element is, therefore, of the order of magnitude of 50 cm^{-1} (14).

(c) Direct population of the first triplet state of the anthracene crystal is possible using intense light sources. Irradiation of an anthracene crystal with a ruby laser ($h\nu = 1.8$ eV) leads to delayed blue fluorescence arising from the transition $^1B_{2u} \rightarrow {}^1A_{1g}$ ($h\nu = 3.42$ eV) (45). Now, the single photon absorption at room temperature is from a vibrationally excited level of the ground state to the triplet state (i.e., the $1 \rightarrow 0$ $^1A_{1g} \rightarrow {}^3B_{2u}$ transition), and the elayed fluorescence (observed on a time scale of milliseconds) depends on the square of the light intensity and follows bimolecular kinetics (45, 46). These results have been interpreted (45) by assuming an annihilation process involving two triplet excitons, with the rate constant for exciton-exciton annihilation estimated to be 2×10^{-11} cm^{-3} sec^{-1}. An alternative method (47) for population of the triplet state in the anthracene crystal (population by radiationless transition from the $^1B_{2u}$ singlet state to the triplet state) leads to the same biexciton annihilation mechanism. Other experiments using (incoherent) continuous light sources (48) also lead to similar conclusions regarding the annihilation of triplet excitons by a bimolecular process. Finally, we note that the excitation spectrum (in the region 14,000 to 20,000 cm^{-1}) has been identified with the $^1A_{1g} \rightarrow {}^3B_{2u}$ absorption spectrum (48).

We now consider the band structure for the first (lowest) triplet exciton states in aromatic crystals. Our results will be used in the later study of triplet exciton dynamics in crystals of aromatic molecules.

As shown in Section II, the structure of an exciton band in a molecular crystal is determined by the interaction energy between the molecules of the crystal. We have already emphasized the importance of using SCF wave functions for the construction of molecular orbitals to provide an improved description of the molecular wave functions at large distances from the molecule. The use of the SCF functions leads to very large changes in the overlap and exchange integrals of importance in the study of triplet excitons. Moreover, it is important to remark that changes in quantitative detail can lead to changes in qualitative interpretation if, as a result of the quantitative changes, the balance is altered between several competing contributions. Therefore we again adopt the SCF carbon atom wave functions and use them for numerical calculations of the triplet exciton band structure in aromatic crystals.

We turn now to the calculation of the J and K integrals for the triplet states of aromatic crystals. As long as spin orbit coupling is neglected the excitation exchange matrix element vanishes (spin orthogonality). An estimate of the integral J can be obtained by considering the spin orbit coupling between the triplet state of interest and the perturbing singlet states. Application of perturbation theory leads to

$$(\varphi^{f}_{m\mu})_1 = (\varphi^{f}_{m\mu})_0 + \sum_s \frac{(H_{so})_{fs}}{E_f - E_s} \varphi^{s}_{m\mu}, \tag{78}$$

where $(\varphi^{f}_{m\mu})_1$ is the perturbed triplet wave function, $\varphi^{s}_{m\mu}$ an unperturbed excited singlet state, E_f and E_s the triplet and singlet energies, and $(H_{so})_{fs}$ the spin orbit coupling matrix element combining the two states. We then find for the coulomb contribution to the excitation exchange interaction

$$J^{f}_{m\mu,n1} = \left\langle \sum_s \frac{(H_{so})_{fs}}{E_f - E_s} \varphi^{s}_{n1} \varphi^{0}_{m\mu} \left| V_{n1,m\mu} \right| \varphi^{0}_{n1} \times \sum_{s'} \frac{(H_{so})_{fs'}}{E_f - E_{s'}} \varphi^{s'}_{m\mu} \right\rangle. \tag{79}$$

In this expression the interaction of the ground state wave function with excited triplet states by spin orbit coupling is not taken into account. For approximate numerical estimates, this omission is not serious.

Spin orbit coupling in aromatic hydrocarbons has been studied by several investigators (37-39). McClure (37) has shown that when only coupling between $\pi \rightarrow \pi^*$ transitions is considered, one and two center contributions to the spin orbit coupling matrix elements vanish, and three center contributions have to be included. Mizishuma and Koida (38) considered the possibility of spin orbit coupling between a triplet $\pi \rightarrow \pi^*$ state and a singlet $\sigma \rightarrow \pi^*$ state, where one center contributions do not vanish. In any case, the spin orbit coupling is extremely weak in these systems, leading to long radiative (phosphorescence) lifetimes. Following McClure (37), we assume that $(H_{so})_{fs} \approx 1\ \text{cm}^{-1}$, and to make a rough estimate of the excitation exchange integral we consider just one perturbing singlet state. Then taking $E_S - E_f = 20{,}000\ \text{cm}^{-1}$, the triplet state coulomb integral, J^f, is related to the perturbing singlet state coulomb integral, J^s, by

$$J^{f}_{m\mu,n1} \approx \frac{|(H_{so})_{fs}|^2}{(E_f - E_s)^2} J^{s}_{m\mu,n1}. \tag{80}$$

The contribution of these coulomb type integrals to the Davydov splitting can be estimated by summing over all translationally inequivalent molecules. For strongly allowed singlet-singlet transitions, $\sum'_{m2} J^{s}_{m2,n1} \approx 10^3$-$10^5$ cm^{-1}, so that the contribution of the coulomb integrals to the Davydov splitting of the triplet state is of the order of 10^{-3} cm^{-1}, and is therefore negligibly small. We conclude that, within the tight binding approximation, the triplet exciton band width arises from intermolecular electron exchange interactions.

We now examine the magnitude of the exchange interaction. Three types of molecular wave functions were used for the calculation of the required exchange integrals: antisymmetrized products of Huckel orbitals, antisymmetrized products of π electron SCF orbitals calculated by Hoyland and Goodman (49), and π electron configuration interaction wave functions computed by Pariser (33). The matrix elements calculated from all three wave functions are nearly equal to one another. The intermolecular exchange integrals for anthracene and naphthalene crystals are displayed in Table VIII. These interactions lead to a total electronic contribution of 32 cm^{-1} for the Davydov splitting in the first triplet exciton state of the anthracene crystal. In the weak coupling scheme, which is obviously appropriate for this case, the contribution of electron exchange interactions to the splitting in the 0-0 vibronic band of the anthracene $^3B_{2u}$ state is about 8 cm^{-1}. A detailed analysis has shown that nonorthogonality corrections are small, and the effects of crystal field mixing are negligible. However, as shown in Section III, the effects of configuration interaction with ion-pair excitons are important.

B. The Band Structure

We have constructed the triplet exciton states in several aromatic crystals in terms of states which involve the symmetric and antisymmetric combinations of molecular wave functions corresponding to the two molecules in the unit cell. The corresponding $\underline{k}$ dependent eigenvalues give the two energy bands (only $\underline{k}$ dependent terms are displayed):

Table VIII. Electron Exchange Intermolecular Integrals in Crystalline Anthracene, Naphthalene and Biphenyl (23)

(Energies in units of 10^{-4} eV)

Location of Molecule	Naphthalene	Anthracene	Biphenyl
$\underset{\sim}{b}$	4.51	4.10	-0.95
$\frac{1}{2}(\underline{a}+\underset{\sim}{b})$	-7.13	-6.83	-0.49

$$\begin{aligned} E'_{+}(\underline{k}) = {} & 2K_2 \cos(\underline{k}\cdot\underline{c}) + 2K_3 \cos(\underline{k}\cdot\underline{b}) \\ & + 2K_4[\cos \underline{k}\cdot(\underline{b}+\underline{c}) + \cos \underline{k}\cdot(\underline{b}-\underline{c})] \\ & + 2K_5 \cos(\underline{k}\cdot\underline{a}) + 2K_6 \cos \underline{k}\cdot(\underline{c}+\underline{a}) \\ & + 2K_7[\cos \underline{k}\cdot(\underline{a}+\underline{b}) + \cos \underline{k}\cdot(\underline{a}-\underline{b})] \\ & + 2K_8[\cos \underline{k}\cdot(\underline{a}+\underline{b}+\underline{c}) + \cos \underline{k}\cdot(\underline{a}-\underline{b}+\underline{c})] \\ & \pm 2K_9[\cos \underline{k}\cdot\frac{(\underline{a}+\underline{b})}{2} + \cos \underline{k}\cdot\frac{(\underline{a}-\underline{b})}{2}] \\ & \pm 2K_{10}[\cos \underline{k}\cdot(\frac{\underline{a}+\underline{b}}{2} + \underline{c}) + \cos \underline{k}\cdot(\frac{\underline{a}-\underline{b}}{2} + \underline{c})] \end{aligned} \tag{81}$$

and the ten K_i are the electron exchange integrals between the reference molecule 1 and i. The band can be readily visualized if we consider now the special cases when $\underline{k}$ is parallel to a reciprocal lattice vector, $\underline{a}^{-1}$, $\underline{b}^{-1}$, or $\underline{c}^{-1}$:

$$E'_{\pm}(\underline{k} \quad \underline{a}^{-1}) = A + B \cos(\underline{k}\cdot\underline{a}) \pm C \cos(\underline{k}\cdot\underline{a}/2) \tag{82}$$

$$\begin{aligned} E'_{\pm}(\underline{k} \quad \underline{b}^{-1}) = {} & D + E \cos(\underline{k}\cdot\underline{b}) \pm F \cos(\underline{k}\cdot\underline{b}/2) \pm \\ & \pm E_{14} \cos(\underline{k}\cdot 3\underline{b}/2) \end{aligned} \tag{83}$$

$$E'_{+}(\underline{k} \parallel \underline{c}^{-1}) = G + H\cos(\underline{k}\cdot\underline{c}),$$
$$E'_{-}(\underline{k} \parallel \underline{c}^{-1}) = I + L\cos(\underline{k}\cdot\underline{c}), \quad (84)$$

where we have defined A through L by the relations

$$\begin{aligned} A &= 2(K_2 + K_3 + 2K_4), \\ B &= 2(K_5 + K_6 + 2K_7 + 2K_8), \\ C &= 4(K_9 + K_{10}), \\ D &= 2(K_2 + K_5 + K_6), \\ E &= 2(K_3 + 2K_4 + 2K_7 + 2K_8), \\ F &= 4(K_9 + K_{10}), \\ G &= 2(K_3 + K_5 + 2K_7 + 2K_9), \\ H &= 2(K_2 + 2K_4 + K_6 + 2K_8 + 2K_{10}), \\ I &= 2(K_3 + K_5 + 2K_7 - 2K_9), \\ L &= 2(K_2 + 2K_4 + K_6 + 2K_8 - 2K_{10}). \end{aligned} \quad (85)$$

The symmetry properties of these energy bands have previously been discussed. For the cases when $\underline{k}$ is parallel to the reciprocal lattice vectors $\underline{a}^{-1}$ and $\underline{b}^{-1}$, the two bands are degenerate at $\pi\underline{a}^{-1}$ and at $\pi\underline{b}^{-1}$ and intersect at these points with equal and opposite slopes. This conclusion is a general result of the effect of time reversal symmetry on the energy bands of crystals. The energy bands are expected to stick together in pairs on the boundary plane of the Brillouin zone that is perpendicular to a twofold screw axis of the reciprocal lattice.

The first triplet exciton band structure of anthracene, naphthalene and biphenyl is displayed in Figure 2. The band structure has been plotted in the three principal reciprocal lattice directions. In Table IX we display the values of the widths of the bands in the $\underline{a}^{-1}$, $\underline{b}^{-1}$ and $\underline{c}^{-1}$ directions. These results are of interest in the study of triplet exciton dynamics.

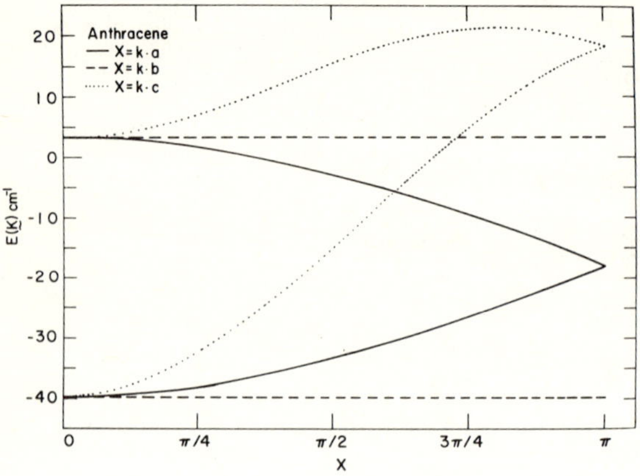

Fig. 2(a). The triplet band structure (neutral exciton) of anthracene.

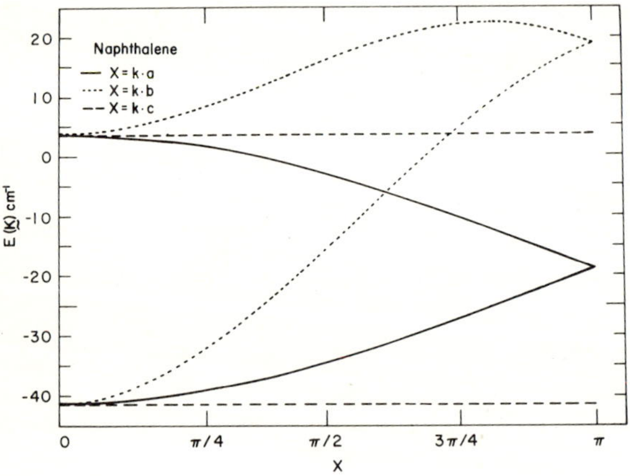

Fig. 2(b). The triplet band structure (neutral exciton) of naphthalene.

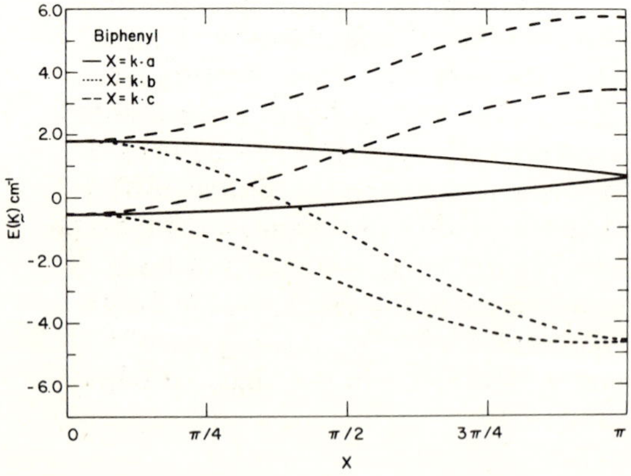

Fig. 2(c). The triplet band structure (neutral exciton) of biphenyl.

Table IX. Electron Exchange Contribution to Triplet Exciton Bandwidths (Energy in Units of 10^{-4} eV)

	Naphthalene	Anthracene	Biphenyl
$(a^{-1})_+$	28.62	27.36	1.105
$(a^{-1})_-$	28.61	27.36	1.212
$(b^{-1})_+$	11.31	11.41	4.922
$(b^{-1})_-$	46.69	43.76	2.775
$(c^{-1})_+$	0.134	0.086	3.679
$(c^{-1})_-$	0.190	0.077	0.334

C. Triplet Exciton Migration

We have demonstrated that in the first excited triplet state of some typical aromatic crystals the electron exchange plus charge transfer interactions between adjacent molecules are of the order of 10 to 50 cm^{-1}. These relatively small interactions determine the rate of migration of triplet excitons. Clearly, the frequency of transfer of triplet excitation will be of the order of magnitude of the frequency of intermolecular lattice vibrations, and much smaller than the frequency of intramolecular vibrations. Under these conditions there will be appreciable scattering of the triplet exciton by the lattice phonons. Two models are useful for the description of the triplet exciton motion. First, we may assume that the scattering of the triplet excitation wave is so strong that the mean free path is of the order of magnitude of the intermolecular separation in the crystal. The motion of the excitation can then be described as a random walk, i.e., a diffusion process. Alternatively, the triplet exciton motion can be described within the framework of the band model, where the limiting case of narrow bands is considered.

The hopping model for triplet exciton motion has previously been considered in some detail (50), and we shall limit ourselves to the basic results. In describing the incoherent motion of the triplet excitation, the transition probability W_i for energy transfer from the reference molecule 1 to molecule i is given by

$$W_{1i} = \frac{2\pi}{\hbar} |M_{1i}|^2 \int F(E) f(E)\, dE, \tag{86}$$

where the matrix element M_{1i} includes ion-pair mixing, and F(E) and f(E) are the singlet-triplet absorption function and the triplet-singlet phosphorescence function, respectively, each normalized to unity on the energy scale. The integral, $\int F(E)f(E)dE$, corresponds to a vibrational overlap factor, the vibrational spectrum being treated as a continuum. In view of the short range of the exchange integrals, K_{1i}, energy transfer only to near neighbors need be considered. The mean lifetime of the triplet excitation on the reference molecule 1, τ_1, is given by

$$(1/\tau_1) = \sum_i W_{1i}, \tag{87}$$

and the mean diffusion coefficient of the triplet excitation is

$$D = \frac{1}{6} \sum_i W_{1i} R_{1i}^2, \tag{88}$$

where R_{1i} is the distance between molecules 1 and i. The diagonal elements of the diffusion tensor along the crystal axes are

$$\begin{aligned} D_{aa} &= a^2/2(\tau_1)_a, \\ D_{bb} &= b^2/2(\tau_1)_b, \\ D_{cc} &= c^2/2(\tau_1)_c. \end{aligned} \tag{89}$$

The vibrational overlap integral $\int F(E)f(E)dE$ for anthracene was estimated from the phosphorescence spectrum in EPA glass (51) and from the reaction spectrum of the anthracene crystal laser induced blue fluorescence (48). We find $\int F(E)f(E)dE = 0.2\ eV^{-1}$, whereupon, with the inclusion of ion-pair mixing we find for the anthracene crystal, $(1/\tau_1) = 3 \times 10^{10}\ sec^{-1}$ and $D = 3 \times 10^{-5}\ cm^2\ sec^{-1}$. It is worthwhile noting that when electron exchange interactions alone are considered, only interactions within the ab plane are of appreciable magnitude. However, the mixing of charge-transfer states into the neutral exciton states leads to an appreciable transition probability from the reference molecule to the one located at $\underset{\sim}{c} + \underset{\sim}{\tau}$, and the motion of the exciton is not confined to a single molecular plane.

The diffusion length for a triplet excitation in the anthracene crystal in the limiting case of small exciton concentration is given by

$\ell = \sqrt{2D\tau_0}$, where $\tau_0 = 10^{-2}$ sec is the triplet lifetime in anthracene (15, 16). Hence, $\ell = 8 \times 10^{-4}$ cm, which is to be compared with the experimental value, $\ell = 10 \pm 5 \times 10^{-4}$ cm as measured by Avakian and Merrifield (52).

An estimate of the rate of encounter of two triplet excitons is of considerable interest with respect to the study of biexcitonic annihilation. Within the framework of the random walk model, the rate of collision of two excitons is (53)

$$\Gamma = 8\pi D \langle R \rangle n_T^2, \tag{90a}$$

where $\langle R \rangle$ is an average crystal spacing and n_T the number of triplet excitons per unit volume. An alternative formalism sets

$$\Gamma = \frac{z}{2\rho\tau_1} n_T^2, \tag{90b}$$

where z is the number of nearest neighbors to a molecule and ρ the number of molecules per unit volume. A comparison between the predictions of (90a) and (90b) sets bounds on the reliability of Γ. The main features of the triplet-triplet annihilation mechanism leading to the formation of an excited singlet can be described in terms of the simple kinetic scheme (54):

$$T + T \underset{k_{-1}}{\overset{k_1}{\rightleftharpoons}} T_iT_j \xrightarrow{k_2} S_1 \xrightarrow{k_3} G + h\nu \tag{91}$$

where G, S_1 and T refer to the ground state, first excited singlet and first excited triplet, respectively. The T_iT_j refers to a state where the two triplets are located on adjacent molecules. The annihilation process may involve some intermediate steps, and will be further discussed later. Under steady state conditions the rate of fluorescence emission is given by

$$\mathcal{R} = \frac{d}{dt}(h\nu) = \frac{k_1 n_T^2}{\left(\frac{k_{-1}}{k_2} + 1\right)}. \tag{92}$$

Two limiting cases are of interest:

(a) When the rate of dissociation of the triplet pair is large compared with the annihilation rate, i.e., $k_{-1} \gg k_2$, the annihilation reaction is expected to be determined by the transition rate, k_2, to the final state, $\mathcal{R} = (k_1 k_2 / k_{-1}) n_T^2$.

(b) If the transition to the final singlet state is fast on the time scale of triplet exciton migration, i.e., $k_{-1} < k_2$, then the annihilation reaction is diffusion controlled. In this case the rate is $\mathcal{R} = k_1 n_T^2$, being determined by the rate of collision of the triplet excitons.

As we shall show later, in the case of triplet-triplet annihilation the rate determining step involves the production of two adjacent excited molecules. It should be noted that if spin conservation is maintained, the conversion of the triplet pair to a singlet state is modified by the probability for the formation of the pair $T_i T_j$ in the singlet state. This leads to a statistical factor of $w = 1/9$ (since two triplet excitons may form a singlet, triplet and a quintet pair). For the case of the diffusion controlled triplet-triplet annihilation, the rate is thus given by

$$\mathcal{R} = w\Gamma \tag{93}$$

The random walk model just considered may underestimate the mean free path, i.e., the mean free path may be larger than the intermolecular spacing. In this case the exciton motion is better described within the framework of the band scheme.

The band structure calculations reported in Section IV-B can be used to estimate the diffusion coefficient of the triplet exciton (within the framework of the band model). We shall adopt the conventional assumption that the rate of exciton migration is limited by scattering arising from interaction with lattice phonons. The most difficult problem faced concerns the treatment of the scattering mechanism. We adopt a phenomenological approach and assume that exciton scattering can be described in the constant relaxation time approximation.

In this approximation, the components of the diffusion tensor are expressed in the form

$$D_{ij} = \tau_s \langle v_i v_j \rangle , \tag{94}$$

where τ_s is a constant isotropic relaxation time, independent of $\underset{\sim}{k}$, and v_i, v_j are the components of the group velocity, $\underset{\sim}{V}(\underset{\sim}{k})$,

$$\underset{\sim}{V}(\underset{\sim}{k}) = \frac{1}{\hbar} \nabla_k E(\underset{\sim}{k}) . \tag{95}$$

The diffusion tensor can also be related to an isotropic constant mean free path, $\Lambda = \tau_s |\underset{\sim}{V}(\underset{\sim}{k})|$, by

$$D_{ij} = \Lambda \left\langle \frac{v_i v_j}{|\underset{\sim}{V}(\underset{\sim}{k})|} \right\rangle . \tag{96}$$

In the following, we employ an orthogonal coordinate system with axes parallel to $\underset{\sim}{a}$, $\underset{\sim}{b}$ and $\underset{\sim}{c}'$ ($\underset{\sim}{c}' \cdot \underset{\sim}{a} = \underset{\sim}{c}' \cdot \underset{\sim}{b} = 0$).

In Eqs. (94) and (96), the angular brackets represent an average over the Boltzmann distribution of excitons in the band,

$$\langle v_i v_j \rangle = \frac{\int \left[\frac{\partial E_+}{\partial k_i} \frac{\partial E_+}{\partial k_j} g(E_+) + \frac{\partial E_-}{\partial k_i} \frac{\partial E_-}{\partial k_j} g(E_-) \right] d^3k}{\hbar^2 \int [g(E_+) + g(E_-)] \, d^3k} \tag{97}$$

where $g(E_i) = \exp(-\frac{E_i}{k_BT})$ is the Boltzmann factor. In the present case the triplet exciton band width, W, is smaller than k_BT (at room temperature), the band states are almost equally populated, and the effective mass approximation cannot be used.

The components of the velocity tensor in the constant free time and constant free path approximation were calculated using the electron exchange interactions previously computed. In this description, the anisotropy of the velocity components along the crystal axes arises only from the anisotropy in the energy bands. Since the off-diagonal components of the velocity tensor are small, the mean diffusion coefficient is approximately

$$D \approx \frac{1}{3} (D_{aa} + D_{bb} + D_{cc}) \tag{98}$$

or

$$D \approx \frac{\Lambda}{3} \left[\left\langle \frac{v_a^2}{|\underset{\sim}{v}|} \right\rangle + \left\langle \frac{v_b^2}{|\underset{\sim}{v}|} \right\rangle + \left\langle \frac{v_c^2}{|\underset{\sim}{v}|} \right\rangle \right]. \tag{99}$$

Now, the existence of a scattering mechanism implies the presence of molecular vibrations. When the band structure in the tight binding approximation is based on the vibronic states of the constituent molecules, the band corresponding to the rigid lattice separates into a number of sub-bands, each determined by the total interaction matrix element multiplied by the vibrational overlap factor $\langle \chi^{o(o)} | \chi^{f(i)} \rangle^2$. For all aromatic molecules of interest, the intramolecular vibrational quanta exceed k_BT at room temperature, so that for our study only the 0-0 vibronic band need be considered. Provided that the triplet band width is smaller than k_BT, the diffusion controlled exciton-exciton annihilation rate now assumes the form

$$\Gamma = \frac{8\pi w}{3} \langle R \rangle \Lambda \langle |\underset{\sim}{v}(\underset{\sim}{k})| \rangle \langle \chi^{o(o)} | \chi^{f(i)} \rangle^2 n_T^2 . \tag{100}$$

In order to reproduce the observed annihilation rate, it is necessary that the mean free path be between 6 Å and 13 Å, corresponding to the two limits of the observed annihilation rate. In this calculation we have used $\langle |\underline{V}(\underline{k})| \rangle \approx 1.3 \times 10^5$ cm/sec as determined from the band structure, and $\langle \chi^{0(0)} | \chi^{f(i)} \rangle^2 = 0.20$, all for the case of crystalline anthracene. From the mean free path we obtain the mean scattering time. Corresponding to the limits on Λ, we find τ_s to be between 0.8×10^{-11} sec and 1.6×10^{-11} sec.

It is the opinion of the authors that when the mean free path of the excitation wave is as short as one or two lattice distances, as is the case in anthracene, then the use of a band scheme is inappropriate. Moreover, the assumption of a constant scattering time, or constant mean free path is so uninformative and so fraught with conceptual difficulties that, before the difficult problem of the nature of exciton scattering is solved for the cases under discussion, no meaningful comparison of band theory and experiment can be made. The very short mean free path leads us to prefer the limiting case of strong scattering, wherein the localized states are dominant and the exciton motion is diffusive.

Consider now the exciton-exciton annihilation rate. We assume that the rate determining step in this reaction is the diffusion of the excitons. When the effect of electron transfer states on the triplet mobility is included, the triplet-triplet annihilation rate is determined by an exciton-exciton collision rate $\Gamma = 4.5 \times 10^{-11} n_T^2$ cm^{-3} sec^{-1} from (90a) and $15 \times 10^{-11} n_T^2$ cm^{-3} sec^{-1} from (90b) so that $0.5 \times 10^{-11} < R < 1.5 \times 10^{-11} n_T^2$ cm^{-3} sec^{-1}. In view of the uncertainties in the calculated electronic matrix elements and the overlap factor $\int F(E)f(E)dE$, our theoretical estimate is uncertain to one order of magnitude. The experimental triplet-triplet [45,46] annihilation rate constant in the anthracene crystal is (2-6) $\times 10^{-11} n_T^2$cm^{-3} sec^{-1}, which is in good agreement with the theoretical value (55).

Thus far in our analysis of triplet-triplet annihilation we have only examined the consequences of the assumption that the rate of annihilation is diffusion controlled. We now turn to an examination of the mechanism of annihilation, in the sense of studying possible intermediate states along the reaction path.

The singlet electronic state involving two triplet excitations on adjacent molecules (a and b) can be constructed using the Wigner angular momentum coupling coefficients (56):

$$\Psi_{TT} = \frac{1}{\sqrt{3}} \sum_{m=0,\pm 1} (-)^{m+1} | T_a^{m} T_b^{-m} \rangle. \tag{101}$$

The simplest wave functions useable in the present discussion involve four electrons occupying the highest bonding and lowest antibonding orbitals on the molecules a and b. These are

$$|T_a^1 T_b^{-1}\rangle = \mathcal{Q}\left[u_a^{N/2+1}\alpha(1)\, u_a^{N/2}\alpha(2)\, u_b^{N/2+1}\beta(3)\, u_b^{N/2}\beta(4)\right],$$

$$|T_a^{-1} T_b^{1}\rangle = \mathcal{Q}\left[u_a^{N/2+1}\beta(1)\, u_a^{N/2}\beta(2)\, u_b^{N/2+1}\alpha(3)\, u_b^{N/2}\alpha(4)\right],$$

$$2\,|T_a^0 T_b^0\rangle = \mathcal{Q}\left[u_a^{N/2+1}\alpha(1)\, u_a^{N/2}\beta(2) + u_a^{N/2+1}\beta(1)\, u_a^{N/2}\alpha(2)\right]\times$$

$$\times\left[u_b^{N/2+1}\alpha(3)\, u_b^{N/2}\beta(4) + u_b^{N/2+1}\beta(3)\, u_b^{N/2}\alpha(4)\right],$$

(102)

with α and β the usual spin functions. Now, the functions displayed in (102) do not represent a stationary state. If the annihilation process leads to the direct formation of the $^1B_{2u}$ singlet excited state (Ψ_{GS_1}), the matrix element which determines the transition probability is

$$\langle \Psi_{TT}|H|\Psi_{GS_1}\rangle = \sqrt{\frac{3}{2}}\left[\langle u_a^{N/2+1}(1)\, u_b^{N/2}(1)|\frac{1}{r}|u_a^{N/2+1}(2)\, u_b^{N/2+1}(2)\rangle\right.$$

$$\left. - \langle u_a^{N/2}(1)\, u_b^{N/2}(1)|\frac{1}{r}|u_a^{N/2}(2)\, u_b^{N/2+1}(2)\rangle\right],$$

(103)

where G and S_1 refer, respectively, to the ground state and first singlet excited state of the molecule. From our preceding analysis, it is seen that (103) is of the order of magnitude of the electron exchange interaction, i.e., 10^{-4} to 10^{-3} eV.

Now, as already mentioned, ion-pair excitons can mix with the neutral triplet exciton band in anthracene and naphthalene (57). It is to be expected, then, that an important contribution to the transition probability may arise from the charge-transfer state, G^+G^-, with the electron and hole on adjacent molecules. The wave function representing G^+G^- has the form

$$\Psi_{G^+G^-} = \frac{\mathcal{a}}{\sqrt{2}}\Big[u_a^{N/2}\alpha(1)\,u_b^{N/2+1}\beta(2)\,u_b^{N/2}\alpha(3)\,u_b^{N/2}\beta(4) - u_a^{N/2}\beta(1)\,u_b^{N/2+1}\alpha(2)\,u_b^{N/2}\alpha(3)\,u_b^{N/2}\beta(4)\Big]. \tag{104}$$

For the four electron system under consideration, it is convenient to take the Hamiltonian in the form

$$H = \sum_{i=1}^{4}\Big[T_i + H_a^c(i) + H_b^c(i) + \sum_{i<j}\frac{1}{r_{ij}}\Big],$$

$$H_a^c = V_a^{NUC} + \sum_{k=1}^{N/2-1}(2J_a^s - K_a^s), \tag{105}$$

where T_i is the kinetic energy operator, H_a^c and H_b^c are the core potentials of molecules a and b and V_a^{Nuc} is the bare nuclear potential of molecule a. Using (105) the transition probability between the states represented by Ψ_{TT} and $\Psi_{G^+G^-}$ is

$$\begin{aligned}\langle \Psi_{TT}|H|\Psi_{G^+G^-}\rangle \cong {} & \langle u_a^{N/2+1}|V_a^{GMS}|u_b^1\rangle \\ & - \sum_{k=1}^{N/2-1}\langle u_a^{N/2+1}u_a^k|\frac{1}{r_{ij}}|u_a^k u_b^1\rangle \\ & + \langle u_a^{N/2+1}u_b^{N/2}|\frac{1}{r_{ij}}|u_b^{N/2+1}u_b^{N/2}\rangle \\ & + \langle u_a^{N/2+1}u_b^{N/2}|\frac{1}{r_{ij}}|u_b^{N/2}u_b^{N/2}\rangle,\end{aligned} \tag{106}$$

when nonorthogonality corrections are neglected. Here V_a^{GMS} is the Goeppert-Mayer-Sklar potential of the neutral molecule a. Because the final density of states is unknown, direct evaluation of (103) or (106) is not, at present, feasible. However, an order of magnitude estimate indicates that $\langle \Psi_{TT} | H | \Psi_{G^+G^-} \rangle$ is one to two orders of magnitude larger than $\langle \Psi_{TT} | H | \Psi_{GS_1} \rangle$. The reason that the two matrix elements differ so much is that the magnitude of (103) is determined by intermolecular exchange integrals, while the magnitude of (106) is determined by coulomb hybrid integrals. Thus, if the vibrational overlap factors do not differ greatly in the two cases cited, the transition probability to the ion-pair state will exceed the transition probability to the excited singlet state by a factor of $10^3 - 10^4$.

We have already cited the estimated energy of the charge-transfer state (3.4 ± 0.5 eV). The reader should note that this estimate places the charge transfer state very close to the first excited singlet state ($^1B_{2u}$) of anthracene. A detailed theoretical study of the spectrum of crystalline anthracene (49) leads to the conclusion that the energy separation between the charge-transfer state and the $^1B_{2u}$ state in anthracene must exceed 500 cm^{-1}. The spectrum is consistent, however, with a limited amount of mixing of the states G^+G^- and GS_1, i.e., a representation of the form $\Psi_{GS_1} + \lambda \Psi_{G^+G^-}$ with $\lambda^2 \approx 0.1$ will not noticeably alter the Davydov splitting. Now, it is not necessary that the final state in the triplet-triplet annihilation be a pure charge transfer state, as previously conjectured by us (50). If configuration mixing of the limited type mentioned above occurs, the transition probability to the mixed state will still be of order of magnitude $10-10^2$ larger than the transition probability to the pure $^1B_{2u}$ state. Under these conditions the annihilation process is still efficient on the time scale of the triplet migration from site to site in the crystal.

There is one indirect corroboration of the suggestion that ion-pair states play a role in the annihilation process. Lindquist (58) has studied the kinetics of the decay in solution of triplet fluorescein generated by flash excitation. It is found that the kinetics cannot be understood unless the decay mechanism includes the reaction between two triplet fluorescein molecules to give as products the oxidized and reduced forms of fluorescein. This reaction is, of course, the direct analogue of the reaction of two triplet excitons to form an ion-pair exciton.

V. THE EXCESS ELECTRON BAND STRUCTURE OF CRYSTALS OF AROMATIC COMPOUNDS

As a final topic in our description of the low pressure behavior of crystals of aromatic compounds, we consider the excess electron and hole band structures and carrier mobility. The experimental work of Kepler (59), LeBlanc (60) and Thaxton, Jarnagin and Silver (61) on the mobilities of excess electrons and holes in crystalline anthracene has shown that: (i) the mobilities are slightly anisotropic and of the order of magnitude of 1 cm^2/sec-v at room temperature, and (ii) the mobilities vary with temperature as T^{-n} with n between 1 and 2. A mechanism of charge carrier transport that involves phonon-assisted hopping from molecule to molecule would imply that the carrier mobility increase with temperature; this is contrary to experiment. On the other hand, a simple band scheme is consistent with experiment. This fact led LeBlanc (62) to attempt a calculation of the excess electron and hole band structures and the mobilities of excess charge carriers in crystalline anthracene using the following assumptions: (a) the crystal is rigid and composed of nonvibrating molecules, (b) the wave function of the excess electron or hole may be represented by a Bloch sum of molecular orbitals, (c) the molecular orbitals are linear combinations (Huckel) of single Slater carbon atom 2p orbitals, (d) intermolecular overlap may be neglected, (e) the potential of a neutral molecule may be taken as the Goeppert-Mayer and Sklar potential, and (f) only two center integrals need be considered. With these assumptions, LeBlanc was able to obtain order-of-magnitude agreement with the experimental observations.

Katz, Choi, Rice and Jortner (63) and later Silbey, Jortner, Rice and Vala (64) extended this calculation, improving on (a), (c), (d), (e) and (f). The analysis properly accounts for the fact that anthracene has two molecules per unit cell by constructing symmetry adapted functions. The best available SCF atomic orbitals for carbon were used in the calculation of the necessary integrals, and many three-center terms were computed. More particularly, the effect of molecular wave functions as the product of an electronic part and a vibrational part. This representation corresponds to the weak-coupling limit of vibronic interaction, i.e., the nuclear part of the Hamiltonian is diagonalized before the crystal field part. Using this approximation, the symmetrized wave functions of the crystal with an excess or deficiency of one electron may be written in the form:

$$\Psi_{\pm}(\underset{\sim}{k}) = \sum_{l=0}^{2N-1} (\pm)^{l} e^{i\underset{\sim}{k}\cdot\underset{\sim}{r}_{ml}} \, \mathcal{Q} u_{l}^{k}(1) \times$$

$$\times \prod_{l=0}^{2N-1} \left(u_i^{1}(i,1) \ldots \bar{u}_i^{a}(i,2a) \, \chi_i \right),$$

(108)

where u^k is the lowest unoccupied molecular orbital of molecule for electrons and the highest occupied orbital for holes, the u_i are the a occupied molecular orbitals of molecule i. As before, the bar over u_i^a means that the electron in that molecular orbital has β spin, and χ_i is the ground state vibrational wave function of the ith molecule. It is assumed that all the vibrational wave functions are the same except for that of the molecule with an excess electron (or hole). This formaiism implies, of course, that every electron or hole band is split into many bands, each characterized by a vibrational wave function. The k = 0 levels of these bands will be separated by the vibrational quantum, which in aromatic molecules is of the order of 0.2 eV. It will be seen later that the band widths are of the order of 0.01 eV and thus the bands are well separated. Finally, in the work of Silbey et al the potential of the nth neutral molecule was taken to be

$$V_n(\underset{\sim}{r}-\underset{\sim}{r}_n) = -\sum_{A} \frac{Ze^2}{R_A} + \sum_{i=1}^{a} (2J_n^{i} - K_n^{i})$$

$$= V_n'(\underset{\sim}{r}-\underset{\sim}{r}_n) - \sum_{i=1}^{a} K_n^{i}.$$

(109)

The Hamiltonian of the crystal may now be written as the sum of two terms, the first of which contains all contributions that refer to the lone electron (H'), and the second of which is the remainder (H-H'). The energy of the crystal, <u>neglecting overlap between molecules</u>, may be written as

$$E_{\pm}(\underset{\sim}{k}) = \mathcal{E}_0 + \sum_{l} (\pm)^{l} \left[\cos(\underset{\sim}{k}\cdot\underset{\sim}{r}_l)\right] \mathcal{E}_l, \qquad (110)$$

where

$$\mathcal{E}_0 = \langle u_0^k | H' | u_0^k \rangle$$

and

$$\mathcal{E}_\ell = \langle u_0^k | V_\ell(\underset{\sim}{r} - \underset{\sim}{r}_\ell) | u_\ell^k \rangle | \langle \chi' | \chi^\circ \rangle |^2. \tag{111}$$

In Eq. (111) χ' represents the vibrational wave function of the positive or negative ion and χ° that of the neutral molecule.

Replacing V_n by $V_n' - \sum_i K_n^i$ one finds:

$$\begin{aligned}\mathcal{E}_\ell &= |\langle \chi' | \chi^\circ \rangle|^2 \langle u^k(\underset{\sim}{r} - \underset{\sim}{r}_\ell) | V_\ell' - \sum_i K_\ell^i | u^k(\underset{\sim}{r}) \rangle \\ &= |\langle \chi' | \chi^\circ \rangle|^2 \Big\{ \langle u^k(\underset{\sim}{r} - \underset{\sim}{r}_\ell) | V_\ell' | u^k(\underset{\sim}{r}) \rangle \\ &\qquad - \sum_{i=1}^{a} \langle u^k(\underset{\sim}{r} - \underset{\sim}{r}_\ell) | K_\ell^i | u^k(\underset{\sim}{r}) \rangle \Big\}. \end{aligned} \tag{112}$$

The second term in the brackets, when expanded in atomic orbitals, becomes:

$$\sum_{i=1}^{a} \sum_{\alpha,\beta,\gamma,\delta}^{M} \langle \omega^\alpha \omega^\gamma | \frac{1}{r_{12}} | \omega^\delta \omega^\beta \rangle c_k^\alpha c_k^\beta c_i^\gamma c_i^\delta, \tag{113}$$

where α, γ, δ number the atoms of molecule ℓ and β the atoms of the reference molecule. Including only two center atoms ($\alpha = \gamma = \delta$) we find:

$$\begin{aligned}\sum_{i=1}^{a} \langle u^k(\underset{\sim}{r} - \underset{\sim}{r}_\ell) | K_\ell^i | u^k(\underset{\sim}{r}) \rangle \\ = \sum_{i=1}^{a} \sum_{\alpha,\beta}^{M} c_k^\alpha (c_i^\alpha)^2 c_k^\beta \langle \omega^\alpha \omega^\alpha | \omega^\alpha \omega^\beta \rangle, \end{aligned} \tag{114}$$

where the index representing the molecules has been suppressed. Since $\sum_{i=1}^{a} (c^i)^2 = 1/2$ by the Coulson-Rushbrooke theorem, we have finally

$$\sum_{i=1}^{a} \langle u^{k}(\underset{\sim}{r}-r_{\ell}) | K_{\ell}^{i} | u^{k}(\underset{\sim}{r}) \rangle = \frac{1}{2} \sum_{\alpha=1}^{M} \sum_{\beta=1}^{M} c_{k}^{\alpha} c_{k}^{\beta} \langle \omega^{\alpha} \omega^{\alpha} | \omega^{\alpha} \omega^{\beta} \rangle . \tag{115}$$

These integrals were computed on an IBM 7094 using the same Hartree-Fock SCF carbon atomic wave functions as described earlier. The hybrid sums for the thirteen molecules considered (2) are listed in Table X. Using these values in Eq. (110), Silbey et al calculated the band structures and mobilities of the excess hole and electron in anthracene and naphthalene for various values of the vibrational overlap. The mobilities were computed in both the constant free time approximation and the constant free path approximation. See Figures 3-6 and Table XI. While the components of the mobility tensor cannot be found without making an assumption for the free time or free path, the ratios of these components may be computed without such an assumption. These results are listed in Table XII. It may be seen that for a vibrational overlap of ~ 0.5, for example, the agreement between calculation and experiment is fairly good.

The calculated band widths are given in Table XI. These are smaller than those calculated by Katz et al (63) and closer to those of LeBlanc (62) for a vibrational overlap of 0.5. It can be seen that the band widths are of the order of 0.02 eV and smaller.

From the calculated components of the mobility tensor and the experimental data, one may estimate the free time and the mean free path of the electrons and holes. Katz et al (63) found the mean free path to be ~ 4 Å for electrons and holes, which is of the order of magnitude of the lattice spacing. The results reported by Silbey et al (64), for vibrational overlap of 0.5, give a mean free path of the order of 10 Å. The calculated free time is such that the band widths are of the order of $\hbar/\tau$, in agreement with the uncertainty principle.

VI. PRESSURE INDUCED STRUCTURAL CHANGES

Because of the sensitivity of the properties of molecular crystals to both the crystal geometry and the magnitude of the overlap, it is necessary to have some measure of the pressure induced intermolecular and intramolecular structural changes. For this reason, we consider first some qualitative arguments which suffice to define the order of

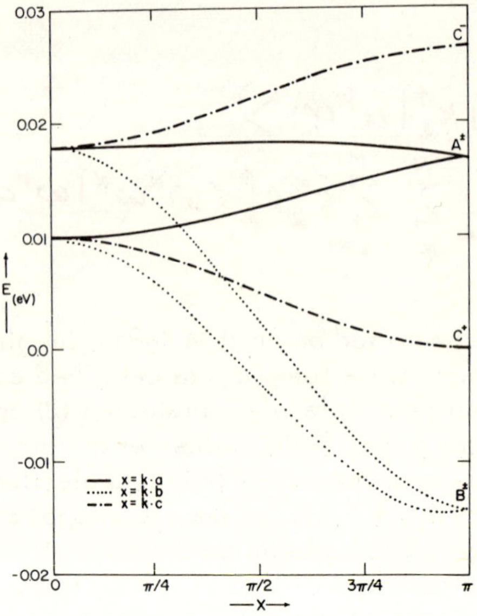

Fig. 3. Shape of excess hole band of naphthalene in the a^{-1}, b^{-1}, c^{-1} directions for vibrational overlap of $(0.2)^{1/2}$.

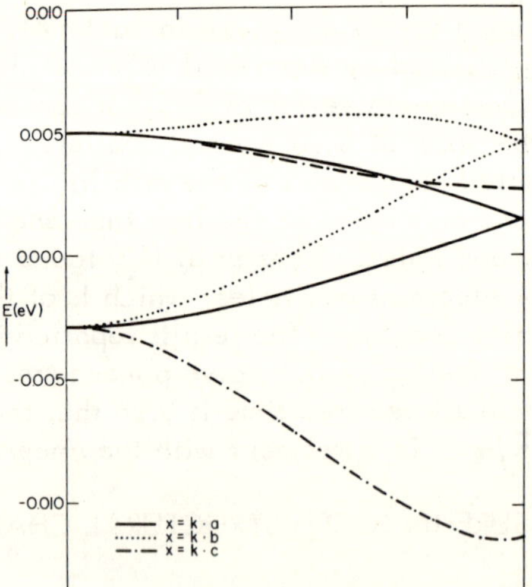

Fig. 4. Shape of excess electron band of naphthalene in the a^{-1}, b^{-1}, c^{-1} directions for vibrational overlap of $(0.2)^{1/2}$.

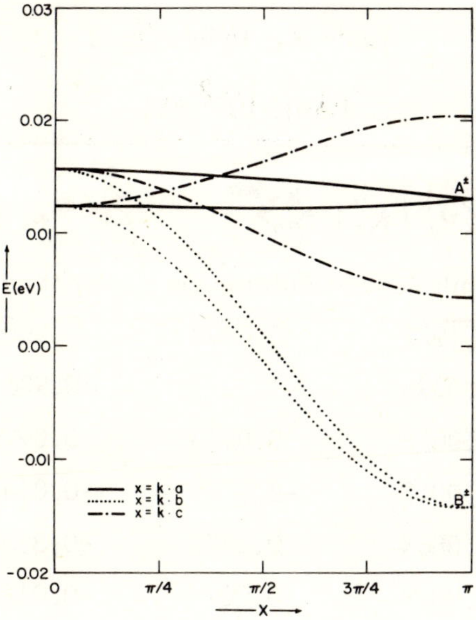

Fig. 5. Shape of excess hole band of anthracene in the a^{-1}, b^{-1}, c^{-1} directions for vibrational overlap of $(0.2)^{1/2}$.

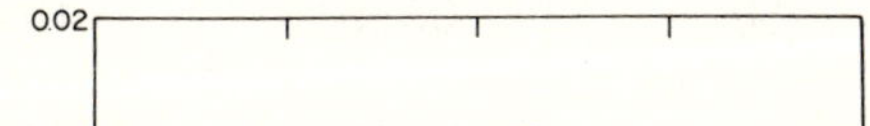

Fig. 6. Shape of excess electron band of anthracene in the a^{-1}, b^{-1}, c^{-1} directions for vibrational overlap of $(0.2)^{1/2}$.

Table X. Hybrid Sums

(Units 10^{-2} eV)

Molecule Coordinates in a, b, c Space*	$\sum_{i=1}^{a} \langle u_0^k \| K_n^i \| u_n^k \rangle^{**}$ Naphthalene	Anthracene	$\sum_{i=1}^{a} \langle u_0^{k-1} \| K_n^i \| u_0^{k-1} \rangle^{**}$ Naphthalene	Anthracene
(0,0,1)	0.1072	0.0408	-0.0001	0.0268
(0,1,0)	-0.5396	-1.529	2.936	3.057
(0,1,1)	-0.0078	0.0013	0.0926	0.0091
(1,0,0)	0.0003	-0.0134	-0.0142	-0.0110
(1,0,1)	-0.0864	0.0435	-0.3850	0.1572
(1,1,0)	0.0024	-0.0020	-0.0171	-0.0023
(1,1,1)	0.0014	-0.0007	-0.0026	0.0011
$(\frac{1}{2},\frac{1}{2},0)$	1.255	2.789	-1.292	2.187
$(\frac{1}{2},\frac{1}{2},1)$	0.0958	-0.0698	1.135	-0.9685
$(-\frac{1}{2},\frac{1}{2},1)$	0.0016	-0.0012	0.0012	-0.0007
(-1,0,1)	0	0	0	0
(-1,1,1)	0	0	0	0
$(\frac{1}{2},\frac{3}{2},0)$	-0.0006	0.0001	-0.0021	-0.0007

*This gives the vector to the center of the molecule, the unit vectors being the lattice vectors of the crystals in the corresponding directions.

**In these, k represents the lowest unfilled orbital of the molecule.

Table XI. Widths of Excess Electron and Hole Bands in Anthracene and Naphthalene

(Units = 10^{-2} eV)

	NAPHTHALENE		ANTHRACENE	
Direction	Hole	Electron	Hole	Electron
a^+	2.6	7.3	7.6	15.1
a^-	1.7	8.1	6.2	14.9
b^+	15.8	4.6	23.6	6.5
b^-	16.8	10.8	10.4	23.7
c^+	10.8	1.2	9.2	0.4
c^-	12.8	0.9	11.2	1.2

(Vibrational overlap squared = 1.0)

Table XII. Ratios of Components of Mobility Tensor in Constant Free Time and Constant Free Path Approximations

NAPHTHALENE	I a	I b	II a	II b	III
HOLE					
μ_{aa}/μ_{bb}	0.07	0.08	0.09	0.10	0.64
$\mu_{c'c'}/\mu_{bb}$	0.19	0.25	0.17	0.21	0.29
$\mu_{c'c'}/\mu_{aa}$	2.5	2.9	1.81	2.11	0.44
ELECTRON					
μ_{aa}/μ_{bb}	2.1	1.92	1.70	1.60	1.0
$\mu_{c'c'}/\mu_{bb}$	0.05	0.01	0.04	0.06	0.57
$\mu_{c'c'}/\mu_{aa}$	0.02	0.04	0.03	0.04	0.57

(continued)

Table XII (continued)

ANTHRACENE					
HOLE					
μ_{aa}/μ_{bb}	0.39	0.44	0.37	0.41	0.5
$\mu_{c'c'}/\mu_{bb}$	0.15	0.19	0.14	0.17	0.4
$\mu_{c'c'}/\mu_{aa}$	0.44	0.44	0.37	0.42	0.8
ELECTRON					
μ_{aa}/μ_{bb}	2.6	2.6	1.87	1.86	1.7
$\mu_{c'c'}/\mu_{bb}$	0.03	0.09	0.013	0.04	0.4
$\mu_{c'c'}/\mu_{aa}$	0.01	0.04	0.01	0.02	0.24

a Constant free time

b Constant free path

I Vibrational overlap squared = 1.0

II Vibrational overlap squared = 0.5

III Experimental results

magnitude of the expected effects. For a system characterized by a constant isothermal compressibility, κ , the isothermal equation of state is $V = V_0 e^{-\kappa p}$, and the compression work W, per molecule is

$$W = \int p\,dV = V_0 \left\{ \frac{1}{\kappa}(1 - e^{-\kappa p}) - p e^{-\kappa p} \right\}, \tag{116}$$

which for $\kappa p < 1$ reduces to $W = \kappa V_0 p^2$. In the case of molecular crystals, we take the average zero pressure volume to be $V_0 = 50$ Å^3 and the compressibility to be of the order of 10^{-6} atm^{-1}. From the data, $(W/eV) \approx 10^{-6}$ $(p/atm)^2$ (for $\kappa p < 1$). This order of magnitude estimate indicates that for pressures up to 10^5 atm., W $\leq$ 0.01 eV, so that only the lattice dimensions will be affected by the compression. In the pressure region p $\geq 3 \times 10^5$ atm., the compression work computed from (116) is W $\geq$ 0.1 eV, i.e., of the order of magnitude of the intermolecular interactions between aromatic molecules. Irreversible transformations involving crosslinking between neighboring molecules may also take place. Only at pressures greater than 10^6 atm. is the compression work expected to be of the order of magnitude of the bond energies: at these pressures intramolecular configuration changes may be expected to occur.

At present there are available no data regarding pressure induced phase transitions in crystals of aromatic molecules. (The well-known graphite-diamond transition is outside the scope of this review). It should be noted that crystalline pyrene is a particularly interesting system for experimental study since pyrene crystallizes in two monoclinic structures containing, respectively, 2 and 4 molecules per unit cell (65).

We have already mentioned possible chemical changes: evidence has been obtained for intermolecular crosslinking in crystalline pentacene, hexacene, and violanthrene (66) at pressures greater than 200 kbar and at temperatures above 200°K. The irreversible crosslinking in pentacene leads to a stable dimer: the absorption spectrum is consistent with a decrease in the size of the π electron system leading to a blue shift of the polycene p band, while the position and the intensity of the β band remains unaltered.

The optical properties of molecular crystals have been studied in the pressure region up to 10^5 bar, and studies of the electrical properties have been carried out up to pressures of 5×10^5 bar. In this pressure range pressure induced intramolecular changes are not expected to occur, and we shall therefore be concerned with the dependence of the manifold of electronic bands of the crystal on the lattice dimensions. To carry out such an analysis, compressibility data are required.

Molecular crystals of aromatic molecules are held together by dispersion forces. Because these intermolecular interactions are relatively weak, these crystals may be expected to be quite compressible in the pressure range up to 100 kbar. Note, however, that because the dispersion forces between conjugated π electron systems are anisotropic, experimental data on the linear compressibilities of these systems are essential. The volume compressibilities of a number of atomatic crystals (i.e., naphthalene, anthracene, biphenyl, terphenyl, and substituted benzenes) have been studied by Bridgman and by Kabalinka (68). Typical volume compressibility data for some molecular crystals are displayed in Table XIII. The experimental data indicate that the compressibilities of all these molecular crystals do not differ by more than 20% up to 50 kbar, the crystals with large molecules being somewhat less compressible, while at higher pressures the deviations between different compounds are even smaller. It is therefore possible (69) to construct an approximate unified compressibility-pressure curve for a generalized molecular crystal of aromatic molecules (Table XIV).

Linear compressibility data for linear polyphenyls (biphenyl, terphenyl, and quaterphenyl) have recently been obtained by Kabalinka (68). These data provide the only available information regarding the changes under pressure of the unit cell dimensions. The organic crystals cited all have a monoclinic unit cell containing two molecules; one at the corner and the other at the center of the ab face. The crystal structure is described by the symmetry classification C_{2h}^5. Naphthalene and anthracene also have the same crystal structure. The crystal structure data (at zero pressure) are displayed in Table XV, while in Table XVI the available linear compressibility data are listed. From these results it is concluded that:

(a) The crystal symmetry is unaffected by the applied pressure, and only the unit cell dimensions are changed.

(b) The absolute values of the changes in the unit cell dimensions are the same (within experimental error) for biphenyl, terphenyl, and quaterphenyl (i.e., $\Delta a = 0.435$ Å, $\Delta b = 0.168$ Å, $\Delta c = 0.2$Å, and $\Delta\beta = 0$ for $p = 10^4$ kg/cm^2).

(c) The linear compressibilities corresponding to the unit cell axes are highly anisotropic.

(d) The small differences between the volume compressibilities of different monoclinic crystals arise from the changes in the linear compressibilities in the c direction.

For a monoclinic crystal the fractional pressure induced volume

Table XIII. Fractional Volume Changes of Some Organic Crystals at Room Temperature

Crystal	10^4 p atm/cm^2	$\Delta V/V$
Anthracene	1	0.0876
	2	0.1336
	3	0.1657
Naphthalene	1	0.105
Biphenyl	1	0.106
Terphenyl	1	0.098

Table XIV. Fractional Pressure Induced Volume Changes for a Model Crystal of Aromatic Molecules. (Extrapolated by Samara and Drickamer (69) from Bridgman's Data).

Pressure 10^3 bar	$\Delta V/V_o$
10	0.095
20	0.133
40	0.190
60	0.227
80	0.255
100	0.276
140	0.303
180	0.319
200	0.324
240	0.328
280	0.330
300	0.331

Table XV. Crystal Constants at Zero Pressure

	a	b	c	β
Naphthalene[1]	8.235 Å	6.003 Å	8.658 Å	122°55'
Anthracene[2]	8.561 Å	6.036 Å	11.163 Å	124°42'
Biphenyl[3]	$8.12_4 \pm 0.02$ Å	$5.63_5 \pm 0.01$ Å	$9.51_3 \pm 0.02$ Å	$95.1° \pm 0.3°$
Terphenyl[4]	8.08 Å	5.60 Å	13.59 Å	91°55'
Quaterphenyl[5]	8.05 ± 0.02 Å	5.55 ± 0.02 Å	17.81 ± 0.05 Å	95.8° (approx)

[1] D. W. J. Cruickshank, Acta Cryst. 10, 504 (1957).

[2] V. C. Sinclair, J. M. Robertson, and A. M. Mathieson, ibid. 3, 251 (1950).

[3] A. Hargreaves and S. Hason Rizvi, ibid. 15, 365 (1962).

[4] L. W. Pickett, Proc. Roy. Soc. (London) A142, 333 (1933).

[5] L. W. Pickett, J. Am. Chem. Soc. 58, 2299 (1936).

Table XVI. Changes in Unit Cell Dimensions of Crystals of Linear Polyphenyls at $p = 10^4$ kg/cm^2

Crystal	$\Delta a/a$	$\Delta b/b$	$\Delta c/c$	$\cot\beta\,\Delta\beta$
Biphenyl	0.054 ± 0.002	0.030 ± 0.002	0.022	0.002 ± 0.002
Terphenyl	0.054 ± 0.002	0.030 ± 0.002	0.014	-0.002 ± 0.002
Quaterphenyl	0.054 ± 0.002	0.030 ± 0.002	0.005	----

changes are given by

$$\frac{\Delta V}{V} = \frac{\Delta a}{a} + \frac{\Delta b}{b} + \frac{\Delta c}{c} + \cot g\,\beta\,\Delta\beta. \tag{117}$$

Assuming that the ratios $\Delta a/a$, $\Delta b/b$, $\Delta c/c$, $\Delta\beta/\beta$ are pressure independent, we can now combine the general volume compressibility data (Table XIV) with the biphenyl linear compressibility data to construct a general linear compressibility curve for monoclinic crystals. These extrapolated data, displayed in Table XVII, should give a reasonable approximation to the change in the unit cell dimensions of monoclinic molecular crystals such as naphthalene, anthracene and the linear polyphenyls, and should serve as a starting point for the computation of pressure effects on the electronic states of these systems.

Table XVII. Linear Compressibility Data
for a Model Monoclinic Crystal

10^3 bar	$\Delta a/a$	$\Delta b/b$	$\Delta c/c$
10	0.054	0.030	0.020
20	0.072	0.040	0.027
40	0.103	0.058	0.038
60	0.123	0.068	0.045
80	0.137	0.076	0.051
100	0.149	0.082	0.055
140	0.164	0.091	0.061
200	0.175	0.097	0.065
240	0.177	0.098	0.066
300	0.178	0.099	0.067

VII. PRESSURE EFFECTS ON RADIATIONLESS TRANSITIONS

Electronic relaxation phenomena are of considerable current interest. Except for a recent study of the effects of pressure on the absorption and emission spectra and excited state lifetime of fluorescein in boric acid glass (70), attention has been confined to the low pressure study of radiationless transitions in aromatic molecules. This is an unfortunate situation, since studies of the variation of lifetimes, etc. with pressure can provide important information on the relative roles of the intermolecular and intramolecular contributions to internal conversion processes.

Gregg and Drickamer (70) have studied the emission spectrum of fluorescein up to pressures of 54 kbar. The fluorescence emission band shows a pressure induced red shift corresponding to the red shift in the absorption spectrum, whereas the position of the phosphorescence band is insensitive to the pressure. These investigators observed no definite change in the intensitives of the two bands but the relative intensities cannot be directly related to absolute intensities (quantum yields). The phosphorescence lifetime was found to decrease by a factor of 3, from 0.90 sec to 0.29 sec. Gregg and Drickamer interpret these observations by assuming a decrease of the natural lifetime of the triplet state arising from an increased spin-orbit coupling between the first singlet and triplet states, which increased coupling is presumably due to a decrease in their energy separation. However, even if these states do mix by virtue of the spin orbit coupling perturbation, it is now fairly well established that the mixing between $\pi \rightarrow \pi^*$ singlet and triplet excited states in aromatic molecules is extremely small. Indeed, the mixing arises from the contributions of three and four center terms only(72), while the intensity of the singlet-triplet transition is mainly borrowed from singlet states polarized perpendicular to the molecular plane (73). An alternative interpretation of the experimental results has been proposed (71) based on the postulate of an increase at high pressures of the rate of the radiationless triplet-ground state singlet intersystem crossing.

Internal conversion rates in complex molecules in a solid are determined by:

(a) An intramolecular process involving tunnelling between the potential energy surfaces corresponding to two electronic states (which generally do not intersect for the energy at which internal conversion takes place). This is followed by

(b) Intermolecular vibrational relaxation, in which the vibrational energy of the final molecular state is transferred to the surrounding medium.

The intramolecular tunnelling rate depends on the product of the square of the electronic matrix element which mixes the two states and a Franck-Condon vibrational overlap factor (74, 75). As we have stated, pressures less than a few hundred kbars are not expected to modify the ground state of the molecule in the solid. It should be noticed, however, that highly excited vibrational states may be strongly perturbed by application of external pressure because of the increase of the nearest neighbor repulsion with increased compression. These "cage effects" may seriously modify the upper vibrational states of the molecule in which intramolecular tunnelling occurs, giving rise to negative anharmonicity effects. This in turn may lead to a modification of the vibrational overlap factors leading to an enhanced probability of radiationless transition. However, at present we cannot rule out the possibility that the increase of the radiationless transition rate is due to the enhancement of vibrational relaxation at higher pressures. More experimental data are obviously required.

IX. A SURVEY OF THE ELECTRICAL PROPERTIES OF ORGANIC CRYSTALS AT HIGH PRESSURE

The electrical resistivities, ρ, of organic crystals lie between 10^{10} and 10^{19} ohm cm at room temperature and zero pressure. The energy gaps in the crystals are large, being of the order of magnitude of 1 to 3 eV. Now, for semiconducting materials, the temperature dependence of the resistivity is given by

$$\rho = \rho_0 e^{E/kT}, \tag{118}$$

while the conductivity σ is given by

$$\sigma = \sigma_0 e^{-E/kT}, \tag{119}$$

and E is the activation energy for conductance. The conductivity is related to the number of charge carriers and to the mobility by

$$\sigma = [n_+\mu_+ + n_-\mu_-]e, \tag{120}$$

where the signs $\pm$ refer to the positive and negative charge carriers, respectively.

Extensive experimental studies (76-78) of the mobilities of excess electrons and holes in crystalline anthracene and naphthalene show that the mobilities increase with decreasing temperature as T^{-n}, with n between 1 and 2. This weak temperature dependence of the mobility implies that the major contribution to the temperature dependence of the band gap for the dark conductivity arises from the temperature dependence of the number of charge carriers. Setting $n = n_+ = n_-$ we can write

$$n \approx n_0 e^{-E/kT}. \tag{121}$$

It should be noted that the equilibrium thermal concentration of charge carriers is extremely small at normal temperatures and pressures, being of the order of one electron per seven cubic feet. There are at present no data requiring as an interpretation the generation of intrinsic charge carriers, and the cited low concentration of charge carriers can be achieved by electron injection processes at the electrodes.

The effects of pressure (up to 10^4 atm) on the resistivity of isoviolanthrone were studied by Inokutchi (79), and extensive studies of the effect of pressure (to several hundred kilobars) on the resistance of tetracene, pentacene, hexacene, and some aromatic quinones (violanthrone, isoviolanthrone pyranthrone) have been published by Drickamer et al (66, 80). These data are summarized in Table XVIII. The following has been learned:

(a) As the pressure is increased, there is a rapid drop in the resistance of powdered samples and of single crystals followed by a leveling off of the resistance at about 200-250 kbar. The relative constancy of the high pressure resistance is apparently associated with the rapid decrease in the volume compressibility at higher pressure. Indeed, a plot of log σ versus the fractional volume change for several compounds is linear approximately up to 300 kbar.

(b) At high pressures (above 200 kbar) the resistivity of a single crystal is lower than that of a powdered sample by about two orders of magnitude.

(c) The limiting resistance obtained at high pressures varies appreciably from compound to compound (Table XVIII). Since the activation energies for all these powdered samples are small in this pressure range, the high pressure mobilities must differ substantially in these compounds.

(d) The activation energy decreases appreciably at high pressures (Table XVIII) for several microcrystalline samples, dropping to a value of about one-sixth of the atmospheric pressure value. Since the activation energies are quite small, trapping processes can be extremely important in microcrystalline samples. These results do not exclude the

Table XVIII. Pressure Effects on the Electrical Resistivity of Polycyclic Aromatic Compounds at Room Temperature (15°C)

Compound	Form	Pressure Atm.	ρ [(a)] ohm cm	ρ_o ohm cm	E eV	Reference
Anthracene	Powder	1	7×10^{15}	30	0.82	(a)
	Crystal	1	1.3×10^{14}	0.5	0.82	(b)
	Powder	3×10^5	10^7	---	---	(c)
Tetracene	Powder	1×10^5	(1×10^{10})	---	---	(c)
	Powder	3.5×10^5	(1.5×10^6)	---	---	(c)
Pentacene	Crystal	1	1×10^{14}	10	0.75	(b)
	Crystal	2.8×10^5	(10)	(10)	0.00	(d)
	Powder	3.6×10^5	(500)	(500)	0.00	(d)
Hexacene	Crystal	1	(10^8)	---	---	(d)
		4×10^5	(4×10^5)	---	---	(d)
Violanthrone	Powder	1	2.3×10^{10}	2.9×10^3	0.39	(b)
	Powder	3×10^2	5×10^9	1.3×10^3	0.375	(b)
	Powder	4.2×10^3	2.6×10^7	22	0.370	(b)
	Powder	8.4×10^3	1.2×10^7	10	0.340	(b)
	Powder	2.5×10^5	(100)	---	0.062	(c)

Table XVIII (continued)

(a) Values in parentheses represent resistance data not converted into specific resistivity.

(b) H. Inokuchi and H. Akamatu, Solid State Physics 12, 93 (1962).

(c) G. A. Samara and H. G. Drickamer, J. Chem. Phys. 37, 474 (1962).

(d) R. B. Aust, W. H. Bentley, and H. G. Drickamer, J. Chem. Phys. 41, 1856 (1964).

possibility of metallic behavior of single crystals at high density.

(e) In one case pseudometallic behavior at high pressures has indeed been observed. The activation energy for a single crystal of pentacene goes to zero at 270 kbar, and the system exhibits "metallic" behavior with a positive temperature coefficient of resistivity. Powdered samples of pentacene become metallic only at much higher pressures (about 360 kbar). In this case trapping and scattering processes at the grain boundaries mask the metallic behavior of the system.

A cursory examination of these experimental data is sufficient to to show that the major contribution to the decrease in the resistance of organic crystals at high pressures arises from the decrease in the activation energy for charge carrier production. Thus, in the case of pentacene, the activation energy drops from 0.75 eV to zero in the pressure range 0-300 kbar. The term exp(0.75/kT) leads to a factor of 10^{12} increase in the conductivity, while the observed increase in passing from zero pressure up to 300 kbar is 10^{13}. The pre-exponential factor, ρ_0, which is inversely proportional to the mobility, decreases by one order of magnitude. We thus conclude that the carrier mobilities increase by only about 1-2 orders of magnitude up to approximately 300 kbar.

Some direct information regarding the pressure dependence of charge carrier mobility in organic crystals is available (81). The results of a study of the mobilities of excess electrons and holes in crystalline anthracene at 3000 atm are presented in Table XIX. The change in the unit cell dimensions is the predominant factor in accounting for the pressure induced increase of the mobility. If the carrier scattering mechanism involves acoustical phonons, the pressure dependence of the Debye temperature of the crystal will affect the relaxation time. However, crude estimates of this effect indicate that it is quite small.

Table XIX. The Pressure Dependence of the Electron and Hole Mobility in Anthracene[a]

Crystal Direction	Electron μ(3000 atm)/ μ(1 atm)	Hole μ(3000 atm)/ μ(1 atm)
a	1.4	1.4
b	1.3	1.4
c'	1.0	1.4

[a]The experimental data are taken from R. G. Kepler, Organic Crystals Conference, edited by J. J. Brophy and J. W. Buttery (McMillan, New York, 1962), p. 1.

X. THE HIGH PRESSURE OPTICAL PROPERTIES OF ORGANIC COMPOUNDS

Experimental studies of the effects of pressure on the visible and near ultraviolet absorption spectra (at room temperature) of several organic cyrstals (i.e., anthracene, phenantrene, tetracene, pentacene, azulene) have been reported (80, 82, 83). The electronic transitions studied were the weak α band (of naphthalene, azulene and phenantrene) and the medium strength p band of the linear condensed hydrocarbons. The polarization of the spectrum at high pressure has not been studied for any compound because of the depolarizing effects of the highly strained optical windows. Indeed, many of the optical measurements have been made on randomly oriented microcrystalline samples diluted by mixture with alkali halides, but single crystals of naphthalene (83) and pentacene (80) have been studied. From these experiments the following general information has been obtained:

(a) The increase in pressure results in a spectral red shift for polycyclic alternant hydrocarbons. Thus, for the 0-0 vibronic component of the p band of crystalline anthracene the shift is 4000 cm^{-1} at 100 kbar, while for crystalline phenantrene at the same pressure the red shift of the 0-0 component of the α band is 1600 cm^{-1}. In the case of azulene, a small blue shift of the α band is observed. This anomalous behavior arises from molecule solvent interactions which change because of the change in the dipole moment of the molecule between the ground and the excited state.

(b) No direct relation has been found between the observed red shift and the oscillator strength of the transition. This is not surprising, since it is now well known that the intermolecular pair interactions in an organic crystal which give rise to the Davydov splitting and to exciton migration display similar nonscaling behavior. Thus, for the p band of anthracene ($f \sim 0.1$) the total Davydov splitting is 400 cm^{-1}, while for the naphthalene α band ($f \sim 10^{-3}$) the Davydov splitting is 186 cm^{-1}. The major contribution to the $\underset{\sim}{k}$ dependent intermolecular interactions arises in the latter case from charge transfer exciton-neutral exciton configuration mixing. On the other hand, typical solvent shifts arising from dispersion forces are determined by two contributions, one depending on the oscillator strength of the transition (see Eq. (124)). The experimental results are in qualitative agreement with the predictions of the Pople-Longuet-Higgins formula (84).

(c) A cursory examination of the experimental data indicates that all the vibronic components of a given electronic transition are red shifted by about the same energy with increasing pressure. However, small differences do exist, and these differences are extremely important for the interpretation of the spectral shifts in the crystal. Thus, in the limit of zero pressure, a careful comparison of the solution and the crystal spectrum shows that not all of the vibronic bands are equally affected by the change of environment. For the anthracene p band, the $0 \rightarrow n$ vibronic components become less displaced in the crystal relative to the solution with increasing quantum number, n, of the a_{1g} vibration of the excited state. This result can be rationalized by the application of exciton theory in the weak coupling limit, and then used for the determination of the relative contributions of solvent shifts and exciton shifts to the crystal excitation energy.

(d) The relative and absolute intensities of the vibronic components seem to be unaffected (with an experimental uncertainty of 20%) up to 100 kbar. This result is consistent with our assumption that there occur no pressure induced intramolecular configuration changes in this pressure range. A considerable increase in the line widths of the vibronic components is observed, resulting from enhanced scattering of the excitons by imperfections and by dislocations.

(e) Interesting experimental information regarding imperfections and dislocations in organic molecular crystals can be obtained from optical studies at high pressures. At present the only evidence for this effect is obtained from studies of the naphthalene crystal. The optical studies of McClure and Schnepp (85) of a thick naphthalene crystal reveal a series of very weak lines on the low energy side of the 0-0 band, and showing the same vibrational spacing as that corresponding to the naphthalene molecule (85). A single crystal of naphthalene at pressures of 20-30 kbar shows a pressure induced long wavelength absorption (located at 335 $m\mu$) which increases dramatically with increasing pressure (83). By way of comparison, a naphthalene solution in lucite does not show this band (83). The authors have speculated that this band arises from a charge-transfer transition, which should show intensity enhancement at high pressures. However, in view of McClure and Schnepp's (84) data for thick crystals at normal pressure, we propose herein to assign this pressure induced band to the absorption of a naphthalene molecule near an imperfection, i.e., in the neighborhood of a vacancy. The role of such imperfections in energy transfer phenomena (i.e., traps for singlet and triplet excitons) may be elucidated by emission studies and EPR studies (of triplet states) at high pressure. The small line width of the vibronic components of the naphthalene α band, when compared with the anthracene p band, makes these pressure induced states amenable to an experimental study in the latter case.

We proceed now to examine the nature of the intermolecular interaction leading to the pressure induced red shifts. We shall limit our discussion to nonpolar alternant hydrocarbons and consider only relatively weak electronic transitions (corresponding to excitations to the α and p states) so that the weak coupling scheme can be applied. Using the formalism outlined in Section II , and neglecting interactions between different one electron excited states, the spectral energy shifts of the vibronic components can be displayed in the form:

$$\Delta_f^{(n)} = D_f + W_f + C_f^{(n)}(\underset{\sim}{q}) \tag{122}$$

where $\Delta_f^{(n)}$ is the environmental shift of the center of gravity of the Davydov components of the n th vibronic band arising from the $0 \rightarrow n$ excitation to the f th excited electronic state, while D_f , as stated earlier, is the uniform shift of the manifold of energy levels

$$D_f = \sum_m \sum_\mu \langle a(\varphi_{n\nu}^f \varphi_{n\nu}^f - \varphi_{n\nu}^0 \varphi_{n\nu}^0 | V_{n\nu, m\mu} | \varphi_{m\mu}^0 \varphi_{m\mu}^0 \rangle \tag{123}$$

Note that for the case of alternant hydrocarbons the dipole term in the multipole expansion of $V_{n\nu, m\mu}$ in Eq. (123) vanishes, so that the D_f term will not exceed 100 cm^{-1}. However, in the case of azulene, a major contribution to the observed blue shift will arise from the D_f term.

W_f in Eq. (122) arises from the contribution of dispersion forces, and can be derived only if two electron excitations are included in the energy expression. An order of magnitude estimate for this solvent shift can be obtained from the Longuet-Higgins-Pople formula (84)

$$W_f \approx \sum_i R_i^{-6} \alpha \left(\mu^2 + \frac{1}{9} \alpha E_f\right), \tag{124}$$

where E_f is the intermolecular separation, α the polarizability, the excitation energy, and μ the transition moment.

The effects on the transition of the ordered crystalline environment are expressed in terms of the sum over intermolecular (coulomb and exchange) interactions between translationally equivalent molecules. The crystal shift is given by

$$C_f^{(n)}(\underset{\sim}{q}) = \sum_{m\iota}{}' (J_{n\iota,m\iota} + K_{n\iota,m\iota}) e^{i\underset{\sim}{q}\cdot(\underset{\sim}{r}_{n\iota}-\underset{\sim}{r}_{m\iota})} \times \langle \chi^{o(o)} | \chi^{f(n)} \rangle^2 . \tag{125}$$

In the weak coupling limit the crystal shift is proportional to the intensity (i.e., the Franck-Condon factor) for that particular transition. This observation makes possible the separation of the environmental shift from the crystal shift. In Table XX we display an analysis of the effect of pressure on the vibronic components of the p band of crystalline anthracene. We define the pressure dependent shift, of the n th vibrational component, $\delta^{(n)}(p)$, in the crystal relative to the solution spectrum by

$$\delta^{(n)}(p) = h\nu(\text{crystal}, p) - h\nu(\text{solution}, 1\text{atm}). \tag{126}$$

It is apparent from Table XX that $\delta^{(n)}(p)$ depends on the vibrational quantum number, converging to a constant value for large n . From Eqs. (122) and (125), it readily follows that $C_f^{(n)} \to 0$ for large n . Assuming that $C_f^{(3)} = 0$, we readily derive the contribution of the exciton shift to the vibronic components by setting

$$C_f^{(n)} \approx \delta^{(n)} - \delta^{(3)}. \tag{127}$$

It should be noted that because the contribution to the crystal shift is obtained from second differences between large numbers, these data are reliable only to about $\pm$ 50 cm^{-1}. In Table XXI we summarize the separate contributions of the solvent shift and of the exciton shift to the red shift of the p band of crystalline anthracene. The solvent shift, W_f , is now defined relative to the same transition in the isolated molecule by taking -700 cm^{-1} to be the red shift relative to the gas

Table XX. Pressure Effects on the p Band of Crystalline Anthracene. (All Energies in Units of cm^{-1})

a) p = 1 atm

Vibronic Component	Relative Intensity $\langle \chi^{0(0)} \mid \chi^{p(n)} \rangle^2$	$h\nu_1$ Solution (Hexane)[86]	$h\nu_2$ Micro-crystalline Sample[82,83]	$h\nu_3$ Single Crystal Davydov Components[87]	$\delta^{(n)}(p)$ $(h\nu_2 - h\nu_1)$	$C_f^{(n)}$
0–0	0.324	26,700	25,200	25213 ($\parallel b$) 25432 ($\parallel a$)	-1500	-330
0–1	0.316	28,210	26,800	26680 ($\parallel b$) 26830 (aa)	-1410	-250
0–2	0.218	29,590	28,275	28200 ($\parallel b$) 28280 ($\parallel a$)	-1315	-155
0–3	0.093	30,960	29,800		-1160	0
0–4	0.050	32,470	31,300		-1170	0

Table XX. (continued)

b) $p = 2 \times 10^4$ atm

Vibronic Component	$h\nu_4$ Crystal	$\delta^{(n)}(p) = h\nu_4 - h\nu_1$	$C_f^{(n)}$
0-0	24,160	-2640	-421
0-1	25,918	-2292	-173
0-2	27,421	-2169	- 50
0-3	28,741	-2119	0

c) $p = 4 \times 10^4$ atm

Vibronic Component	$h\nu_5$ Crystal	$\delta^{(n)}(p) = h\nu_5 - h\nu_1$	$C_f^{(n)}$
0-0	23,231	-3569	-431
0-1	24,962	-3248	-210
0-2	26,556	-3034	0
0-3	27,922	-3038	0

Table XXI. Spectral Shifts (Relative to the Gas Phase) for Crystalline Anthracene

p atm	$C_f^{(0)}$ cm^{-1}	$\sum_n C_f^{(n)}$ cm^{-1}	W_f cm^{-1}	$(V_0/V)^2$
1	-330 ± 50	-735 ± 200	-1860	1.00
2×10^4	-421 ± 50	-623 ± 200	-2819	1.33
4×10^4	-431 ± 50	-641 ± 200	-3738	1.53

Table XXII. First Order Contributions to the Exciton Spectral Shift in Crystalline Anthracene at Zero Pressure

Dipole-Dipole to 55 Å	-457 cm^{-1}
Higher Multipoles	+117 cm^{-1}
Exchange Terms	+ 8 cm^{-1}
Long Range Dipole-Dipole Interactions	-250 cm^{-1}
First Order Contribution	-582 cm^{-1}
Experimental	-735 ± 200 cm^{-1}

phase of the spectrum of anthracene in heptane solution. The total first order contribution to the exciton shift is obtained in the weak coupling limit by summation over all the exciton shifts corresponding to the individual vibronic components. From these data the following conclusions follow:

(a) The major contribution to the red shift arises from the solvent shift. However, the exciton shift is not negligible. It is impossible to interpret the effects of pressure on the optical spectra of organic crystals considering only solvent shifts (83) or only exciton shifts (82).

(b) The solvent shift, W_f, which arises from dispersion forces, shows a marked pressure dependence. This is attributed to the decrease in the intermolecular spacing at increasing pressure. According to Eq. (124), the solvent shift depends on the inverse sixth power of the intermolecular separation so that in the case of uniform isotropic crystal it is expected that the solvent shift will depend linearly on the square of the density. The data of Table XXI show that for crystalline anthracene there is an approximate linear dependence of the solvent shift on the square of reciprocal fractional volume (relative to zero pressure). The zero pressure solvent shifts relative to the gas phase are $-700\ cm^{-1}$ in heptane solution (85), $-980\ cm^{-1}$ in lucite glass (83), and $-1860\ cm^{-1}$ in crystalline anthracene. These results are consistent with the large polarizability of the anthracene molecule ($\sim 25\ Å^3$), which, according to Eq. (124), should lead to a large solvent shift.

(c) The exciton shifts in crystalline anthracene do not exhibit a marked pressure dependence in the region $0 - 4 \times 10^4$ kbar. It should be noted that in view of the considerable error associated with the determination of these exciton shifts, it is impossible to eliminate the possibility of a 25% change.

How are the exciton shifts accounted for within the framework of exciton theory? As in the case of the calculation of the Davydov splitting in the anthracene crystal, we must consider the contribution of dipole-dipole interactions summed over a distance of the order of magnitude of the wave length of the exciting light. To this must be added the contribution of short range higher multipole terms and of electron exchange interactions. The contributions of charge-transfer interactions are of no importance for the anthracene p band. These components of the intermolecular interactions between translationally equivalent molecules (for zero pressure) are listed in Table XXII. They lead to a first order contribution to the exciton shift which is in reasonable agreement with the experimental data.

A more refined theoretical study of the exciton contribution to the spectral shifts in crystalline anthracene should include the effects of crystal field mixing between the p and β exciton states. Now, it is necessary to consider the separate contributions of the A_u and B_u Davydov components. These spectral shifts, listed in Table XXIII, show that the A_u state is only slightly affected while the B_u state is shifted to lower energies by crystal field mixing. It is fair to conclude from these data cited that the observed spectral shifts in the anthracene crystal are adequately accounted for in terms of exciton theory.

How may the exciton shift be expected to change with increasing pressure? The long range dipole-dipole interactions depend on the unit cell volume, which decreases by 19% at 4×10^4 atm. The contribution of the short range dipole-dipole terms to 55 Å will scale approximately by the same factor. Thus, the total contribution of the dipole-dipole terms (in first order) is expected to increase at 4×10^4 atm from $-707\ cm^{-1}$ at 1 atm to $-841\ cm^{-1}$. The calculations presented in Section XI show that the contribution of exchange interactions increases from $+8\ cm^{-1}$ to $+25\ cm^{-1}$ in this pressure range. The effect of the compression of the unit cell on the short range contribution of higher monopoles has not been studied. If these terms increase to $+200\ cm^{-1}$ in this pressure range, the estimated exciton shift at 4×10^4 atm is $-700 \pm 200\ cm^{-1}$. Thus, the theory of exciton states predicts only a slight change in the pressure induced exciton shifts, in agreement with the experimental data. Just as in the case of the application of exciton theory to the study of Davydov splittings in molecular crystals, we conclude that the observed spectral shifts represent only the end effects of very subtle balances between different interactions, all of which must be carefully considered.

XI. PRESSURE EFFECTS ON INTERMOLECULAR EXCHANGE INTERACTIONS

It is now well established that a substantial contribution to the triplet exciton band width and to the Davydov splitting in triplet exciton states in crystals of organic molecules arises from electron exchange interactions. The dynamical properties of triplet excitons, as experimentally demonstrated by electron paramagnetic resonance, optical studies of triplet energy transfer, and by triplet exciton-exciton annihilation processes, are determined by intermolecular electron exchange interactions with an additional contribution from charge-transfer interactions. The electron exchange interactions are extremely sensitive to the detailed form of the molecular wave function at large intermolecular separation. In the calculations performed by us (88), a linear combination of carbon atom SCF atomic orbitals has been employed to represent the molecular wave function. This procedure is obviously only the first step in obtaining a good representation of the

Table XXIII. First and Second Order Contribution of Exciton Shifts to the Anthracene p Band at Zero Pressure. (All Energies in cm^{-1} Units)

	Davydov Component		Mean Energy of Two Components	Experiment
	A_u	B_u		
First order	-922	-251	-587	
Second order (p-β mixing)	-929	-560	-744	$-735 \pm 200\ cm^{-1}$

molecular wave functions at large distances from the molecule. Our calculations lead to an estimate of the Davydov splitting in triplet exciton states which has been found to be in good agreement with some very recent experimental studies of crystals of dibromonaphthalene (see Section IV). The predicted triplet exciton diffusion length and triplet exciton-exciton annihilation rate constant, calculated on the basis of a random walk model for triplet exciton migration, have also been found to be in excellent agreement with the experimental data. A further test of the adequacy of the approximations inherent in our calculations can be obtained from a study of the dependence of electron exchange interactions on the internuclear spacing. These interactions are extremely sensitive to the unit cell dimensions, varying approximately exponentially with the intermolecular spacing. An experimental study of the effect of pressure on the Davydov splittings of a set of triplet exciton states is too difficult at present, but studies of the pressure dependence of triplet exciton dynamics in organic crystals can be performed and will yield valuable information regarding the dependence of the short range intermolecular interactions on the unit cell dimensions and should provide a significant test of the theoretical studies of the electron exchange interactions.

We turn now to the numerical calculation of the pressure dependence of the K integrals. The procedure employed was the same as has been previously described. The molecular wave functions employed in the calculations were the Hoyland-Goodman semiempirical SCF molecular orbitals, and the pressure dependence of the lattice parameters was taken from Section VI. Typical results for crystalline anthracene are displayed in Table XXIV. We have only included the largest electron exchange interactions for molecules located at $\underset{\sim}{b}$ and at $(1/2)(\underset{\sim}{a} + \underset{\sim}{b})$ since all the other integrals were found to be two orders of magnitude smaller. In Table XXIV we have also included the components of the group velocity tensor, calculated in the constant mean free path (isotropic) approximation, within the framework of the band model. From these results we conclude that:

(a) Intermolecular electron exchange interactions increase by about a factor of 4 in the pressure region $1-10^5$ atm. Hence, the Davydov splitting in the 0-0 band (using a vibrational overlap factor of 0.25) increases from $-11\ cm^{-1}$ at 1 atm to $-38\ cm^{-1}$ at 10^5 atm. The contribution of electron transfer interactions increases somewhat more slowly over this same pressure range. For the case of anthracene, the charge transfer contribution to the Davydov splitting increases from $-4\ cm^{-1}$ at 1 atm to $-9\ cm^{-1}$ at $10^5\ cm^{-1}$.

(b) Using a random walk, strong scattering, model for triplet exciton diffusion, we predict that the diffusion coefficient will in-

Table XXIV. Pressure Effects on Intermolecular Exchange

Interactions in Crystalline Anthracene

p atm	$K(\underset{\sim}{b})$ 10^{-4} eV	$K(\frac{\underset{\sim}{a}+\underset{\sim}{b}}{2})$ 10^{-4} eV	$\left\langle \frac{V_{ae}^2}{\lvert \underset{\sim}{V} \rvert} \right\rangle$ 10^5 cm/sec	$\left\langle \frac{V_{bb}^2}{\lvert \underset{\sim}{V} \rvert} \right\rangle$ 10^5 cm/sec	$\langle \lvert \underset{\sim}{V}(\underset{\sim}{k}) \rvert \rangle$ 10^5 cm/sec
1	4.10	- 6.83	0.605	0.520	1.13
10000	6.24	-11.52	1.035	0.850	1.88
20000	7.20	-12.14	1.084	0.932	2.02
40000	13.82	-17.23	1.488	1.608	3.10
60000	10.95	-20.07	1.814	1.523	3.33
100000	13.81	-24.02	2.169	1.877	4.05

crease by a factor of 10 in the pressure range 1-10^5 atm. The diffusion length will therefore increase by a factor of 3 provided that there is no decrease in the lifetime of the triplet state.

(c) The exciton-exciton annihilation rate constant, herein interpreted in terms of a diffusion controlled reaction, is predicted to increase by one order of magnitude in the pressure range 1-10^5 atm.

(d) A study of triplet exciton migration within the framework of the narrow band model, with a scattering length of the order of intermolecular spacing, predicts that the diffusion coefficient and the annihilation rate constant increase by a factor of 4 in the pressure range 1-10^5 atm, since the velocity depends linearly on the band width. Thus, both the strong and weak scattering models predict an enhanced rate of triplet exciton migration at high pressure. However, the enhancement is larger for the strong scattering model than for the band model.

XII. PRESSURE EFFECTS ON THE ELECTRON AND HOLE MOBILITY IN ORGANIC CRYSTALS

A theoretical investigation of the electrical properties of organic crystals at high pressures requires the calculations of the pressure dependence of the excess electron and hole band structures and of the carrier mobilities. The band structure (Eq. (110), Section V) was calculated using the method of Silbey et al (89) and the crystal structure data of Section VI. It should be noted at this point that the excess charge carrier band structure is mainly determined by intermolecular hybrid integrals, which are less affected by a decrease in unit cell dimensions than are the intermolecular electron exchange integrals considered in Section XI. We therefore expect that the effects of pressure on the excess electron and hole bands will be somewhat smaller than the corresponding triplet exciton bands. In Table XXV we display the band widths, relative to zero pressure, calculated using a vibrational overlap integral of 0.5. From these data we conclude that increasing pressure results in an increase of the band width, the effect on the narrow bands being more pronounced than on the wider bands. Thus the a-electron band width increases from 7.5×10^{-4} eV at 1 atm to 15×10^{-2} eV at 10^5 atm, while the b-electron band width increases from 12×10^{-2} eV to 25×10^{-2} eV in this same pressure range. The pressure dependence of the components of the electron and hole velocity tensors are presented in Table XXVI. Assuming that the relaxation time, or the mean free path, is independent of the external pressure, these data represent the cal-

Table XXV. Pressure Dependence of the Excess Electron and Hole Band in Crystalline Anthracene
(Calculated for Vibrational Overlap Factor of 0.5)

(Units 10^{-2} eV)

Dir.	$p = 1$ atm		$p = 10^4$ atm		$p = 2 \times 10^4$ atm		$p = 4 \times 10^4$ atm		$p = 10^5$ atm	
	Electron	Hole	Electron	Hole	Electron	Hole	Electron	Hole	Electron	Hole
a^+	7.58	3.82	10.22	3.74	11.03	3.47	12.81	2.90	15.17	1.39
a^-	7.46	3.12	10.08	2.88	10.89	2.57	12.64	1.94	14.95	0.37
b^+	3.26	11.82	4.31	14.58	4.53	15.44	5.07	17.15	5.60	19.18
b^-	11.84	5.22	16.12	8.19	17.60	9.57	20.69	12.41	25.18	17.43
c^+	0.22	4.61	0.31	6.20	0.59	6.84	1.25	7.87	2.76	9.70
c^-	0.60	5.59	0.80	7.79	0.90	8.71	1.11	10.32	1.55	13.25

Table XXVI. Pressure Dependence of the Components of Mobility Tensor in Constant Free Time and Constant Free Path Approximation for Excess Charge Carriers in Crystalline Anthracene

	p = 1 atm		$p = 10^4$ atm	
	I	II	I	II
$\langle v_a^2/\lvert\mathbf{v}\rvert\rangle^a$	6.09	13.56	5.29	15.86
$\langle v_b^2/\lvert\mathbf{v}\rvert\rangle^a$	14.99	7.29	19.23	7.97
$\langle v_{c'}^2/\lvert\mathbf{v}\rvert\rangle^a$	2.54	0.26	2.55	0.58
$\langle v_a^2\rangle^b$	180.11	375.84	174.50	512.59
$\langle v_b^2\rangle^b$	484.46	201.26	716.30	262.37
$\langle v_{c'}^2\rangle^b$	66.76	2.60	81.99	7.78
	$p = 2 \times 10^4$ atm		$p = 4 \times 10^4$ atm	
	I	II	I	II
$\langle v_a^2/\lvert\mathbf{v}\rvert\rangle^a$	4.88	16.30	4.16	17.24
$\langle v_b^2/\lvert\mathbf{v}\rvert\rangle^a$	20.90	8.21	23.99	8.63
$\langle v_{c'}^2/\lvert\mathbf{v}\rvert\rangle^a$	2.47	0.80	2.19	1.35
$\langle v_a^2\rangle^b$	165.03	544.87	145.03	619.35
$\langle v_b^2\rangle^b$	814.78	283.60	1016.89	325.30
$\langle v_{c'}^2\rangle^b$	85.82	12.08	86.06	25.17
	$p = 6 \times 10^4$ atm		$p = 10^5$ atm	
	I	II	I	II
$\langle v_a^2/\lvert\mathbf{v}\rvert\rangle^a$	3.92	17.61	3.44	17.74
$\langle v_b^2/\lvert\mathbf{v}\rvert\rangle^a$	25.59	8.94	27.66	9.46
$\langle v_{c'}^2/\lvert\mathbf{v}\rvert\rangle^a$	2.09	1.90	2.05	9.23
$\langle v_a^2\rangle^b$	137.28	658.98	117.92	698.62
$\langle v_b^2\rangle^b$	1142.67	354.40	1324.62	397.74
$\langle v_{c'}^2\rangle^b$	87.56	39.92	91.61	71.13

Table XXVI (continued)

a Constant free path approximation units 10^5 cm/sec

b Constant free time approximation units 10^{10} cm/sec

I Hole

II Electron

Vibrational overlap squared = 0.50

---

culated pressure dependence of the electron and hole mobilities. The carrier mobilities in anthracene are thereby predicted to increase by a factor 2-3 in the pressure range $1-10^5$ atm. These results are in qualitative agreement with Kepler's mobility data for anthracene at 3000 atm (see Table XIX).

From these calculations the following negative conclusions may be reached:

(a) The contribution of the change in mobility to the large increase of the conductivity at high pressures is negligible.

(b) The pressure dependence of the activation energy for charge carrier production, which decreases by about 0.5-1 eV in the pressure range $1 - 3 \times 10^5$ atm, cannot be attributed to an increase in the excess electron and hole band widths with increasing pressure. Indeed, in the case of the anthracene crystal, the bottom (i.e., lowest energy state) of the excess electron band is located at -2.53×10 eV at one atmosphere and at -4.96×10^{-2} eV at 10^5 atm (these data are presented for the $\underset{\sim}{k}$ dependent terms in Eq. (110), calculated for a vibrational overlap of 0.5).

What is, then, the reason for the decrease in the activation energy for charge carrier production which leads to the dramatic increase in the electrical conductivity at high pressure? We propose that the decrease in the energies of the excess electron and hole states should be assigned to the $\underset{\sim}{k}$ independent term (i.e., ε_0 in Eq. (110)). This term contains the electronic polarization energy of the lattice by the excess charge (90). The polarization energy, P, of the anthracene lattice in interaction with an anthracene negative ion can be calculated using the known anisotropic polarizability of the anthracene molecule (90): The result is P = 2.0 eV at zero pressure. Since the polarization energy depends on the inverse fourth power of the intermolecular separation, it is expected to increase (in absolute value) by a factor of $(\rho/\rho_0)^{4/3}$ (where ρ is the density) with increasing pressure. From the data displayed in Table XIV we conclude that at 3×10^5 atm the crystal polarization energy should increase by 45%, i.e., to $P = -2.9$ eV. It should be noted that our rough estimate of 0.9 eV for the decrease of the activation energy for charge carrier production in the pressure range $1 - 3 \times 10^5$ atm is in good agreement with the experimental value obtained for pentacene in this pressure range. This analysis provides strong support for the idea that the carrier generation process leading to dark conductivity in organic crystals is due to extrinsic electron injection from the electrodes. The increase in the absolute value of the polarization energy with increasing pressure will lead to a decrease in the energy difference between the Fermi level of the electrode and the bottom of the excess charge carrier conduction band. The "metallic" behavior of the pentacene crystal at high pressure is not due to a drastic change in the band structure but just to the matching of the Fermi level of the electrode with the narrow excess electron band of the crystal, thereby leading to an efficient charge injection process at high pressures.

The authors do not believe that the red shift in the position of the first singlet exciton band, caused by the contribution of dispersion forces to the solvent shift, is relevant to the interpretation of the decrease in the energy gap for dark conductivity with increasing pressure. No evidence is available at present for an intrinsic process involving the dissociation of the first singlet exciton in anthracene to yield charge carriers. This negative evidence is consistent with general theoretical considerations regarding the energy balance in this system. The photoconductivity induced in the first exciton band of the anthracene crystal is due either to surface dissociation of a singlet exciton, or to multiexciton processes.

XIII. NEW EXPERIMENTS

We hope that the exhausted and blurry-eyed reader is convinced at this stage that the study of pressure effects on the electronic states of organic crystals may lead to interesting and valuable information regarding the band structure and exciton dynamics. Most of the experiments we would like to suggest are by no means easy. However, they appear to us to be worth performing, and we suggest the following:

1. Examination of the enhancement of the transition between the ground state and charge-transfer states using optical absorption techniques at low temperatures.
2. Examination of triplet exciton dynamics using laser excitation.
3. Direct studies of electron and hole mobility using essentially the same techniques as have been employed at low pressures.
4. Examination of the dependence of the Davydov splitting of singlet exciton states at high pressures. This experiment requires that

the windows of the cell be strain-free so that polarization measurements can be made.

It is fairly obvious that many more experiments could be suggested, and these should be obvious from the text of this review. With a sigh of relief, we therefore end this review.

XIV. ACKNOWLEDGMENTS

This research was supported by the Directorate of Chemical Sciences of the Air Force Office of Scientific Research and the United States Public Health Service. We have also benefited from facilities provided by the Advanced Research Projects Agency for materials research at the University of Chicago.

REFERENCES

1. J. Frenkel, Phys. Rev. 37, 17 (1931).
2. R. Peierls, Ann. Physik 13, 905 (1932).
3. (a) A. S. Davydov, Theory of Molecular Excitons (McGraw-Hill, 1962).
   (b) D. Fox and O. Schnepp, J. Chem. Phys. 23, 767 (1955).
   (c) D. S. McClure, Solid State Physics 8, 1, Academic Press, New York (1959).
   (d) D. P. Craig and S. H. Walmsley, "Physics and Chemistry of the Organic Solid State", ed. M. M. Labes, D. Fox and A. Weissberger (John Wiley and Sons, New York, 1963), Vol. I, p. 585.
4. G. Wannier, Phys. Rev. 52, 191 (1937).
5. R. S. Knox, Theory of Excitons (Academic Press, New York, 1959).
6. L. Meyer, S. A. Rice and J. Jortner (to be published).
7. (a) D. P. Craig and S. H. Walmsley, Mol. Phys. 4, 113 (1961).
   (b) T. A. Claxton, D. P. Craig and T. Thirunamachandran, J. Chem. Phys. 35, 1525 (1961).
8. R. Silbey, J. Jortner and S. A. Rice, J. Chem. Phys. (in press).
9. D. P. Craig and J. R. Walsh, J. Chem. Soc. 1958, 1613.
10. M. H. Cohen and F. Keffer, Phys. Rev. 99, 1128 (1955).
11. G. G. Hall, Proc. Roy. Soc. A270, 285 (1962).
12. A. T. Amos, Mol. Phys. 6, 393 (1963).
13. W. T. Simpson, Rad. Res. 20, 87 (1963).
14. H. C. Bolton, W. Fawcett and D. G. Gurney, Proc. Phys. Soc. (London) A80, 199 (1962).
15. (a) W. Heller and A. Marcus, Phys. Rev. 84, 809 (1951).
    (b) D. Fox and S. Yatsiv, Phys. Rev. 108, 938 (1957).
16. M. Born and E. Wolf, Principles of Optics (Pergamon Press, London, 1959).
17. B. G. Anex and W. T. Simpson, Rev. Mod. Phys. 32, 466 (1962).
18. E. F. McCoy and I. G. Ross, Aust. J. Chem. 4, 573 (1962).
19. M. S. Brodin and S. V. Manisova, Optics and Spectroscopy 10, 242 (1961).
20. R. W. Brandon, R. E. Gerkin and C. A. Hutchison, Jr., J. Chem. Phys. 37, 447 (1962).
21. G. C. Nieman and G. W. Robinson, J. Chem. Phys. 37, 2150 (1962).
22. J. C. Slater and W. S. Shockley, Phys. Rev. 50, 705 (1936).
23. E. Wigner, Group Theory (Academic Press, New York, 1959).

24. The main source of the intensity of the CT transition may be intensity borrowing: see Reference 25.
25. L. E. Orgel and R. S. Mulliken, J. Am. Chem. Soc. 79, 4839 (1957); J. L. Murrell, J. Am. Chem. Soc. 81, 5037 (1959).
26. L. E. Lyons, J. Chem. Soc. 1957, 5001.
27. R. S. Berry, J. Jortner, J. Mackie, E. Pysh and S. A. Rice, J. Chem. Phys. (in press).
28. R. Hilsch and R. W. Pohl, Z. Physik 48, 384 (1928); 57, 145 (1929); 59, 812 (1930).
29. I. M. Dykman, Zh. Eksperim. i Tior. Fiz. 26, 307 (1954).
30. R. Daudel, R. Lefevre and C. Moser, Quantum Chemistry (Interscience Publishing Company, New York, 1959).
31. R. Silbey, J. Jortner, M. T. Vala and S. A. Rice, J. Chem. Phys. (in press).
32. D. P. Craig, L. F. Lyons and J. R. Walsh, Mol. Phys. 4, 97 (1961).
33. R. Pariser, J. Chem. Phys. 24, 250 (1956).
34. This is a consequence of the pairing of molecular orbitals in these approximate theories, and would be true for any molecular wave function preserving the pairing of orbitals in the state.
35. E. Konijnenberg, Thesis, University of Amsterdam (1964).
36. D. P. Craig, M. Hollas, M. F. Kedies and S. C. Wait, Phil. Trans. Roy. Soc. (London) A253, 543 (1961).
37. D. S. McClure, J. Chem. Phys. 20, 682 (1952).
38. M. Mizushima and S. Koide, J. Chem. Phys. 20, 765 (1952).
39. H. F. Hameka and L. J. Oosterhoff, Mol. Phys. 1, 358 (1958).
40. R. M. Hochstrasser (private communication).
41. N. Hirota and C. A. Hutchison, Jr., J. Chem. Phys. (in press).
42. M. A. El-Sayed, M. T. Wauk and G. W. Robinson, Mol. Phys. 5, 205 (1962).
43. H. Sternlicht, G. C. Nieman and G. W. Robinson, J. Chem. Phys. 38, 1326 (1963).
44. G. C. Nieman and G. W. Robinson, J. Chem. Phys. 38, 1928 (1963).
45. R. G. Kepler, J. C. Caris, P. Avakian and E. Abramson, Phys. Rev. Letters 10, 400 (1963).
46. J. L. Hall, D. A. Jennings and R. M. McClintock, Phys. Rev. Letters 11, 364 (1963).
47. S. Z. Weisz, A. B. Zahlan, M. Silver and R. C. Jarnagin, Phys. Rev. Letters 12, 71 (1964).
48. P. Avakian, E. Abramson, R. G. Kepler and J. C. Caris, J. Chem. Phys. 39, 1127 (1963).
49. J. R. Hoyland and L. Goodman, J. Chem. Phys. 36, 12 (1962).

50. J. Jortner, S. Choi, J. L. Katz and S. A. Rice, Phys. Rev. Letters 11, 323 (1963).
51. S. P. McGlynn and M. Kasha, J. Chem. Phys. 24, 588 (1956).
52. P. Avakian and R. E. Merrifield, Phys. Rev. Letters 13, 541 (1964).
53. S. Chandrasekhar, Rev. Mod. Phys. 15, 1 (1943).
54. We consider here the limiting case of high exciton concentration where the bimolecular mechanism predominates over monomolecular decay processes.
55. It should also be noted that the reduction by 1/9 may overestimate the effects of spin selection rules, since it is conceivable that radiationless transitions could occur from the compound intermediate state to S_1. All such effects tend to raise .
56. E. U. Condon and G. H. Shortley, The Theory of Atomic Spectra (Cambridge University Press, 1935), p. 73.
57. S. Choi, J. Jortner, S. A. Rice and R. Silbey, J. Chem. Phys. 41, 3294 (1964).
58. L. Lindquist, Arkiv Kemi 16, 79 (1961).
59. R. G. Kepler, Phys. Rev. 119, 1226 (1960).
60. O. H. LeBlanc, J. Chem. Phys. 33, 626 (1960).
61. G. D. Thaxton, R. C. Jarnagin and M. Silver, J. Phys. Chem. 66, 2461 (1962).
62. O. H. LeBlanc, J. Chem. Phys. 35, 1275 (1961).
63. J. L. Katz, S. I. Choi, S. A. Rice and J. Jortner, J. Chem. Phys. 39, 1683 (1963).
64. R. Silbey, J. Jortner, S. A. Rice and M. T. Vala, J. Chem. Phys. 42, 733 (1965).
65. J. Tanaka, Bull. Chem. Soc. Japan 36, 1237 (1963).
66. R. B. Aust, W. H. Bently and H. G. Drickamer, J. Chem. Phys. 41, 1856 (1964).
67. P. W. Bridgman, Proc. Am. Acad. Arts Sci. 76, 9, 71 (1945); ibid. 76, 1, 19 (1945); ibid. 76, 3, 83 (1948).
68. G. S. Kabalinka, Soviet Phys.--Solid State 4, 2288 (1963).
69. G. A. Samara and H. G. Drickamer, J. Chem. Phys. 37, 474 (1962).
70. D. W. Gregg and H. G. Drickamer, J. Chem. Phys. 35, 1780 (1961).
71. M. A. El-Sayed, J. Chem. Phys. 37, 1568 (1962).
72. D. S. McClure, J. Chem. Phys. 20, 682 (1952).
73. M. A. El-Sayed, Nature 197, 401 (1963).
74. G. W. Robinson and R. P. Frosch, J. Chem. Phys. 37, 1962 (1962).
75. G. R. Hunt, E. F. McCoy and I. G. Ross, Aust. J. Chem. 15,

591 (1962).
76. R. G. Kepler, Phys. Rev. 119, 1226 (1960).
77. O. H. LeBlanc, J. Chem. Phys. 33, 626 (1960).
78. G. D. Thaxton, R. G. Jarnagin and M. Silver, J. Phys. Chem. 66, 2461 (1962).
79. H. Inokutchi and H. Akamata, Solid State Physics 12, 93 (1963).
80. G. A. Samara and H. G. Drickamer, J. Chem. Phys. 37, 474 (1962).
81. R. G. Kepler, Organic Semiconductors Conference, edited by J. J. Brophy and J. W. Buttery (McMillan Company, New York, 1962), p. 1.
82. S. Wiederhorn and H. G. Drickamer, J. Phys. Chem. Solids 9, 330 (1959).
83. H. W. Offen, "Optical Spectra of Polyacenes under Pressure", Ph.D. Thesis, University of California, Los Angeles (1963), University Microfilms, Inc., Ann Arbor, Michigan 63-7884.
84. H. C. Longuet-Higgins and J. A. Pople, J. Chem. Phys. 27, 192 (1957).
85. D. S. McClure and O. Schnepp, J. Chem. Phys. 23, 1575 (1955).
86. M. Orchin and H. H. Jaffe, Theory and Applications of Ultraviolet Spectroscopy (Wiley, New York, 1962).
87. M. S. Brodin and S. V. Marisova, Opt. Spectry. 10, 242 (1961).
88. J. Jortner, S. A. Rice, J. L. Katz and S. I. Choi, J. Chem. Phys. 42, 309 (1965).
89. R. Silbey, J. Jortner, S. A. Rice and M. T. Vala, J. Chem. Phys. 42, 733 (1965).
90. R. S. Berry, J. Jortner, J. C. Mackie, E. S. Pysh and S. A. Rice, J. Chem. Phys. 42, 1535 (1965).

SESSION B

ELECTRONIC PROPERTIES II

Chairman: H. BROOKS

SOME REMARKS ON CURRENT THEORIES OF THE TRANSPORT PROPERTIES OF METALS, WITH REFERENCE TO THE EFFECTS OF PRESSURE

J. M. Ziman: H. H. Wills Physics Laboratory
University of Bristol, England

Abstract

The electrical properties of ordinary metals depend on the shape of the Fermi surface, on the electron-phonon interaction, and on the phonon spectrum. In recent theories of electronic structure these may all be derived from the effective potential (or "pseudo-potential") of the ions as seen by the conduction electrons. This, in its turn, depends on the atomic volume. Thus, in principle, it ought to be possible to make fairly complete calculations of properties such as the electrical resistivity of sodium and aluminum, and to predict their variation with pressure.

The experimenter has a decisive tactical advantage over the theorist: he can just go into his laboratory, and in a day or two measure a quantity that will take a year or two to compute. That is not to say that high pressure measurements are all that easy - but they do generate a mass of data that are not to be interpreted without a great deal of hard work.

The present theoretical situation is that we are just beginning to understand how to calculate the electronic transport properties of very simple metals at their normal volume: it will take a little time to make similar calculations for the same properties at high pressures. Nevertheless, the outlook is much more hopeful than it was five years ago. There has been a palace revolution, in which we have expelled a lot of the Old Guard of ideas, and a new and simpler outlook now reigns. This lecture will be a brief account of this new point of view, and its implications.

I am going to keep to the simplest basic problem - the electrical resistivity of a cubic non-transition metal - say Na or Al. And of course I shall have to say many things without full documentation. (But see (1)(2)(3)). So let us start at once, with a formula

$$\rho = \frac{3\,\hbar^2 q_D{}^2 kT}{2\hbar e^2 N M v_F{}^2 k^2\ \Theta^2} \int |\,U(K)|^2\ [K^2/q^2]$$
$$\times\ (1-\cos\,\theta)\,d\Omega. \qquad (1)$$

This is rather a hybrid, but it is the outcome of the following line of argument.

First, we treat the conduction electrons as if they were perfectly free. This is never quite true, but so long as the Fermi surface is not too distorted - that is, so long as it can be sketched out, in an extended zone scheme, as effectively spherical, - it may be a plausible assumption. So we think of a scattering process in which the electron wave vector changes by K, and its velocity changes by the angle θ, into the solid angle $d\Omega$. The factor $(1-\cos\,\theta)$ is familiar in the elementary theory of the relaxation time.

The probability of scattering is usually calculated from the theory of the electron-phonon interaction. That is all right - only it tends to get rather elaborate. I am supposing that we are thinking about what happens at ordinary room temperature, where we may assume that the electron scattering is essentially elastic. The problem is then much simpler; it is just diffraction of the electron wave by the slightly disordered lattice of ions. Most of the diffraction just gives rise to band structure, and can for the moment be ignored. We are interested in the extra scattering caused by the thermal disarray of the ions, and this comes out proportional to kT, inversely proportional to Θ^2, etc., which introduces a lot of factors into the front of the formula.

It also introduces a thoroughly nasty geometrical factor, which I have written, rather blandly, $[K^2/q^2]$, under the integral sign. Most of this factor comes from our old friends, the Umklapp processes: we know that the phonons responsible for scattering through the vector $\underline{K}$ have wave vector $\underline{q}$, whether either $\underline{q} = \underline{K}$ (as in an N-process), or where $\underline{q}$ differs from $\underline{K}$ by a vector of the reciprocal lattice. The job of calculating the ratio $[K^2/q^2]$, and averaging it over all processes where K is the same, is elementary in principle, but tedious in practice. There is also hidden here a factor for the weights of transverse and longitudinal modes, since these interact differently with electrons, and also have different velocities.

However, this discussion of the geometry and dynamics of the lattice vibration is probably familiar. I am not really going beyond Electrons and Phonons (4) in this part of the formula. The new ideas are in the factor that I have written $|U(K)|^2$. In the conventional theory of the electron-phonon interaction this is a rather elaborate function derived from the Bloch theory, in which one thinks about moving an ion, and then treating the changed potential as a perturbation on the electron states. The matrix element does not come out clearly, as if quite separate from the derangement factors to which I have just referred; we have to do some work to get these out at all.

But from the diffraction point of view $|U(K)|^2$ is just the matrix element of the potential of an ion for the transition $\underline{k}^1 - \underline{k} = \underline{K}$, i.e., it is the Fourier transform

$$U(\underline{K}) = N\int U(\underline{r})e^{i\underline{K}\cdot\underline{r}}\, d\underline{r}. \qquad (2)$$

Well, that is not quite correct; what we have now learned is that it is the matrix element of the screened pseudo-potential of an ion.

Here again it would take too long to go into

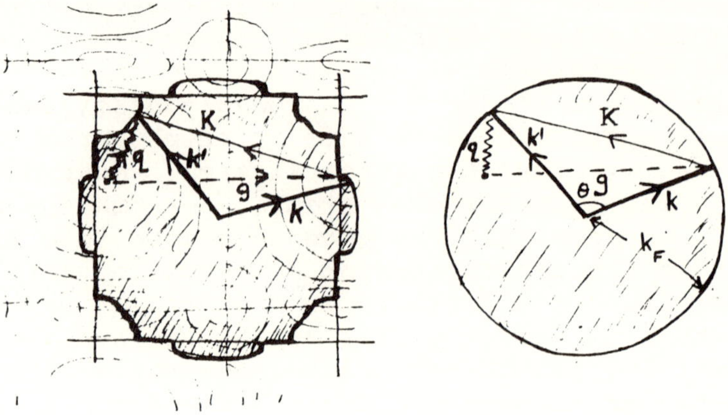

(a) (b)

Fig. 1. Idealization of Electron-Phonon Interaction.

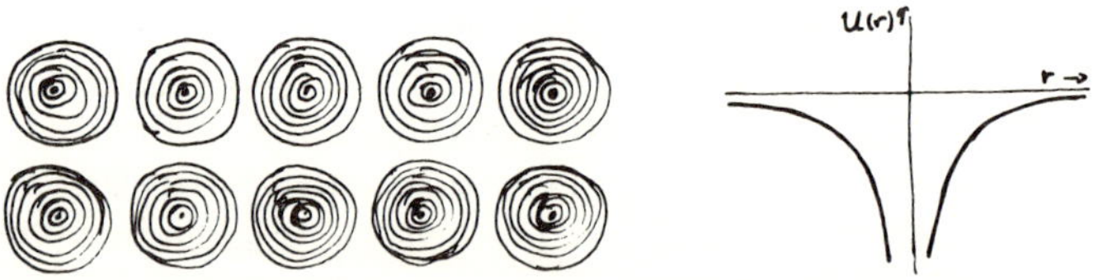

(a) Bare ions

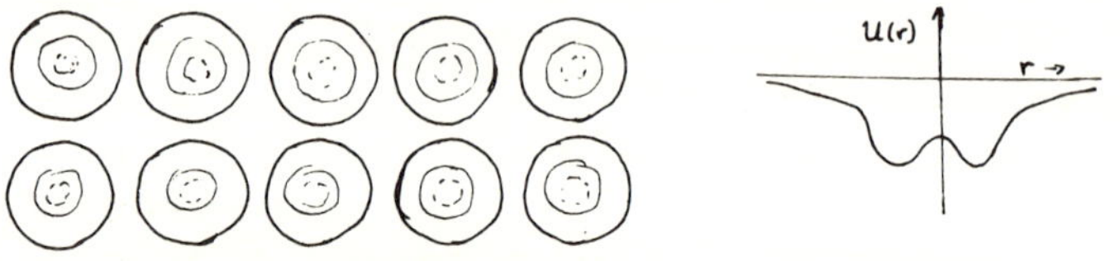

(b) Bare Pseudo potentials

(c) Screened Pseudo potentials.

Fig. 2.

details. But the general argument is as follows. First think of the potential of a bare ion. This would be much too strong to be treated in Born approximation for the scattering of an electron at the Fermi surface. So we do some tricks with it and replace it by a "pseudo-potential", or "effective potential", which would give the correct scattering cross section if put into a Born formula. Then we remember that the "other" conduction electrons also have their effect on the electron, and that these tend to form a screening cloud round each "bare" ion. This we allow for by dividing the matrix elements of the effective potential of the bare ion by an appropriate dielectric function. The result is U(K), to be used for our general formula. Actually the Bardeen formula for the electron-phonon interaction is a special case of such a function - but it took us 25 years to realize this.

The actual form of U(K) is now known for a number of metals. It is always something like Figure 2. Because the field of a bare ion is a Coulomb field outside the core, we always get $-2/3\ E_F$ for the value of U(K) near K = 0. But for larger values of K the matrix element becomes quite small - a fraction of an electron volt - and may change sign, or oscillate a bit. Of course we are only interested in its value out to the limit $2\ k_F$, corresponding to scattering from one side of the Fermi sphere to the other.

We can now see what needs to be known in order to calculate the resistivity of a metal. We need all the information we can get about the lattice spectrum - it is surprising how sensitive the calculation can be to, say, the anisotropy of velocity of the transverse modes - and we need to know U(K). To do a good job, we also need to know more about the shape of the Fermi surface, since there are important corrections to be made where this cuts the zone boundaries (as in a polyvalent metal), or where it is somewhat distorted from a sphere. It is not difficult to get a value for ρ within a factor of two (there is a formula in <u>Electrons and Phonons</u> which does this), but from then on the calculation is complicated, and requires much more information than we usually have. That is why there are only a few such calculations, and none to which one should trust ones life or scientific reputation.

So it looks extremely difficult to "predict" the effect of pressure on the electrical resistivity of even so simple a metal as I have assumed here, for we should have to know what the effect would be for all these different sub-contractors pseudo-potentials, lattice spectrum, and Fermi surface - before we could name our price. It looks as if this calculation could only come after these subsidiary changes had all been carefully measured - for example by measuring the velocity of sound at high pressures, or noting the change of Θ with volume, or determining anew the shape of the Fermi surface.

But our new point of view gives us a way of estimating these subsidiary changes <u>from the change of U(K) alone</u>. This function does not merely govern the electron-phonon interaction; it also determines the shape of the Fermi surface, and the parameters of the lattice spectrum. In principle, given U(K), we can construct all the other properties, and hence all the quantities that occur in the formula for the resistivity. I am stating this categorically to give it emphasis, although really it is only an approximation. At the moment, anyway, it is the best approximation we have, and is worth treating with respect.

The idea that U(K) should govern the shape of the Fermi surface is obvious enough. Indeed, it was out of the theory of band structure that the pseudo-potential formulation arose. All that we need do is to take seriously the diffraction of the conduction electrons by the regular ordered lattice array, which we threw away previously. Putting this back, we get the usual band gaps; for example, the zone boundary that bisects the reciprocal lattice vector $\underline{g}$, gives rise to an energy gap of width $2|U(\underline{g})|$. Thus, the values of $U(\underline{g})$ for various reciprocal lattice vectors $\underline{g}$ are the parameters in a nearly-free-electron calculation of the band structure and the shape of the Fermi surface. Since we can estimate these parameters quite accurately from Fermi surface studies (5), this can even be a check on our calculation of U(K) from atomic potentials, etc. And then, if we knew how U(K) varied with volume, we could estimate, with fair reliability, the consequent changes in the shape of the Fermi surface.

The connection between the pseudo-potential and lattice dynamics is not so evident - and indeed, has not been proved rigorously. The idea is that if we can say that an electron "sees" a screened ion as if it had a potential with Fourier transform U(K), then another ion, also, will see it with this same potential. Thus, if the field of the first ion at large distances looks like

$$U(R) = -\frac{Ze^2}{R}\exp(-k_s R) \tag{3}$$

to an electron, then this will also be the field seen by the second screened ion at this distance. If the valence charge of this ion is also Ze, then there will be a mutual potential energy interaction of the form

$$E(R) = \frac{Z^2e^2}{R}\exp(-k_s R) \tag{4}$$

which we should put into a calculation of lattice dynamics.

This is easy enough - but for near neighbors we have to recall that the ion is not a point charge but is an extended charge cloud of complicated structure. The trick (and this has not really been proved to be correct) is to calculate, by Poisson's equation, what charge distribution would have given rise to the pseudo-potential of this <u>second</u> ion, and to treat this as the charge cloud that is in the field of the first one. One gets quite a simple formula, e.g.

$$E(R) = \int \frac{|U(K)|^2 K^2}{4\pi e^2 \epsilon(K)} e^{i\underline{K}\cdot\underline{R}}\, d\underline{K} \tag{5}$$

where $\epsilon(K)$ is the dielectric function in the screening theory.

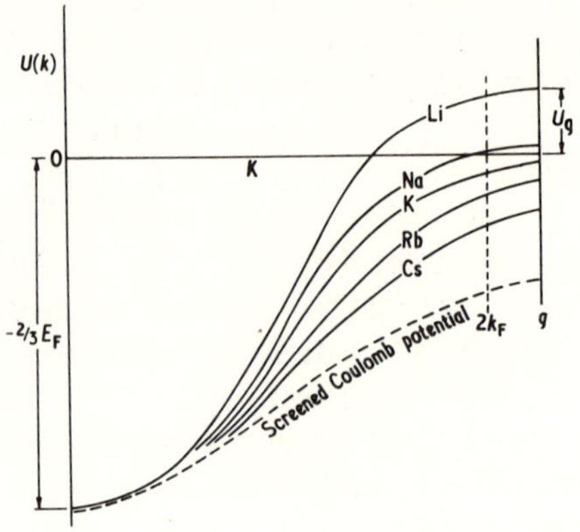

Fig. 3. Pseudo potentials for alkali metals.

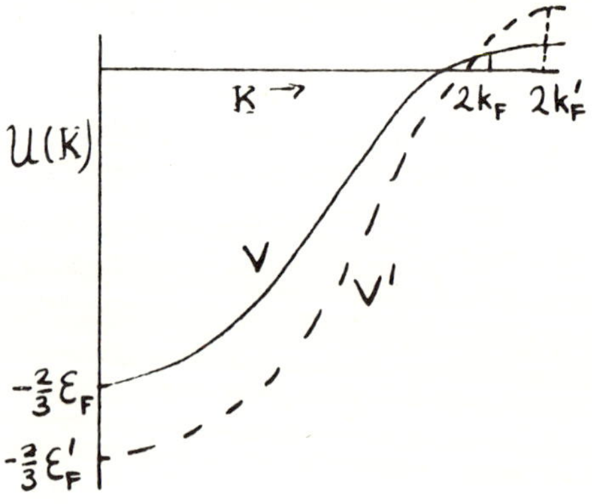

Fig. 4. Volume-dependence of Pseudo potential.

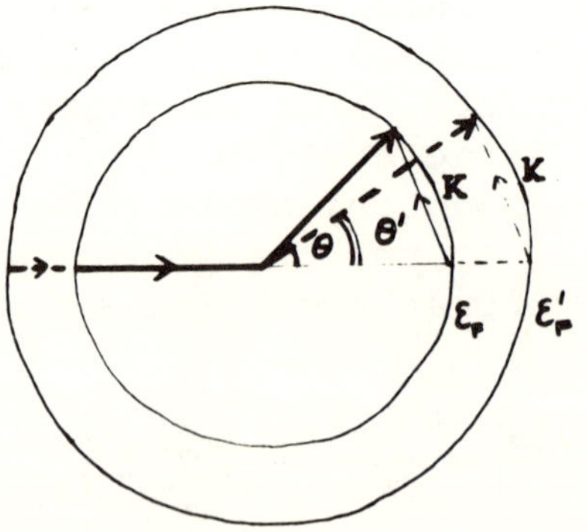

Fig. 5.

To treat the whole of the lattice spectrum we need to know a little more than this; volume changes, as in a longitudinal wave, will change the average Fermi energy of the conduction electrons. But it has been verified by Cochran (6) that the spectrum of the transverse modes in Na can be derived quite accurately from a formula such as (5): in principle such an analysis gives a way of estimating the form of U(K) directly. Perhaps the most striking verification of the general idea that interatomic forces are related to the pseudo-potentials of the ions is the observation, in both liquid and solid alkali metals (7) (8) of long range oscillations in these forces, exactly as predicted by the theory.

Once we have satisfied ourselves that the lattice spectrum is calculable from U(K), then we can estimate how it too varies with volume - so that we can put the correct variation of Θ, and of the factor $[K^2/q^2]$, into our formula for the resistivity. Other quantities in this formula will also depend on the volume - for example k_F, V_F, q_D, N, etc. - but these can be allowed for by elementary arguments. The heart of the question must be the variation of U(K) itself.

One thing is easily proved. Since

$$U(K) \rightarrow -2/3\ E_F \qquad (6)$$

as $K \rightarrow 0$ the contribution to the resistivity from small values of K (i.e., small-angle scattering) is readily estimated. The variation of this contribution with volume is also easy to compute; this is just the sort of calculation that was attempted in Electrons and Phonons (§ 9·15), and compared carefully with observation by Paul (9). In fact, it throws all the variation of resistivity on to the variation of Θ, so that

$$\frac{\partial \ln \rho}{\partial \ln V} = -2\frac{\partial \ln \Theta}{\partial \ln V} = 2\gamma \qquad (7)$$

where γ is the Grüneisen parameter.

We now see that this is quite unsound. A great part of the resistance of most metals comes from the large angle scattering, which is very heavily weighted in the integral, both by the factor $(1 - \cos\theta)$ and by the geometrical factor $[K^2/q^2]$, which may become quite large for U-processes. We must look carefully at the behavior of U(K) near the upper limit.

There are several effects to consider here. Compression of the solid increases the density of the electron gas, which modifies the screening parameter, so that the screened potential itself is altered. This we have seen above in (6); the value of U(K) at K = 0 is altered automatically to be equal to the new value of $-2/3\ E_F$. But for large values of K this effect is probably not very important. At short wavelengths the screening by the conduction electrons is not so great. The pseudopotential is then determined largely by the properties of the atomic potential within the core, and is not altered much by changing the density of the electron gas in which the ion is immersed.

But decreasing the volume increases the radius of the Fermi sphere, so that we take a new range of the variable K into the integral. If U(K) is varying rapidly with K near $K = 2k_F$, then we may see a significant contribution to $\partial\rho/\partial V$.

We must also remember that changing the Fermi level can change the scattering power of the ions in a more direct way. It is essential to keep in mind that the pseudo-potential is not a real localized potential; it is a function that simulates the scattering amplitude of the ion <u>for electrons of a given energy</u>. If we change that energy, by moving the Fermi level, we alter the function U(K).

This point seems to worry some people. Let me put it another way. Suppose we represent the pseudo potential by the formula for the scattering amplitude in terms of phase shifts:

$$U(K) = -\frac{2\pi\hbar^2 N}{mk_F}\sum_{\ell=0}^{\infty}(2\ell + 1)\sin\eta_\ell e^{i\eta_\ell}P_\ell(\cos\theta). \qquad (8)$$

We know that the phase shifts, η_ℓ, are functions of the energy, and in general the right hand side of this expression is not just a function of the magnitude of the scattering vector, $K = 2k_F \sin \frac{1}{2} \theta$. Suppose, for example, that the Fermi level were not far from a resonance energy of the ion. Then one of the phase shifts would be changing rapidly with energy and the whole function U(K) would be very sensitive to the exact value of E_F. This, in its turn, would contribute heavily to the change of resistivity with volume.

I hope that it will now be clear that an elementary semi-qualitative, semi-quantitative, "interpretation" of the effect of pressure on the electrical resistivity of a given metal is not to be achieved easily. It really has to be worked out. But such a computation looks quite straightforward. One might start, for example, with the "model potential" of Heine and Abarenkov (1964), which has already been very successful as the basis of calculations of the electrical properties of Al and Pb (11), and of the liquid alkali metals (12). One might then recalculate the potential for a different Fermi level, and hence compute the shape of the Fermi surface and the lattice spectrum in the compressed metal. It would be wise to check these computations by other experiments - for example, by direct observations of the velocity of sound under pressure. One might then be able to put all the numbers into an integral such as our formula (1) and calculate the change in the resistivity.

Such calculations do indeed seem quite feasible - and I hope very much that we may soon be able to attempt them, for they would add a further dimension to our understanding of the properties of the simple metals and would, I trust, confirm the new methods which we are now using in the analysis. The electron-phonon interaction is a surprisingly complicated phenomenon that cannot be summed up in a few parameters; the mathematical connections can only be unravelled with the aid of a variety of experimental observations. In a cubic metal we do not really have very many essentially different "bits" of evidence to help us, so that the dependence of the resistivity may well turn out to be an important clue.

All this is perhaps still very cautious and limited. What about the Hall constant and thermoelectric power? What about non-cubic metals? What happens in transition metals? What about alloys? The answer in all cases is "we do not know".

In fact we have almost no direct, reliable computations of these properties in simple metals at ordinary temperatures. For example, why is the thermoelectric power of Li positive? There have been various conjectures on this subject - but no convincing calculation. Why is the Hall constant of Pb small and positive? Again, we might suggest a number of contributory mechanisms - hole-like regions of the Fermi surface, anisotropy of relaxation time, etc. - but we have no quantitative explanation. It seems to me to be pointless to speculate qualitatively about why or how this or that changes with volume until we understand better why and how it is as it is.

Perhaps we should put it this way. The observed variations in the electrical transport properties of metals with pressure cannot yet be explained, except in vague principle, by our present theories. But as a student of the theory of metals one warmly welcomes these experiments, especially where they are reliable, systematic and comparative, for the clues that they afford, and the circumstantial evidence that they provide, in the trial of new models and new theories. Long may the experimenters exploit their advantages, and keep flitting ahead of us as we plod along.

References

1. J. M. Ziman, Adv. Phys. 13, 89 (1964)

2. J. M. Ziman, Principles of the Theory of Solids, (Cambridge University Press, 1964)

3. L. J. Sham and J. M. Ziman, Solid State Physics 15, 223, (1963).

4. J. M. Ziman, Electrons and Phonons, (Oxford: Clarendon Press, 1960).

5. N. W. Ashcroft, Phil. Mag. 8, 2055 (1963)

6. W. Cochran, Proc. Roy. Soc. A276, 308 (1963)

7. M. D. Johnson, P. Hutchinson and N. H. March, Proc. Roy. Soc. A. 282, 283, (1964).

8. S. H. Koenig, Phys. Rev. 135, A 1693, (1964).

9. W. Paul, in High Pressure Physics and Chemistry (ed. R. S. Bradley)(Academic Press Inc., N.Y., 1963), pp. 299 - 354.

10. V. Heine and I. Abarenkov, Phil. Mag. 9, 451 (1964).

11. N. W. Ashcroft, to be published.

12. Lorna J. Sundström, Phil. Mag. (in press).

THE USE OF PRESSURE IN THE STUDY OF IMPURITY STATES IN SEMICONDUCTORS*

H. FRITZSCHE
Department of Physics and
Institute for the Study of Metals
The University of Chicago, Chicago, Illinois 60637

Since most electrical and optical properties of solids depend on the details of the electronic band structure, the change of the band parameters under hydrostatic pressure and shear stresses leads to interesting effects which can give valuable information about the band structure and also about the electronic process studied. The covalently bonded semiconductors offer the additional advantage that the relevant deformation potentials are relatively large and that strains as large as 1 percent can be applied at room temperature and below without plastic deformations.

The presence of large piezoresistance coefficients (1) for certain orientations of the applied shear indicates whether the semiconductor is of the multi-valley type like Si and Ge, and it determines the orientations of the conduction band valleys. Further details (2) of the structure of shallow donor states are observed when the electrical measurements are carried to shear stresses high enough so that the valleys moved up by the shear have transferred essentially all electrons to the lower valleys.

Germanium, for instance, which has the lowest conduction band minima at the point L in the Brillouin zone (at the BZ boundary in the [111] directions), can be changed from a 4-valley to a 2-valley semiconductor by [110] uniaxial stress and into a 1-valley semiconductor by [111] compression. This is shown in Figs. 2 and 4 below.

Let us consider the different effects of uniaxial stress on the resistivity of germanium at

*Research sponsored by Air Force Office of Scientific Research through Grant No. AF-AFOSR 148-63.

various temperatures and donor concentrations.

1. Extrinsic Range

Because of the mobility anisotropy $K = \mu_\perp/\mu_\parallel$ of each valley (here $\mu_\perp$ and $\mu_\parallel$ are the mobilities perpendicular and parallel to the [111] valley symmetry axis) the resistivity in general changes when electrons are transferred from some valleys to others and the electron concentrations in the valleys are made unequal by the stress. [110] stress X (positive for compression) splits the valleys symmetrically by an amount 2ϵ where

$$\epsilon = E_2 S_{44} X/6. \quad (1)$$

$S_{44} = 1.47 \times 10^{-3}$ kbar^{-1} is the elastic shear constant and E_2 is the shear deformation potential. For current flow parallel to the [110] stress direction the piezoresistance due to the electron transfer effect is

$$\Delta\rho/\rho_X = [(K-1)/(2K+1)] \tanh(\epsilon/kT). \quad (2)$$

This dependence on stress is shown by the solid curve in Fig. 1 for T = 6.5°K. The discrepancy between this curve and the experimental data points for Sb- and As-doped germanium is the effect of interest here. It is caused by the following (2): Corrections to the effective mass treatment of shallow donors (3), due to the breakdown of the dielectric constant and effective mass concept and due to imperfect screening near and in the lattice cell of the impurity, are generally accounted for by adding to the Coulomb potential of the donor ions a short range potential which depends on the donor ion. This central cell potential splits the 1s hydrogenic ground state by an amount $4\Delta_c$ called valley-orbit splitting. [110] stress causes a shift of these donor levels which is different from the shifts of the valleys as shown on the right-hand side of Fig. 2. The ground state shifts with stress only in second order and comes closer to the lowest valleys by $2\Delta_c$ at large stresses. For $kT < 4\Delta_c$ this amounts to a decrease of the donor ionization energy and hence results in an increase in total carrier concentration for both compression and tension. When this effect is taken into account Eq. (2) is replaced by

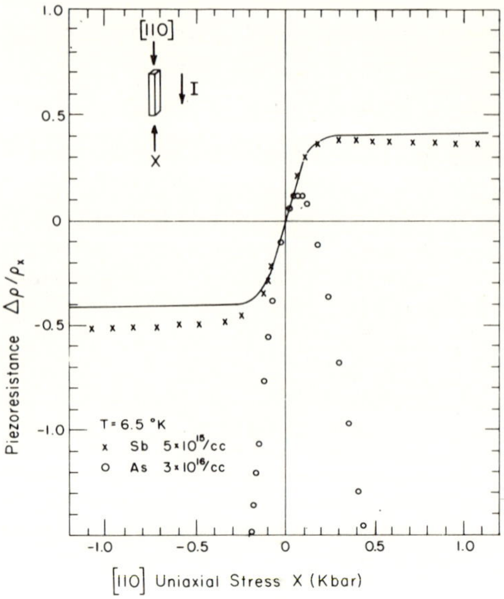

Fig. 1. Longitudinal piezoresistance of Sb- and As-doped germanium in the extrinsic temperature range.

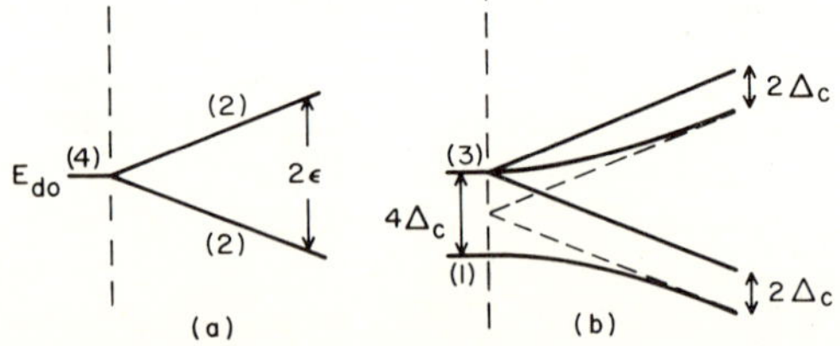

Fig. 2. Splitting of the conduction band states and of the shallow donor states in germanium due to [110] uniaxial stress (a) in the effective mass approximation and (b) with a valley-orbit splitting $4\Delta_c$.

$$\Delta\rho/\rho_X = 1 - \frac{n}{n_o}\left[1 - \frac{K-1}{2K+1}\tanh\left(\frac{\epsilon}{kT}\right)\right] \qquad (3)$$

where $$n/n_o = \exp(E_F/kT)\cosh(\epsilon/kT) \qquad (4)$$

is the ratio of the total electron concentration in the presence of stress to that without stress. By fitting Eq. (3) to the experimental curves at various temperatures and calculating E_F as a function of stress from the donor level splitting of Fig. (2), one obtains (2) the shear deformation potential $E_2 = 19.2 \pm 0.4$ eV/strain, $4\Delta_c(\text{As}) = 4.1 \pm 0.15$ meV, $4\Delta_c(\text{Sb}) = 0.57 \pm 0.03$ meV, and a mobility anisotropy which for this impurity concentration and temperature is K = 7.8. Measurements of the stress dependence of the electron spin resonance hyperfine splitting (4), of the Lyman absorption spectrum of the donors (5), and of the low temperature thermal conductivity (6) show interesting effects of the valley-orbit splitting and have verified their magnitudes given above. The corrections to the effective mass approximation are negligible for Sb donors for which $4\Delta_c$ is only 6% of E_{do}, the ionization energy calculated in that approximation. For As donors $4\Delta_c$ is nearly 50% of E_{do} and the central cell potential is large. This causes some striking differences between Sb and As donors which will be discussed below.

Interesting transport measurements have also been carried out on p-type germanium (7) which has two valence bands degenerate at the zone center. One can lift this degeneracy sufficiently by means of uniaxial stress to obtain a simple ellipsoidal surface of constant energy ("split p-germanium").

2. Impurity Conduction

As the temperature is lowered below 6°K the carrier concentration in the conduction band becomes negligible and conduction due to a charge exchange between neutral and ionized donors predominates. This so-called impurity conduction (8) depends sensitively on the wave function overlap of neighboring donor sites. A change of the donor wave functions due to stress, therefore, gives rise to a large change in resistivity (9). This is shown in Fig. 3 for current flow parallel to the [111] compressional stress at 1.2°K. Sb-doped Ge shows a dramatic

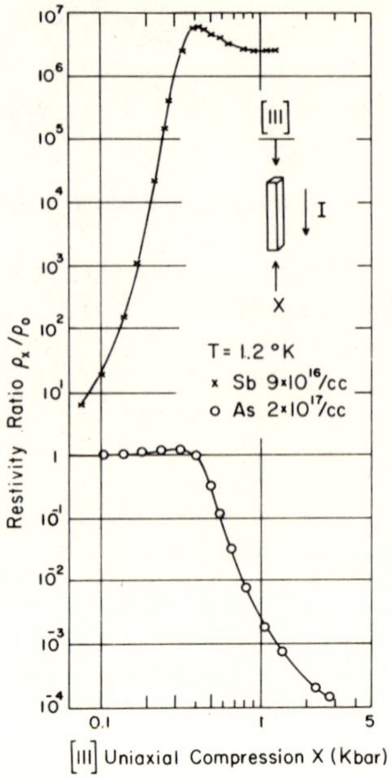

Fig. 3. Change of resistivity with [111] compressional stress in the range of impurity conduction for Sb- and As-doped germanium.

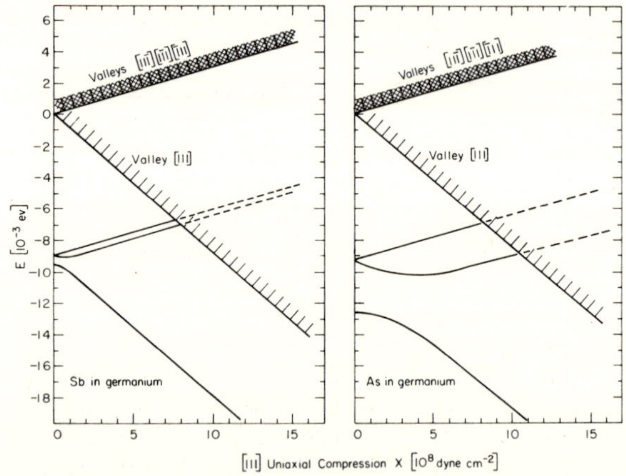

Fig. 4. Splitting of the conduction band states and of the 1s-like donor states with [111] compression for Sb and As donors in germanium.

increase of resistivity with stress whereas for As-doped Ge the resistivity decreases by several orders of magnitude. The Sb donors behave as expected from the effective mass treatment of the donor states. At zero stress the ground state wave function is a linear combination of the bound states formed by the four individual conduction band valleys. With [111] compressional stress the contributions of the valleys which are raised to the donor ground state decrease. At large stresses the gound state wave function is formed solely by the lower conduction band valley. It is oriented in such a way that the overlap between neighboring donors is greatly reduced in the [111] direction. This results in the large increase in resistivity which is expected to saturate when the demixing is complete.

Why is a decrease in resistivity observed in the case of As donors? Here the same demixing of the individual valley contributions to the donor ground state takes place as for Sb donors. In addition, however, the size of the donor wave function or the effective Bohr radius increases considerably with stress because the donor ionization energy decreases as a result of the large valley-orbit splitting of the As donors. This effect can be seen in Fig. 4 where the splitting of the valleys and of the donor states is shown for [111] compression for Sb and for As donors. The valley-orbit splitting energies can again be obtained from these experiments. The large resistivity changes observed demonstrate the sensitivity of impurity conduction to small stress-induced changes of the impurity wave functions. This can be used as a tool for determining whether impurity conduction prevails in certain temperature ranges in crystals for which the type of conduction is uncertain.

The stress-induced change in overlap and hence in the interaction between neighboring impurity states has been observed to result also in a change of the motional narrowing of the spin resonance lines (4).

The effect of the stress-induced splitting of the valence bands on impurity conduction and on the acceptor wave functions in p-germanium has been studied by F. Pollak (10).

3. Degenerate Region

As the impurity concentration is increased the

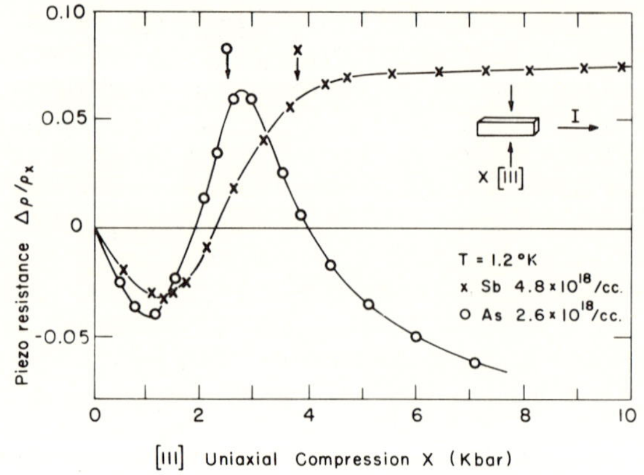

Fig. 5. Transverse piezoresistance of degenerate Sb- and As-doped germanium.

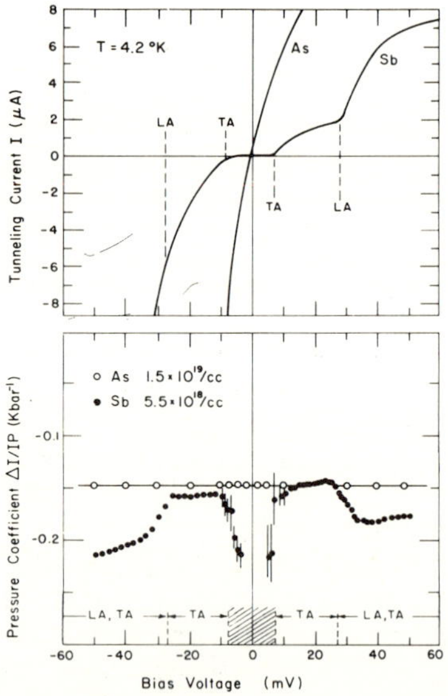

Fig. 6. Tunnel current and tunnel pressure coefficient in narrow p-n junctions for phonon-assisted tunneling (Sb-doping) and impurity-assisted tunneling (As-doping).

impurity ionization energy reduces to zero because of screening and the Fermi level penetrates the conduction band. In this high concentration or degenerate region quite different stress effects are observed. We quote here only one example from a more complete study (11). Fig. 5 shows the piezoresistance at 1.2°K of Ge highly doped with Sb and As for compressional stress along [111] and current flowing perpendicular to this direction. For this orientation the stress causes transfer of electrons from three valeys of lower mobility to one valley of higher mobility. Hence one expects a negative piezoresistance with a monotonic increase in magnitude until saturation is reached at a stress beyond which the upper valleys are empty. This saturation stress is indicated by a vertical arrow for each sample in Fig. 5. In contrast to this a positive saturation value is reached by the Sb-doped sample. This additional positive piezoresistance is due to the fact that the density of states at the Fermi level is less for the one-valley case (large stress) than for the four-valley case (zero stress) and hence the screening decreases as the electrons collect in the lowest valley.

As-doped germanium again behaves differently. At small [111] compressions the piezoresistance is negative because of electron transfer and the mobility anisotropy of the valleys. At higher stresses the piezoresistance becomes positive because of the decrease in screening. Beyond the expected saturation stress a further decrease to negative values is observed. This last behavior is not yet explained. It might be due to a transfer of electrons from low mobility tail states which are associated with the upper valleys and which are raised by the stress above the Fermi level. Such tail states extending from the conduction band into the gap are expected to be more pronounced in the case of As donors because of this large central cell potential.

4. Interband Tunneling

Interband tunneling (12) is observed in narrow p-n junctions (Esaki diodes) formed with degenerate p- and n-material. Here again the difference between Sb and As donors gives rise to interesting effects. In the case of Ge the initial and final tunneling states are localized near the center of the

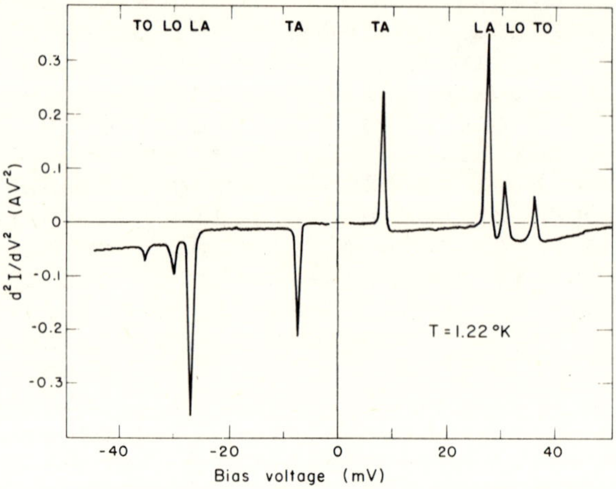

Fig. 7. Second derivative of tunneling current with respect to bias voltage of phonon-assisted tunneling in germanium.

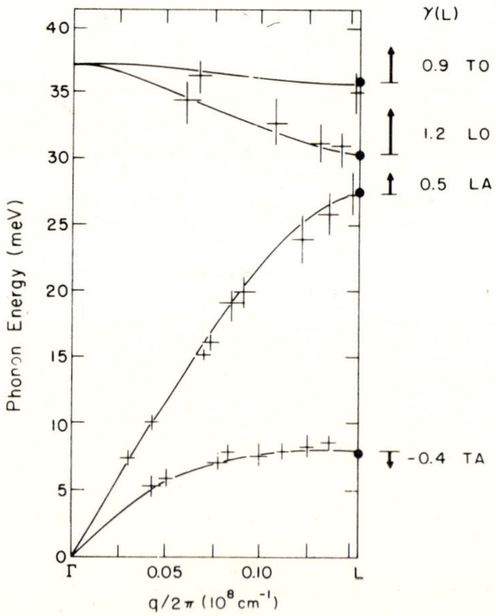

Fig. 8. Phonon spectrum of germanium along the [111] direction of the Brillouin zone. Crosses represent data of Brockhouse and Iyengar; full dots are data by Payne who determined the Grüneisen constants γ (L) listed on the right.

zone (valence band) and near the points L (the valleys of the conduction band), respectively. In the ideal case momentum can be conserved in the tunneling process only when an appropriate phonon is simultaneously being emitted or absorbed (13). The I-V curves of Sb-doped junctions show at low temperatures, where phonon absorption is excluded, the characteristic steps of phonon-assisted tunneling. See Fig. 6. Current can only flow when the bias raises the initial state above the final electron state by at least the energy of the phonon to be emitted. The I-V curves of As-doped junctions do not show this structure. It is believed that the strong localized central cell potential of the As ions introduces components into the electron wave functions from large parts of the Brillouin zone including all of the valleys and the region around the zone center. Thus no phonons are needed to conserve momentum.

By changing the relevant band gaps and effective masses, stress strongly affects the tunneling current and further details of the tunneling process are revealed (14). The lower half of Fig. 6 shows the relative change of the phonon-assisted tunneling current (Sb donors) with hydrostatic pressure. The observed structure in the bias dependence coincides with the phonon structure in the I-V curve. The analysis of the Sb data shows that different pressure coefficients have to be attributed to the tunneling currents assisted by different phonons. This means that the present tunneling theories have to be modified to take proper account of the characteristic differences of the phonons such as the selection rules for the phonons which can connect different Bloch states in the Brillouin zone (15).

The pressure coefficient of the impurity-induced tunneling in As-doped diodes on the other hand (shown in the lower half of Fig. 6) does not show any structure. This is consistent with the absence of phonon assistance.

The energies of the phonons which have a wave number equal to the wave number difference of the initial and final tunneling states can be determined very accurately by measuring d^2I/dV^2 versus bias voltage V (16). Such a plot is shown in Fig. 7 for Sb-doped diodes at 1.22°K. The half-width of these second derivative peaks is given by the phonon dispersion and kT broadening (17). By measuring the shift of these peaks under hydrostatic pressure Payne (17) determined the Grüneisen constants of the various phonon modes at the point L in the zone

$$\gamma(L) = K^{-1} d\ln E(L)/dP,$$

where K is the volume compressibility and E the energy of the transverse and longitudinal acoustic and optic phonons, TA, LA, LO, and TO, respectively. The results are shown in Fig. 8 together with the neutron scattering data for the phonon spectrum of Brockhouse and Iyengar (18). The negative sign of γ (L) of the TA phonon is the first verification of Daniels' explanation (19) of the anomalous negative volume coefficient of thermal expansion in diamond-like semiconductors at low temperatures (20). Because of the difficulty of applying large hydrostatic pressures at helium temperatures, R. T. Payne applied uniaxial compression of the order of 10 kbar to the tunnel diodes. By changing the orientation of the stress both the shear and the hydrostatic pressure deformation potentials of the phonon spectrum at the point L could be obtained (17).

The experiments discussed here are only a few examples of the use of pressure and uniaxial stress in semiconductor research. The most important contributions to this field have been reviewed recently by Paul and Warschauer (21). Some very elegant work on the effect of uniaxial stress on the excitation spectrum of impurities in germanium and silicon should also be pointed out (22).

References

1. R. W. Keyes, in Solid State Physics, edited by F. Seitz and D. Turnbull (Academic Press, Inc., New York, 1960), Vol. II, p. 149.
2. H. Fritzsche, Phys. Rev. 115, 336 (1959); 120, 1120 (1960).
3. P. J. Price, Phys. Rev. 104, 1223 (1956); W. Kohn, in Solid State Physics, edited by F. Seitz and D. Turnbull (Academic Press, Inc., New York, 1957), Vol. 5, p. 257.

4. D. K. Wilson, Phys. Rev. 134, A265 (1964).
5. G. Weinreich and H. G. White, Bull. Am. Phys. Soc. 5, 60 (1960).
6. R. W. Keyes and R. J. Sladek, Phys. Rev. 125, 478 (1962).
7. S. H. Koenig and J. J. Hall, Phys. Rev. Letters 5, 550 (1960).
8. For a review on this subject see: N. F. Mott and W. D. Twose, Advances in Physics, edited by N. F. Mott (Taylor and Francis, Ltd., London, 1961), Vol. 10, p. 107.
9. H. Fritzsche, Phys. Rev. 125, 1552 (1962); 125, 1560 (1962).
10. F. Pollak, Phys. Rev. (to be published).
11. M. Cuevas and H. Fritzsche, Phys. Rev. 137, A1847 (1965); and to be published.
12. L. Esaki, Phys. Rev. 109, 603 (1958).
13. H. Holonyak, Jr., I. A. Lesk, R. N. Hall, J. J. Tiemann, and H. Ehrenreich, Phys. Rev. Letters 4, 456 (1960).
14. H. Fritzsche and J. J. Tiemann, Phys. Rev. 130, 617 (1963); Proceedings of the International Conference on the Physics of Semiconductors, Paris, 1964, p. 599.
15. J. J. Tiemann and H. Fritzsche, Phys. Rev. 137, A1910 (1965).
16. A. G. Chynoweth, R. A. Logan, and D. E. Thomas, Phys. Rev. 125, 877 (1962).
17. R. T. Payne, Phys. Rev. Letters 13, 53 (1964); Phys. Rev. (to be published).
18. B. N. Brockhouse and P. K. Iyengar, Phys. Rev. 111, 747 (1958).
19. W. B. Daniels, Proceedings of the International Conference on Semiconductor Physics (Institute of Physics, London, 1962), p. 482.
20. D. F. Gibbons, Phys. Rev. 112, 136 (1958); R. D. McCammon and G. K. White, Phys. Rev. Letters 10, 234 (1963).
21. W. Paul and D. M. Warschauer in Solids Under Pressure, Ed. W. Paul (McGraw-Hill, 1963), p. 179.
22. R. L. Aggarwal, P. Fisher, V. Mourzine, and A. K. Ramdas, Phys. Rev. 138, A882 (1965).

CONTACT POTENTIAL OF A SHOCKED METAL*

George E. Duvall
Washington State University
Department of Physics
Pullman, Washington

Robb M. Thomson
University of Illinois
Department of Metallurgy
Urbana, Illinois

Abstract

The motion of electrons in a mechanically shocked solid is examined theoretically, using as a basis the assumptions that the electrons have a fermi distribution of velocities, that they accelerate classically under the influence of internal electric fields, and that their interactions with the solid lattice are adequately described by a relaxation process. In an acoustic approximation the potential difference between two pressure-free surfaces of a shocked slab is found to be zero.

I. Introduction

A shock front in a metal is a fast-moving boundary between shocked material, which has been accelerated, compressed, heated and deformed, and relatively undisturbed material ahead. The thickness of the shock front is not well known. For weak shocks it has been found to be the order of a millimeter in some cases. For strong shocks it could conceivably approach atomic dimensions.

---

*This work was supported primarily by the AFOSR. Partial support was provided by Washington State University and by Stanford Research Institute. Work in progress at SRI provided motivation for the problem.

Shock velocities in metals are characteristically somewhat greater than acoustic velocities.

The rise in temperature resulting from shock heating is proportional to the cube of the compression for weak shocks. The magnitude of the temperature rise for 10% compression is shown for several metals in Table 1, along with other shock parameters. Maximum densities achievable in shock compression are limited by heating to about twice normal density.

Casual consideration leads to the tentative conclusion that significant electrical effects should accompany the passage of a shock wave through a metal. The compression should lead to a change in the fermi energy of the valence electrons and in the energy of the bottom of the valence band. Imperfect coupling between the valence electrons and the lattice phonons should lead to some charge imbalance and differences in temperature between shocked and unshocked material should do likewise.

More careful consideration shows that the problem is indeed a complicated one. We present here a first attempt at both qualitative description and quantitative estimates of the electrical effects which may be produced under such conditions. In order to simplify the calculations, the initial state is assumed to be at $0^{o}K$ and shock heating is ignored.

II. Effects of Compression on Band Structure and Fermi Energy

In a real experimental situation a metal undergoing shock consists of an undisturbed region in which the relation between neighboring atoms is that of the normal crystal, the shock transition zone in which pressure rises to its peak value, and a region of pressure decline in which pressure is reduced to ambient value through a rarefaction following the shock. Thus the atomic separation ahead of the shock is constant and normal, in the transition zone it is decreasing rapidly, and behind the shock front it is increasing more slowly. The gradient of pressure in the rarefaction may be kept as low as 10 kb/cm in a carefully designed experiment; for the purposes of preliminary discussion we take it to be zero and consider two uniform states joined by the shock zone.

The energy of compression goes primarily to change the electron levels in the crystal; in

Table 1. Shock Parameters for 10% Compression (1)

Metal	Shock pressure, Kilobars	Increase of internal energy, ev/atom	Temperature increase, ^{o}C
Na	10.4	.013	43
Al	105	.054	75
Cu	195	.072	100

Table 2. Energy Changes Resulting From 10% Volume Compression

Metal	ϵ_{f_o}, ev	$\Delta\epsilon_f$, ev	ΔE_o,* ev	ΔE_f,* ev
Na	2.5	0.167	- 0.10	0.067
Al	11.8	0.79	- 0.47	0.314
Cu	7.0	0.467	- 0.48	-0.013

*Assuming that total energy remains constant.

discussing its effects, two cases are to be distinguished. In the first only the valence electrons are affected by the change in the lattice spacing; lower lying levels are so deep that they do not participate in the resistance of the metal to compression. In the other case this is not true and the ion cores interact repulsively. Consider the latter case and let the total energy of the crystal be

$$E = E_o + \bar{\mathcal{E}}_f + E_c, \tag{1}$$

where E_o is the energy level at the bottom of the valence band, $\bar{\mathcal{E}}_f$ is the mean kinetic energy of the valence electrons, and E_c is the interaction energy of the ion cores. Then for small changes from the equilibrium lattice spacing, ΔE vanishes to first order and

$$\Delta E_o = - \Delta\bar{\mathcal{E}}_f - \Delta E_c \tag{2}$$

For a free electron gas, $\bar{\mathcal{E}}_f$ varies as the two-thirds power of the number density of electrons, N. N is, in turn, proportional to the mass density ρ if the number of electrons per atom is constant and charge equilibrium prevails. Then

$$\Delta\bar{\mathcal{E}}_f = 2\bar{\mathcal{E}}_f \Delta\rho/3\rho = 2\mathcal{E}_f \Delta\rho/5\rho \tag{3}$$

where $\mathcal{E}_f$ is the fermi energy. Then

$$\Delta E_o = -(2\mathcal{E}_f \Delta\rho/5\rho) - \Delta E_c \tag{4}$$

The total energy of the most energetic electrons is $E_f = E_o + \mathcal{E}_f$ and

$$\Delta E_f = (2\Delta\mathcal{E}_f/5) - \Delta E_c \tag{5}$$

For Na, Mg, and Al it is not unreasonable to assume that $\Delta E_c = 0$; then the relations between energy levels before and after compression are as shown in Fig. 1. The total energy of the valence electrons is unchanged to 1st order in $\Delta\rho$, as stated earlier.

For copper the interaction of ion cores may be important, and it is suggested by Callaway (2) that one of the principal contributing interactions may be the van der Waals energy, amounting to perhaps one electron volt per atom. In this case, since the van der Waals energy is proportional to the inverse square of the atomic volume, we have

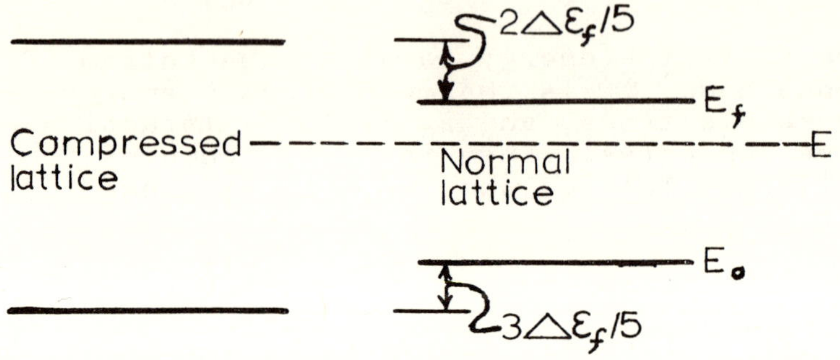

Fig. 1. Effect of Compression on Energy Levels in Alkali Metals

$$\Delta E_c \simeq 2E_c \, \Delta\rho/\rho = 2 \, \Delta\rho/\rho \text{ , ev.} \tag{6}$$

Numerical values of ΔE_f and ΔE_o for 10% compression are shown in Table 2.

III. Static Pressure Couple

Before considering the real physical situation of a moving shock front, we examine a hypothetical case in which a compressed sample is brought into contact with a normal sample of the same metal and contact potential differences are established, as for two dissimilar metals. Just as in that case, we expect the fermi levels to adjust themselves by charge transfer until they have the same absolute energy. Then

$$\Delta E_o = \Delta \mathcal{E}_f = 2 \mathcal{E}_f \, \Delta\rho / 3\rho$$

or

$$dE_o/dx = (2\mathcal{E}_f/3\rho) \, d\rho/dx \tag{7}$$

in the transition region. From Table 2 we see that for Na, and Al, the uncompressed material has the lower fermi energy so it will acquire an excess negative electric charge. Because of the ion core interaction, Cu shows a very small effect of opposite sign.

Now the boundary is, in fact, moving at a speed of approximately 5×10^5 cm/sec. Since this velocity is small compared to that of fermi electrons, it is reasonable to discuss the electron distribution for a static couple and to see what happens to it in a finite volume as the boundary is quasi-statically moved from one side to the other. The electrostatic potential jumps to a negative value as the shock front enters the metal and remains at that value until the entire sample is compressed. The contact potential arising from such a couple is, however, unobservable. The pressure must return to zero at any surface of observation; during de-compression the electric potential returns to its initial value; and the net external potential is nil.

Then in order to relate any theory of electron motion in a shocked metal to experimentally observable quantities, the entire sequence of shock compression followed by rarefaction must be considered. Two modes of observation are possible: short circuit and open circuit. In the former case a net current might be expected; in the latter, a difference of potential. We consider the open circuit case. In the static approximation the fermi level remains constant, as shown in Fig. 2b, whatever the density and the variation of E_0. Consequently there is no difference between the potential measured on the r.h.s. of Fig. 2 and that on the left. Here the shock and rarefactions have been separated by a uniform state and the compression and rarefaction zones are assumed to have linear variations of density with distance. These simplifications will help to clarify the following discussion.

IV. Dynamic Effects

Motion of the shock wave through the lattice provides an opportunity for the energy levels depicted in Fig. 2 to deviate from equilibrium. Gradients existing in the rarefaction are subject to experimental control and can be made very small. Gradients in the shock front are characteristic of the compression ratio and the material and are very large.

Under these conditions it is plausible to suppose that the potential in the uniform region and rarefaction following the shock will have its equilibrium value and that any departure from equilibrium will occur in the shock front. This will produce a change in the fermi level, ΔE_f, which will appear externally as a contact potential of the same magnitude. We proceed in the remainder of this paper to estimate the magnitude of this change across the shock front.

Our discussion is restricted to the case where the mean free path for electron-phonon collisions is small compared with the shock transition zone length. The model of A. B. Pippard (3) is used in which electrons are assumed to be free and independent, with degenerate fermi distribution. Only electrons at the fermi surface are considered, and they are divided into groups traveling in direction θ with respect to the direction of shock propagation; their interactions with the lattice are represented by relaxation terms in their equations of motion, and they are accelerated by electric fields resulting from compression and deviations from charge neutrality.

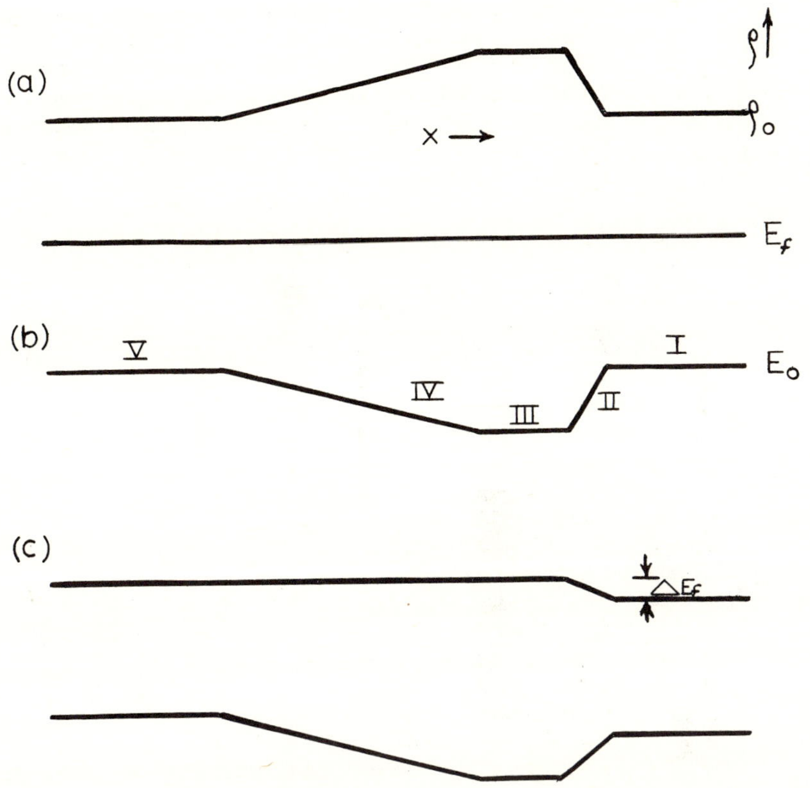

Fig. 2. Schematic Representations of Density and Energy Distributions in a Shock Wave:

(a) Density variation

(b) Bottom of Valence Band for Equilibrium

(c) Hypothetical situation in which shock compression provides isolation between normal and compressed lattice but equilibrium exists through the rarefaction.

The equations to be satisfied are

i) the equations of motion for the electrons (Eq. (8))

ii) equations of continuity for both ions and electrons (Eqs. (9) and (10))

iii) Maxwell's equations for quasi-stationary fields (Eqs. (11) and (12)).

The shock wave is assumed to be a plane wave traveling with speed a in the positive z-direction. Compressions are assumed to be sufficiently small that acoustic approximations are valid, and mass density, ρ, is taken as a known function. The pertinent equations become:

$$d\vec{v}/dt = e\vec{F}/m - (\vec{v} - \vec{v}_{eq})/\tau \qquad (8)$$

$$\partial\rho/\partial t + \partial \rho u/\partial z = 0 \qquad (9)$$

$$\partial Ne/\partial t + \partial J_e/\partial z = 0 \qquad (10)$$

$$\partial F/\partial t = - J/\mathcal{E} \qquad (11)$$

$$\partial F/\partial z = q/\mathcal{E} \qquad (12)$$

Where
- m = electron mass
- $\vec{v}$ = local velocity of an electron at the fermi surface
- $\vec{v}_{eq}$ = velocity of an electron at the top of the fermi surface corresponding to local mass density and velocity
- e = electronic charge = -1.6×10^{-19} coulombs
- $\vec{F}$ = $\vec{a}_z\, F$ = electric field
- τ = relaxation time for electron-phonon collisions
- u = mass velocity
- N = local electron density
- J_e = electrical current corresponding to electron flow
- J = total electrical current
- q = net charge per unit volume

Since ρ, u, $\vec{F}$, J, J_e, and q are functions of z - at alone, the transformation x = z - at leads to the following forms for Eqs. (8) - (12):

$$d\vec{v}/dt = (d\vec{v}/dx)(dx/dt) = (v\mu - a)\, d\vec{v}/dx$$

$$= e\vec{F}/m - (\vec{v} - \vec{v}_{eq})/\tau \qquad (13)$$

$$d\rho(u - a)/dx = 0, \quad (u - a)\rho = -\rho_0 a \qquad (14)$$

$$d(J_e - Nea)/dx = 0, \quad J_e - Nea = -\rho_o ae/M \quad (15)$$

$$J = (N - \rho/M)ea = J_i + J_e \quad (16)$$

$$dF/dx = (N - \rho/M)\, e/\epsilon = J/\epsilon a \quad (17)$$

Here, M is the mass of the metal ion, J_i is the ion current, subscript "o" refers to the state ahead of the shock, and $\theta = \arccos\mu$ is the angle between the velocity v and the direction of propagation. For simplicity it has been assumed that each atom is singly ionized.

These equations are to be supplemented by the expression for J_e in terms of $\vec{v}$, (Pippard, Eq. (5)):

$$2J_e = 3N_o e \int_{-1}^{1} \mu(v - v_o)\, d\mu \quad (18)$$

where $N_o = \rho_o/M$ and v_o is the fermi speed in the undisturbed material. The relations between v_o, v_{eq} and v are illustrated in Fig. 3. The scalar product of Eq. (13) with unit vector in the direction $\vec{v}$ yields

$$(\mu v - a)dv/dx = \mu e F/m - (v - v_{eq})/\tau \quad (19)$$

Setting $v = v_o$ in the first bracket, this becomes

$$(\mu - \alpha)dv/dx + v/\lambda = \mu e F/mv_o + v_{eq}/\lambda \quad (20)$$

where $\alpha = a/v$ and $\lambda = v_o\tau$. Changes in mass density and acceleration of the lattice cause v_{eq} to be triaxial,

$$v_{eq} = v_o + \mu u + v_o\, \delta N/3N_o \quad (21)$$

From Eq. (14), $u = a\delta\rho/\rho \cong a\, \delta\rho/\rho_o$. Consequently Eq. (20) becomes

$$(\mu - \alpha)dv/dx + v/\lambda = \mu(eF/mv_o + a\, \delta\rho/\lambda\rho_o)$$

$$+ v_o\, \delta N/3\lambda N_o + v_o/\lambda \quad (22)$$

Solutions to this equation are sought which lead to an expression for the unknown field F(x). Let

$$v = \sum_{n=o}^{\infty} B_n(x)\, P_n(\mu)$$

$$dv/dx = \sum_{n=o}^{\infty} B_n'(x)\, P_n(\mu) \quad (23)$$

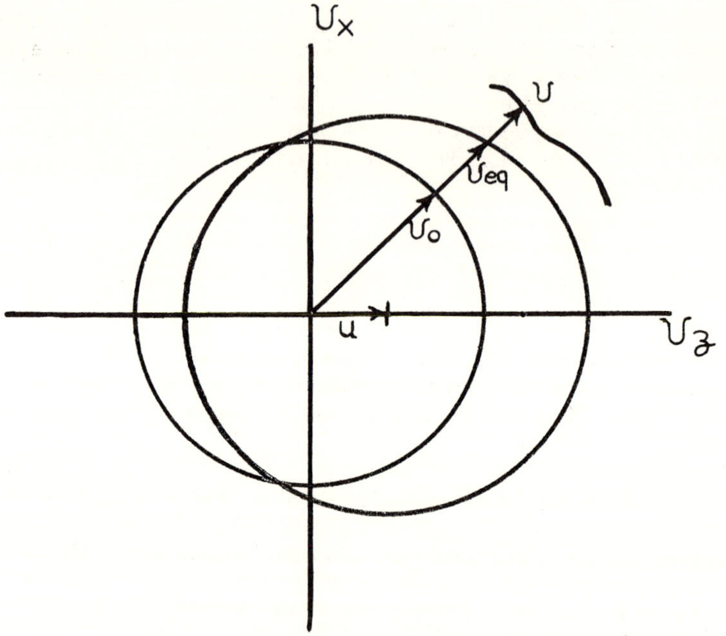

Fig. 3. Fermi Surfaces for Normal, Equilibrium Compressed, and Transient Distributions.

where $P_n(\mu)$ is the Legendre polynomial of degree n. Substitution of this series into Eq. (22) leads to a set of equations for the coefficients B_n:

$$B_1'/3 - \alpha B_0' + B_0/\lambda = (v_0/\lambda)(1 + \delta N/3N_0) \qquad (24)$$

$$B_0'/1 + 2B_2'/5 - \alpha B_1' + B_1/\lambda = eF/mv_0 + \alpha v_0 \, \delta\rho/\lambda\rho_0 \qquad (25)$$

$$2B_1'/3 + 3B_3'/7 - \alpha B_2' + B_2/\lambda = 0 \qquad (26)$$

etc.

The right hand side of all the higher order equations is zero. Since the series of Eq. (23) is expected to converge, a suitable procedure for solution of the set of equations (24) is to terminate the sequence at an arbitrary number, set the highest derivative equal to zero, and solve the remaining equations. If two of the B's are known, the remaining ones can be eliminated to yield an equation for F(x). B_1 can be immediately related to dF/dx through Eqs. (16), (17), and (18), as follows. From Eq. (23),

$$2B_0 = \int_{-1}^{1} v d\mu$$
$$2B_1 = 3\int_{-1}^{1} \mu v d\mu \qquad (27)$$

Thus Eq. (18) becomes $J_e = N_0 e B_1$; the positive ion current is $J_i = -\rho e u/M = -\alpha e v_0 \delta\rho/M$. Then $J = J_e + J_i = N_0 e B_1 - \alpha e v_0 \delta\rho/M$

$$= \alpha\epsilon v_0 \, dF/dx,$$

from Eq. (17). Or

$$B_1 = \alpha v_0 \, \delta\rho/\rho_0 + (\alpha\epsilon v_0/N_0 e) dF/dx \qquad (28)$$

by Eq. (17). To determine B_0 we calculate the volume enclosed by the local, non-equilibrium, fermi surface and equate it to $Nh^3/2$:

$$N = (2/h^3)\int_{-1}^{1} d\mu \int_0^{p(\mu)} 2\pi p^2 dp$$

$$= (4\pi/3h^3)\int_{-1}^{1} p^3(\mu) d\mu$$

Here p is the electron momentum and $p(\mu)$ is the momentum of an electron at the fermi surface traveling in the direction arcosμ with respect to the z-axis. Let $p(\mu) = p_o + \delta p$, where $p_o = mv_o$. Then

$$p^3 = p_o^3 + 3p_o^2\,\delta p$$

and

$$N = (3N_o/2v_o)\int_{-1}^{1} v\,d\mu - 2N_o,$$

or

$$B_o/v_o = 1 + \delta N/3N_o \tag{29}$$

B_o can also be related to dF/dx, as follows. From Eqs. (16) and (17)

$$J/ea = N - N_o - (\rho/M - N_o)$$

$$= \delta N - \delta\rho/M$$

$$= (\epsilon/e)dF/dx$$

or

$$\delta N = \delta\rho/M + (\epsilon/e)dF/dx.$$

Then

$$B_o/v_o = 1 + \delta\rho/3\rho_o + (\epsilon/3N_oe)dF/dx \tag{30}$$

Comparison of Eqs. (28) and (30) shows that

$$B_1 = 3\alpha(B_o - v_o) \tag{31}$$

Substitution of Eqs. (28) and (30) into (24) yields an identity. Eq. (25) is the first one to yield an approximation to F. Except for B_o, the even coefficients are expected to be small; i.e. variations in v should be approximately odd in μ. An electron traveling in the positive z direction should experience approximately the negative of the acceleration experienced by one going in the negative z direction. Accordingly, setting $B_2' = 0$ in Eq. (25) should give a reasonable approximation to F. Neglecting B_1' with respect to B_o', Eq. (25) leads to the following differential equation for F:

$$(\epsilon v_o/3N_oe)\,d^2F/dx^2 + (\alpha\,\epsilon v_o/\lambda N_oe)\,dF/dx - eF/mv_o = -(v_o/3\rho_o)d\rho/dx. \tag{32}$$

It is convenient at this point to introduce a new function

$$Z = -eF + (mv_o^2/3\rho_o)\, d\rho/dx. \qquad (33)$$

Since $\frac{1}{2} mv_o^2 = \mathcal{E}_{fo}$, $Z = 0$ when F has the equilibrium value given by Eq. (7). Moreover, the displacement of the fermi level in any interval (x_1, x_2) is given by the integral

$$\Delta E_f = \int_{x_1}^{x_2} Z dx \qquad (34)$$

Using Eq. (33), Eq. (32) becomes

$$d^2Z/dx^2 + (3\alpha/\lambda)\, dZ/dx - (3N_oe^2/\epsilon mv_o^2)\, Z$$
$$= \left[d^3\rho/dx^3 + (3\alpha/\lambda)\, d^2\rho/dx^2\right] mv_o^2/3\rho_o \qquad (35)$$

Taking sodium as a metal to which these calculations may apply, we have approximately

$$\begin{aligned}
E_{fo} &= 2.5 \text{ electron volts} \\
v_o &= 10^6 \text{ meters/sec} \\
a &= 3 \times 10^3 \text{ meters/sec} \\
\alpha &= 3 \times 10^{-3} \\
N_o &= 2.5 \times 10^{28} \text{ electrons/meter}^3 \\
\lambda &= 10^{-7} \text{ meters} \\
\epsilon &= 8.85 \times 10^{-12} \text{ coul}^2\text{/joule meter} \\
m &= 9.1 \times 10^{-31} \text{ kg}
\end{aligned}$$

Then

$$3N_oe^2/\epsilon mv_o^2 \simeq \frac{10^{21}}{4} \text{ meter}^{-2}$$

$$(3\alpha/\lambda)^2 \simeq 10^{10} \text{ meter}^{-2}$$

Consequently any disturbances by large values of ρ'' and ρ''' will die out in a distance the order of 10^{-10} meters. Considering the density distribution in Fig. 4a, we will then find that $Z = 0$ except near the corners A and B where ρ' changes discontinuously. To see what these may contribute to the shift of the fermi level, integrate Eq. (35) directly over a small interval including the point A:

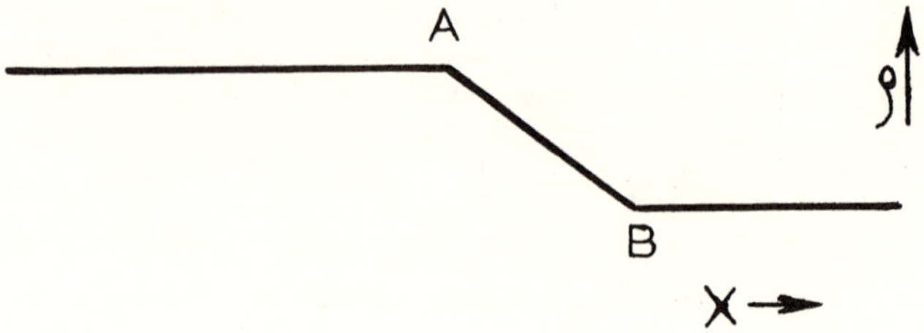

Fig. 4a. Assumed Density Variation in the Shock Zone

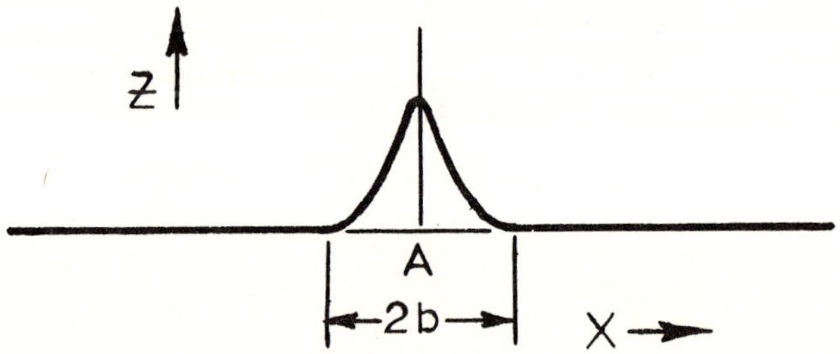

Fig. 4b. Deviations From Equilibrium Field.

$$(dZ/dx)_{A+b} - (dZ/dx)_{A-b} + (3\alpha/\lambda)(Z_{A+b} - Z_{A-b})$$

$$-(3N_o e^2/\epsilon m v_o^2)\int_{A-b}^{A+b} Z dx = (mv_o^2/3\rho_o)(\rho''_{A+b} - \rho''_{A-b}$$

$$+ (mv_o^2/3\rho_o)(\rho'_{A+b} - \rho'_{A-b}) \qquad (36)$$

According to the above discussion, we may expect Z to vary as shown in Fig. 4, where the non-zero region is very narrow. Then if b is taken outside the non-zero region,

$$Z_{A\pm b} = (dZ/dx)_{A\pm b} = \rho''_{A\pm b} = 0$$

Then Eq. (36) reduces to

$$(\Delta E_f)_{A-b}^{A+b} = -(\epsilon m v_o^2/3N_o e^2)(2\epsilon_{fo}/3) \cdot$$

$$\cdot (\rho'_{A+b} - \rho'_{A-b})/\rho_o$$

$$\simeq -3 \times 10^{-21}\, \epsilon_{fo}\, \rho'_{A+b}/\rho_o$$

For $\delta\rho/\rho_o = 0.1$ and a shock transition zone length of 10^{-6} meters,

$$\Delta E_f = +3 \times 10^{-16}\, \epsilon_{fo}$$

Small as this is, it is, nonetheless, cancelled by an equal and opposite contribution at B.

V. Conclusions

If we calculate the potential generated around a circuit of a wire containing a pressure pulse (Fig. 2) from zero pressure, through the shock front, out through the rarefaction and finally back to the zero pressure environment (where any instruments must operate), we find the EMF generated is zero. This result is tentative since the model investigated here operates under the following restrictive assumptions.

1) The temperature is set at zero, and no account is taken of the heating which is always observed behind the shock front. Thus all thermoelectric effects are neglected.

2) The approach to equilibrium by the electrons in the shock front is described in terms of a simple relaxation time. The mean free path of the electrons is assumed to be short compared to the width of the shock front and the deviation from equilibrium is supposed to be small.

3) The acoustic approximation is applied to the shock front, leading to a form of pressure pulse which moves uniformly through the solid.

4) The series development (Eq. 24) is terminated at the first nontrivial order equation which yields dynamic effects.

It is certainly conceivable that in certain systems the shock front is short compared with an electron mean free path. Then a completely different discription is necessary. The work indicated for this is in progress. This case is especially interesting since the shock front must be treated by a long mean free path model, while the rarefaction will always correspond to a short mean free path. In such circumstances, the chances for interesting observable dynamic effects will be difficult but is being attempted.

References

1. "Compression of Solids by Strong Shock Waves", M. H. Rice, R. G. McQueen, and J. M. Walsh, Solid State Physics, edited by F. Seitz and D. Turnbull (Academic Press, Inc., New York, 1958), Vol. 6.
2. "Electron Energy Bands in Solids", Joseph Callaway, Solid State Physics, edited by F. Seitz and D. Turnbull (Academic Press, Inc., New York, 1958), Vol. 7.
3. "Ultrasonic Attenuation in Metals", A. B. Pippard, Phil. Mag., 46, 1104 (1955).

EFFECT OF HYDROSTATIC PRESSURE ON THE NÉEL TEMPERATURE AND ELECTRICAL RESISTIVITY ANOMALY IN CHROMIUM METAL*

Tadayasu Mitsui†
Department of Physics, University of Arizona
Tucson, Arizona

Abstract

The electrical resistivity of high-purity polycrystalline chromium was measured as a function of temperature in the range of hydrostatic pressures from atmospheric pressure to 8 kbar. The Néel temperature, which manifests itself as a resistivity minimum, was observed to decrease linearly with pressure at a rate of $-5.1 \pm 0.2°C/kbar$. This value is in agreement with the one estimated by the thermodynamical relation of the second order phase transition. The electrical resistivity anomaly, which is associated with magnetic energy gaps introduced by the periodicity of antiferromagnetic spin arrangement, was found to be pressure insensitive.

Introduction

Many experiments (1-8) have been carried out in recent years in an effort to clarify the nature of the magnetic behavior of metallic chromium. It now appears that chromium can be considered as a typical example of a collective electron antiferromagnet. Several theoretical studies have been made to explain the magnetic, electrical, and other properties of chromium. Lidiard (9) first suggested that the antiferromagnetism of chromium must be associated with the conduction electrons and showed that the unusually small anomalies in the specific heat and magnetic susceptibility are explained on

*This work was sponsored in part by the U.S. Air Force Office of Scientific Research through Grant No. AF 105-63.
†On leave of absence from Department of Physics, University of Hokkaido, Sapporo, Japan.

the basis of the collective electron model. Overhauser (10) proposed the spin density wave theory in which the electrical resistivity anomaly, magnetic properties and the specific heat can be interpreted semiquantitatively as a consequence of magnetic energy gaps at the Fermi surface for a spin density wave ground state. Tachiki and Nagamiya (11) calculated the total energy of spin density waves as a function of the wave vector, based on a formal theory proposed by Yosida (12). However, the strength of the exchange interaction and its relation to the Néel point have not yet been established.

It is then of interest to determine how the Néel point changes with increasing pressure, since it would appear that the change in the Néel point is primarily due to the increase in the Fermi energy and the decrease in the pseudo-exchange interaction with decreasing volume.

Some information on the pressure dependence of the Néel point of chromium is available from Bridgman's data (13) on electrical resistivity *versus* pressure. These data, however, do not provide precise information for the determination of the pressure dependence of the Néel point of chromium, since the measurements covered only six temperatures between 75 and -80°C. These measurements gave a Néel point at atmospheric pressure about 25°K lower than the 310°K which was recently obtained by many measurements (2,3,4,5,7,8).

The purpose of this experiment was to determine the effect of pressure up to about 8 kbar on the Néel point and on the electrical resistivity anomaly of chromium in order to elucidate further its magnetic behavior. In the present experiment, electrical resistivity was employed as a measure of the Néel point. The existence of a cusped minimum in the curve of the electrical resistivity *versus* temperature indicated conveniently the location of the Néel point as shown in the works of Arajs (7) and of Marcinkowski and Lipsitt (14).

Experimental Work

Sample

The chromium specimen, supplied by Professor R. Street of Monash University in Melbourne, consisted of polycrystalline chromium of approximate dimensions 58 mm by 0.5 by 0.5 mm. It was manufactured from ARL chromium ingots which contained

0.08 wt % nitrogen and 0.03 wt % oxygen. Other impurities were below the limit of spectrographic detection. After being ground and formed, the specimen was annealed and recrystallized at 1250°C for approximately two hours in order to remove excess dislocations.

Apparatus and General Procedure

The oil pressure system used was a typical Bridgman-type system which is similar to the one described by Emrick (15). Dow Corning 200 silicone fluid was used as a pressure medium. The pressure was measured with a coil of manganin wire calibrated against the freezing point of mercury at 0°C (7640 kg/cm^2) and was controlled to within ± 1%. The pressure vessel made of Cu-Be alloy was mounted inside an oil or dry ice bath to obtain any temperature between -30°C and 70°C. Current and potential leads (copper) were spot-welded to the chromium specimen which was mounted to a specimen holder attached to a Cu-Be plug fitted with electrical lead-ins.

Electrical leads and thermocouple wires were led out from the vessel through steel, chromel or alumel cones seated in the pressure plug and insulated by conical insulators made of unfired lava A (American Lava Corporation). The temperature was measured by using chromel-alumel thermocouples because of the small pressure effect on the emf output. The temperature inside the vessel was kept constant within ± 0.2°C along the long axis of the specimen.

Fig. 1 shows a cross-section of the assembled pressure vessel and the specimen holder showing arrangement of specimen, electric leadwires, and thermocouple locations.

The electric resistivity was measured by the standard dc potentiometric techniques. The voltages across a constant length of the specimen were determined with a Rubicon six-dial thermofree potentiometer in conjunction with a Keithley milli-microvoltmeter as the null indicator. The electric current through the sample was determined with a Rubicon high precision type B potentiometer across a resistance standard and was kept constant within 0.01%.

The accuracy of the measurement was estimated to be within ± 0.05 μV. This voltage spread is due

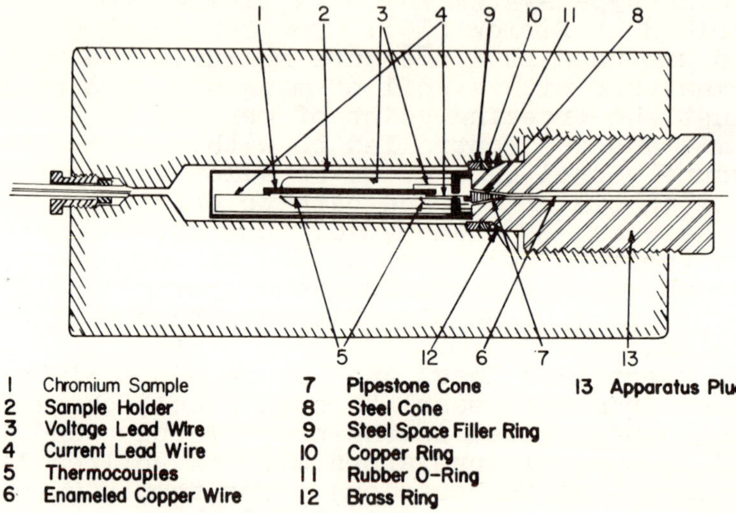

Fig. 1. Cross section of the assembled pressure vessel showing arrangement of specimen and thermocouples. For clarity, only one electrical lead through the plug is shown.

to the difficulty in eliminating thermal emf's from the circuit. Data were taken while the temperature of the specimen was raised and the pressure was held constant. Warm-up time rate was always about 0.1° 0.1°C/min or less.

Results and Discussions

Fig. 2 shows the relative resistivity of chromium as a function of temperature in the range of pressure from atmospheric pressure to about 8 kbar. Here the relative resistivity was taken as ρ_T/ρ_0, where ρ_T and ρ_0 are observed values of resistivity at T°C and 0°C, respectively. The sharp minimum in the resistivity *versus* temperature curves observed at 38.0°C ± 0.2°C for atmospheric pressure is in good agreement with the determinations by many investigators (2,3,4,5,7,8).

In order to determine if there was any hysteresis in the resistivity *versus* temperature curves between cooling and warming, data were also taken during cooling at constant pressure for four runs (0, 0.98, 1.96, and 5.30 kbar).

The measurements indicated clearly a hysteresis in the neighborhood of the Néel point as shown in Fig. 2. However, the temperatures corresponding to the minimum points in curves were found to be independent of the hysteresis within experimental errors. Similar hysteresis in the temperature range in the neighborhood of the Néel point has been reported by Arajs et al (7). Such a hysteresis might be attributable to antiferromagnetic domain structure and also to the residual short range order in the paramagnetic phase.

Fig. 3 shows that the Néel point of chromium is depressed linearly with pressure at a slope of $dT_N/dp = -5.1 \pm 0.2$°C/kbar. Recently the result of neutron diffraction obtained by Litvin and Ponyatovski showed that the Néel point of chromium varied with a pressure coefficient of -6×10^{-3}°C/atm. If the magnetic transition of chromium is treated as the second order phase transition, we can estimate the change of Néel point with pressure by the thermodynamical relation (Ehrenfest relation), $dT_N/dp = 3T_N\, \Delta\alpha/\Delta Cp^\circ \rho$, where $\Delta\alpha$ and ΔCp are the abrupt changes in the temperature coefficient of thermal expansion and in the specific heat (per unit mass) at the Néel point, respectively. ρ is the density at the Néel point. Using $\Delta\alpha \cong -7 \times 10^{-6}$ per deg(5),

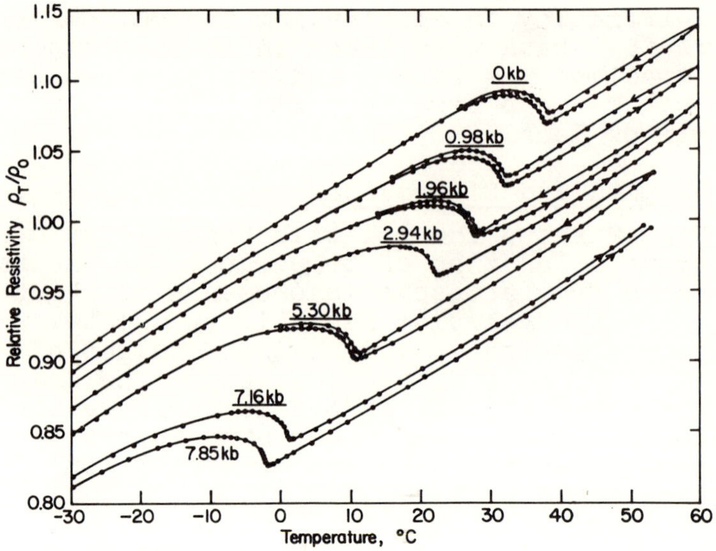

Fig. 2. The relative electrical resistivity of chromium as a function of temperature and pressure. ρ_0 at 0°C and atmospheric pressure was found to be 1.27×10^{-5} Ω cm.

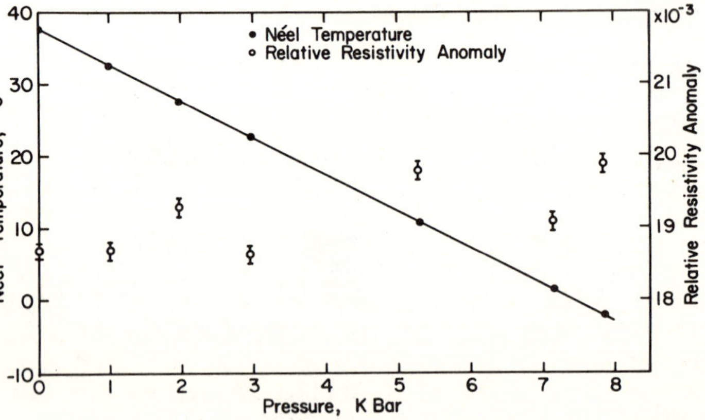

Fig. 3. Change of Néel temperature and relative resistivity anomaly versus hydrostatic pressure for chromium.

$\Delta C_p \cong 0.15$ cal/g-atom deg (4), $\rho = 7.20$ g/cm^3, and $T_N = 311°K$, one obtains $dT_N/dp = -5.8°C$/kbar which agrees reasonably well with the measured value of $dT_N/dp = -5.1°C$/kbar.

In order to determine if there was any electrical resistivity drift or any shift in the Néel point after high pressure cycling or both, the following procedure was carried out. The specimen was left at 7 kbar pressure for 20 hours at room temperature after measuring the resistivity at 7.85 kbar. The pressure was then released and resistivity measurements were again taken with increasing temperature at 0 kbar after establishing thermal equilibrium at 25°C in about 20 minutes. In this run, there was no shift in the Néel point between initial and final runs within experimental error. However, a small reduction of about 0.08% was observed between runs over the whole range from 25°C to 50°C. Such a pressure hysteresis might be understood by a strain sensitivity of the polarization direction of a transverse linear spin density wave, as was proposed by Overhauser (10). The same theory could account for the temperature hysteresis effect observed near the Néel point even in well annealed crystals or under hydrostatic pressure.

It is well known that the electrical resistivity anomaly below the Néel point is one of the significant phenomena associated generally with magnetic energy gaps which are introduced into the Brillouin zone structure when the periodicity of the antiferromagnetic spin arrangement differs from the lattice periodicity. In the case of chromium, Overhauser (10) used the Néel point to estimate a fractional area of the Fermi surface truncated by the spin density wave potential and showed that this estimate is in agreement with that derived from the observed resistivity anomaly. In terms of the Overhauser theory one can express the pressure dependence of the Néel point as a decrease in the magnetic energy gaps by 1.6% per kbar. In this connection it is interesting to determine whether or not there is any pressure dependence of the electrical resistivity anomaly.

In Fig. 3 the pressure dependence of the relative resistivity anomaly, $(\rho_{max} - \rho_{min})/\rho_0$, was also plotted. No evidence for a change in the resistivity anomaly was detected in the temperature range below the Néel point. This anomaly seems to be sensitive to the hysteresis effects associated with antiferromagnetic domain structure and to the rate at which an apparent equilibrium is reached. Further experimental data are required to settle this point using a strain-free crystal of very high purity.

Acknowledgments

The author wishes to express his sincere gratitude to Professor C. T. Tomizuka for his valuable advice. He also wishes to acknowledge the many valuable discussions with Dr. H. Sato. For generously supplying the prepared and well-annealed specimen of high-purity chromium, the author wishes to thank Professor R. Street. He also wishes to thank Mr. R. Lowell for his help in the operation of the pressure system.

References

1. M. K. Wilkinson, E. O. Wollan, W. C. Koehler, and J. W. Cable, Phys. Rev. 127, 2080 (1962).
2. G. Shirane and W. J. Takei, J. Phys. Soc. Japan 17, Suppl. B-III, 35 (1962).
3. E. W. Collings, F. T. Hedgcock, and A. Siddiqui, Phil. Mag. 6, 155 (1961).
4. R. H. Beaumont, H. Chikora, and J. A. Morrison, Phil. Mag. 5, 188 (1960).
5. D. I. Balef and J. de Klerk, Phys. Rev. 129, 1063 (1963).
6. R. G. Barnes and T. P. Graham, Phys. Rev. Letters 8, 248 (1962).
7. S. Arajs, R. V. Colvin, and M. J. Marcinkowski, J. Less-Common Metals 4, 46 (1962).
8. T. I. Kostina, T. N. Kolova, and E. I. Kondorski, Zh. Eksperim. i Theor. Fiz. 45, 1352 (1963); [English transl.: Soviet Phys. - JETP 18, 931 (1964)].
9. A. B. Lidiard, Proc. Phys. Soc. London A66, 1188 (1953).
10. A. W. Overhauser, Phys. Rev. 128, 1437 (1962); J. Appl. Phys. Suppl. 34, 1019 (1963); A. Arott, in Magnetism, edited by H. Suhl and G. Rado (Academic Press, Inc., New York, to be published), Vol. 2.
11. M. Tachiki and T. Nagamiya, Physics Letters 3, 214 (1963).
12. K. Yosida, Prog. Theor. Phys. (Kyoto) 28, 759 (1962).

13. P. W. Bridgman, Proc. Am. Acad. Arts Sci. 68, 27 (1932-33).
14. M. J. Marcinkowski and H. A. Lipsitt, J. Appl. Phys. 32, 1238 (1961).
15. R. M. Emrick, Phys. Rev. 122, 1720 (1961).
16. D. F. Litvin and E. G. Ponyatovski, Doklady A. N. SSSR 156, 69 (1964) [English Translation: Soviet Physics - Doklady 9, 388 (1964)].

Discussion: ELECTRONIC PROPERTIES

G. K. WHITE
Bell Telephone Laboratories, Murray Hill, New Jersey
(Permanent address: C.S.I.R.O., Sydney, Australia)

I hope it is not heresy at this conference to point out that measurements of $(dV/dT)_P$, that is thermal expansion, may give similar information about interaction energies in solids as is given by high pressure measurements. In some instances one type of measurement may be technically simpler than the other. I believe that Drs. Swenson and Smith will be discussing later the complementary nature of measurements using pressure, volume and temperature as variables in relation to lattice properties. In this session on electronic properties I wish to mention that recent measurements at low temperatures have given values for the electronic contribution to the expansion coefficient (β_e). This is a measure of the pressure or volume dependence of the density of states, $N(E_F)$, at the Fermi surface just as the corresponding term in the heat capacity C_e gives a measure of the density of states. Data from the E.T.H. (Zurich), C.S.I.R.O. (Sydney) and Iowa State University have shown that the electronic Gruneisen parameter $\gamma_e = \beta_e VK/C_e = d\ln N(E_F)/d\ln V$ (V = molar volume, K = bulk modulus) is close to its ideal free-electron value of 2/3 for copper but is appreciably larger for Al(1.8), Mg(1.4), Nb(1.5), Pb(ca 1.7), Pd(2.1), Pt(2.4), Ta(1.3), V(1.7), Fe(2.1), Ni(2.0); for W, it is close to zero (see review by Collins and White, Prog. Low Temp. Phys. IV, 450, 1964). These data may be interpreted in terms of the separate derivatives $(dN(E_F)/dV)_{E_F}$ and $(dN(E_F)/dE_F)_V$ (Varley, Proc. Roy. Soc. A237, 413, 1956) but only represent an averaged behavior as do heat capacity data. Undoubtedly for the noble metals more information on the behavior of the individual parts of the Fermi surface is given by de Haas-van Alphen measurements under pressure. For some other elements, including transition elements, detailed information may also be obtained by studying changes in dimensions with magnetic field, i.e.

magnetostriction, in addition to thermally induced changes.

Discussion: MAGNETISM

Thermal expansion measurements at low temperatures may provide useful data complementary to pressure studies of magnetically interacting systems, as in the case of electronic and lattice properties. For example, we have observed in the antiferromagnets chromium, α-manganese, FeMn and in f.c.c. iron-nickel alloys close to the invar composition (ca 35 at % Ni), a relative large negative contribution β_m, to the expansion coefficient (White, Proc. Phys. Soc., Lond., in press). This contribution is proportional to temperature and predominates below ca. 20°K. If it is magnetic in origin and the analogous term in the specific heat is magnetic (C_m), then values can be calculated for the "magnetic Gruneisen" parameter, $\gamma_m = \beta_m VK/C_m = - d\ln J/d\ln V$ (J = exchange interaction) of $\gamma_m = -20$ for Fe + 35 %Ni, $\gamma_m = -11$ for Fe + 40 at % Ni, and $\gamma_m \simeq -7$ for Cr and α - Mn. This behavior contrasts sharply with that in nickel and b.c.c. iron where a relatively small positive term (proportional to T) is observed, comparable in magnitude to those in Pd, Pt, and Ta, which are regarded as electronic in origin. If the magnetic interpretation for Invar, Cr, α-Mn is correct then $d\ln J/d\ln V$ is large and positive and similar in magnitude and size to that obtained from the pressure dependence of the magnetic ordering temperature, $d\ln\theta/d\ln V$ (e.g. Patrick, Phys. Rev. 93, 384, 1954).

Again in some rare earths, there is a parallel behavior between negative expansion coefficients observed at low temperatures (Andres, Phys. Kondens. Mater 2, 294, 1964) and negative values of $d\theta/dP$ (Patrick, also McWhan and Stevens, Phys. Rev. in press). Presumably, however, γ_m (from expansion at low temperature) and the analogous parameter (from $d\theta/dP$ measurements) cannot be equated exactly unless the saturation magnetization at 0°K is volume independent. Further data on the pressure dependence of the magnetization or volume magnetostriction at different temperatures are required to fully describe the magnetic equation of state.

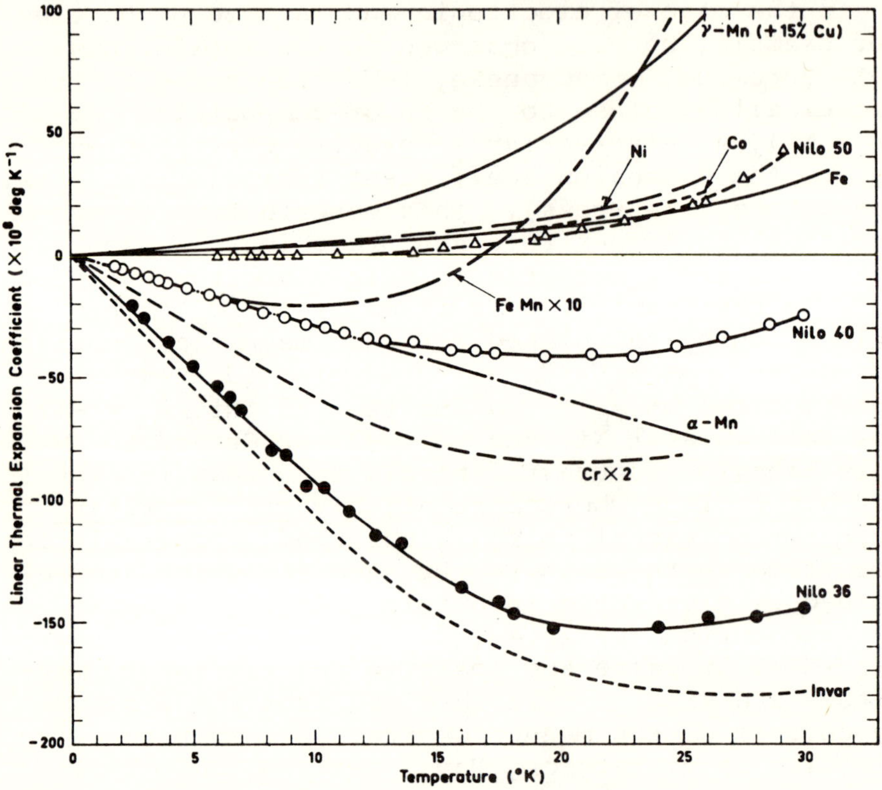

Fig. 1. Linear coefficient of thermal expansion ($\alpha \times 10^8$ degK^{-1}). Nilo alloys are Fe + Ni alloys, e.g. Nilo 36 is Fe + 36wt%Ni, similar to Invar. Curve marked Invar is from P. Meincke (University of Toronto, Thesis 1963).

SUPERCONDUCTIVITY AND HIGH PRESSURES

B. T. Matthias
University of California
La Jolla, California
and
Bell Telephone Laboratories
Murray Hill, New Jersey

High pressures have contributed a great deal to the investigations of superconductivity. In order to represent these results in an orderly fashion, two subdivisions are perhaps a logical aproach:

A. The application of high pressures has changed elements and compounds from normal conductors into superconductors, thus far in the temperature range above 1°K.

B. The application of high pressures to known superconductors has changed their transition temperatures considerably, sometimes to higher but usually to lower temperatures.

The combination of both approaches has increased our knowledge of superconductivity quite considerably. Let us consider first the search for new superconductors.

Thus far, three otherwise normal elements have become superconducting with application of high pressures. They are Bi, (1) Sb, (2) and Te. (3) It is indicative of the very recent impact of high pressure techniques that, while the superconductivity of Bi had been discovered as early as 1953, the superconductivity of Te and Sb was found only during the last year. This was in spite of the fact that the superconducting phase of InSb had been found two years earlier. (4) InSb was a beautiful example of the general occurrence of superconductivity among the nontransition elements.

Its high pressure structure, isomorphous to that of white tin, illustrated furthermore that averaging over the periodic system is possible in the same way as for the transition elements. Since then quite a number of normal compounds have been found to become superconducting under pressure, but since they all contain at least one normally superconducting element, I chose to limit this discussion to the pure elements. The general occurrence of superconductivity in the periodic system was, of course, even more drastically illustrated by the discoveries of superconducting antimony and tellurium. It seems very likely that the remaining nonsuperconductors among the nontransition elements, such as Ge, Si, As and black phosphorus will also eventually become superconducting under pressure, i.e., upon a change of their crystal lattice. Another element that may become superconducting eventually is cerium. (5) It is well known that under pressure it is possible to squeeze out most of its 4f electrons. Should it be possible to do so entirely, then cerium will have a transition temperature slightly above that of thorium at 1.4°K. The importance of a change in symmetry, lattice spacing or electron configuration then brings us to the second part.

Very strong variations for the transition temperatures of the superconducting elements have been observed here. Very often thin films condensed at liquid helium temperatures also have signaled the effects later found under pressure. This was the case for bismuth and for gallium. (6) These films formed many new but metastable lattices, sometimes with several different transition temperatures for the various modifications. Bismuth and beryllium films condensed at liquid helium temperatures all became superconducting (7) and the transition temperature of gallium was raised in this way by almost an order of magnitude. (8) Quite frequently there was even more than one superconducting modification of the same element, as was the case for bismuth and gallium. Beryllium is an exception since it has not yet become superconducting under high pressure, but undoubtedly will do so eventually. Once the structure of the superconducting beryllium films is known, it may be possible to predict the necessary pressures.

All the results described so far, however, are due to a change of either lattice symmetries or electron configuration. When only the lattice dimensions are changed, the results are generally not quite so drastic. The usual volume change at the superconducting transition is positive (9) and thus transition temperatures usually drop with increasing pressures. The depression of the transition temperature is very large in cadmium, thus indicating from extrapolation that the superconducting state may eventually be suppressed entirely from the accessible temperature range. (10) The c/a ratio of cadmium is abnormally large, 1.89 as compared to 1.6333, for a hexagonal close packing. Under high pressures this ratio is reduced towards the ideal value. (11) It seems tempting to speculate that the transition temperatures of the hexagonal alkaline earths with c/a ratios close to 1.6 would, therefore, be at present below the accessible temperature range, as well as those for yttrium and scandium.

As had been pointed out a few years ago, (12) there may be a close correlation between the pressure effects and isotope effects in superconductors. This correlation became obvious when a distinct difference in behavior of transition temperature vs. pressure was found (13) between the transition elements and the nontransition elements. This same difference had been found earlier also in the isotope effect of the transition temperatures. Thus, high pressure data will give information or prediction of the isotope effect even if this cannot ever be measured in reality because of lack of purity or because no isotopes exist. This has been confirmed again just recently from high pressure data on zirconium (14) and corresponding isotope measurements. (15) The isotope effect of zirconium is zero, the same as in ruthenium. Again the difference between transition and nontransition elements is shown. Also shown again is the symmetric behavior of superconductivity in transition elements with respect to the half filled d-shell, since zirconium has four valence electrons, the half filled d-shell at molybdenum has six and ruthenium with eight valence electrons on the symmetrically opposite side, corresponds to zirconium.

Recently, a very large positive pressure effect in lanthanum has been observed. (16) This is quite different from the behavior of all other superconducting transition elements. The reason for this is to be found in the different interactions causing superconductivity. In this case it is the interaction with the virtual 4f electrons or the approaching 4f level (17) that causes the large variation of the transition temperature with a decreasing lattice constant.

Whether superconductivity is the normal behavior for metals once their purity is sufficiently high and their temperature sufficiently low is, of course, still an open question. High pressure techniques will help considerably to reach an answer to this question, which I think will be in the affirmative.

ACKNOWLEDGMENTS

I would like to thank D. B. McWhan and T. H. Geballe for many valuable discussions.

Part of this work was supported by the Air Force Office of Scientific Research.

REFERENCES

1. P. F. Chester and G. O. Jones, Phil. Mag. 44, 1281 (1953).
2. T. R. R. McDonald, E. Gregory, G. S. Barberich, D. B. McWhan, T. H. Geballe and G. W. Hull, Jr., Phys. Letters 14, 16 (1965).
3. B. T. Matthias and J. L. Olsen, Phys. Letters 13, 202 (1964).

4. H. E. Bommel, A. J. Darnell, W. F. Libby, B. R. Tittman and A. J. Yencha, Sci. 141, 714 (1963).
5. T. F. Smith, Phys. Rev. 137, A1435 (1965).
6. R. Hilsch, Proceedings of the International Conference on Low Temperature Physics, edited by R. Bowers (Oxford, 1951), p. 119.
7. B. G. Lazarev, A. I. Sudovtsev and A. P. Smirnov, Soviet Phys.----JEPT 6, 816 (1958).
8. W. Buckel and W. Gey, Z. Physik 176, 336 (1963).
9. D. Shoenberg, Superconductivity (Cambridge University Press, 1960).
10. N. E. Alekseevskii and I. U. P. Gaidukov, Soviet Phys.----JEPT 2, 762 (1956).
11. D. B. McWhan, J. Appl. Phys. 36, 664 (1965).
12. J. L. Olsen, E. Bucher, M. Levy, J. Muller, E. Corenzwit and T. H. Geballe, Rev. Mod. Phys. 36, 168 (1964).
13. E. Bucher and J. L. Olsen, Proceedings of the Eighth International Conference on Low Temperature Physics, edited by R. O. Davies (Butterworths, Washington, 1963), p. 139.
14. N. B. Brandt and N. I. Ginzberg, Zh. Eksperim. i. Teor. Fiz. 46, 1216 (1964).
15. E. Bucher, J. Muller, J. L. Olsen and C. Palmy, Physics Letters, April 15 (1965).
16. T. F. Smith and W. E. Gardner, Bull. Am. Phys. Soc. 10, 519 (1965).
17. D. C. Hamilton and M. A. Jansen, Phys. Rev. Letters 11, 205 (1963).

DISCUSSION

By R. L. Orbach

It is possible, even within the weak coupling B.C.S. theory, to obtain an anomalous isotope effect if the width of the conduction electron band is less than or of the order of the Debye energy. This would never be the case for s electrons. However, a transition metal may well exhibit peaks in the density of states with widths of this order (e.b. V_3X compounds). These peaks must lie close to the Fermi level to be effective, and, as is well known, may be detected by a temperature change in the specific heat γ and the Pauli susceptibility. This argument assumes the d electrons play the dominant role in these materials, a point which seems to be substantiated by the striking dependence of T_c on the number of d electrons.

SESSION C

MAGNETISM

Chairman: L. F. VERESCHAGIN

STABILITY OF MAGNETIC MOMENTS UNDER PRESSURE IN RARE-EARTH METALS

A. Blandin, B. Coqblin and J. Friedel
Faculté des Sciences
Physique des Solides
Orsay (Seine & Oise)
France

I. Experimental Introduction.

In most cases, rare-earth metals can be described with an ionic model. The ions have the same angular momentum $\vec{J}$ as the free trivalent ions R^{+++}. These ions interact with the conduction electrons by an s-f exchange mechanism and many properties in the ordered and disordered magnetic states are fairly well understood with this model.

There are three exceptions in the whole series: Cerium, Europium and Ytterbium. The peculiarities of these rare-earth metals can be seen in macroscopic constants such as atomic radii (Fig. 1) (1), thermal expansion coefficients and compressibilities (2) and valency and magnetic moments.

Cerium has a phase diagram (Fig. 2) (3) with four phases. We shall be interested in the α- and γ-phases. Both of them are face-centered cubic. There is only a change of the lattice parameter (or atomic radius): at normal pressure, the atomic radius for γ-Ce is nearly on the same curve as the radii of the other metals. The atomic radius of α-Ce is smaller. At low pressure, the valences of γ- and α-Ce have been estimated to be 3.06 and 3.67 (2). These figures can be criticized, as they are derived mainly from Hall measurements, but there certainly remains the qualitative fact that the number of electrons in the conduction band (6s and 5d) are very different in γ- and α-Cerium.

Moreover, α-Cerium is nonmagnetic and γ-Cerium has a magnetic moment giving rise to a Curie-Weiss susceptibility at normal pressure; the magnetic moment per atom is 2.40 (4) or 2.52 (2) Bohr magnetons, a value which is nearly the theoretical value

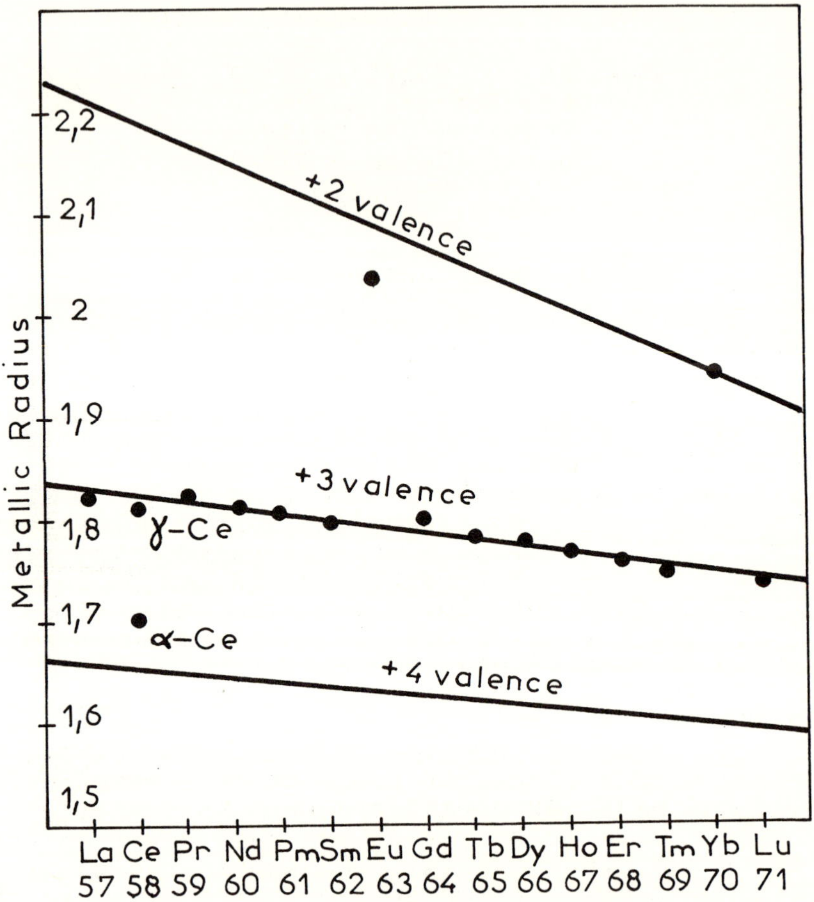

Fig. 1. Metallic radii of rare-earth metals (H. Tracy Hall, J. Dean Barnet and L. Merrill), Science **139**, 111, (1963).

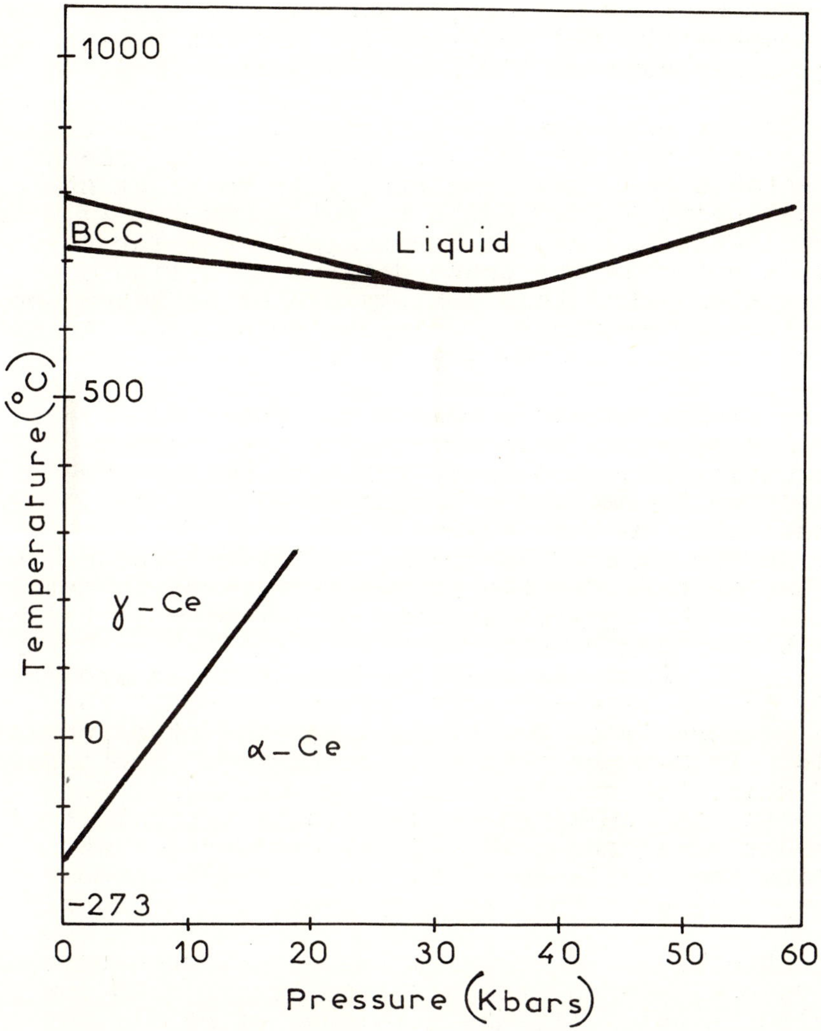

Fig. 2. Phase diagram of cerium (A. Jayaraman, Phys. Rev. **137**, A179, 1965).

for the Ce^{+++} ion. At low temperature, these moments may order so there exists a magnetic hexagonal β-phase. We shall discuss only the α- and γ-phases.

The discontinuous volume change in the α ⟷ γ transition goes to zero at about 630°K and 20 kbar (5). The heat of transition vanishes at nearly the same temperature and pressure (6) and the resistance change disappears at 550°K and 17.5 kbar (Fig. 3) (3).

Due to experimental difficulties, these figures are not precise and vary somewhat. It seems well established, nevertheless, that there exists a critical point P ($T \simeq 550°K$ to $600°K$ and $p \sim 17.5$ to 20 kbar) which looks, at first sight, similar to a critical point in the phase diagram of a liquid vapor system. But this analogy cannot be taken too far: turning around the critical point, one can go from the α-phase to the γ-phase with a continuous change in the atomic volume. But during this process, an atomic magnetic moment has appeared. Thus, there should exist another line in the diagram representing a second order transition from the nonmagnetic to the magnetic Cerium metal. We shall comment later on this point.

In the case of Europium, the atomic volume is much larger than for the trivalent rare-earth metals (Fig. 1). The compressibility of Europium is two to three times larger than for the "normal" rare-earth metals (2). Thus the relative variation in atomic volume is 0.2 between normal pressure and 40 kbar, to be compared with 0.1 or less for the other normal metals. This means that at a pressure of the order of 40 kbar, the atomic volume (or radius) of Europium metal is coming close to a value typical for a trivalent metal (Fig. 1). It is estimated that Europium has a valency of two and magnetic properties typical of a full spin up shell as in Gadolinium.

Ytterbium, as Europium, at normal pressure and temperature, has an atomic volume larger than for trivalent metals. It has a valency of nearly two (2) and is nonmagnetic (temperature independent Knight shift (7)). At room temperature there is a rapid increase with pressure in the resistivity from 25 to a few hundred micro-ohms between zero and 40 kbar (8,9,10,11). At lower temperatures, the maximum of the resistivity increases very much and

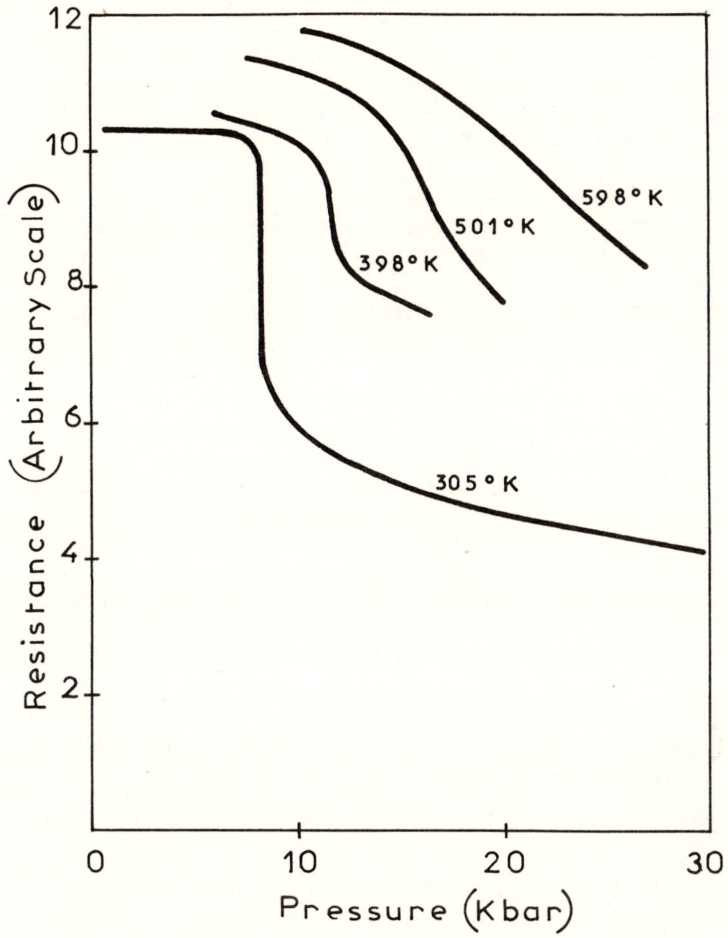

Fig. 3. Resistance of fcc-cerium (A. Jayaraman, Phys. Rev. **137**, A179, 1965).

is shifted to higher pressures. There are large disagreements with regard to the magnitude of the peak at low temperature. It appears, nevertheless, clear that Ytterbium has a behavior which looks more like that of a semiconductor or a semi-metal in that range of low temperatures. At room temperature and above 40 kbar, Ytterbium undergoes a phase transition to a body-centered cubic phase, the resistivity decreases abruptly to a value typical of a metal and the atomic radius comes back near the curve of the normal trivalent metals.

II. Theoretical discussion.

Clearly, the strange behavior of Cerium, Europium and Ytterbium seems to be related to a nearly empty or nearly full f shell in Cerium and Ytterbium, respectively, and a nearly full spin up shell in Europium. One knows from atomic data and chemical considerations that a configuration $4f^1\ 5d^1\ 6s^2$ has an energy close to $5d^2\ 6s^2$ or $5d^1\ 6p^1\ 6s^2$ in the case of Cerium and similar results hold for Europium and Ytterbium. Thus we would like to have a model capable of explaining the general features of the series.

The theoretical model which we shall use is an extension of the virtual bound state model or localized magnetic states model of Anderson and Friedel (12,13). The approach of Anderson is mathematically more appropriate for this discussion so we shall use it although the results can be derived from scattering formalism. A similar model has been used by Rocher (4) but this work represents a large extension of his paper.

A. Model for non-degenerate orbital states. (13)

Let us consider conduction electrons with ϵ_k and a localized f state with energy E_0 which falls inside the conduction band. This represents typically the case of an impurity. There exist small matrix elements V_{kf} which provide an admixture of the conduction and localized states. We have thus a resonant scattering state. If we take into account the Coulomb energy between the two spin up and down f states, the Hamiltonian in the Hartree-Fock approximation is

$$H = \sum_{k,\sigma} \epsilon_k a^+_{k\sigma} a_{k\sigma} + E_0 \sum_\sigma a^+_{f\sigma} a_{f\sigma} + U n_{f\sigma} n_{f-\sigma} \tag{1}$$
$$+ \sum_{k,\sigma} V_{fk}(a^+_{k\sigma} a_{f\sigma} + a^+_{f\sigma} a_{k\sigma}).$$

By use of Green's functions, the shape of the resonant scattering state (or virtual bound state) is easily obtained:

$$\rho_\sigma(E) = -\frac{1}{\pi} \operatorname{Im} \frac{1}{E - E_\sigma - \Gamma + i\Delta}$$
$$= \frac{1}{\pi} \frac{\Delta}{(E - E_\sigma - \Gamma)^2 + \Delta^2} \tag{2}$$

$$\text{where} \quad \Gamma - i\Delta = \sum_k \frac{| V_{kf} |^2}{E - \epsilon_k + i\Delta} \tag{3}$$

Γ is the shift and Δ the width of the state. We shall assume that Γ and Δ are constant, independent of E; and we can take into account Γ by shifting the unperturbed energy E_0. E_σ is given by

$$E_\sigma = E_0 + U n_{-\sigma}$$
$$\text{where} \quad n_\sigma = \int^{E_f} \rho_\sigma(E)\, dE \tag{4}$$

(and the same for the $-\sigma$ spin).

The two Equations (4) represent a set of self-consistent equations to be satisfied. With the Lorentzian shape, Eq. (2), we have

$$\operatorname{cotg} \pi n_\sigma = \frac{E_0 + Un_{-\sigma} - E_f}{\Delta}. \tag{5}$$

$\rho_\sigma(E)$ represents the extra density of states due to the presence of the virtual state. This term contributes to the density of states and represents electrons which are localized in space on the impurity center.

The set of Equations (5) has solutions which can be magnetic or not. If $E_0 - E_f$ is large, the solution is nonmagnetic ($\rho_+ = \rho_-$). Then, if $E_o - E_f$ decreases, we obtain a magnetic solution when

$$U \rho(E_f) = 1. \tag{6}$$

This condition is very similar to the condition of Stoner for the appearance of ferromagnetism in a band picture; $E_o - E_f$ decreases, the state becomes nearly full and we have again a nonmagnetic solution. The total number of f electrons and the magnetic moment are functions of E_o; they are given by Figures 4A and 4B.

The transition from the nonmagnetic to the magnetic case is a second-order one if by some means (pressure, for example) we move E_o relative to the Fermi energy E_f.

B. Extension of the model to degenerate orbital states.

The extension of the model to degenerate orbital states is easy (13,14). For each state labelled by m (orbital quantum number) and σ (spin quantum number), we have

$$\rho_{m\sigma}(E) = \frac{1}{\pi} \frac{\Delta}{(E - E_{m\sigma})^2 + \Delta^2} \tag{7}$$

where
$$E_{m\sigma} = E_o + \sum_{m'\sigma'} U_{mm'} n_{m'\sigma'} - U_{mm} n_{m\sigma} - \sum_{m' \neq m} J_{mm'} n_{m'\sigma} \tag{8}$$

where the first term is the Coulomb interaction term, the second one the exchange within the same orbital state and the last one the exchange between different orbital states.

The $U_{mm'}$ are much larger than the $J_{mm'}$ and fairly constant. The $J_{mm'}$ are not constant, but for simplicity we shall take both $U_{mm'}$ and $J_{mm'}$ constant, which we shall call U and J, respectively. For simplicity also, we shall consider for a qualitative discussion the case of a doubly-degenerate orbital state.

We have thus a set of self-consistent equations:

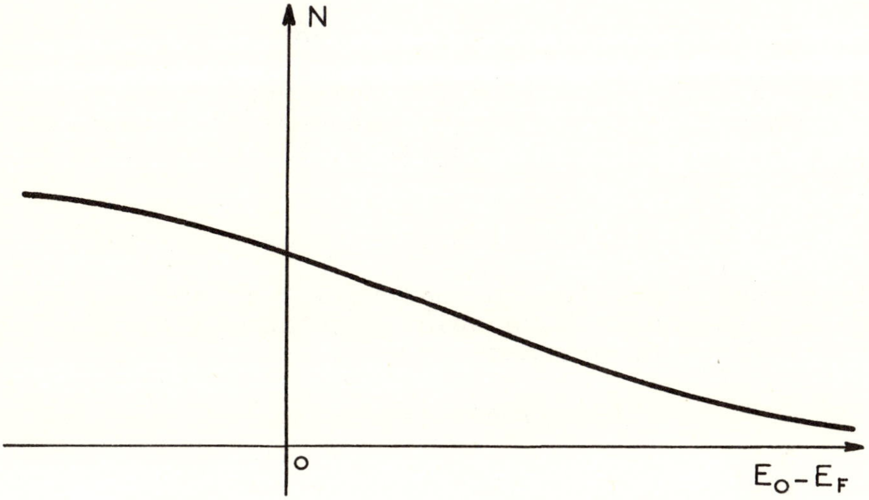

Fig. 4A. Total number N of electrons.

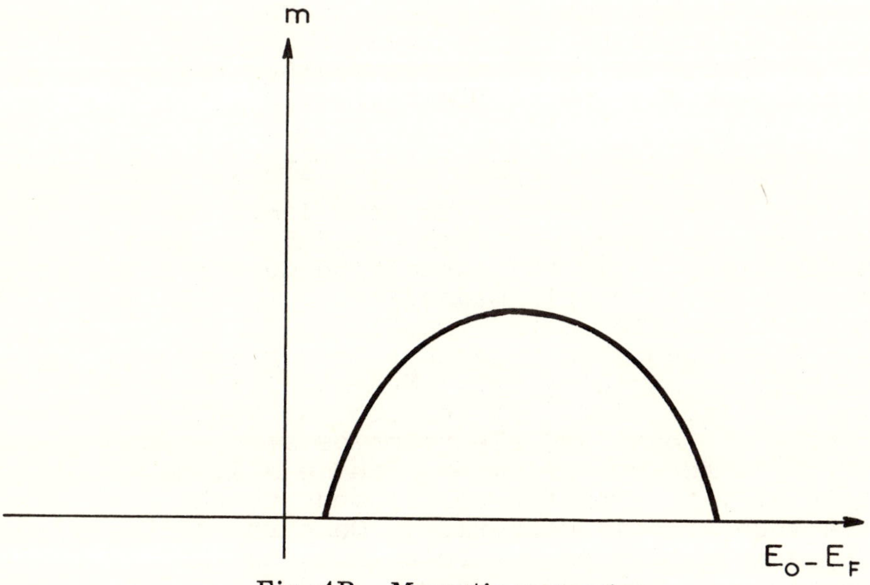

Fig. 4B. Magnetic moment m .

$$
\begin{aligned}
E_o + U\,N - Un_{1+} - Jn_{2+} &= \Delta \cot g\, \pi\, n_{1+} \\
E_o + U\,N - Un_{2+} - Jn_{1+} &= \Delta \cot g\, \pi\, n_{2+}
\end{aligned} \tag{9}
$$

and two similar equations for down spin.

When $E_o - E_f$ is large, there is no decoupling (Fig. 5A). If $E_o - E_f$ decreases, the appearance of magnetic moment occurs when

$$(U + J)\ \rho(E_f) = 1\ . \tag{10}$$

This is the condition for decoupling of spin up and down states and we have spin magnetism (Fig. 5B). Then, if $E_o - E_f$ decreases again,

$$\rho_+(E_f) \neq \rho_-(E_f)\ .$$

The condition for the appearance of orbital magnetism is

$$(U - J)\ \rho_+(E_f) = 1\ . \tag{11}$$

We have then the situation of Fig. 5C. If $E_o - E_f$ decreases again, the state $|\,1_+ >$ crosses the Fermi surface; then state $|\,2_+ >$ crosses it; then state $|\,2_- >$; then $|\,1_- >$. The intermediate steps represent nearly the configurations f^0, f^1, f^2, f^3, ...

Evidently different situations may occur, depending on the parameters U/Δ and J/Δ. The complete discussion would be very long. We have described here the case when U and J are much greater than Δ. If this condition does not hold, one may have only spin magnetism. (This is the case of transition metal impurities.)

The discussion which has been given does not bring in anything new. Nevertheless, a new effect has appeared. If we plot the total number N of localized electrons as a function of $E_o - E_f$ for $U >> \Delta$, $J >> \Delta$ and $U > J$, we obtain Fig. 6. Point A corresponds to the appearance of spin magnetism. On curve AB there is spin magnetism only. Point B marks the appearance of orbital magnetism and curve BCD describes states with both spin and orbital magnetism.

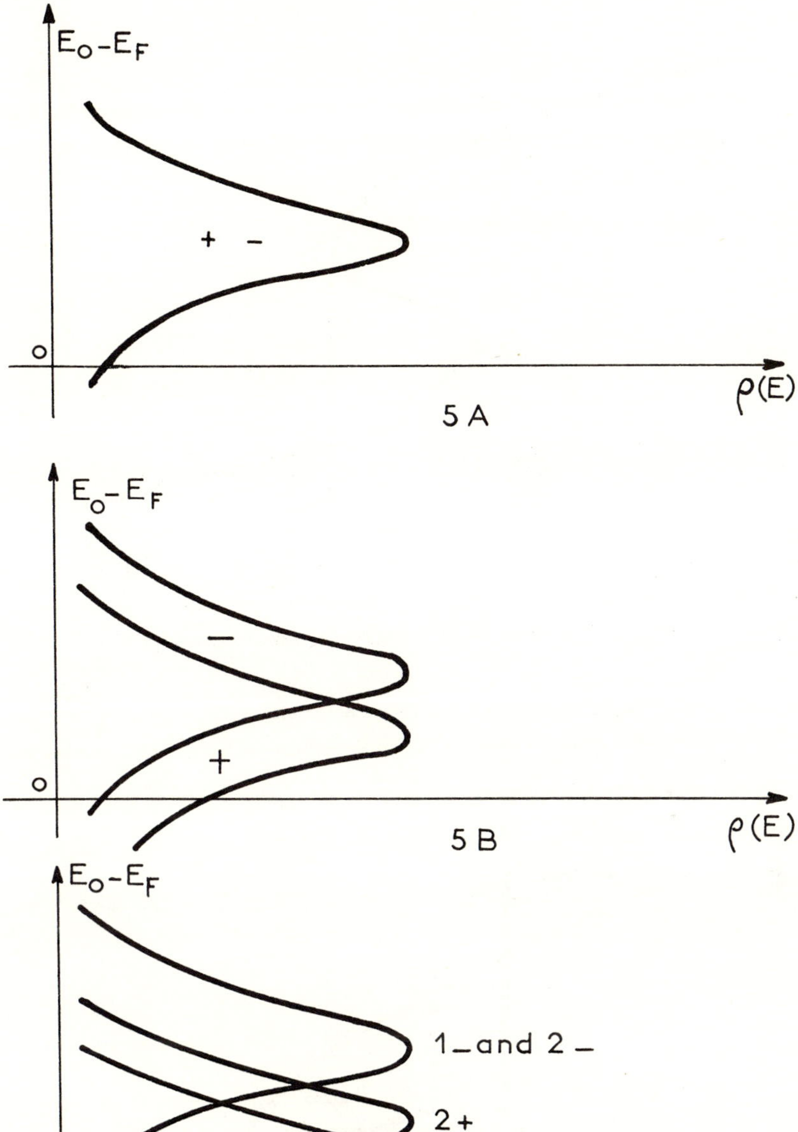

Fig. 5A, B, C. Density of states.

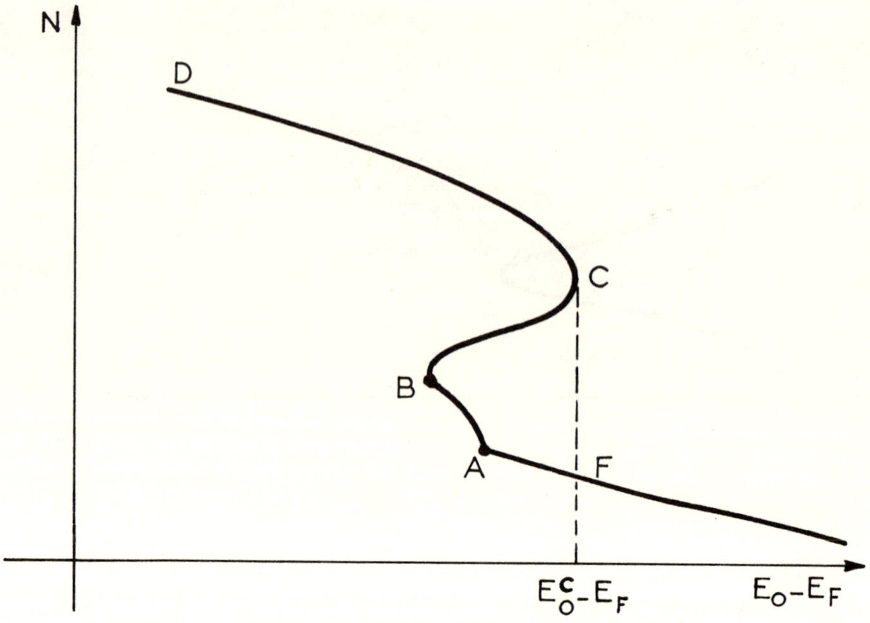

Fig. 6. Total number N of localized electrons.

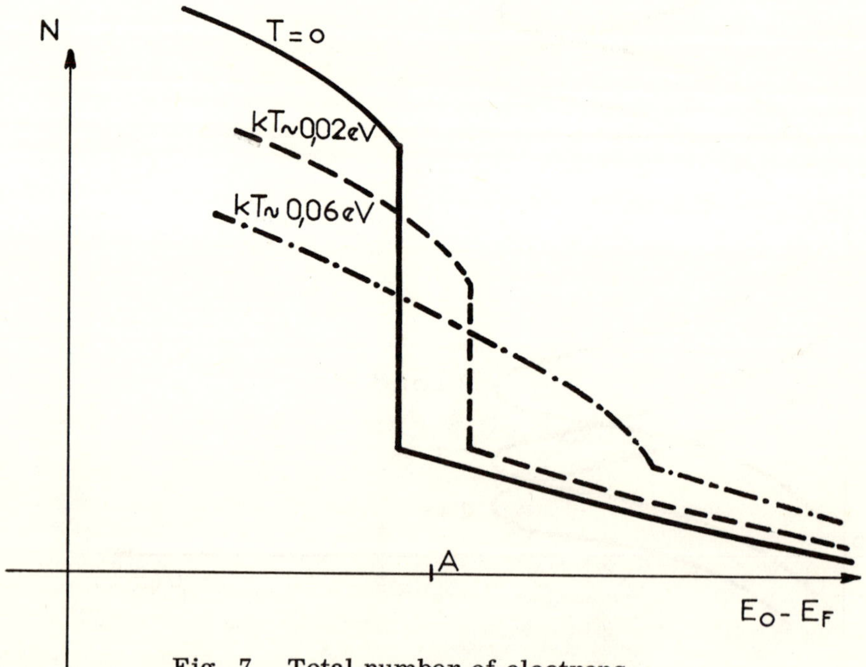

Fig. 7. Total number of electrons.

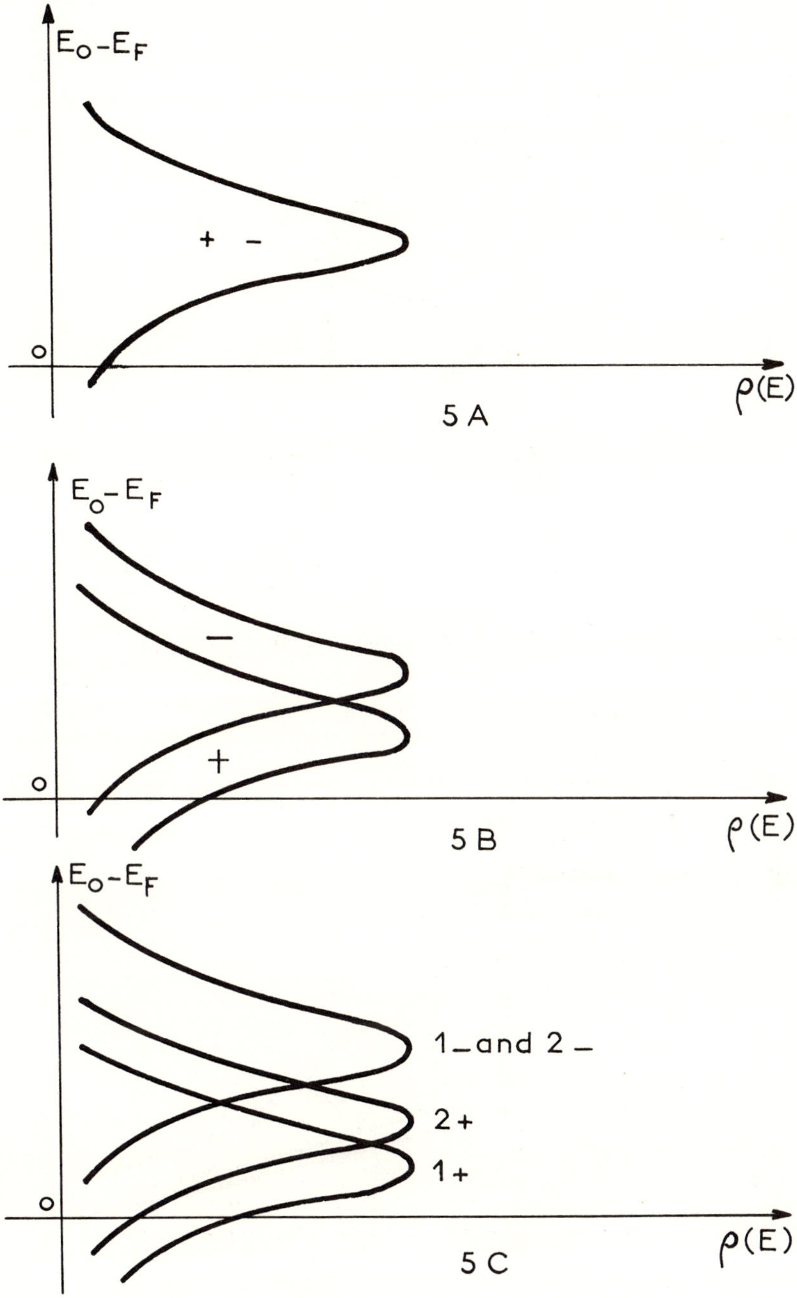

Fig. 5A, B, C. Density of states.

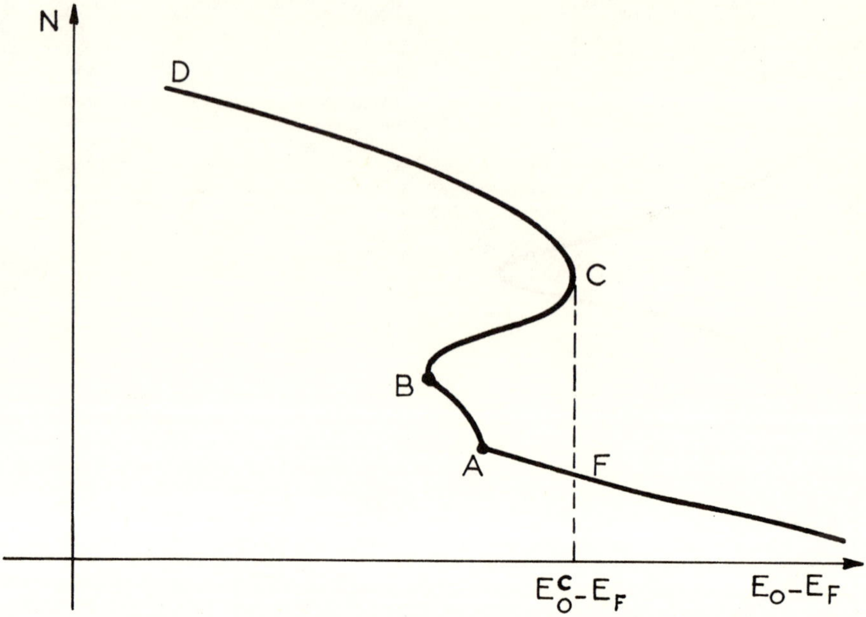

Fig. 6. Total number N of localized electrons.

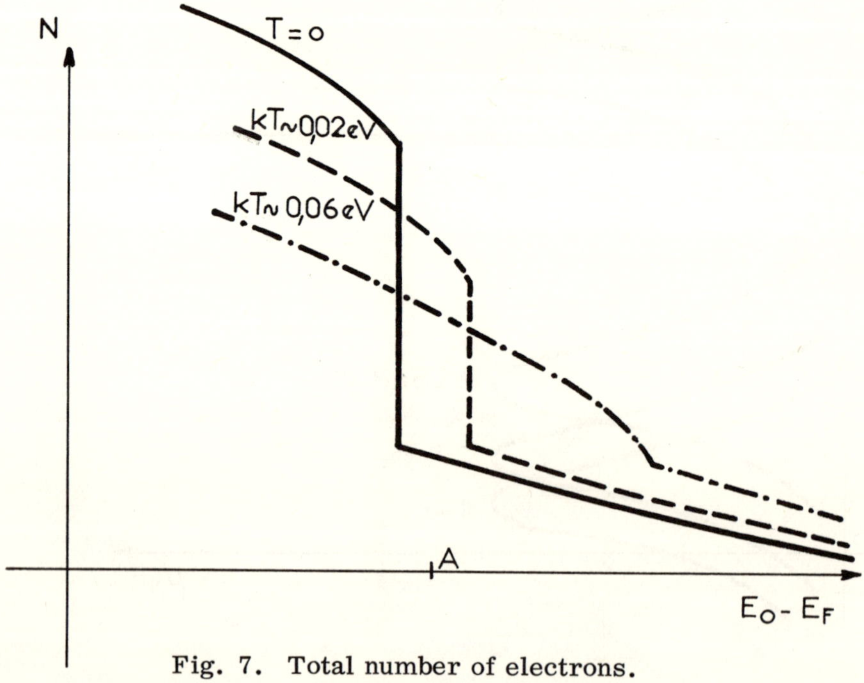

Fig. 7. Total number of electrons.

Let C be the point where the tangent is vertical. This corresponds to a critical value of E_o, viz. E_o^c. The tangent crosses the nonmagnetic portion of the curve at F. The states which are described by the curve FABC are unstable. Thus, when $E_o - E_f$ decreases through the value $E_o^c - E_f$ there is a jump to the magnetic solution with both orbital and spin magnetism. We have here a description of a first-order transition with an abrupt variation in the magnetic moment and in the number of localized electrons. The f electrons are governed by the conduction band. As $E_o - E_f$ continues to decrease, successive jumps occur as a state passes abruptly through the Fermi surface.

In rare-earth metals the direct overlap of f electrons on adjacent atoms is certainly very small since the f electrons are fairly tightly bound. We shall neglect this overlap. What is usually called the f band (2) is probably a set of magnetic states localized on each atom. We shall consider the pure metal as a set of impurities, neglecting the collective effect. Thus, if each atom is magnetized, we discuss mainly the paramagnetic region.

III. Application to the Rare-Earth Series.

In the case of rare-earths, the orbital degeneracy is seven-fold so that numerical calculations become difficult. For the beginning of the series, conditions (10) and (11) become:

(for the appearance of spin magnetism)

$$\rho(E_f)\,(U + 6J) = 1\ ; \tag{12}$$

(for the appearance of orbital magnetism)

$$\rho_+(E_f)\,(U - J) = 1\ . \tag{13}$$

U and J in the Hartree-Fock computations are of the order of 25 eV and 1 to 2 eV, respectively, not including the effect of correlation in the atom. These values may be reduced to 12 eV for U and 0.5 to 1 eV for J by including this effect. Even these values may be reduced again by the correlation which occurs when a virtual bound state exists. Δ is probably very small compared with 1 eV. We shall come back to these points later.

A. Normal Metals.

For most of the rare-earth metals the situation is such that the value of E_0 is far from a critical value. Thus the number of f electrons is nearly an integer and the magnetic moment is what one expects for a $4f^n$ configuration of the ion.

We may also consider the effect of the magnetic coupling of the moments on different atoms when the states are nearly empty or nearly full. Here, the indirect interaction via the conduction band by the admixture mechanism becomes negligible in comparison with the Rudermann-Kittel-Yosida s-f exchange mechanism (15).

B. The Case of Cerium.

In this case one is tempted to apply the theory which has been developed above.

The critical point of Fig. 2 is at a temperature of 600°K and a pressure of 20 kbar. Such a temperature corresponds to 0.05 eV. Under such pressures the displacements of levels in solids are of the order of 0.05 eV to 0.2 eV. We need a virtual state of the order of one or two tenths of an electron-volt above the Fermi surface and of the order of a few hundredths of an electron-volt wide. This virtual state would correspond to $m = -3$ and $\sigma = \frac{1}{2}$, $m = -3$ being favored by the L-S coupling.

Under these conditions, one can make a temperature dependent theory. The only change which must be made is to replace

$$n_{m\sigma} = \int^{E_f} \rho(E)\, dE \quad \text{in Eq. (4) by}$$

$$n_{m\sigma} = \int_{-\infty}^{+\infty} \rho(E) \frac{dE}{1 + \exp\left(\frac{E-E_f}{kT}\right)}$$

The effect of temperature will be important if kT is not too small compared with $E - E_f$.

Now one expects the following results. When kT increases, the magnitude of the discontinuity at $E_o{}^c - E_f$ decreases. The reason is that $\rho(E)$ must be replaced by an effective $\rho(E)$ which is smooth and has a larger width (Fig. 7). The discontinuity disappears at T = Tc.

We can then try to explain the behavior of Cerium by the following mechanism. E_o is such that at T = 0 and atmospheric pressure, we are at point A on Fig. 7, i.e., to the right of the discontinuity of the curve for T = 0. Then, when T increases, the curve changes and is displaced to the right and crosses A. We thus have a first-order transition to the γ-phase. Now, at a given temperature, if we increase the pressure, E_o will stay approximately constant but E_f will move. The value of E_f relative to the energy at the bottom of the band E_b will increase. However, the energy E_b usually decreases much more rapidly (16) so that the absolute value of E_f decreases, the point A goes to the right and we have a transition to α-Cerium. If T > Tc, the transition is a second-order one; a line should be added to the phase diagram of Fig. 2 describing a second-order transition with a continuous appearance of a magnetic moment.

The qualitative results which we have described are obtained when U and J are large compared with Δ. The discontinuity for the total number of electrons is of the order of 0.7 to 1 electron per atom at T = 0. The transition temperature Tc appears to be too large with the parameters actually used. Moreover, one should take into account the change of the Fermi energy as electrons jump from the conduction band into the f shell. This fact tends to decrease the jump at T = 0 and favors the appearance of the virtual point at lower temperatures.

As a final point for Cerium, we would like to discuss the optical absorption observed in γ-Cerium and reported by Gschneidner (2). An optical absorption is observed at room temperature at 0.076 eV with a width of the order of 0.02 eV. This evidently can be explained in this model: this absorption comes from transitions from the virtual bound state ($m = -3$, $\sigma = \frac{1}{2}$) to the conduction band. The parameters are of the right order of magnitude for this explanation.

The resistivity measurements (3) with pressure can be fairly well explained with this model. The only point to comment on is the behavior above Tc. A simple minded argument would lead one to think that one should have a big maximum in resistivity as the virtual bound state goes continuously across the Fermi surface. In fact, due to the temperature broadening of the virtual bound state, this maximum would be smaller. It would be important to check this point experimentally.

The theory could also be checked mainly by making pressure dependent magnetic studies. This could be a fundamental test of this theory.

C. Europium and Ytterbium.

Little will be said of Europium. Nevertheless, the fact that it looks like a divalent metal leads to the conclusion that under pressure or higher temperatures it could undergo a transition to a trivalent normal metal.

Let us make some comments on the case of Ytterbium. Rocher, (4) knowing only the resistivity data at room temperature, had tentatively described the resistivity maximum as due to a continuous crossing of a virtual bound state through the Fermi level. The atomic radius of Ytterbium in the BCC phase at 40 kbar suggests that the metal is trivalent and magnetic.

On the other hand, at lower temperatures in the FCC phase the resistivity data, which are not very precise, cannot be explained by this mechanism. It is also not clear why the transition should be a second-order one in that case. It would be very useful to look for the magnetic properties both in FCC and in BCC phases and to clarify that situation regarding the resistivity.

IV. Concluding Remarks.

The model which has been described suffers a number of deficiencies:

A. It is an Hartree-Fock model with the choice of a particular axis which breaks the symmetry of the problem. As a consequence, the calculation of magnetic moments, Curie constants, . . . cannot be made in a clear way.

B. We have neglected crystalline field effects. This is not very important and could be added easily to the model. In fact, it seems that the order of magnitude of this effect is not large.

C. We have neglected spin-orbit coupling. This can be more important. The introduction of spin-orbit coupling may suppress the intermediate step with only spin magnetism and may favor the direct onset of orbital and spin magnetism.

D. Another point is not well understood. Why is the virtual bound state in α-Cerium 0.1 to 0.2 eV off the Fermi energy? From the energies of the different configurations of which we spoke at the beginning of this paper, one would think that this energy would be of the order of one or two electron-volts. Is it accidental in the case of Cerium? Then, one would think that there is no reason for similar behavior to occur for Europium and Ytterbium. But it is clear that there is something in common between these metals. Then one is tempted to believe that there is a fundamental reason. Perhaps the abilities of conduction and f electrons to screen the nuclear charge are in severe competition and lead to this surprising behavior.

Before ending, we would like to emphasize that this paper reports work which is in progress. A number of things remain to be discussed in more detail, both in the theoretical approach and in the applications to rare-earth metals and alloys. It is our hope to present a detailed paper in the near future.

REFERENCES

1. H. Tracy Hall, J. Dean Barnett, Leo Herrill Science 139, 111 (1963).

2. K. A. Gschneidner, Rare Earth Research III, Proceedings of the Fourth Conference on Rare-Earth Research, Leroy Eyring, Ed., (Gordon and Breach, 1964).

   K. A. Gschneidner, Jr. and R. Smoluchowski, J. Less-Common Metals 5, 374 (1963).

3. A. Jayaraman, Phys. Rev. 137, A179 (1965).

4. Y. A. Rocher, Advances in Phys. 11, 233 (1962).

5. R. I. Beecroft and C. A. Swenson, J. Phys. Chem. Solids 15, 234 (1960).

6. E. Ponyatovskii, Dokl. Akad. Nauk. SSSR 120, 1021 (1958) [Soviet Physics - Doklady 3, 498 (1958)].

7. A. C. Gossard, V. Jaccarino and J. H. Wernick, Phys. Rev. 133, A881 (1964).

8. P. W. Bridgman, Proc. Am. Acad. Arts Sci. 83, 1 (1954).

9. R. A. Stager and H. G. Drickamer, Science 139, 1284 (1963).

10. H. T. Hall and L. Merrill, Inorg. Chem. 2, 618 (1963).

11. P. C. Souers and G. Jura, Science 140, 481 (1963).

12. J. Friedel, Can. J. Phys. 34, 1190 (1956).

    A. Blandin and J. Friedel, J. Phys. Radium 19, 573 (1958).

13. P. W. Anderson, Phys. Rev. 124, 41 (1961).

14. Kei Yosida, Ayao Okiji and Sōshin Chikazumi, Prog. Theor. Phys. 33, 559 (1965).

15. B. Caroli, A. Blandin, J. Friedel (to be published).

16. N. F. Mott and H. Jones, Metals and Alloys, Oxford (1936).

Microwave Resonance in Metals at High Pressures

Glen E. Everett and A. W. Lawson
Department of Physics
University of California
Riverside, California

Previous resonance measurements (1,2) at high pressures have been based on feeding r.f. power to a resonant system by means of a coaxial line. The coaxial method will not operate at sufficiently high frequencies to be useful for some of the broad, low field lines in ferromagnetic materials. It also suffers from the fact that the resonant frequency of the cavity is greatly affected by the pressure dependence of the dielectric constant of the pressure transmitting medium.

A waveguide coupling system has been developed in this laboratory which overcomes the frequency limitations of the coaxial techniques. A schematic diagram of this system is shown in Fig. 1 and a more complete description of constructional details has been given elsewhere (3). The essential feature of the design is a sapphire cone which acts both as a resonant cavity and a pressure seal. The cone resonates as a consequence of the impedance mismatch on the low pressure side and the presence of a metal sample placed on the surface of the cone on the high pressure side. The pressure, generated by standard techniques, is transmitted to the Be-Cu bomb via non-magnetic capillary tubing; the system is mounted between the poles of an electromagnet and the measurements are made using standard microwave techniques.

There are several features of this system that should be noted. 1. The resonant frequency of the sapphire cone is only very slightly pressure dependent ($\Delta\nu/\nu$=.01%/kbar). Also, the Q of the system does not depend on the pressure. This is very important and convenient for ferromagnetic systems since the experiments can be carried out at constant frequency. This

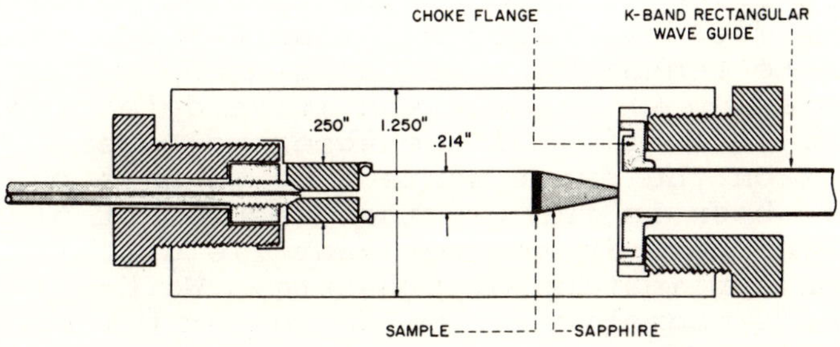

Fig. 1

Microwave bomb showing the sample terminating the sapphire cavity. Not shown is the spring holding the sample in place.

eliminates the need for the large corrections due to changes in operating frequency characteristic of the coaxial system. 2. The coupling, being directly from standard rectangular waveguide, eliminates the need for impedance matching sections and also permits the easy control of the polarization of the microwave fields. 3. The small diameter of the bomb makes the attainment of the necessary magnetic fields possible using standard laboratory electromagnets. While this technique has been applied exclusively to work on metallic systems, it should be possible to extend it to insulators.

For the work undertaken so far, we have used two methods of detection. For strong signal systems such as iron, the detected microwave crystal voltage is applied to a differential voltmeter, nulled at zero field, and the change in voltage vs. field is recorded on an X-Y recorder. The alternate technique also being used is that of field modulation with phase sensitive detection to measure the derivative of the absorbed power. The limitations on the experiments that we have performed have been in the precise measurement of the very small changes in the resonant fields that result upon application of pressure. We have constructed a magnetic field controller using the derivative of the power absorption as an error signal to lock the magnetic field very precisely to the point of resonant absorption. Based on the work of Benedek (4) this will permit a determination of the magnetic field to the very high degree of precision needed.

Our work has been on the ferromagnetic resonance in iron and nickel and on the paramagnetic resonance in gadolinium and cerium. We will now describe the results of some of these experiments.

Gadolinium. Liu (5) has attempted to explain the results of Patrick (6) on the change of Curie temperature vs. pressure in gadolinium. Liu pointed out that on the basis of an indirect exchange theory of ferromagnetism, the change in Curie temperature, T_C, is related to both the change in the magnitude of the exchange interaction, $|J|$, and in the density of states at the Fermi surface of the conduction electrons, $N(\varepsilon_F)$.

Furthermore, he showed that the g-factor, or more properly, the departure of g from the free electron value, Δg, is also related to these same two parameters. The logarithmic derivatives vs. volume for T_c and Δg can be written as

$$\frac{\partial \ln T_c}{\partial \ln V} = \frac{\partial \ln N(\varepsilon_F)}{\partial \ln V} + \frac{\partial \ln |J|^2}{\partial \ln V} \qquad (1)$$

$$\frac{\partial \ln \Delta g}{\partial \ln V} = \frac{\partial \ln N(\varepsilon_F)}{\partial \ln V} + \frac{\partial \ln |J|}{\partial \ln V} \qquad (2)$$

From these equations it can be seen that a determination of the pressure dependence of Δg used in conjunction with the results of Patrick for T_c, will permit a determination of the volume dependence of $|J|$ and $N(\varepsilon_F)$ separately. An estimate of the magnitude of the expected change in the resonant field vs. pressure has been made and is about 1 gauss/kbar at 8,000 gauss. Our initial result of 70 gauss/kbar arises from the large contribution of the pressure dependent susceptibility. The resonant frequency ω is related to the applied field H in the paramagnetic region by the relation

$$\omega = g\,\beta_o H \sqrt{1+4\pi\,\chi} \qquad (3)$$

where β_o is the Bohr magneton and χ is the paramagnetic susceptibility. This equation applies to the case of an infinite plane with the magnetic field applied parallel to the plane. The rapid dependence of ω on χ is a consequence of the proximity of room temperature to the Curie temperature in gadolinium. The effect of χ can be made smaller by working at elevated temperatures (100°C). We are at present measuring the susceptibility vs. temperature and pressure. These data will be used in connection with a much more precise measurement of the change in the resonant field to determine Δg.

Iron. We have measured the ferromagnetic resonance in iron to 6 kbar using both the direct and derivative techniques. Typical data are shown in Fig. 2 and Fig. 3. The samples in the form of very thin disks (.220" dia. x .005" thick) were prepared by spark cutting and planing 99.99% pure iron. The samples were carefully annealed in vacuo and subsequently electropolished. For the ferromagnetic case, the resonant condition is given by

$$\omega = g\beta \sqrt{BH} \qquad (4)$$

where B is the magnetic induction. For the frequencies used B can be written as

$$B = H + 4\pi M_s \qquad (5)$$

where M_s is the saturation magnetization. The most recent measurement of M_s vs. pressure (7) gives a change in the magnetic moment per atom of $\Delta\sigma/\sigma = -3.04\text{x}10^{-7}/\text{kg/cm}^2$. An estimate of the contribution of the change in M_s to the change in the resonant field is approximately 0.1 gauss/kbar. This is about 10 times smaller than that expected from the change in the g-values. We are at the present able to determine the position of the resonant field to $\pm$ 10 gauss. With the incorporation of the field control system, locking to the resonant field, we will be able to measure to better than the $\pm$0.1 gauss necessary for this experiment.

Nickel. The experiments that have been performed on nickel are of a much more preliminary nature. The ferromagnetic resonance was observed as a function of pressure. At 5 kbar the equipment was operated in the derivative mode, the magnetic field set to give the zero derivative corresponding to resonant absorption, the gain of the system was greatly increased and subsequently the pressure was slowly reduced. From the absence of any change in the derivative signal we can infer in a very qualitative way, that the pressure dependence in nickel will be very small. As in the case of both iron and gadolinium a measurement of any change will necessitate the use of the field locking system described earlier.

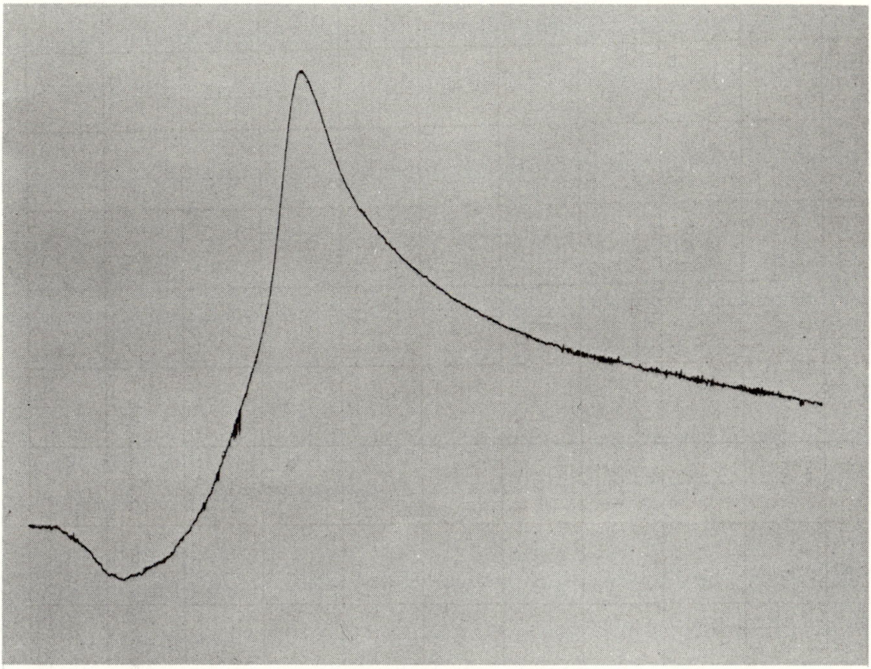

Fig. 2

Ferromagnetic resonance absorption in iron taken at 5 kbar. The field range is from 0 to 9,100 gauss. The microwave frequency is 22.5 Gc/sec.

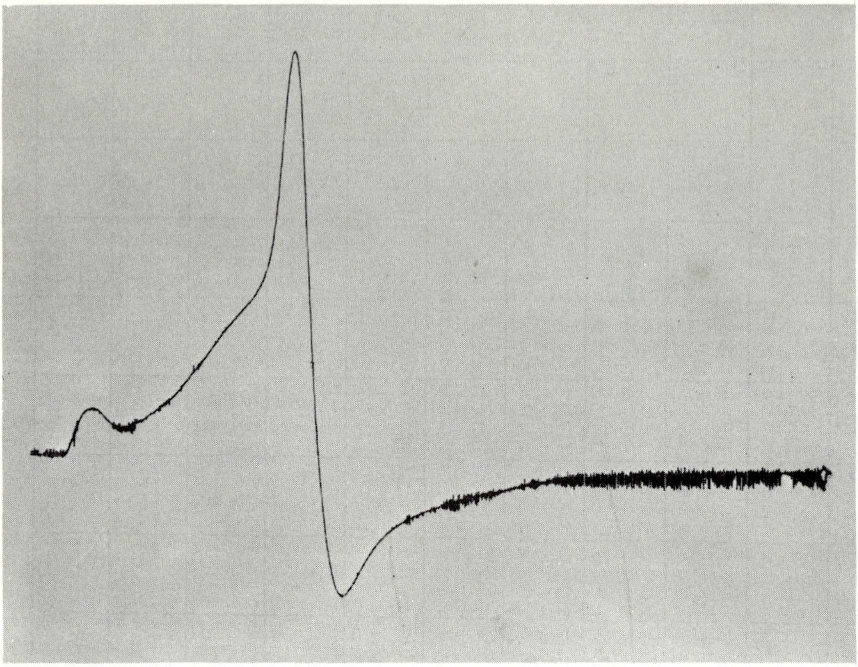

Fig. 3

The derivative of the ferromagnetic resonance absorption in iron taken at atmospheric pressure. The field range is from 0 to 9,100 gauss. The microwave frequency is 22.5 Gc/sec.

Cerium. The γ-α phase transition in cerium at 7-9 kbar and at room temperature is from a f.c.c. to a f.c.c. structure with a 20% decrease in volume. The origin of the transition is now generally assumed to be associated with the promotion of the single 4f electron of the γ-phase into the conduction band. Efforts to substantiate this on the basis of resistivity and Hall effect measurements are inconclusive in that each effect depends not only on the number of conduction electrons but also on the relaxation times. Our intention of measuring the E.P.R. spectrum through the transition and thereby determining the relative number of 4f electrons in both phases has been delayed by the necessity of a careful study and interpretation of the zero pressure spectrum. This work has now been completed (8). The signals encountered in this experiment are much more complex than was originally supposed on the basis of previous experiments (9). In particular, rather than just a single isotropic transition associated with the ground state doublet of the γ-phase, it has been necessary to include the effects of a trigonal distortion on this line as well as exchange interactions with ions in the quartet state. While the investigation of the nature of the γ-α transition remains as a primary interest, the possibility of determining the pressure dependence of the exchange interaction and of the crystal field splitting is most attractive. The signal strengths encountered in cerium, while much weaker than in, say, gadolinium, will present no limitation on these investigations owing to the immediate availability of signal averaging equipment that conveniently increases the signal-to-noise ratio by a factor of 10 to 100.

In summary, the necessary tools for the precise determination of microwave resonance in metals under pressure appear to be at hand. The authors gratefully acknowledge the support of the Atomic Energy Commission.

References

1. W. M. Walsh, Jr., and N. Bloembergen, Phys. Rev. 108, 904 (1957).
2. G. Benedek, Chap. 9, p. 251, Solids under Pressure, edited by William Paul and Douglas Warschauer, McGraw Hill Book Co., Inc. (1963).
3. R. G. Goodrich, Glen E. Everett and A. W. Lawson, Rev. Sci. Instr. 35, 1596 (1964).
4. G. Benedek and T. Kushida, J. Chem. Phys. Solids 5, 241 (1958).
5. S. H. Liu, Phys. Rev. 127, 1889 (1962).
6. L. Patrick, Phys. Rev. 93, 384 (1954).
7. E. Tatsumoto, H. Fujiwara, H. Tange and Y. Kato, Phys. Rev. 128, 2179 (1962).
8. R. G. Goodrich and G. E. Everett, to be submitted to the Physical Review.
9. S. G. Salikhov, Zh. eksper. theor. Fiz. 26, 439 (1954).

Hydrostatic Pressure Effect on the Magnetic Properties of Ferromagnetic Metals and Alloys

Hiroshi FUJIWARA, Tetsuhiko OKAMOTO
and Eiji TATSUMOTO

Department of Physics, Hiroshima University,
Hiroshima, Japan

Introduction

The pressure effect on the magnetic moment σ_s at any temperature may be a consequence of the pressure effect not only on the intrinsic magnetic moment σ_{so} at 0°K but also on the Curie temperature T_c. So far the measurement of the pressure effect on σ_s has been made on ferromagnetic metals at some definite temperatures such as 4.2°K and room temperature (1). However, the pressure effect on both σ_{so} and T_c may be estimable from the temperature dependence of the pressure effect on σ_s observed over a wide temperature range. In our previous paper (2), the measurement of the hydrostatic pressure effect on σ_s has been made on Fe and Ni over a range from -73°C to 100°C and recently by Tange on a 24 at. % Cu-Ni alloy (3). Compressibility is one of the basic quantities in high pressure research. In ferromagnetics it would be expected to have an anomaly around T_c.

In this report, the present state of the activities of the high pressure group at Hiroshima University on the measurement of the compressibility as well as the hydrostatic pressure effect on σ_s and T_c will be briefly described. The measurements have been made mostly on Cu-Ni alloys for the two reasons that T_c lies in an available temperature range and no systematic experimental investigation has been done over the whole composition range.

Experimental

General

The standard Bridgman press has been employed to generate hydrostatic pressures (4). Petroleum ether was used as the transmitting medium and the highest pressure applied was 15 kbar. The pressure was determined with manganin wire which was calibrated by the help of the freezing pressure of mercury at 0°C (5). The specimens used were polycrystalline rods of Fe, Ni and Cu-Ni alloys and each element was commercial electrolytic, 99.8 % in purity. For Cu-Ni alloys careful heat treatments were done (3), so as to homogenize the composition. It was possible to control the temperature to within ±0.1°C in all measurements.

Pressure Effect on the Magnetic Moment per Unit Mass σ_s

The pressure effect on σ_s is derived from the observed pressure effect on the saturation flux. The experimental procedures for the observation of the pressure effect on the saturation flux have already been reported briefly (4) and will briefly be described here again. The measurement of the pressure effect on the saturation flux has previously been done by an electromagnet, whose pole gap and pole surface were 80 mm and 100 mm, respectively. Sample and dummy pressure bombs, 25 mm in outer diameter and 42 mm in length, were made of precipitation-hardened beryllium copper. Two specimens, 5.5 mm in diameter and 14.5 mm in length, were separately inserted in both bombs and the pressure was applied only in the sample bomb. Two search coils of the same size were separately disposed on both bombs. They were electrically connected in series and in the opposite sense so that, in the absence of pressure, no resultant flux could be observed with a ballistic galvanometer and lamp-and-scale. The field applied lay in the lengthwise direction of the specimen and the intensity was sufficient to magnetize the specimen to saturation. The change in the saturation flux, which was caused by the increase in pressure, was obtained by observing the deflection of the galvanometer. The highest pressure applied was 15 kbar and the measurements were done over a range from -73°C to 100°C.

Pressure Effect on T_c

The pressure effect on T_c has been obtained in two ways: (i) by the extrapolation of the temperature dependence of the pressure effect on σ_s near T_c (3) and (ii) by the direct observation of the shift of a peak appearing at T_c in the magnetoresistance vs. temperature curve, which is caused by an applied pressure (6). The magnetoresistance has been measured in an electromagnet with a water jacket, the pole gap and the pole diameter being 55 mm and 110 mm, respectively. In this measurement, a cylindrical pressure bomb made of beryllium copper, 8 mm in inner diameter, 30 mm in outer diameter and 120 mm in length, was employed. The specimen used was 1 mm in diameter and 40 mm in length and was placed in the bomb so as to be subjected to a field perpendicular to the lengthwise direction of the specimen. The magnetoresistance, i.e., the change in resistance, ΔR, caused by the increase in field, ΔH, from 8,000 oe to 18,000 oe, was measured by the use of a galvanometer and lamp and scale. The highest pressure applied was 8 kbar.

Measurements of the Linear Compressibility

A new technique has been developed by Tatsumoto, et al., utilizing an Advance wire which is usually used as a strain guage. A guage wire elastically stretched is fixed by gluing both its ends on a specimen. The relative change in resistance of the wire, $(\Delta R/R)_s$, which is caused by the application of pressure P, is given by

$$(\Delta R/R)_s = (\Delta R/R)_f - (K_s - K_w)GP. \qquad (1)$$

Here, $(\Delta R/R)_f$ indicates the relative change in resistance of the wire itself with a pressure P; K_s and K_w are the linear compressibilities $(-\Delta \ell/\ell \Delta p)$ of the specimen and the wire, respectively; G is the gauge factor of the wire. Accordingly, K_s can be obtained by substituting the observed values of $(\Delta R/R)_s$, $(\Delta R/R)_f$, K_w and G into Eq. (1). Recently an improvement was made which permitted repeated use of the same strain gauge wire. A specimen 1.7 mm in diameter and 38 mm long fitted with the wire was inserted into a hollow brass cylinder 11 mm in diameter and 50 mm long. Measurements could be made over the temperature range of -73°C to 150°C at pressures up to 12 kbar.

Experimental Results and Discussion

Pressure Effect on σ_s and T_c

The change in the saturation flux Φ_s', $\Delta\Phi_s'$, observed at a fixed temperature varies almost linearly with pressure for Cu-Ni alloys as well as for Fe and Ni as reported previously (2,4). Thus, a constant gradient $\Delta\Phi_s'/\Delta p$ has been obtained. Here, Φ_s' indicates the flux picked up by a search coil, the diameter of which is larger than that of the specimen. The pressure coefficient of σ_s, i.e., $\partial\sigma_s/\sigma_s\partial p$, is given by

$$\partial\sigma_s/\sigma_s\partial p = \partial\Phi_s/\Phi_s\partial p - k/3. \qquad (2)$$

Here, Φ_s is the flux picked up by a coil with the same diameter as that of the specimen and k is the volume compressibility. However, Eq. (2) is not applicable to the present experiment since the pressure coefficient observable is not $\partial\Phi_s/\Phi_s\partial p$ but $\partial\Phi_s'/\Phi_s'\partial p$. In the present measurement, the discrepancy between the results from $\partial\Phi_s/\Phi_s\partial p$ and $\partial\Phi_s'/\Phi_s'\partial p$ can not be neglected. However, it may be corrected and the correction factor is estimated experimentally. The compressibility in Eq. (2) has been measured exactly with the device previously mentioned for the specimens used.

The pressure coefficients of σ_s thus obtained are plotted as a function of the reduced temperature T/T_c for Fe and Ni in Fig. 1. The values in this figure are not appreciably different from the previous values (2) in which the correction mentioned above was not made. The values at 4.2°K obtained by Kondorskii (7) are also cited in this figure.

The pressure effect on σ_s has also been measured on Cu-Ni alloys, the copper contents of which are 4, 13, 18, 25 and 29 at. % Cu. The pressure coefficients $\partial\sigma_s/\sigma_s\partial p$ are plotted as a function of composition in Fig. 2, where the values for Ni given in Fig. 1 and for 24 at.% Cu due to Tange (3) are also cited. Each curve changes monotonically with composition. The present experiment shows that the pressure coefficient isotherm at 200°K remains negative while those at 300°K and 350°K become positive as Cu content increases. The triangles in this figure indicate the values at 4.2°K obtained by substituting the values of $\partial\Phi_s/\Phi_s\partial p$ due to Kondorskii (7) and those of the compressibilities estimated by the present experiment in Eq. (2). As an example, the temperature dependence of the pressure coefficient $\partial\sigma_s/\sigma_s\partial p$ for 25 at.% Cu is shown in Fig. 3. This curve suggests that σ_{so} decreases, while T_c increases, with pressure.

The change of Curie temperature with pressure, $\Delta T_c/\Delta p$, has been estimated by determining the Curie temperature both at atmospheric pressure and at a pressure p by two methods: (i) the curves of σ_s^2 and $(\sigma_s + \Delta\sigma_s)^2$ vs. temperature near the Curie temperature are extrapolated ($\Delta\sigma_s$ can be estimated from the pressure effect on σ_s); (ii) the peaks of the magnetoresistance vs. temperature curves taken at atmospheric pressure and at pressure p are used. The values of $\Delta T_c/\Delta p$ have been obtained for 24 at.% (3) and 25 at.% Cu from (i) and for 22 and 28 at.% Cu from (ii), respectively. For Ni and Monel, $\Delta T_c/\Delta p$ has been measured by Patrick (8) by the use of an A.C. transformer method. The experimental results including those of Patrick are plotted as a function of composition in Fig. 4. It may be assumed that $\Delta T_c/\Delta p$ is positive and decreases

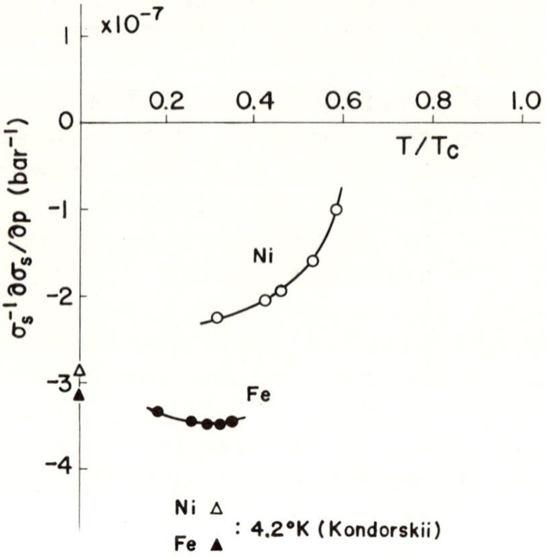

Fig. 1. A plot of $\partial\sigma_s/\sigma_s\partial p$ as a function of reduced temperature T/T_c.

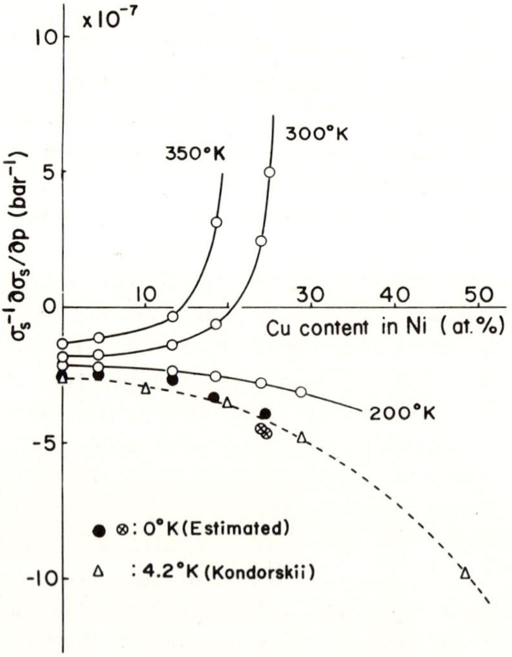

Fig. 2. A plot of $\partial\sigma_s/\sigma_s\partial p$ as a function of Cu content.

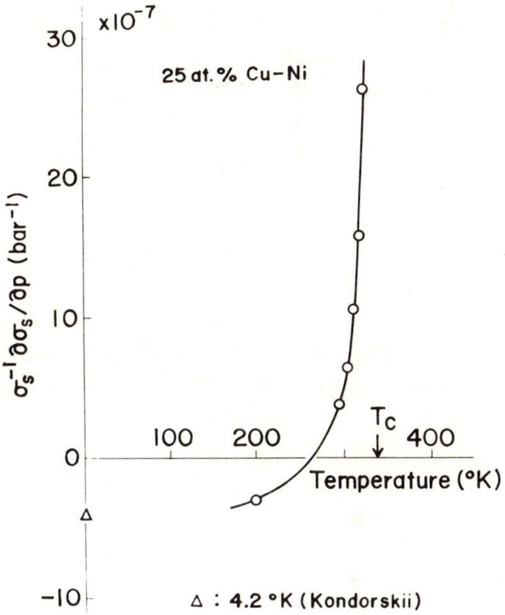

Fig. 3. A plot of $\partial\sigma_s/\sigma_s\partial p$ as a function of temperature for 25 at.% Cu-Ni alloy.

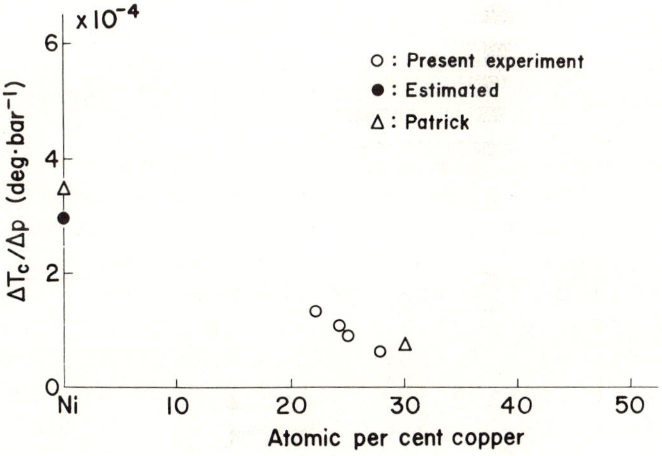

Fig. 4. A plot of $\Delta T_c/\Delta p$ as a function of Cu content.

monotonically with increasing copper concentration.

In the present report, the analysis of the results will be done on the basis of the equation (9)

$$\partial T_c/T_c\partial p = \frac{\partial\sigma_{so}/\sigma_{so}\partial p - \partial\sigma_s/\sigma_s\partial p}{T(\partial\sigma_s/\sigma_s\partial T + 3\alpha\partial\sigma_s/k\sigma_s\partial p)} \qquad (3)$$

(i) The pressure coefficients, $\partial\sigma_{so}/\sigma_{so}\partial p$ and $\partial T_c/T_c\partial p$, and thus $\Delta T_c/\Delta p_o$ may be estimated by the use of the values of $\partial\sigma_s/\sigma_s\partial p$ and $\partial\sigma_s/\sigma_s\partial T$ measured at various temperatures. For Ni, the values estimated are -2.7×10^{-7} bar^{-1} for $\partial\sigma_{so}/\sigma_{so}\partial p$ and 3.0×10^{-4} deg·bar^{-1} for $\Delta T_c/\Delta p$, where the values of $\partial\sigma_s/\sigma_s\partial T$ were referred to Weiss, et al. (10). These values are in fairly good agreement with -2.9×10^{-7} bar^{-1} due to Kondorskii at 4.2°K (7) and 3.5×10^{-4} deg·bar^{-1} due to Patrick (8), respectively. These estimated values for Ni are given in Fig. 3 and Fig. 4, respectively. The values of $\partial\sigma_{so}/\sigma_{so}\partial p$ estimated for Cu-Ni alloys are plotted with closed circles in Fig. 2. (ii) The pressure coefficient of $\partial\sigma_{so}/\sigma_{so}\partial p$ for 24 and 25 at.% Cu is also evaluated by substituting the observed values of $\partial\sigma_s/\sigma_s\partial p$, $\partial\sigma_s/\sigma_s\partial T$ and $\partial T_c/T_c\partial p$ in Eq. (3), since the value of $\Delta T_c/\Delta p$ has been obtained. The values estimated are indicated by a circle with a cross in Fig. 2. It is to be noted in Fig. 2 that the estimated values of $\partial\sigma_{so}/\sigma_{so}\partial p$ and those by Kondorskii (7) appear to lie on the same curve.

The present experiment thus appears to lead one to the conclusion that for Cu-Ni alloys of 0-30 at.% Cu the application of pressure depresses σ_{so} while it raises T_c. However, a more reliable conclusion will be reported in the near future, since the same observations are still being continued on Cu-Ni alloys with larger Cu concentration. Furthermore, direct observation of the pressure effect on T_c is also in progress by the use of the magnetoresistance effect.

Linear Compressibility

For Fe and Ni, the linear compressibility was measured in a range from -73°C to 120°C and it was found that the temperature coefficient is positive and nearly constant for both Fe and Ni. The compressibilities at room temperature are in good agreement with those due to Bridgman (11). From Bridgman's data (12) for Ni rich Cu-Ni alloys and for pure Cu, the compressibility in the ferromagnetic region appears to be different from that in the paramagnetic region. In the present experiment, the linear compressibility was measured at room temperature over the whole composition range of Cu-Ni alloys and plotted as a function of composition in Fig. 5. The linear compressibility observed increases almost linearly with increasing Cu concentration in the ranges from Ni to 20 at.% Cu and from 50 at.% Cu to Cu. However, an anomalous increase is found in a composition range where T_c is in the neighborhood of room temperature. This result suggests that at a temperature T, the linear compressibility of a specimen in the ferromagnetic state is appreciably different from that in the corresponding paramagnetic state where the spontaneous magnetization would disappear. This effect may be due to the appearance of the spontaneous magnetization. Thus the difference in linear compressibility would be given by the difference between the extrapolation of the dotted line and the curve plotted on the Ni rich side in Fig. 5. As a result of an empirical analysis, K_T, the linear compressibility at T, in a ferromagnetic state appears to be expressed adequately by the formula

$$K_T = K_{T,\sigma_s=0}(1 - A(\sigma_s/\sigma_{so})^{\frac{1}{2}}). \qquad (4)$$

Here A=0.09 and $K_{T,\sigma_s=0}$ indicates the linear compressibility at the same temperature in the assumed paramagnetic state and σ_s and σ_{so} are the magnetic moments at T and 0°K, respectively. In Fig. 6 the linear compressibility is plotted as a function of temperature for the 27 at.% Cu alloy which exhibits a similar anomaly around T_c. The dotted line in

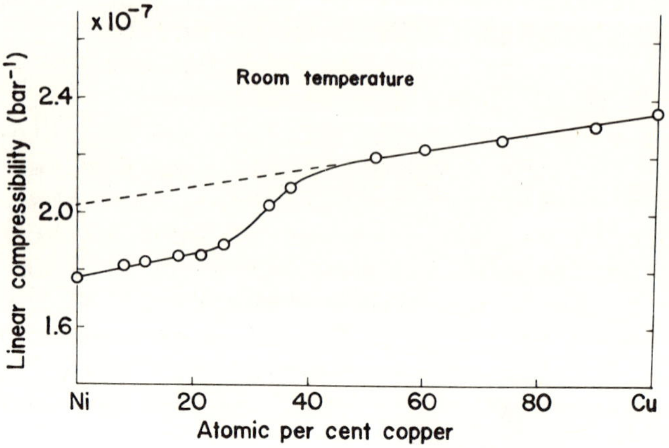

Fig. 5. A plot of linear compressibility as a function of Cu content.

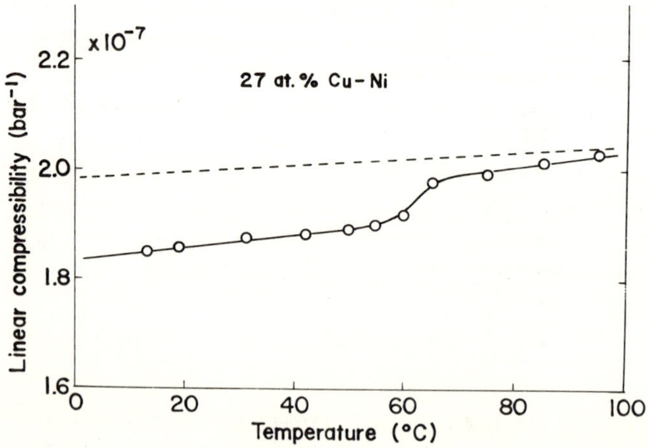

Fig. 6. A plot of linear compressibility as a function of temperature for 27 at.% Cu-Ni alloy.

this case is drawn to satisfy Eq. (4). Thus the slope of the line is almost the same as that of the 50 at.% Cu-Ni alloy. Further details of this investigation will appear in the near future.

References

1. J. S. Kouvel, Solids under Pressure (ed. by W. Paul and D. M. Warschauer, McGraw-Hill Book Co., Inc., New York, 1963), Chapt. 10; Metallurgy at High Pressures and High Temperatures (ed. by K. A. Gschneidner, Jr., M. T. Hepworth and N. A. D. Parlee, Gordon and Breach Science Publishers, New York, 1964), Chapt. 8.
2. E. Tatsumoto, H. Fujiwara, H. Tange and T. Hiraoka, J. Phys. Soc. Japan 18, 1348 (1963).
3. H. Tange, J. Sci. Hiroshima Univ., Ser. A-II, 29 (1965) in press.
4. E. Tatsumoto, T. Kamigaichi, H. Fujiwara, Y. Kato and H. Tange, J. Phys. Soc. Japan 17, 592 (1962).
   E. Tatsumoto, H. Fujiwara, H. Tange and Y. Kato, Phys. Rev. 128, 2179 (1962).
5. P. W. Bridgman, The Physics of High Pressure (G. Bell and Sons, Ltd., London, 1958).
6. N. P. Grazhdankina, L. G. Gaidukov, K. P. Rodionov, M. I. Oleinik and V. A. Shchipanov, Soviet Physics - JETP 13, 297 (1961).
7. E. I. Kondorskii and V. L. Sedov, Soviet Physics JETP 11, 561 (1960).
8. L. Patrick, Phys. Rev. 93, 384 (1954).
9. J. S. Kouvel and R. H. Wilson, J. Appl. Phys. 32, 435 (1961).
10. P. Weiss and R. Forrer, Ann. Phys. (Paris) 5, 153 (1926).
11. P. W. Bridgman, Proc. Am. Acad. Arts Sci. 84, 131 (1957).
12. P. W. Bridgman, Proc. Am. Acad. Arts Sci. 77, 187 (1949).

Magnetic Behavior of Rare Earth Metals at High Pressures

Lawrence Baylor Robinson,
Frederick Milstein,* and Swie-In Tan*

Department of Engineering
University of California
Los Angeles

and

K. F. Sterrett
Space Physics and Chemistry Laboratories,
Northrop Corporation, Los Angeles, California

I. INTRODUCTION

It is known that cooperative magnetic effects (e.g., ferromagnetism or antiferromagnetism) are exhibited only by certain transition metals and not by others. This is largely because the exchange interaction, which is responsible for the alignment of the elementary magnetic dipoles, depends critically upon the interatomic distance in the crystal. The exchange interaction in a ferromagnetic or antiferromagnetic material would therefore be expected to vary when the material is subjected to high pressure. This variation would be manifested by a change of the Curie or Néel temperature of the material.

Experimental investigations of the effect of hydrostatic pressure on the magnetic transition temperatures of the ferromagnetic rare earth metals are presently underway. The purpose of this paper is to present the results of these investigations for gadolinium (Gd)(1), terbium(Tb), and dysprosium (Dy). Some important contributions have been achieved in this research. They are: (1) the observation of the disappearance of the Curie point of an element with increasing pressure, (2) direct magnetic measurement of pressure effect on Néel temperature to 25 kilobars (kbar), and (3) measure-

*Contribution supported in part by the NASA Space Science Center at UCLA, Project Number NSG 237-67 (W. F. Libby).

ment of magnetic transitions at 40°K and 25 kbar simultaneously.

II. EXPERIMENTAL WORK

A. Procedure

The magnetic properties of the rare earth metals under pressure were correlated with the secondary voltage from a small toroidal transformer, the core of which was the rare earth metal itself. As the sample changes from a nonferromagnetic to a ferromagnetic state, a large increase in permeability and hence in secondary voltage is observed (the Curie point). A transition from a paramagnetic state to an antiferromagnetic state (the Néel point) is indicated by a cusp in the secondary voltage curve.

For the work with Gd, pressures up to about 40 kbar were obtained by use of a piston-cylinder high pressure press located in Professor G. C. Kennedy's Laboratory. A similar double-ram press was used for Tb and Dy; the lower temperatures needed were achieved by means of a new 4-wall stainless steel dewar, designed and constructed by Dr. K. F. Sterrett and made available at Northrop Space Laboratories by Dr. J. T. Thomas. Figure 1 shows a cross section of this dewar and die stack assembly. To improve the insulation a piece of single crystal mica, about 1mm thick was inserted between the top of the stack and the dewar lid; similarly, fused alumina or mica was inserted between pistons. For the terbium runs, temperatures down to 77°K were obtained by filling the dewar slowly with liquid nitrogen. Heating was accomplished by a heater wound around the platen (or high pressure cylinder). Both heating and cooling were carefully controlled as not to exceed 2.5°K per minute at all times. The heating and cooling rate around the transition temperatures were usually maintained at about 1.5°K per minute. The lower temperatures were obtained with liquid helium. During these determinations the larger piston was sprayed with liquid nitrogen to reduce heat leakage.

Figure 2 shows a cross section of the sample

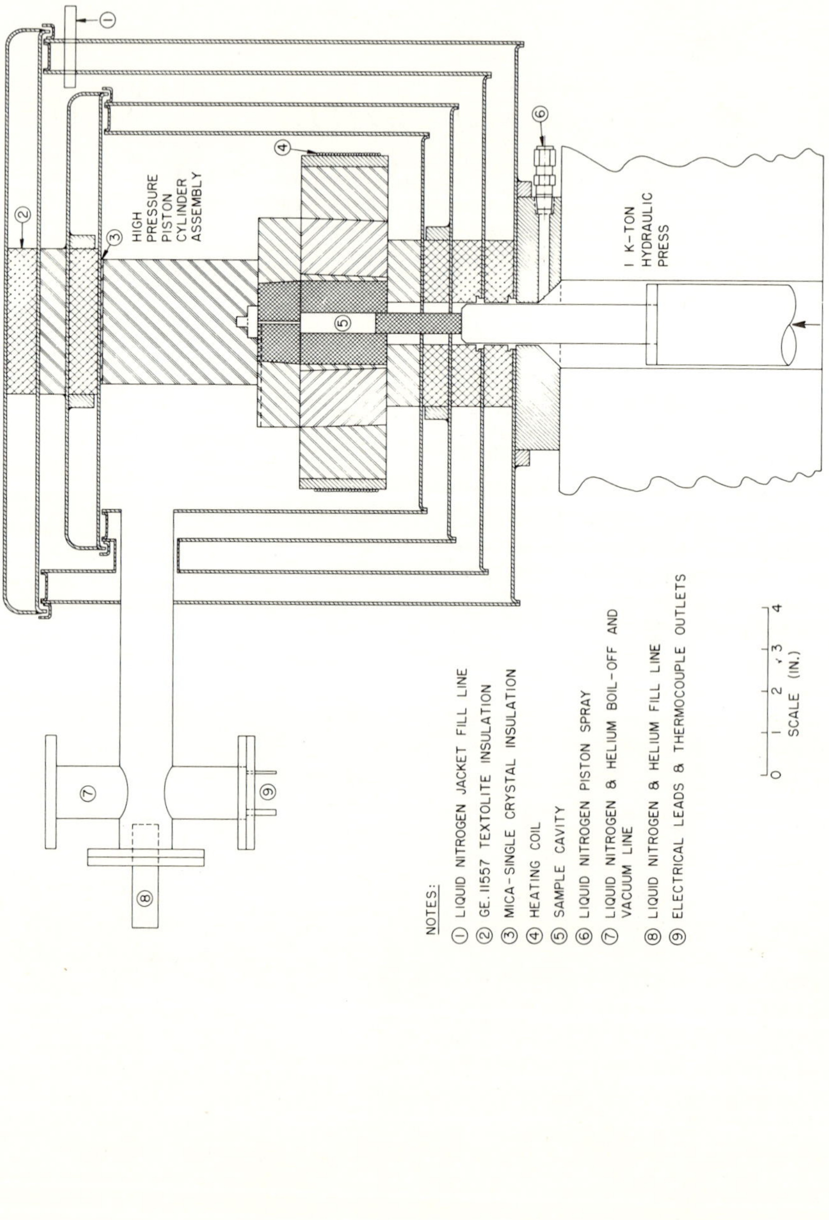

Figure 1. High Pressure Dewar and Piston-Cylinder Assembly

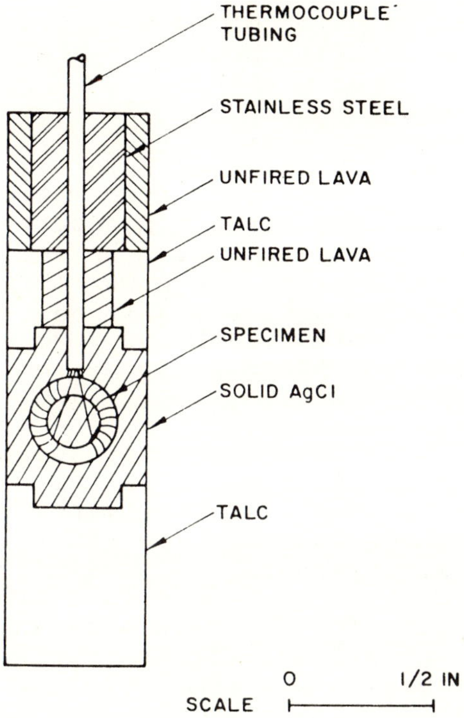

Figure 2. Cross Section of Sample Capsule Assembly

capsule assembly. The rare earths used were certified by the vendors to be 99.9% pure. The specimens were machined in the form of small toroids which were coated with epoxy and covered with a layer of masking tape to prevent the windings from shorting and to prevent the sample from reacting with the silver chloride.

B. Results

1. Gadolinium

The interesting results in the investigation of Gd were:

(a) The Curie point of Gd decreased linearly with increasing pressure with a slope of -1.60 ± 0.05°C/kbar, up to a pressure of 21.5 kbar.

(b) At pressures greater than 26.7 kbar the expected Curie temperature disappeared, indicating that a transformation to a non-ferromagnetic state had occurred. The phase transformation occurred between 21.5 and 26.7 kbar. This state was temporarily retained after the pressure was released. The total pressure range investigated was atmospheric pressure to 38 kbar.

The secondary voltage versus temperature isobars for Gd are shown in Fig. 3. The ordinate is such that one unit corresponds to about 8μV. Figure 4 shows the Curie temperature of the Gd specimen as a function of pressure, as determined by the displacement of these isobars. The shift of the Curie temperature with pressure was determined by two techniques. One was to define the inflection point of the curves in Fig. 3 as the apparent Curie point. The other was to extrapolate these curves so that they would intersect the temperature axis, and to equate this intersection to the apparent Curie temperature. The pressure coefficient of Gd was -1.62 and -1.58°C/kbar as calculated by these two techniques, respectively. It should be pointed out that the measuring technique employed is not suitable for a precise determination of the Curie point itself, although it is accurate for determining the shift of the Curie point. The Curie point of Gd is usually listed in the literature at from 16° to 18°C.

The effect of pressure on the Curie point of Gd

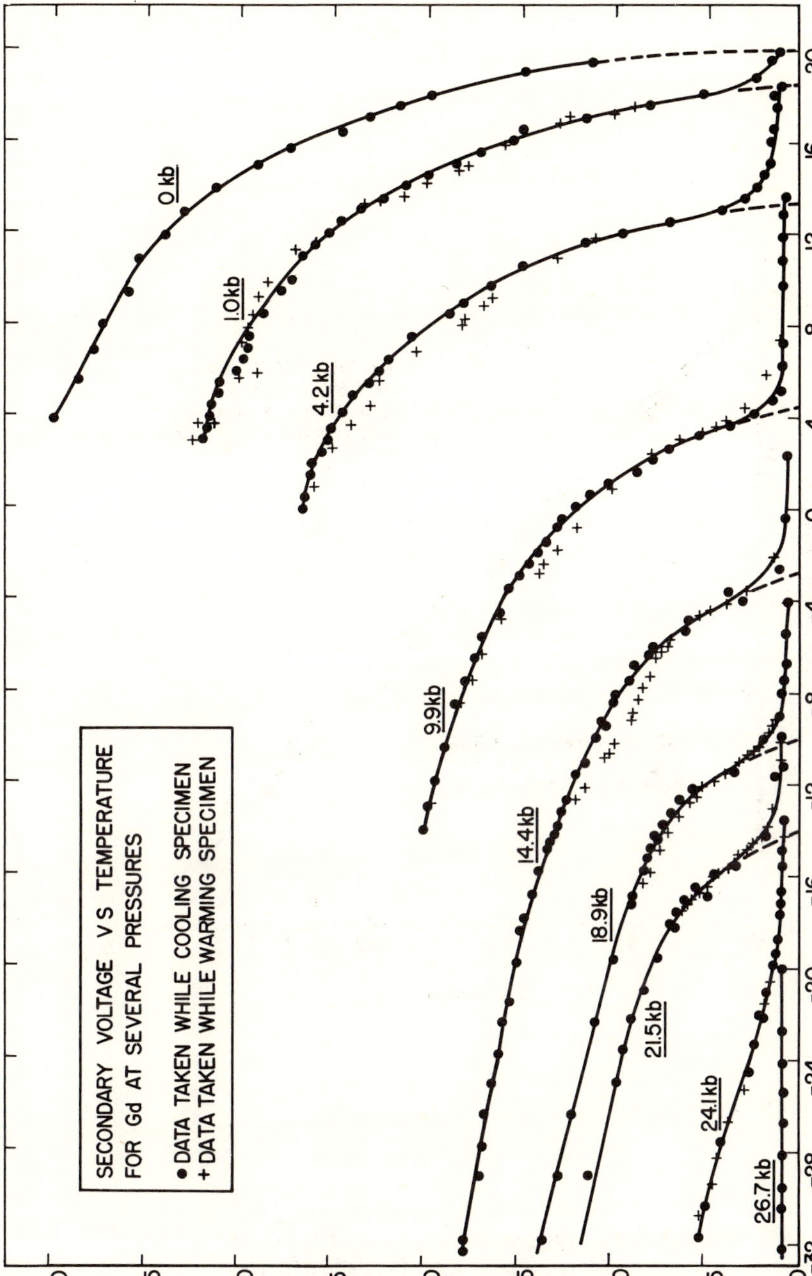

Figure 3. Secondary Voltage versus Temperature Isobars for Gadolinium

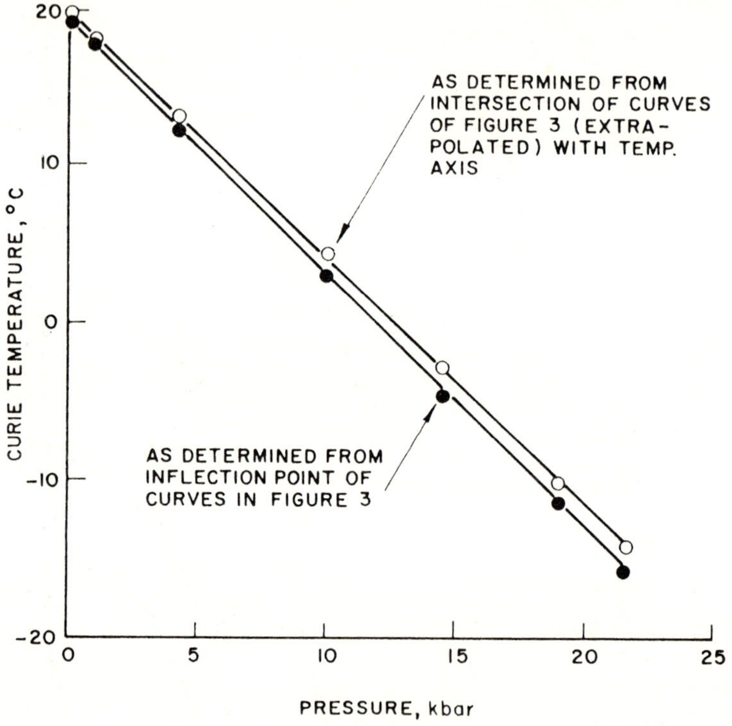

Figure 4. Apparent Curie Point versus Pressure for Gadolinium

was first measured by Patrick (3) in the pressure range to 8 kbar, and later by Bloch and Pauthenet (4) in the range to 6 kbar. They reported a decrease in the Curie point with pressure at rates of -1.2 and -1.53°C/kbar, respectively. The results reported by Bloch and Pauthenet are in fair agreement with those reported herein. Neither Patrick nor Bloch & Pauthenet offered an explanation for the observed decrease of the Curie temperature of Gd with pressure.

Bridgman (5) investigated the effects of pressure on the electrical resistivity of Gd and has reported "an instability of resistance beginning at approximately 20,000 Kg/cm^2. Below this, resistance is steady under constant pressure. Above 20,000 resistance at constant pressure drifts rapidly downward. This drift occurs in the entire region between 20,000 and 25,000 Kg/cm^2.

This anomaly in resistance observed by Bridgman was evidently the transition to a non-ferromagnetic state observed in the present work. Jayaraman and Sherwood (6) performed X-ray studies of a Gd sample, after pressurizing it to above the transition pressure and then removing the sample from pressure. They reported finding the presence of a samarium type structure with layers of atoms arranged in the sequence ABABCBCACA. Their work was performed subsequent to that presented herein.

2. Terbium

The magnetic behavior of terbium at different pressures is shown in Figs. 5 and 6 as isobaric plots of secondary voltage vs. temperature. At atmospheric pressure, Tb exhibits both a Néel temperature and a Curie temperature at about 230 and 220°K, respectively (7). Between these two temperatures, the magnetic ordering is known to be that of a helical spin configuration. The large increase in secondary voltage with decreasing temperature for the curves of Figures 5 and 6 is characteristic of the onset of ferromagnetic ordering. The anomalous higher temperature peak in the 0 + 0.2 kbar run is evidently the result of the presence of the helical spin state. It was not present in any of the runs at higher pressure. Thus the shift of the Néel point

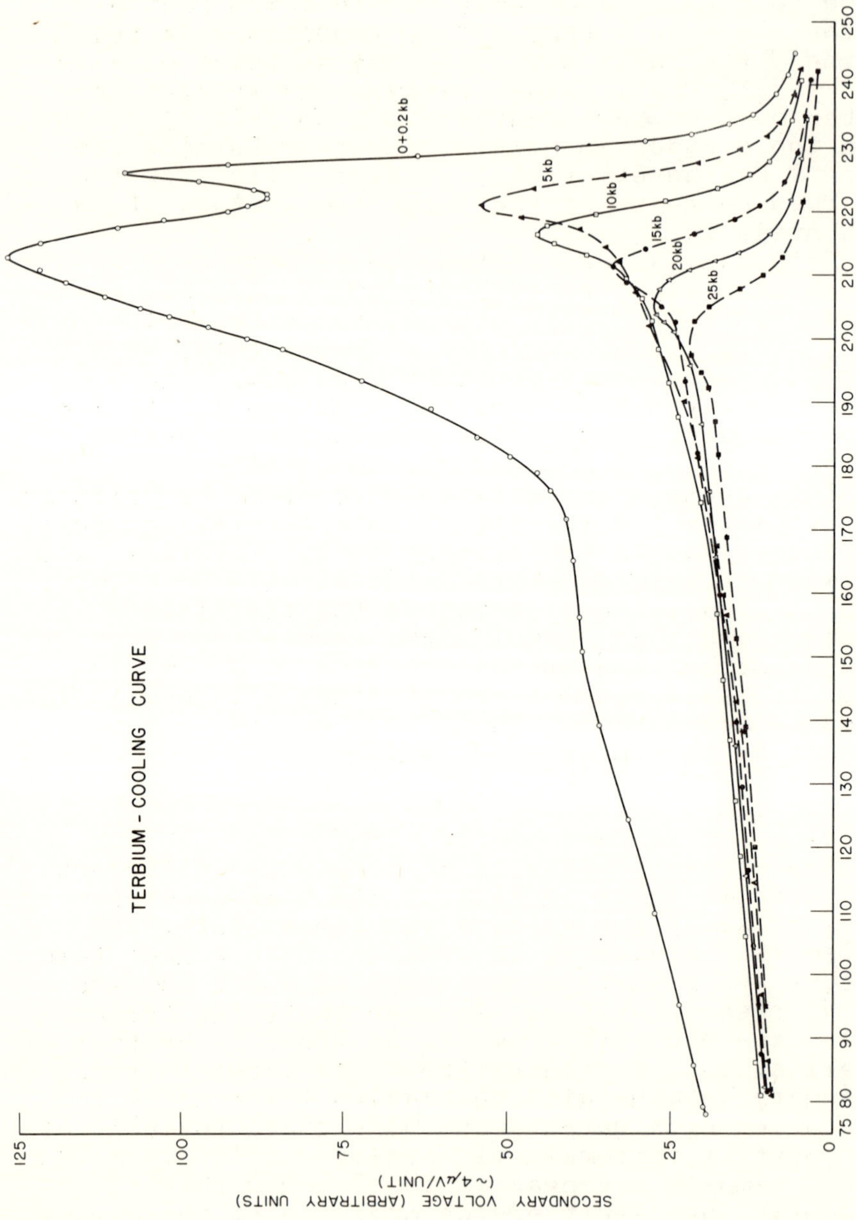

Figure 5. Secondary Voltage Versus Temperature Isobars for Terbium.

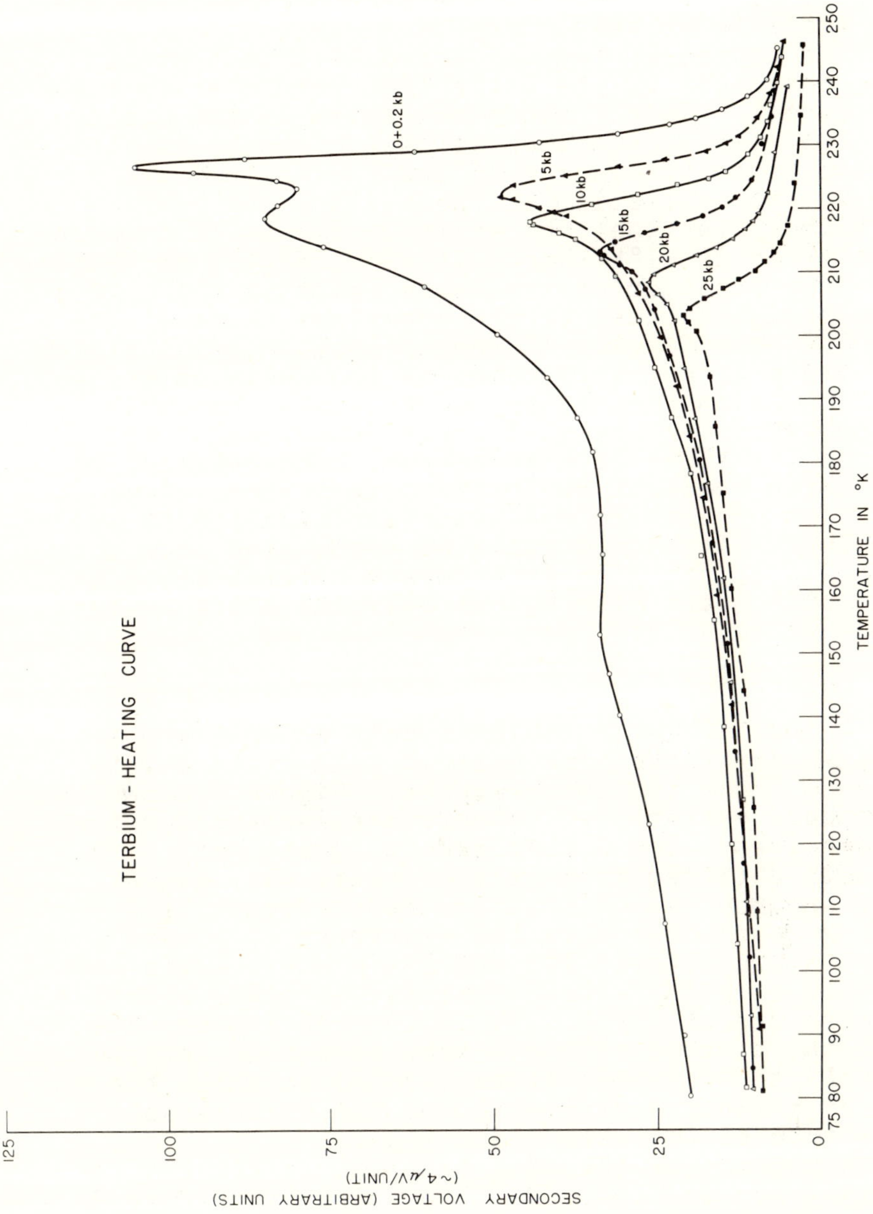

Figure 6. Secondary Voltage Versus Temperature Isobars for Terbium.

of Tb with pressure could not be determined. For the higher pressure runs it is difficult to say whether the helical spin state was actually removed, or whether its effect on the secondary voltage was too small to be seen.

The curves in Figs. 5 and 6 will be taken to give the shift in the Curie temperature of Tb with pressure. Figure 7 shows the shift in the Curie point of the Tb specimen as determined by the shift in the inflection point of the curves in Figs. 5 and 6. An evaluation of the heating and cooling data indicates that the shift in the Curie temperature of Tb is -1°K/kbar. It is difficult to determine the behavior of the Curie point in the 0 to 5 kbar region because of the anomalous shape of the 0 + 0.2 kbar curve. If the inflection point of the 0 + 0.2 kbar curve is taken as the Curie point, as with the higher pressure curves, then the Curie temperature vs. pressure curve would appear to change its slope in this region. The behavior of the Curie temperature in the 0 to 5 kbar region is still an unresolved question; additional data are required. It should again be emphasized that this method does not give an accurate determination of the Curie point itself, but only the shift in the Curie point.

3. Dysprosium

Dysprosium also exhibits both a Curie and a Néel temperature. The helical spin state exists between these two temperatures. Figure 8 shows the secondary voltage vs. temperature isobars for the Dy specimen. The 0 kbar result agrees well with results obtained from heat capacity (8), magnetic (9), and neutron diffraction measurements (10). Such measurements show the existence of a Néel temperature between 175 and 179°K and a Curie temperature at 87°K. The 0 kbar isobar gives a Néel temperature of 177°K. It is interesting to note that the minimum of the 0 kbar isobar corresponds closely to the temperature (140°K) at which the onset of a minor modification of the helical spin structure of Dy takes place (10).

A summary of the Néel temperature results for Dy is given in Figure 9. The points in Figure 9

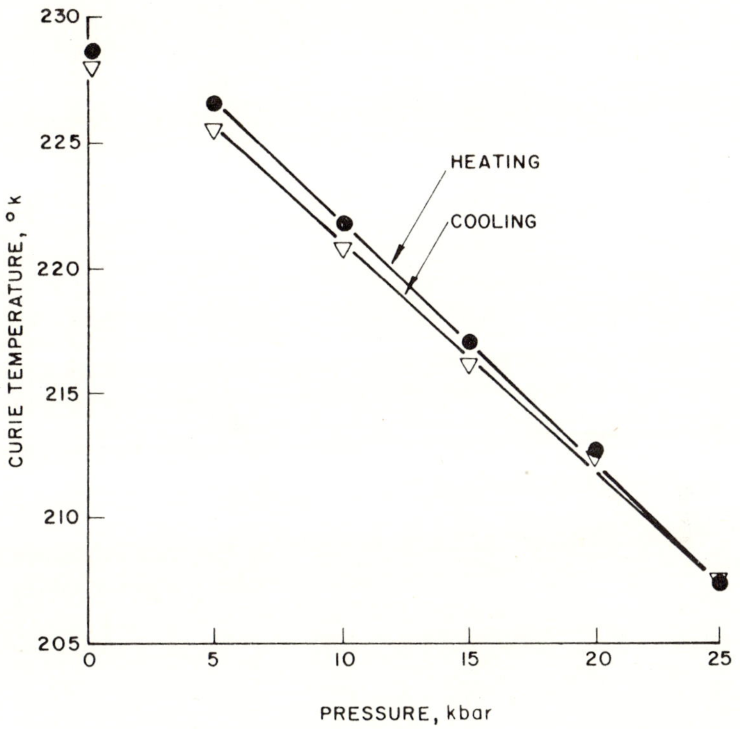

Figure 7. Apparent Curie Point versus Pressure for Terbium

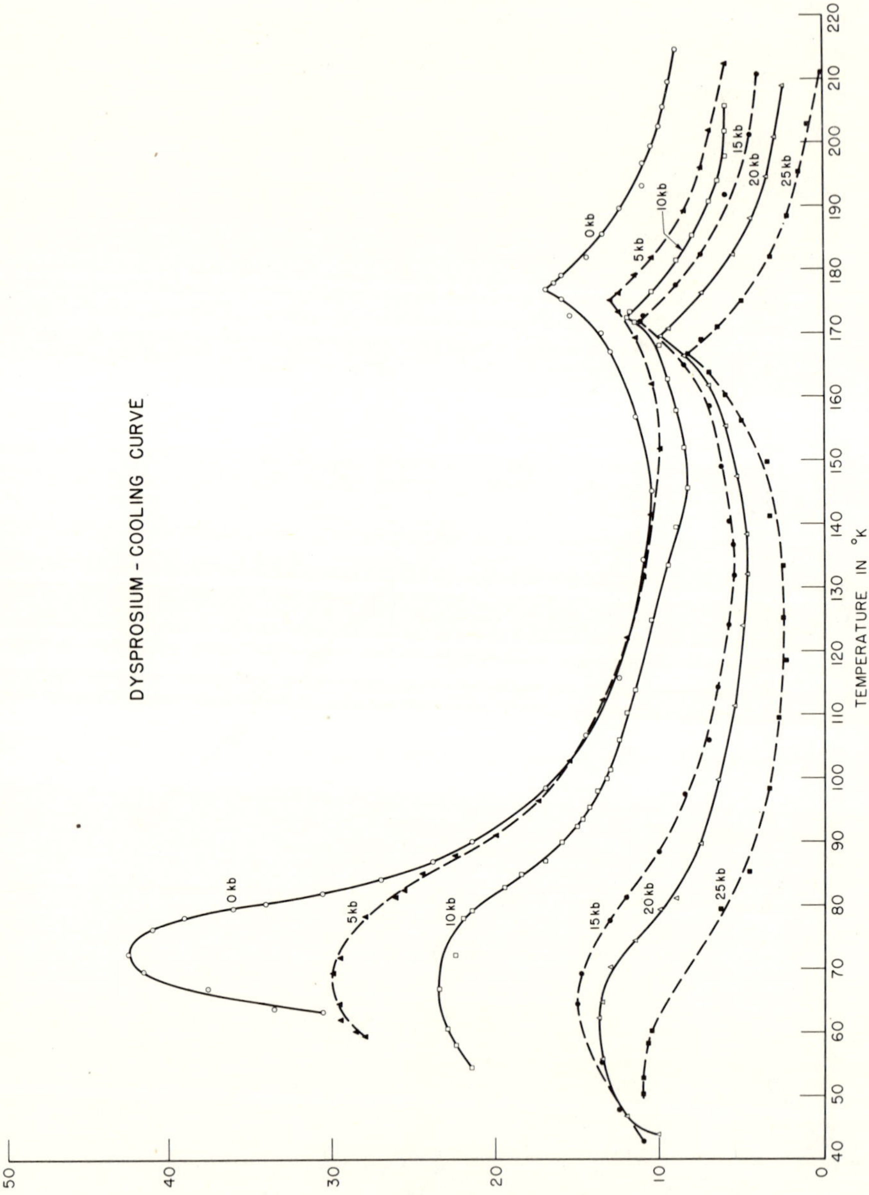

Figure 8. Secondary Voltage versus Temperature Isobars for Dysprosium

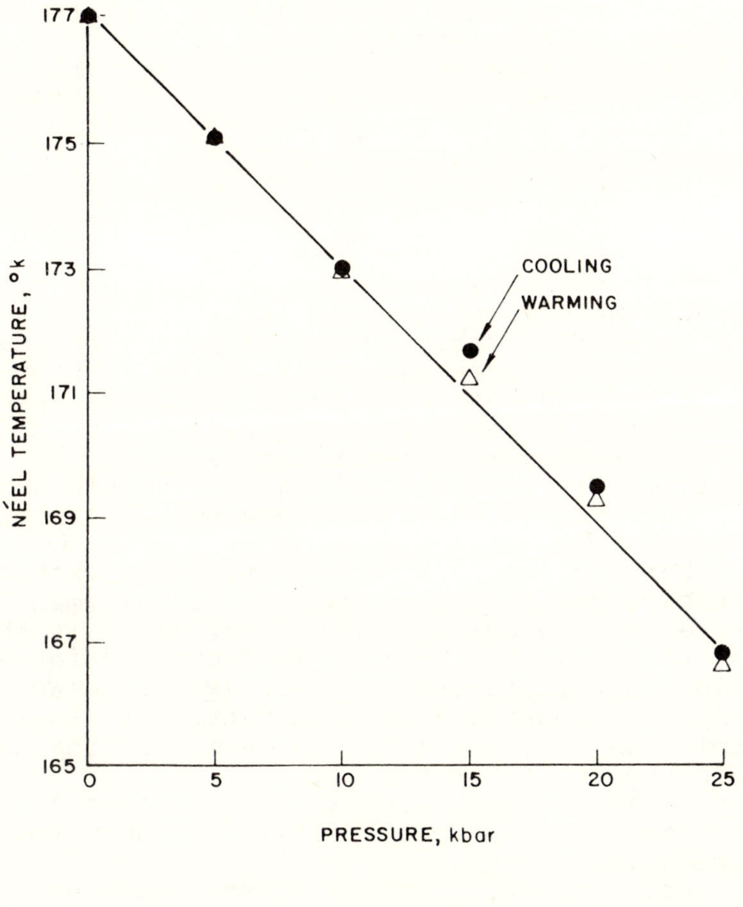

Figure 9. The Shift of the Neel Temperature of Dysprosium as a Function of Pressure

represent cusps in the isobars of Fig. 8. The pressure coefficient of the Néel temperature is -0.4°K/kbar.

It is somewhat more difficult to interpret the variation of the Curie temperature with pressure, since the characteristic increase in secondary voltage due to the Curie temperature does not vary with pressure in as regular a manner as for the cases of Gd and Tb. If, as with Gd and Tb, the inflection point is taken as the Curie temperature, then the Curie temperature appears to increase at first and then to decrease under the influence of pressure. Fig. 10 shows the apparent behavior of the Curie temperature of Dy with pressure. The results for the Curie temperature of Dy will have to be taken as tentative, pending further experimental investigation.

III. CONCLUSIONS

The effect of pressure on the Curie points of ferromagnetic materials is often interpreted in relation to exchange interaction curves. The ordinate for these curves is some parameter of the exchange integral (or energy) J_e, related to the Curie temperature in terms of the Weiss molecular field constant. The abscissa is a parameter relating the interatomic distance in the crystal to the radius of the unfilled electron shell in the atom. The interaction curve, suggested by Bozorth (11) is shown in Fig. 11. For the ordinate Bozorth took the exchange energy to be equal to $k\Theta_c/z$, where k is Boltzmann's constant, Θ_c is the Curie temperature, and z is the number of near neighbor atoms. Another form for an interaction curve, proposed by Néel (12), is shown in Fig. 12. λ is the Weiss field constant, D is the interatomic spacing, and R is the radius of the unfilled electron shell in the atom.

The rare earths must lie far to the right of the peak on these curves due to their large interatomic spacing relative to their radius of the unfilled 4f electron shell. Thus, in terms of these curves, the Curie temperature of a ferromagnetic rare earth element would be expected to increase with pressure. The experimental results presented herein, as well as those of Patrick and of Bloch and Pauthenet, have shown it is not proper to include the rare earths on an interaction curve constructed for the elements with an unfilled 3d electron shell (i.e., the iron

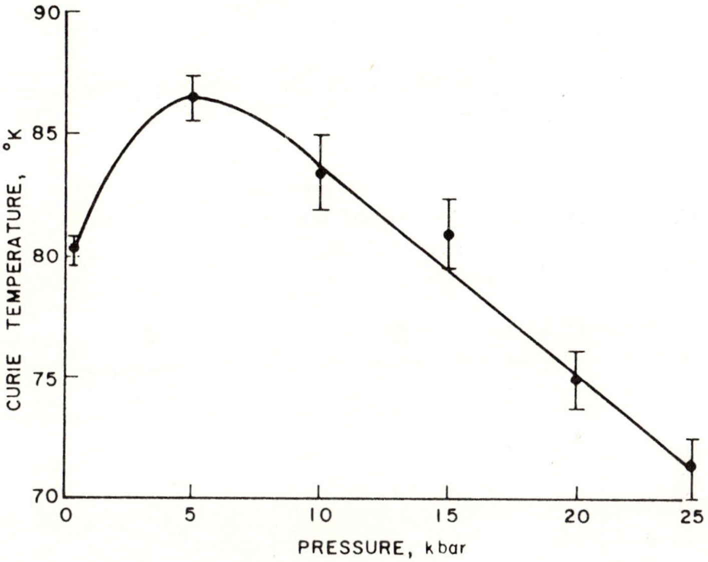

Figure 10. Apparent Curie Point versus Pressure for Dysprosium (Tentative Results)

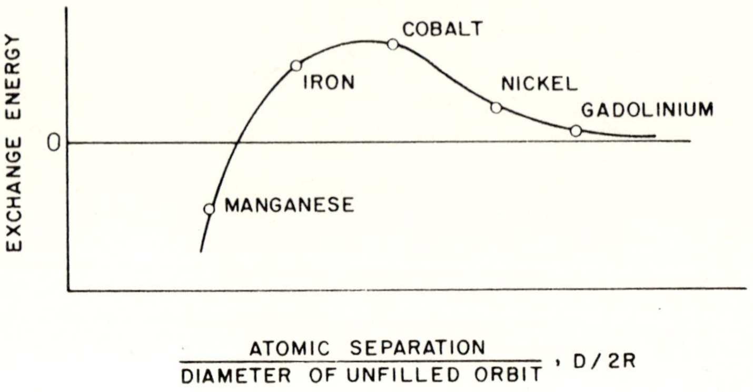

Figure 11. Bozorth Form of the Slater-Sommerfeld-Bethe Interaction Curve

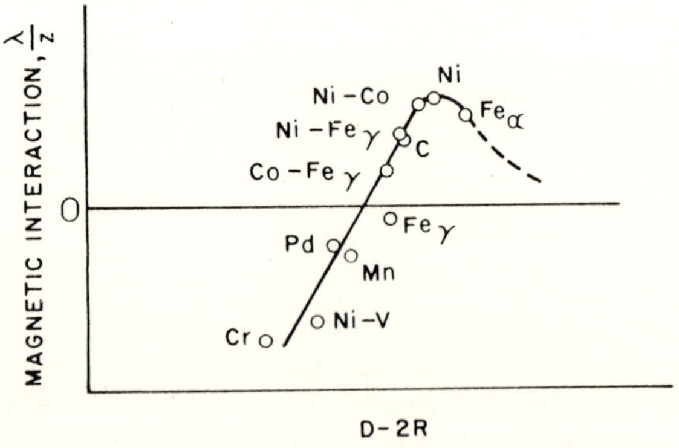

Figure 12. Neel Interaction Curve

group.

Following the experimental work with Gd, Robinson and Milstein suggested an exchange interaction curve limited to the ferromagnetic rare earths alone(1). The relation between the exchange energy J_e and the Curie temperature θ_c which was used in the construction of this exchange curve was first derived by Néel(13)in 1938. Néel assumed that the magnetic moments of neighboring atoms were coupled through the spin vectors only, giving a coupling energy of $-2\ J_e\ z\ S^2$. He equated this to the energy of interaction of the magnetic moment of an atom $g\ J\mu_b$ with the Weiss molecular field λM,

$$-2\ J_e\ z\ S^2 = -g\ J\mu_b\ \lambda M. \tag{1}$$

S and J are the spin and total quantum numbers of an atom, respectively, g is the Landé g-factor, μ_b is the Bohr magneton, and the magnetization M is given by

$$M = N\ g\ J\mu_b\ . \tag{2}$$

In terms of the Weiss molecular field theory, the susceptibility of a ferromagnetic material above its Curie point, where the solid is in the paramagnetic state, is

$$\chi = \frac{N\mu^2/3k}{T - N\mu^2/3k}\ . \tag{3}$$

N is the atomic density, T is the temperature and quantum mechanically the magnetic moment μ is given by

$$\mu = g\sqrt{J(J+1)}\mu_b\ . \tag{4}$$

Experimentally the susceptibility has the form

$$\chi = \frac{C}{T - \theta_c}\ . \tag{5}$$

Thus, from (3), (4), and (5), the Weiss constant λ is

$$\lambda = \frac{3k\theta_c}{Ng^2\ J(J+1)\mu_b^{\ 2}} \tag{6}$$

The expression derived by Néel is obtained by substituting Eqs. (2) and (6) into Eq. (1) and solving for J_e,

$$J_e = \frac{3k\Theta_c J}{2 z S^2(J+1)} \quad . \qquad (7)$$

The essential point is that the exchange energy is directly proportional to $\Theta_c J/S^2(J+1)$, since the rest of the expression is constant for the rare earths. Even if Eq. (7) does not give good numerical values for J_e, as long as the proportionality holds, the equation is useful for present purposes.

In the Weiss molecular field model of a magnet, when $T \leq \Theta_c$, the material is ferromagnetic and Eq. (5) is meaningless. For real materials the ferromagnetic Curie temperature Θ_{cf} (i.e., the temperature at which the material actually becomes ferromagnetic) is normally a few percent less than the paramagnetic Curie temperature Θ_{cp} (i.e., Θ_c in Equation (5)). The Weiss theory does not distinguish between Θ_{cf} and Θ_{cp}. This usually would not make very much of a difference, and either Θ_{cp} or Θ_{cf} could be used in an expression such as Eq. (7). For the ferromagnetic rare earths (other than Gd), however, the situation is complicated by the presence of the helical spin state which exists in the temperature range between the ferromagnetic and the paramagnetic states. (See Fig. 13.) The Curie point to use in finding J_e by Eq. (7) is the paramagnetic Curie point Θ_{cp}; the ferromagnetic Curie point could be used only when it is the temperature of a transformation from the paramagnetic state to the ferromagnetic state. This latter point would be given by Θ_c in Fig. 13 were it not for the presence of the screw state.

The rare earth interaction curve suggested by Robinson and Milstein is shown in Fig. 14. The ordinate J_e/k is determined by Eq. (7), using known Curie temperatures. The abcissa is the interatomic distance divided by the diameter of the 4f electron shell (calculated by the Slater orbital method (14)). Two curves are drawn, one based on the paramagnetic Curie temperature Θ_{cp} and the other on the ferromagnetic Curie temperature Θ_{cf}. The point for Tb on the Θ_{cf} curve was estimated as follows: For Gd, which does not possess the helical screw state, Θ_{cf} is 5% less than Θ_{cp}. For Tb, the temperature to use on the Θ_{cf} curve must be between Θ_1 and Θ_2 which are 230 and 220°K, respectively. For Tb, Θ_{cp} is 236°K (15); 5% less than this is 224°K, which is almost midway between Θ_1 and Θ_2 for this element.

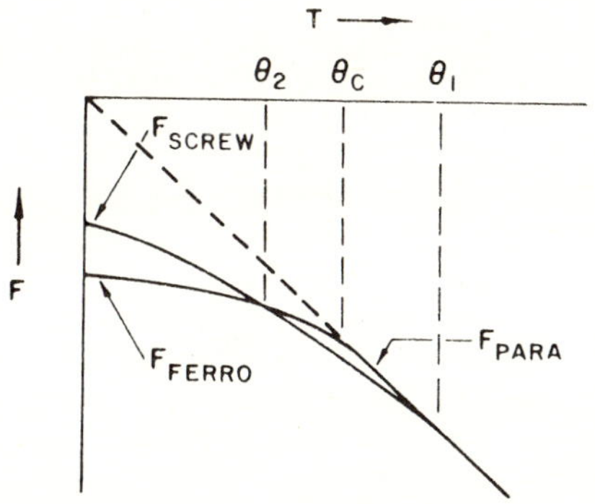

Figure 13. Free Energy versus Temperature Curves for the Ferromagnetic, Screw, and Paramagnetic States (after Yosida and Miwa) [K. Yosida and H. Miwa, J. Appl. Phys. 32, Suppl. 3, 85 (1961)]. Between 0 and θ_2, the Material is Ferromagnetic, Between θ_1 and θ_2 the Spin Structure Exists, and Above θ_1 the Substance is Paramagnetic.

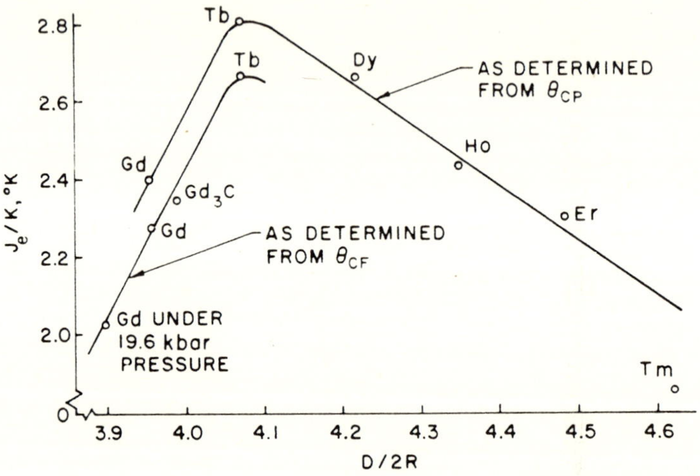

Figure 14. Exchange Interaction Curve for the Ferromagnetic Rare Earth Metals

Thus, for Tb, Θ_{cf} may be taken as 224°K in Eq. (7). It has been possible to estimate Θ_{cf} for Tb because Θ_1 and Θ_2 are relatively close to each other, which is not the case for the elements dysprosium through thulium.

The effect of pressure on the Curie point of Gd was found to be consistent with this interaction curve, as is shown by the point labeled "Gd, under 19.6 kbar pressure." If this curve is valid, one should be able to predict the change of the Curie temperature with pressure for Tb. From the interaction curve, the slope $d(J_e/k)/d(D/2R)$ is the same for Gd as for Tb. If it is assumed that the compressibilities of Gd and Tb are the same, and that $\Delta D \approx \frac{1}{3}\frac{\Delta V}{V} D$ where V is volume, then the pressure coefficient for the Curie temperature of Tb is given by

$$\left[\frac{d\Theta_c}{dP}\right]_{Tb} = \frac{\left[\frac{S^2(J+1)}{J} \frac{D}{2R}\right]_{Tb}}{\left[\frac{S^2(J+1)}{J} \frac{D}{2R}\right]_{Gd}} \left[\frac{d\Theta_c}{dP}\right]_{Gd} \qquad (8)$$

For Gd, $d\Theta_c/dP$ is -1.60 to ±0.05°C/kbar. Thus, the predicted slope for Tb is

$$\left[\frac{d\Theta_c}{dP}\right]_{Tb} \approx \frac{10.5 \times 4.067}{15.75 \times 3.952} \times (-1.60)$$

$$\approx -1.1\ °C/kbar \qquad (9)$$

This is in fair agreement with the measured value of -1°C/kbar.

The Curie temperature measured experimentally is represented by Θ_2 in Fig. 13, not Θ_c. For Tb, these two are close together, but for Dy they are not (Θ_2 and Θ_{cp} are 87 and 151°K (15) respectively for Dy). Thus, the best one could hope for would be qualitative agreement for Dy. The tentative experimental result, yielding a positive pressure coefficient in one pressure range and a negative pressure coefficient in a higher pressure range is in good agreement with the proposed exchange curve.

REFERENCES

(1) Robinson, L.B., F. Milstein and A. Jayaraman, Phys. Rev. 134, A187 (1964).

(2) Kennedy, G.C. and P. N. La Mori, Progress in Very High Pressure Research, Proceedings, edited by F. P. Bundy, W. R. Hibbard, Jr., and H. M. Strong (John Wiley & Sons, Inc., New York, 1960), p. 304 ff.

(3) Patrick, L., Phys. Rev. 93, 384 (1954).

(4) Bloch, D. and R. Pauthenet, Compt. Rend. 254, 1222 (1962).

(5) Bridgman, P. W., Proc. Am. Acad. 82, 83 (1953).

(6) Jayaraman, A. and R. C. Sherwood, Phys. Rev. Letters 11, W535 1-2 (1964).

(7) Koehler, W. C., H.R. Child, E. O. Wollan, and J. W. Cable, J. Appl. Phys. 34, 1335 (1963).

(8) Griffel, M., R. E. Shockdopole and F.H. Spedding, J. Chem. Phys. 25, 75 (1956).

(9) Behrendt, D.R., S. Levgold, and F.H. Spedding, Phys. Rev. 109, 1544 (1958).

(10) Wilkinson, M.K., W.C. Koehler, E.O. Wollan, and J. W. Cable, J. Appl. Phys. Supp. 32, 48S (1961).

(11) Bozorth, R.M., Bell Syst. Tech. J. 19, 1 (1940).

(12) Néel, L., Le Magnetisme (University of Strasbourg, 1939), Vol. 2, p. 65; J. Phys. et radium 1, 242 (1940).

(13) Néel, L., Compt. Rend. 206, 49 (1938).

(14) Slater, J. C., Phys. Rev. 36, 57 (1930).

(15) Arajas, S. and V. Colvin, Rare Earth Research, Seminar, Macmillan, New York, 1961.p. 178.

DISCUSSION

By D. B. McWhan

I would like to make several remarks with regard to the effect of pressure on the magnetic properties of the rare earth metals.

(1) The curves of secondary voltage versus temperature for gadolinium published by Robinson et al showed two new peaks above 25 kbars. They were apparently associated with the high pressure phase which according to the work of Jayaraman and Sherwood has the samarium type structure. A similar transition also occurs in terbium, dysprosium, and holmium, and typical curves of secondary voltage versus temperature for dysprosium are shown in Figure 1. The occurrence of two peaks probably implies that there is a different ordering temperature for each crystallographic site in the Sm type structure. A similar situation occurs in neodymium according to the neutron diffraction studies of Moon, Cable, and Koehler.

(2) The shape and amplitude of the peaks observed in the high pressure phase of Gd are similar to those observed for Dy at the Neel temperature. This suggests that the transition from the hexagonal close packed to the Sm type structure in Gd is accompanied by a change from a ferromagnetic to an antiferromagnetic state at low temperatures.

(3) In Figure 2 the data for Gd, Tb, Dy, and Ho are shown. The position of the first peak in the high pressure phase is also shown. In each metal the high pressure phase orders at a lower temperature. Finally, in Figure 3 the available data in $\frac{d \ln T}{d \ln V}$ in the rare earth metals is presented. For all but Gd the points represent Néel temperatures. The sign of the point for Sm is inferred from the work of Jayaraman. The Ce and Nd points were obtained from thermal expansion measurements by Andres. It is interesting to note that

the points fall onto two distinct curves with a break between Eu and Gd.*

*This work was presented at the A.P.S. Meeting March 1965.

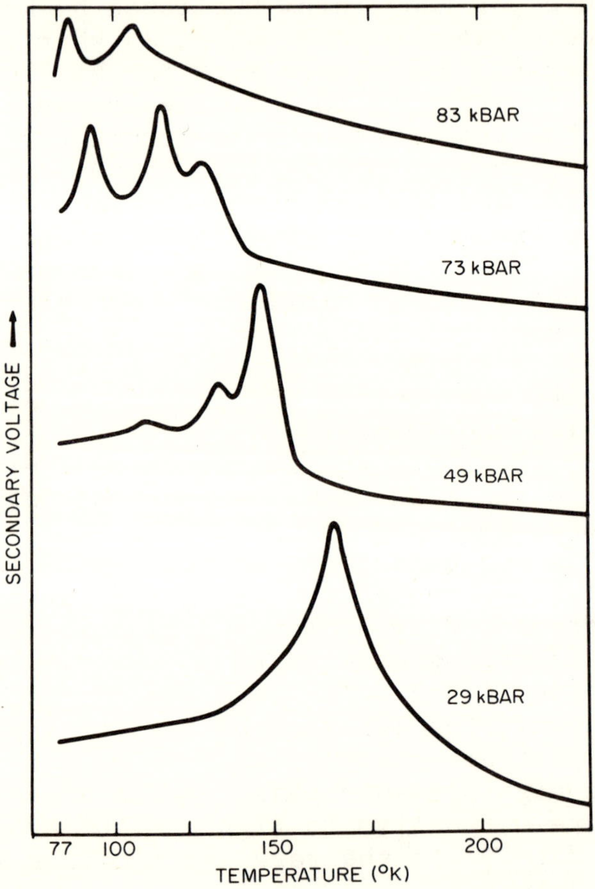

Fig. 1. Secondary voltage versus temperature for dysprosium.

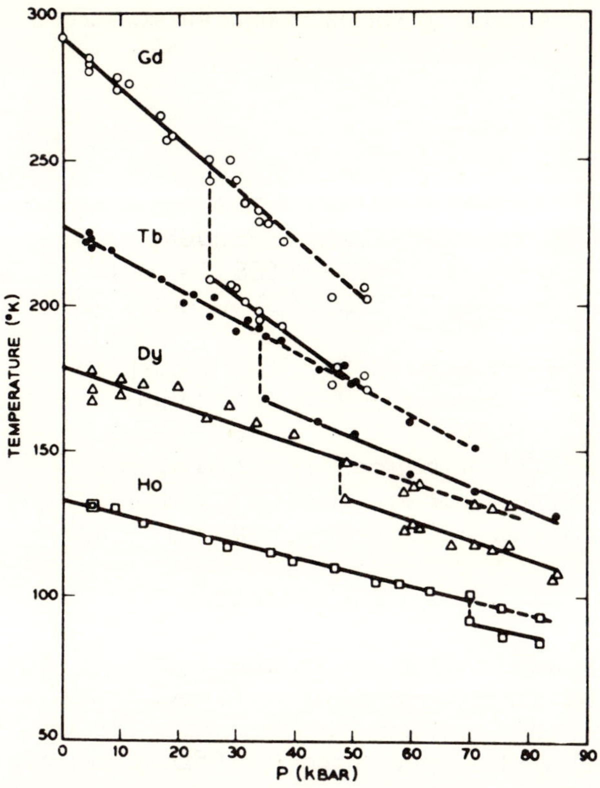

Fig. 2. Data for Gd, Tb, Dy and Ho. Position of first peak in high pressure phase is also shown.

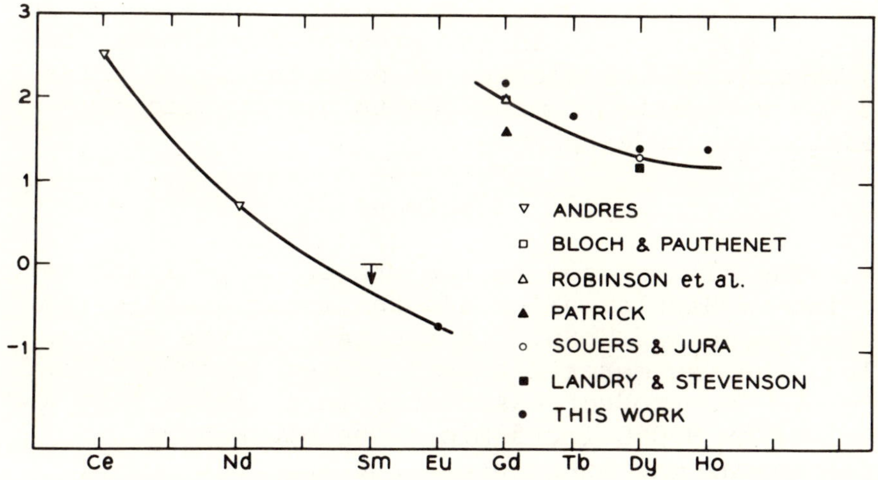

Fig. 3. $\dfrac{d\ \ell n\ T}{d\ \ell n\ V}$ for rare earth metals.

ON THE PRESSURE DEPENDENCE OF MAGNETIC PROPERTIES IN SOME MANGANESE INTERMETALLIC COMPOUNDS

by Tokutaro Hirone, Takejiro Kaneko and Kazuo Kondo

The Research Institute for
Iron, Steel and Other Metals
Tohoku University, Sendai, Japan

1. Introduction

It is well known that the Curie point as well as the saturation moment of a ferromagnet changes under hydrostatic compression. As a result of the measurement of the pressure coefficients of these two quantities, it may be possible to find the pressure dependence of the coupling constants between a pair of spin moments in a ferromagnet. At present, however, very few measurements of both these coefficients have been made on the same ferromagnetic substance. In the present experiment, the measurement of the changes of both the Curie point and the saturation moment due to hydrostatic compression are undertaken.

2. Specimens

The alloys used in the present experiment were the indium Heusler alloy and the intermetallic compounds Au_4Mn and MnSb. In the case of the Heusler alloy, the raw materials were melted at 1300°C for 10 min. The product was placed in a silica tube and remelted at 950°C for 5 min. The block thus obtained was machined into a rod and annealed at 650°C for 1 hr. The raw materials for the Au_4Mn specimens were sealed in a silica tube together with argon gas and melted at 1300°C for 5 min. The product was remelted and then cast into a silica tube. The rod thereby obtained was annealed at 800°C for 50 hrs in an argon atmosphere. It was then heated at 350°C for 70 hrs to promote the ordered arrangement of atoms in this alloy. The raw materials for the MnSb specimen were melted at 1200°C and cast in a silica tube.

X-ray diffractograms of these alloys were taken. In the case of the Heusler alloy, well defined order lines could be seen in the diffractograms. The diffractogram of the Au_4Mn alloy revealed a tetragonal deformation with $\Delta c/a = -0.013$. This result shows that the ordered arrangement found by Watanabe (1) is developed in the present alloy.

3. Experimental Results and Discussion

To determine the change of the Curie point with hydrostatic pressure, isothermal magnetization curves for these alloys were taken. Measurements under hydrostatic compression were made near the Curie point up to magnetic field intensities of 3500 Oe. The results of these measurements were well represented by the law of approach:

$$\sigma = \sigma_S(1 - \frac{b}{H^2}) + \chi H , \qquad (1)$$

where σ and σ_S are the specific and saturation magnetization, respectively, and b and χ are constants independent of the magnetic field H. Then plots of $\sigma_S^2(T)$ versus temperature just below the Curie point were made. Such a plot for $MnAu_4$ is shown in Fig. 1. The plot shows a fairly good linear relationship, which makes it possible to determine by extrapolation the Curie point under hydrostatic compression. The results of the measurement of the Curie point shift by hydrostatic compression are shown in Table 1.

The measurement of the change of the absolute saturation moment due to hydrostatic compression

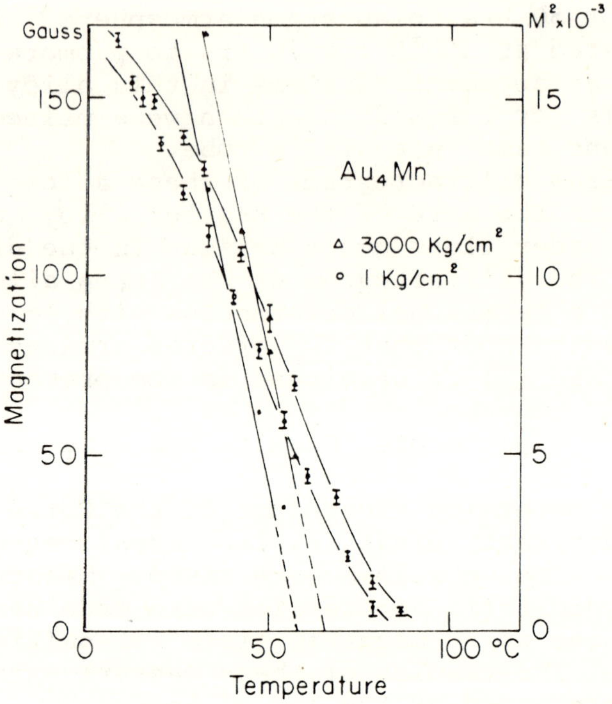

Fig. 1. Magnetization vs. Temperature Curve for Au_4Mn.

Table 1.

	θ (°C) (1 kg/cm²)	θ (°C) (3000 kg/cm²)	dθ/dP deg/cm² dyne⁻¹
$MnAu_4$	58.7 ± 0.9	66.7 ± 1.0	2.7×10^{-9}
$Cu_2Mn_{1.2}In_{0.8}$	176.0 ± 0.5	180.6 ± 0.8	1.5×10^{-9}
MnSb	283°	276°	-3.2×10^{-9}

imposes difficulties which Kondorskii (2) once overcame by means of a freezing ice technique in closed vessels at low temperatures. However, in the present experiment a thermodynamical relation

$$\frac{1}{\rho}\cdot\left(\frac{d\sigma}{dP}\right)_H = -\frac{d\omega}{dH} \qquad (2)$$

$$\omega = dV/V \qquad (3)$$

is used instead of measuring directly the change of magnetic moment due to compression. In this equation ρ is the density and $d\omega/dH$ is the forced volume magnetostriction coefficient. Instead of measuring the quantity on the left hand side of Eq. 2, the forced magnetostriction coefficient was measured by a strain gauge technique from liquid air temperature to room temperature in a field range up to about 15 kOe. In the weak field range the forced magnetostriction is masked by the form and crystal effect in magnetostriction. Hence, in the present experiment, the value of the magnetostriction after magnet saturation was reached was used to deduce the forced magnetostriction constant.

Fig. 2 shows the magnetostriction data for Au_4Mn. As seen in the figure, the value of $d\ell/\ell dH$ from the linear part in the strong field region takes a negative sign at room temperature. However, there is a change in the sign of $d\ell/\ell dH$ at about -10°C and this slope increases monotonically as the temperature is lowered.

Based on the results in Fig. 2, the tempera-

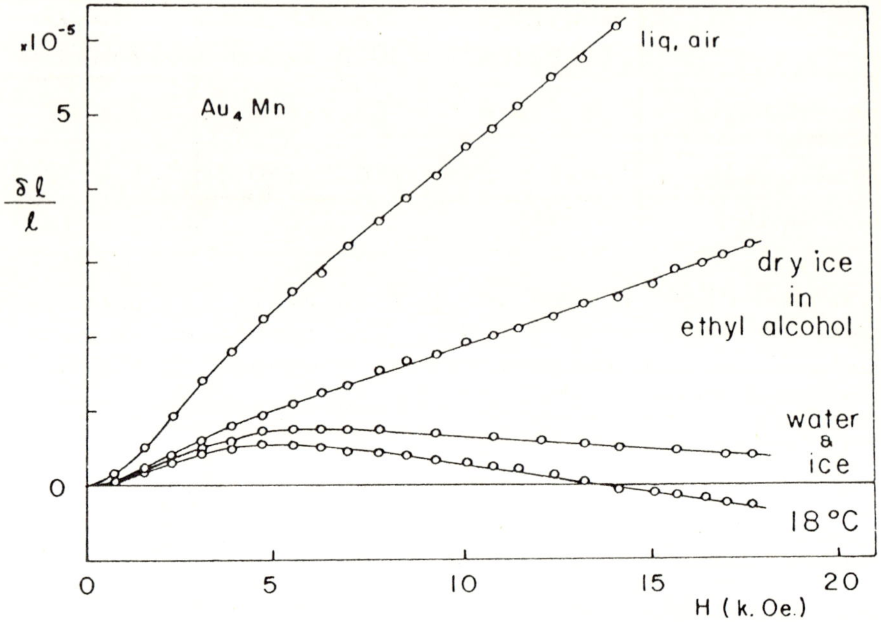

Fig. 2. Magnetostriction Curve for Au_4Mn.

ture dependence of the forced magnetostriction coefficient for $MnAu_4$ can be obtained (Fig. 3). The value extrapolated to the absolute zero of temperature reaches a value of 17×10^{-9}. Also in Fig. 4 and 5 the same results for the MnSb alloy are shown. The extrapolated values of the forced magnetostriction constants are summarized in Table 2. By putting these results in Eq. 2, the change of

Table 2.

	$\frac{d\omega}{dH}$	ρ	σ_0	$\frac{1}{\sigma_0} \cdot \frac{d\sigma_0}{dP}$ $cm^2/dyne$
Au_4Mn	17×10^{-9}	16.7	27.5	-3.70×10^{-11}
Cu_2MnIn	very small	8.24	74.2	very small
MnSb	4.5×10^{-9}	6.86	112	-0.95×10^{-11}

saturation moment at absolute zero due to compression can be calculated. The results of this calculation are also shown in Table 2.

Now the Curie point of the ferromagnet is, according to simple molecular theory, given by

$$\theta = \frac{2zJS(S+1)}{3k} . \qquad (4)$$

In this expression J represents the coupling coefficient of the spins in ferromagnets and S is the mean spin moment of magnetic atoms or ions in the ferromagnet.

Now the above expression can be rewritten as

$$J \sim \frac{\theta}{S(S+1)} \qquad (5)$$

or in the classical limit,

$$J \sim \theta S^{-2} . \qquad (6)$$

If we put

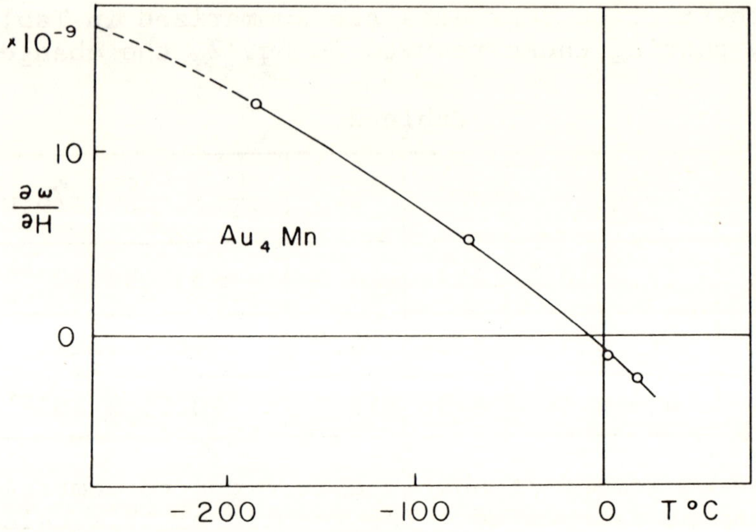

Fig. 3. Forced Magnetostriction Constant vs. Temperature for Au_4Mn.

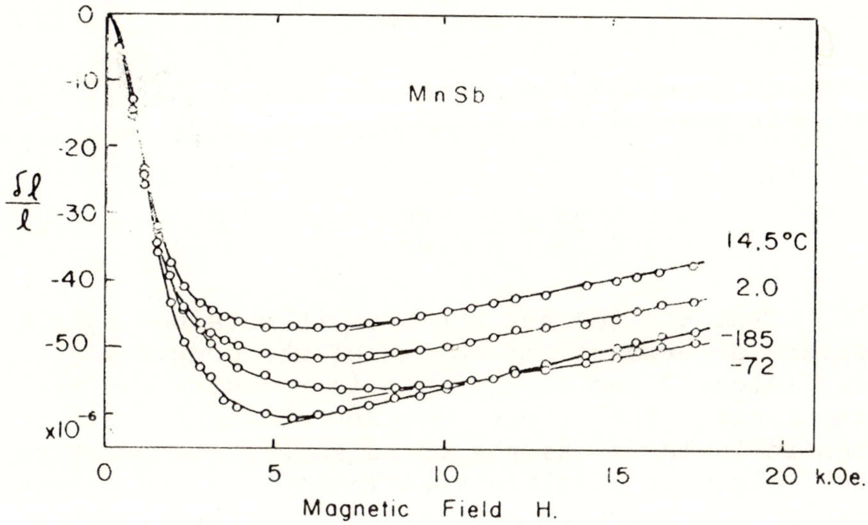

Fig. 4. Magnetostriction Curve for MnSb.

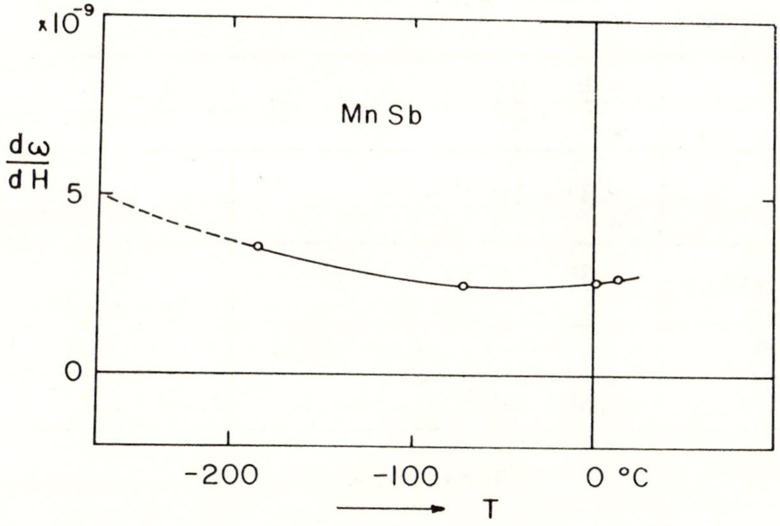

Fig. 5. Forced Magnetostriction Constant vs. Temperature for MnSb.

$$\sigma_0 = nS \tag{7}$$

in which n means the number of elementary magnets per unit mass of the ferromagnet, it then follows that

$$\frac{1}{J} \cdot \frac{dJ}{dP} = \frac{1}{\theta} \cdot \frac{d\theta}{dP} - \frac{2}{\sigma_0} \cdot \frac{d\sigma_0}{dP} \, . \tag{8}$$

From Eq. (8) and the results obtained by the present experiment the change of the coupling coefficient J due to hydrostatic pressure can be calculated. The results of the calculation can be seen in Table 3. As the table shows, the pressure coefficient of coupling coefficients in the manganese alloys reported here are all positive, which seems to suggest a special mechanism of exchange interaction between spins in the ferromagnetic manganese alloys.

Table 3.

	$\frac{1}{\theta} \cdot \frac{d\theta}{dP}$	$\frac{1}{\sigma_0} \cdot \frac{d\sigma_0}{dP}$	$\frac{1}{J} \cdot \frac{dJ}{dP}$
Au_4Mn	0.81×10^{-11}	-370×10^{-11}	$8.2 \times 10^{-11}\ \frac{cm^2}{dyne}$
$Cu_2Mn_{1.2}In_{0.8}$	-0.33×10^{-11}	——	0.33×10^{-11}
Cu_2MnIn	——	very small	
MnSb	-0.58×10^{-11}	-0.59×10^{-11}	0.60×10^{-11}

References

1. D. Watanabe, J. Phys. Soc. Japan, 15, 1251 (1960).
2. E. I. Kondorskii and V. L. Sedov, J. Phys. radium 20, 185 (1959);
J. Exptl. Theoret. Phys. (U.S.S.R.) 35, 845

(1958);
J. Appl. Phys. 31, 331 (1960);
J. Exptl. Theoret. Phys. (U.S.S.R.) 11, 561 (1960).

DISCUSSION

G. A. Samara
Sandia Laboratory, Albuquerque, New Mexico

and

A. A. Giardini
University of Georgia, Athens, Georgia

We have recently reported (1) on the effect of pressure, to over 50 kilobars (kbar), on the Curie points of the ordered hexagonal manganese compounds MnAs, MnSb, and MnBi. Our findings complement the results presented by Dr. Hirone and will be presented very briefly here.

These compounds crystallize in the hexagonal NiAs-type structure. At atmospheric pressure the Curie points are: MnAs = 45°C, MnSb = 313°C, and MnBi = 360°C. A plot of these Curie points versus the nearest neighbor Mn-Mn distance (½ the c axis), is shown in Fig. 1. The interesting feature of this curve is its shape which resembles very closely that of the well-known Bethe-Slater interaction curve and similar correlations. These are plots of the exchange integral J (or Curie point) versus interatomic distance as was described in the previous paper (C4) by Robinson et al. If such a correlation is valid, it might be applicable to the present case since all three compounds have the same crystal structure. From Fig. 1 one would then predict that the Curie point of MnAs should decrease sharply with increasing pressure, that of MnSb should decrease at a slower rate, and that of MnBi should exhibit a very small change, possibly an increase.

For MnAs we find that the shift of the Curie point, dT_c/dP, is -12.3°C/kbar. This is in good agreement with the result of Bean & Rodbell (2) who reported a value of -12°C/kbar in measurements up

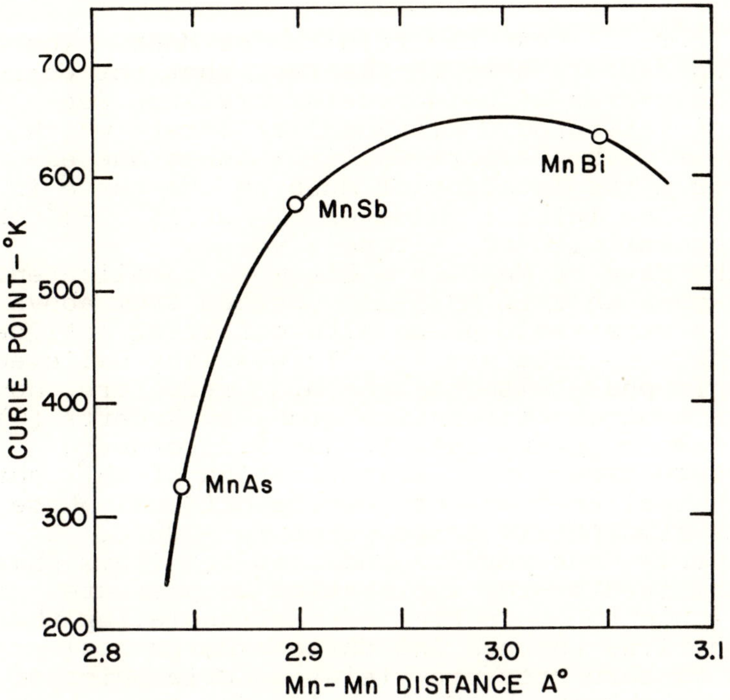

FIG. I CURIE POINT VS DISTANCE BETWEEN NEAREST NEIGHBOR Mn ATOMS to ~1 kbar. The slope is constant over the short pressure range covered (~2.5 kbar in this case). Above 2 kbar the Curie point occurs below room temperature.

Figure 2 shows the results for MnSb. The striking feature here is the fact that the Curie point decreases with increasing pressure (decreasing interatomic distance) at a rate which increases with pressure. This is what one expects from the location of MnSb on the curve in Fig. 1. The initial value of dT_c/dP is ~ -3°C/kb in agreement with Dr. Hirone's value.

The case of MnBi is a bit more complicated. It is known that if MnBi is quenched from above 380°C, a metastable phase with different lattice parameters is obtained (3). It was long believed that this phase also had the NiAs structure; however, a neutron diffraction study by Roberts (4) definitely suggests that it has a disordered $NiIn_2$-type structure. Magnetization of this phase is much smaller than that of the original phase and its atmospheric Curie point is ~200°C. Unfortunately, our results indicate that this phase is stabilized by the application of pressure. The results are shown in Fig. 3. Extrapolating the straight line through the three experimental points to zero pressure yields a Curie point of 200°C. Although the Curie point for this phase does not fall on the curve in Fig. 1, it is interesting to note that dT_c/dP is very small, namely -0.7°C/kbar.

In conclusion, it appears that the decrease of the Curie points of MnAs and MnSb are in good qualitative agreement with the location of these two compounds on the interaction curve (Fig. 1). More work is needed to determine the pressure range of the stability of the normally stable room temperature phase of MnBi and to determine dT_c/dP for this phase.

References:

1. G. A. Samara and A. A. Giardini, Bull. Am. Phys. Soc. 9, 635 (1964). The work was preformed at the U.S. Army Electronics Laboratories, Fort Monmouth, New Jersey.

2. C. P. Bean and D. S. Rodbell, Phys. Rev. 126, 104 (1962).
3. R. R. Heikes, Phys. Rev. 99, 446 (1955).
4. B. W. Roberts, Phys. Rev. 104, 607 (1956).
5. The data in Figs. 2 and 3 have been corrected for the pressure effect on the thermal emf of the Pt-Pt 13% Rh and Ch-Al thermocouples used. The corrections are given by R. E. Hanneman and H. M. Strong, J. Appl. Phys. 36, 523 (1965).

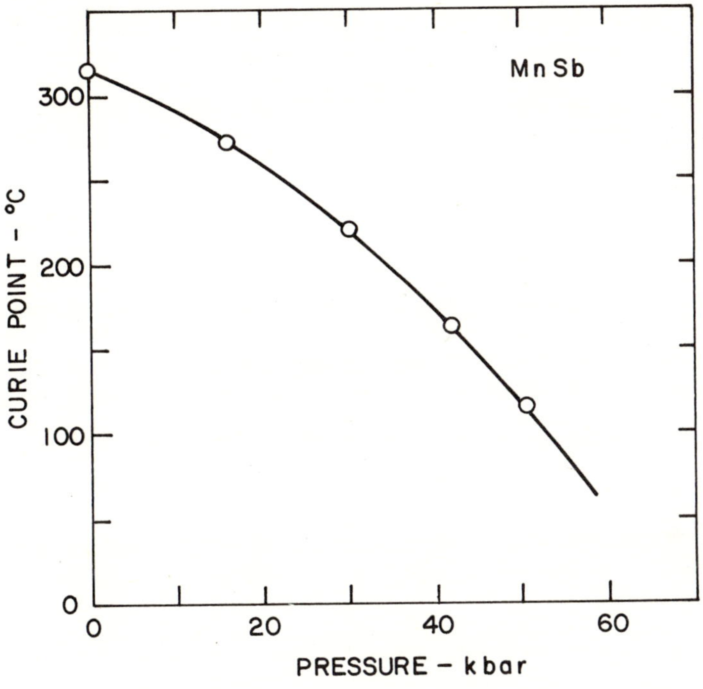

FIG. 2 SHIFT OF THE CURIE POINT OF MnSb WITH PRESSURE

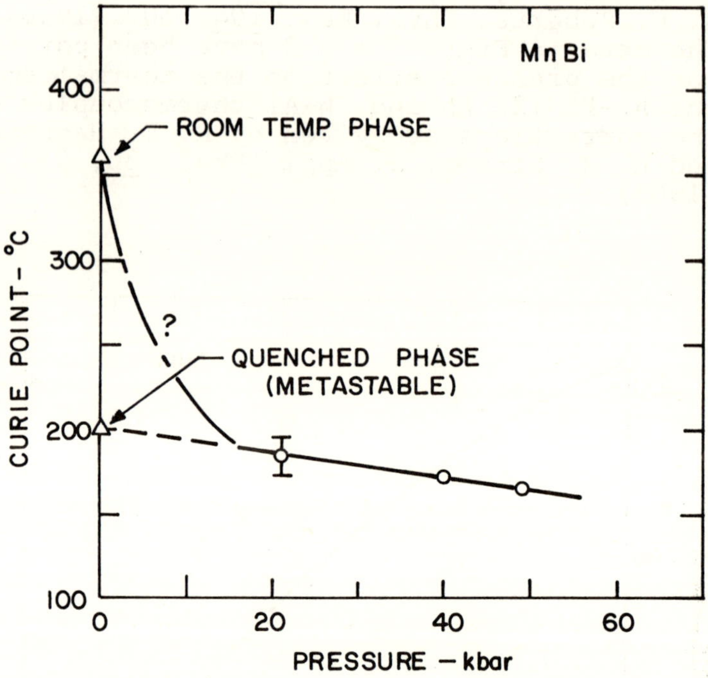

FIG. 3 SHIFT OF THE CURIE POINT OF MnBi WITH PRESSURE

THE MÖSSBAUER EFFECT AT HIGH PRESSURE

by
H. G. Drickamer, R. L. Ingalls, and C. J. Coston
Department of Chemistry and Chemical Engineering,
Department of Physics and
Materials Research Laboratory
University of Illinois, Urbana, Illinois

The purpose of this paper is to describe the results of certain current experiments in our laboratory involving the use of the Mössbauer effect in Fe^{57} to obtain information concerning the relationship between interatomic distance and electronic structure in alloys and compounds. At this early stage of the project no attempt is made to synthesize any detailed theoretical discussion of the results, but a variety of data is presented to illustrate the scope of the method as a means of elucidating electronic structure.

The use of recoilless resonant radiation to study electronic structure in solids has been discussed in a multitude of papers and a number of excellent reviews (1, 2, 3). We shall here consider the types of information obtainable under three categories; the s electron density at the nucleus (the isomer shift), the internal magnetic field, and the electric field gradient at the nucleus.

The isomer shift measures the difference in energy between the nuclear transitions in the source and absorber. These nuclear transitions are slightly, but measurably, perturbed by changes in the environment of the outer electrons. In the first place, the outer s electron wave functions have a finite probability density at the nucleus and in the second place the core s electrons are polarized by interaction with the outer d electrons. From the change in isomer shift with pressure one can infer something about the state of the outer electrons as a function of interatomic distance.

The change in isomer shift (i.e., of s electron density at the nucleus) of Fe^{57} with change of environment or pressure is due to changes in the 4s and 3s wave functions. The 3s wave functions are affected by changes in shielding by the 3d electrons. It seems reasonable to presume that in dilute alloys the 4s electrons of the Fe^{57} are in the conduction band of the metal. The situation with respect to the 3d electrons is not so clear, but it would seem that they are sufficiently itinerant to interact with the 3d bands of the host.

The 3d electrons in compounds are, of course, strongly affected by the crystal field provided by the ligands. The examination of these interactions as a function of interatomic distance forms a very promising field of study which will be only lightly touched upon here.

The ground state of iron has spin one half and the first excited state spin three halves. In the presence of a magnetic field these split into two and four states respectively. The selection rules allow six transitions, and therefore in a magnetic field, one obtains a six-line spectrum. The splitting is a measure of the magnetic field at the nucleus, which again can be obtained as a function of interatomic distance.

In an environment of less than cubic symmetry the interaction between the nuclear quadrupole moment and the electric field gradient can give rise to a splitting of the Mössbauer gamma-ray spectrum. The magnitude of the splitting depends on the gradient of the EFG at the nucleus (i.e., corrected for shielding by the electrons) and the nuclear quadrupole moment Q. Since Q should not be affected by pressure one thus has a measure of the change in EFG with interatomic distance.

The experimental techniques used for these high pressure Mössbauer studies has been previously described (4). Results are discussed under the headings of Isomer Shift, Magnetic Field, and Quadrupole Splitting.

Isomer Shift

Figure 1 shows the isomer shift for Fe^{57} in iron (4). At low pressures iron has the bcc structure and is, of course, ferromagnetic. At 130 kilobars, there is a first order phase transition to the hcp structure which is not ferromagnetic.

Walker, Wertheim and Jaccarino (5) developed a semi-empirical relationship between the isomer shift and the s electron density at the nucleus. If one assumes that the increase in s electron density scales with the decrease in volume, the isomer shift should change with volume according to the relationship:

$$\frac{\partial \epsilon}{\partial (V/V_0)} = 0.14 \text{ cm/sec} \qquad (1)$$

The solid line in Figure 1 represents the above equation.

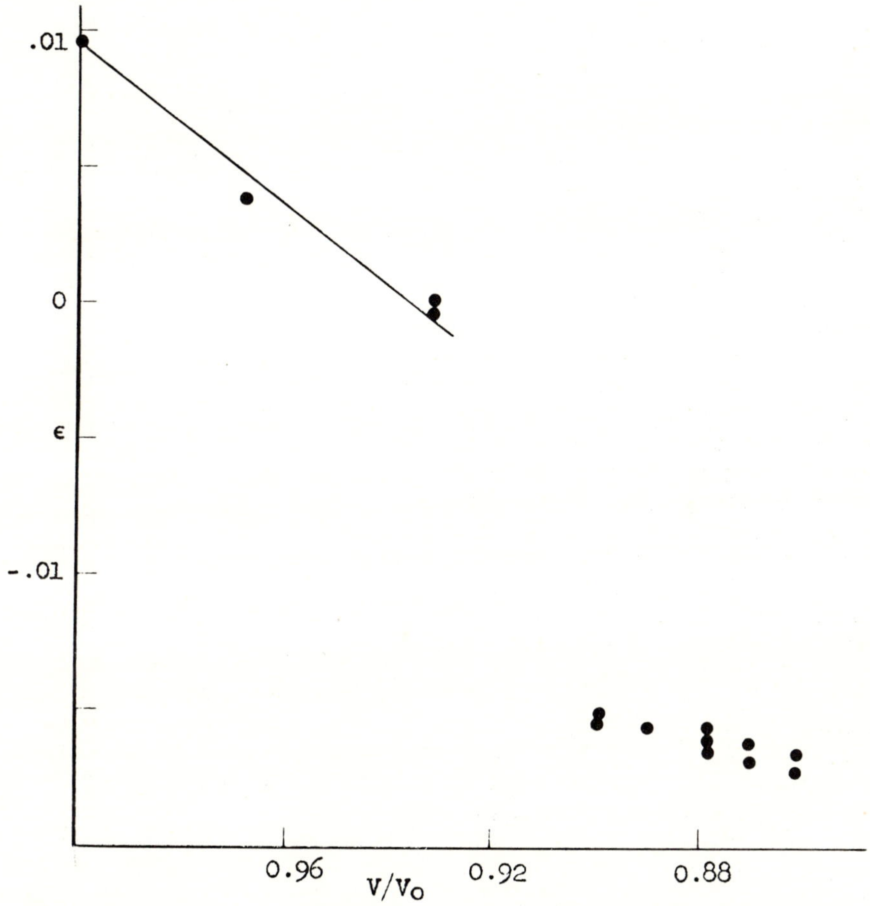

Fig. 1. ϵ vs V/V_o - Iron

For bcc iron one can say that the 4s electron density increases with density by compression with no shift in relative energies of the bands arising from atomic s and d states. There is a very large change in isomer shift at the transition, much larger than can be accounted for by the density change or the fact that the high pressure phase is non-ferromagnetic. A six percent increase in the s character of the valence electrons would account for the change. Apparently the broad s band is lowered in energy relative to the rather narrow d bands in going to the hcp phase. The relatively small change in isomer shift with density in the hcp phase is consistent with the behavior of close packed systems as discussed below.

Figures 2 and 3 show the change of isomer shift with V/V_0 for Fe^{57} in dilute solution in Cu, Ni, Pd, Ti and Co (6). The volumes were taken from shock wave data (7) extrapolated to 25°C. Data for the hcp phase of iron are also included. Here the fiducial volume and isomer shift are taken at 150 kilobars. These very recent results have not yet been fully analyzed, but certain generalizations can be made.

(1) For Fe^{57} dissolved in metals with close packed structures (fcc or hcp) the isomer shift is always considerably less than predicted by Eq. (1) and remarkably similar for all the dilute alloys studied to date. The dotted line shown in Figure 2 and 3 represents the average change in isomer shift for the fcc alloys. It can be seen that it is a very reasonable fit for the hcp alloys also. Although there are obviously significant differences in the electronic structure of these six close-packed metals, the change in isomer shift with density is remarkably similar in all six. In all cases the overriding feature is a decrease in s electron character of the valence electrons which partially counteracts the increase due to compression. Apparently, the 4s band tends to increase in energy with decreasing volume at a more rapid rate than the more tightly bound 3d electrons. The essentially quantitative agreement among these close packed metals is nevertheless surprising.

(2) Figures 4 and 5 show the change in isomer shift with V/V_0 for Fe^{57} in dilute solution in V, Nb, and Mo, all of which have the bcc structure. Data for bcc iron are also included. Here one finds significantly different behavior from metal to metal. As mentioned above, the s electron density scales with volume for iron. For vanadium the change in s electron density with V/V_0 is slightly less than the "scaling line" at low pressure and deviates further at higher density. At the highest density it appears actually to reverse direction, but it is difficult to be certain so near the upper limit of our pressure range.

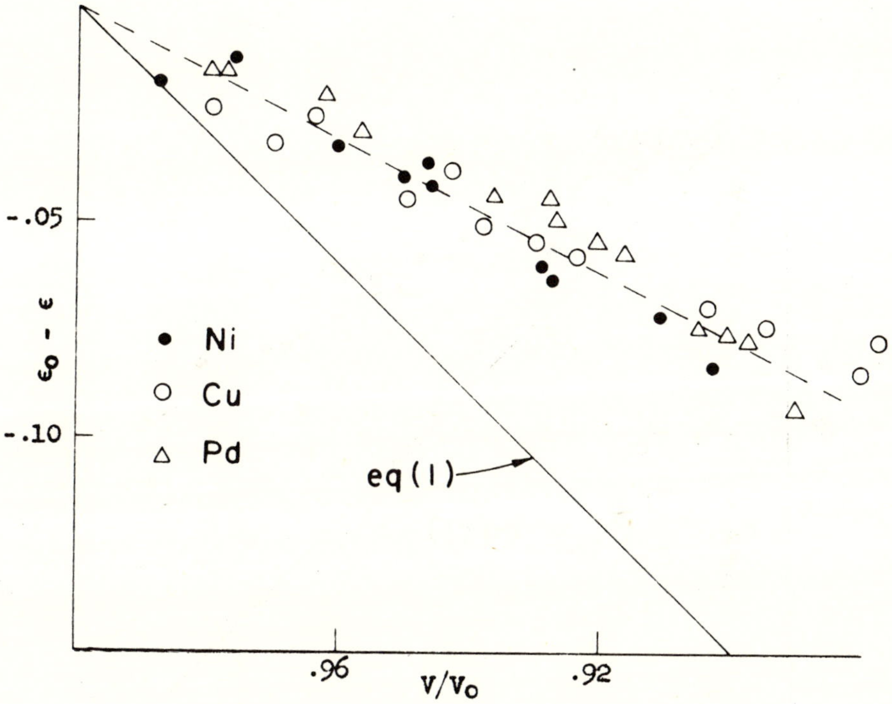

Fig. 2. $\epsilon_0 - \epsilon$ vs V/V_0 - Fcc Phases

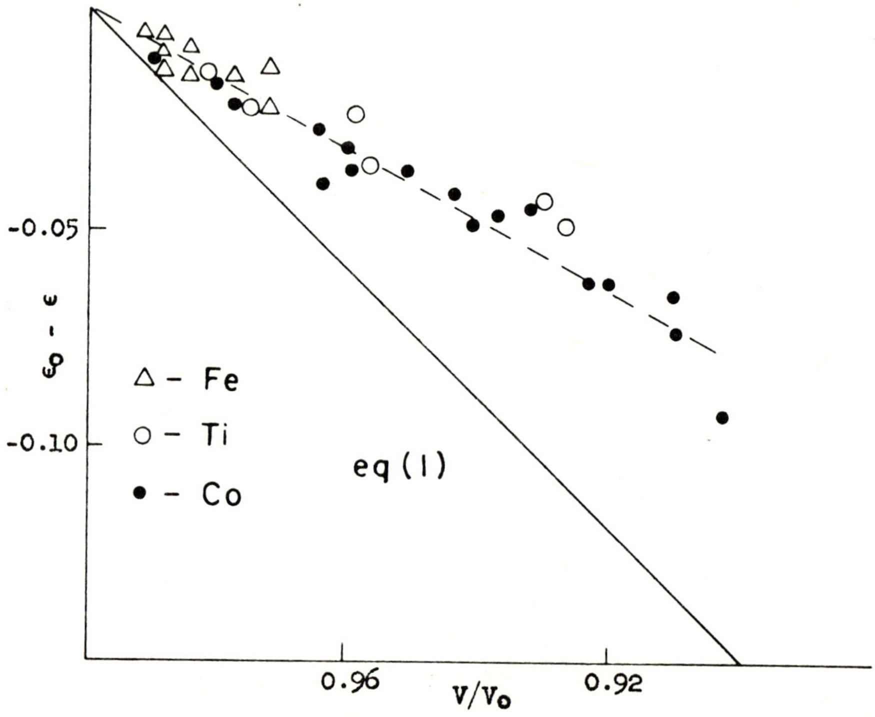

Fig. 3. $\epsilon_o - \epsilon$ vs V/V_o - hcp Phases

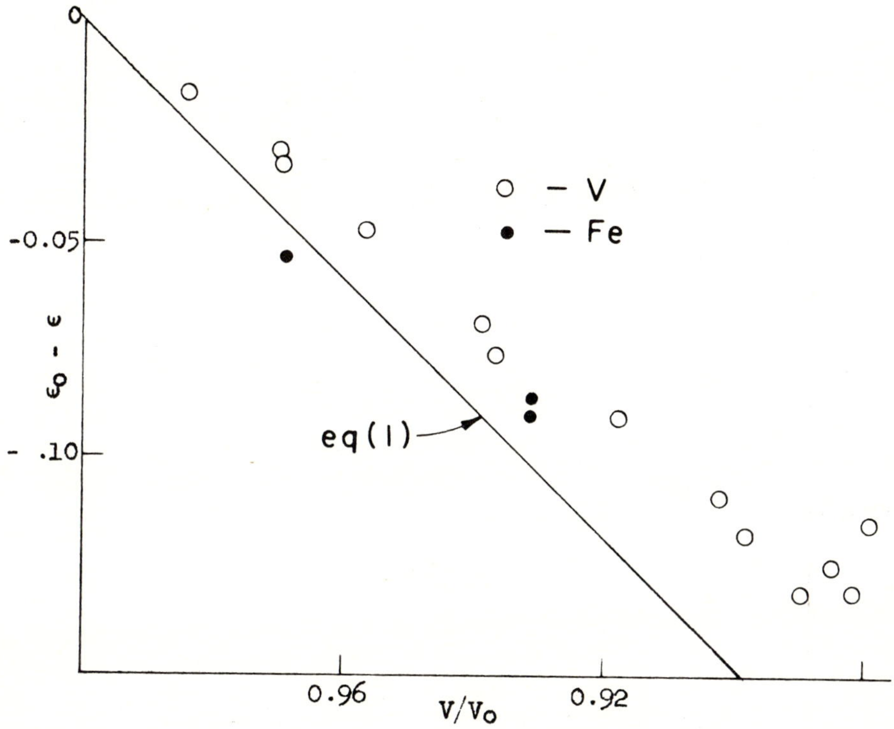

Fig. 4. $\epsilon_0 - \epsilon$ vs V/V_0 - Bcc Phases (3d)

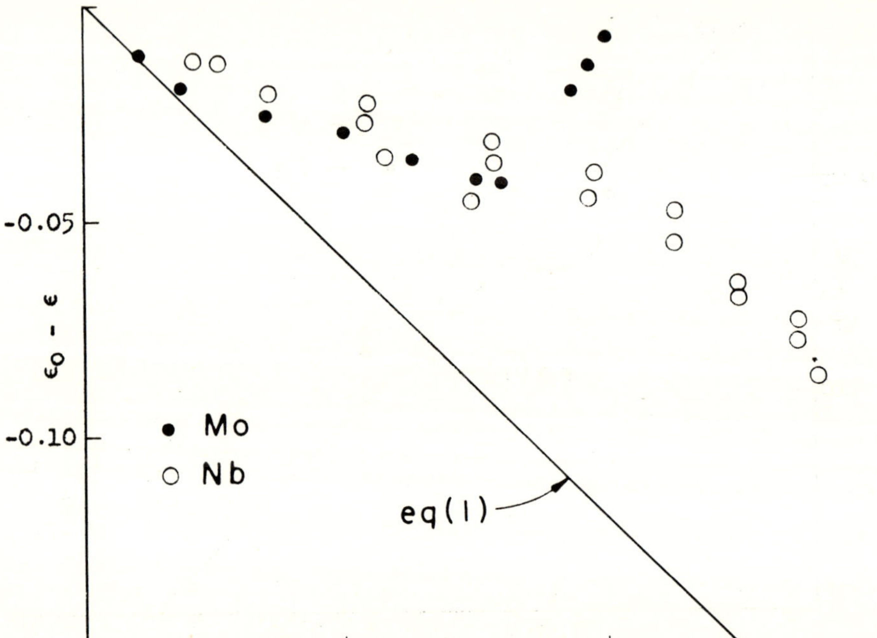

Fig. 5. $\epsilon_0 - \epsilon$ vs V/V_0 Bcc Phases (4d)

Niobium shows a very small increase in s electron density with density at low pressure, but the shift increases rapidly at higher pressures and at $V/V_0 = 0.91$, the s electron density is essentially scaling with density.

For Mo the isomer shift appears to scale with density at very low density but slows down rapidly with increasing density and at $V/V_0 = 0.93$ there is a distinct reversal of sign.

These systems of more open structure show strong variations from metal to metal. It is possible that the d bands are narrower and more typical in shape of the individual metal in these systems. Although no satisfactory treatment is available, it appears that these results should form the basis for some interesting theoretical work. Experiments in other bcc transition metals, e.g., Cr, Ta, and W, would also be important.

Magnetic Field

To date measurements have been made of the magnetic field at an Fe^{57} atom in bcc iron, in nickel, and in hcp cobalt. The results are shown in Figure 6. The magnetic field in iron decreases with increasing pressure at a significant rate. The solid line to 65 kilobars represents the zero field NMR measurements of Litster and Benedek (8). The agreement with Mössbauer results is excellent. This is particularly interesting since NMR measurements are sensitive primarily to nuclei in the domain walls, while the Mössbauer effect does not discriminate by location.

The results for Fe^{57} in nickel show a small but distinct decrease in field with pressure. The data for Fe^{57} in cobalt show an apparent maximum at low pressure and a small decrease thereafter; but further experiments are desirable here. While no theory is yet available to explain these results, it is of interest to note that NMR measurements for Ca^{59} in cobalt (9) and Ni^{61} in nickel (10) show a distinct increase in field with pressure. One could assume that the field at an iron nucleus in these metals represents a balance between the field for iron in iron and the field at a Co or Ni nucleus in these pure metals.

Quadrupole Splitting

The studies of quadrupole splitting with pressure are in their initial stages and only very preliminary results will be mentioned. Figure 7 shows the quadrupole splitting for Fe^{57} in cobalt. At first the interaction increases with

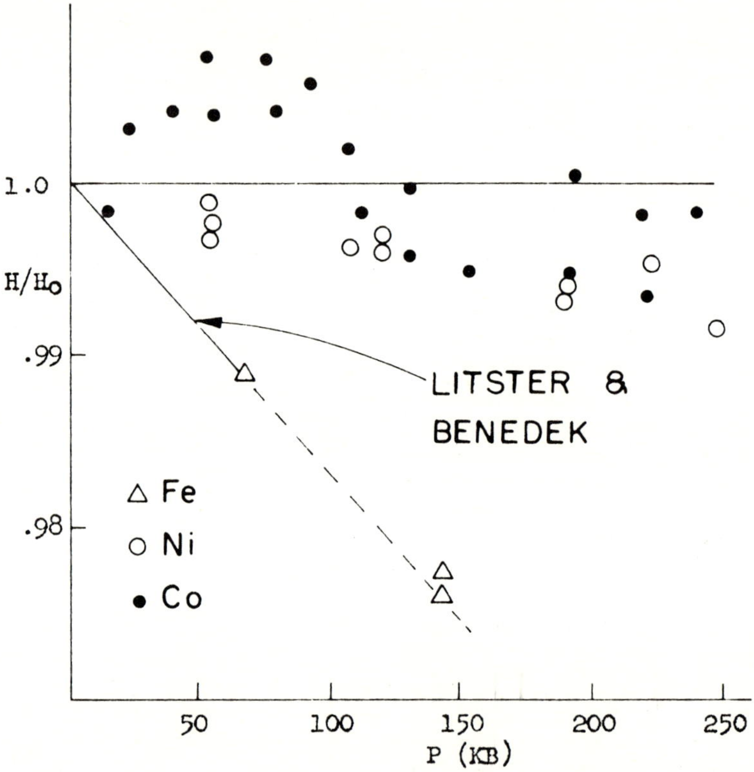

Fig. 6. H/H_0 vs Pressure

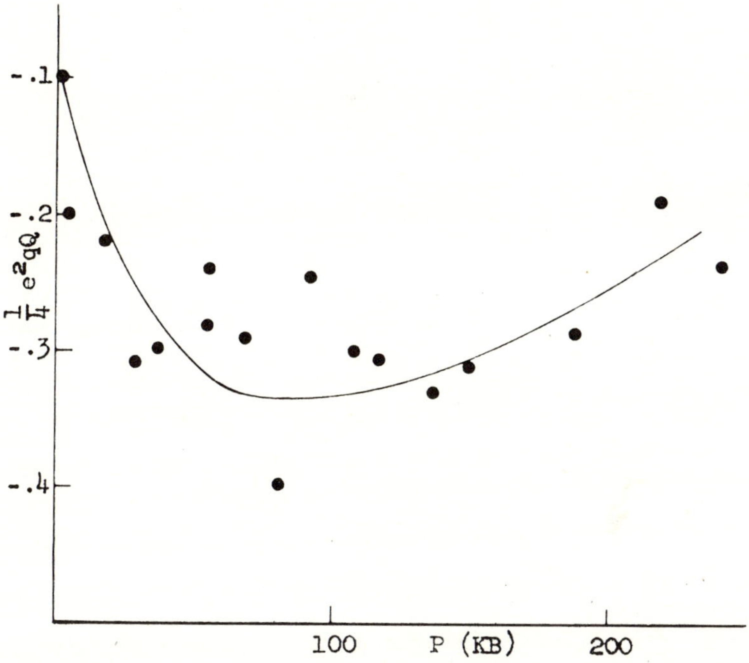

Fig. 7. Quadrupole Splitting vs Pressure - Co

pressure and then decreases at high pressure. It would be of interest to correlate these results with changes in c/a ratio for the material.

Figure 8 exhibits the change in quadrupole splitting with pressure for $CoCl_2$ which has a rhombohedral structure. The shape of the curve compares very closely with the predictions of Ingalls (11) who calculated the change in splitting with change in crystal field energy Δ. For the rhombic splitting Δ changes as $1/R^3$ where R is the ion ligand distance. Ingalls then predicts an increase in quadrupole splitting with decreasing R which levels for high values of Δ and ultimately decreases slightly.

The development of this new tool opens up a wide variety of possible experiments including the change in magnetic field with density near the Curie point or Néel point of alloys and compounds, the effect of distortions of the crystal field on electronic energy levels, and changes in covalency. Extensions in technique to permit working with a moving source outside the cell and to allow measurements at liquid nitrogen temperatures open up the possibility of studies of tin compounds and alloys, of rare earths, and of iodine compounds.

It is a pleasure to acknowledge the assistance of the United States Atomic Energy Commission in this work.

References

1. H. Frauenfelder, The Mössbauer Effect, W. A. Benjamin, New York (1962).
2. A. F. J. Boyle and H. E. Hall, Repts. Prog. Phys. 25, 441 (1962).
3. G. K. Wertheim, The Mössbauer Effect, Academic Press, New York (1964).
4. D. N. Pipkorn, C. K. Edge, P. Debrunner, G. de Pasquali, H. G. Drickamer, and H. Frauenfelder, Phys. Rev. 135, A1604 (1964).
5. L. R. Walker, G. K. Wertheim, and V. Jaccarino, Phys. Rev. Letters 6, 98 (1961).
6. Data for Cu, Ti, and V appear in a paper by C. K. Edge, R. L. Ingalls, P. Debrunner, H. G. Drickamer, H. Frauenfelder, Phys. Rev. in press .
7. M. H. Rice, R. G. McQueen, and J. M. Walsh, Solid State Physics, Vol. 6, F. Seitz and D. Turnbull, eds., Academic Press, New York (1958).
8. J. D. Litster and G. B. Benedek, J. App. Phys. 34, 688 (1963).
9. D. Anderson, Sandia Corp. , private communication .

10. T. Kushida, quoted in Magnetic Resonance at High Pressure, by G. B. Benedek, Interscience, New York (1963) p. 23.
11. R. L. Ingalls, Phys. Rev. 133, A787 (1964).

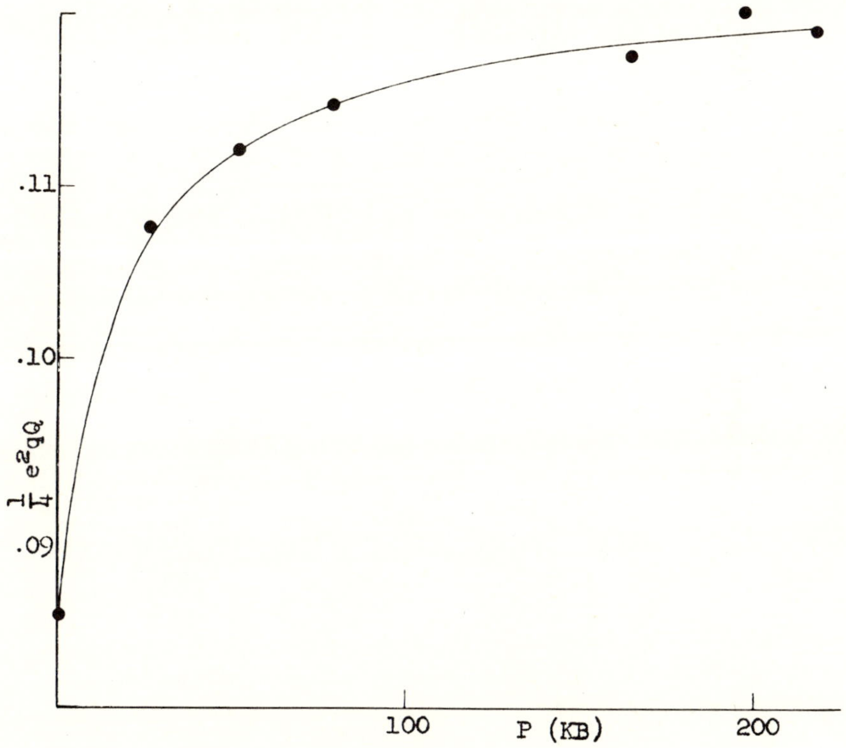

Fig. 8. Quadrupole Splitting vs P $CoCl_2$

SESSION D

LATTICE DEFECTS

Chairman: A. D. LE CLAIRE

EFFECT OF PRESSURE ON THE DIFFUSION RATE IN METALS*

R. H. Dickerson,† R. F. Bonanno‡
and C. T. Tomizuka
Department of Physics, University of Arizona
Tucson, Arizona

Various physical properties of vacancies in metals, especially in noble metals, are known at present. For example, by careful comparison of x-ray lattice parameter and bulk dilatation due to thermal expansion, Simmons and Balluffi (1,2,3) succeeded in determining the vacancy concentration as a function of temperature in a number of metals and thus the energy to form a vacancy, E_f, in these metals. By quenching in vacancies from high temperatures it is also possible to determine E_f and, by observing the annealing rate of the quenched-in vacancies, one can determine the energy to move a vacancy, E_m, (4,5) except that a unique interpretation of the latter quantity is somewhat in doubt due to the recent work of Emrick (6) and of Kauffman and Meshii (7).

The volume of an individual vacancy can be estimated indirectly by combining the results of

*This work was originally sponsored in part by the U.S. Air Force Office of Scientific Research under Contract No. AFOSR 49(638)-790 and later by the U.S. Atomic Energy Commision under Contract No. AT(11-1)-1041.

†Present address: Department of Physics, University of Illinois, Urbana, Illinois.

‡Present address: Air Force Materials Laboratory, Wright-Patterson AFB, Ohio.

two or more experiments (3,8). A more direct determination is possible by employing pressure instead of temperature as a single variable. Nachtrieb, Lawson and their co-workers succeeded in determining the activation volume, ΔV, for self-diffusion in sodium (9) and other low melting point metals (10, 11, 12). The molar activation volume is defined as

$$(\partial \Delta G/\partial p)_T = \Delta V,$$

where ΔG is the molar Gibbs free energy for the self-diffusion process. ΔV is to be interpreted as the volume of a mole of vacancies in motion. ΔV then is the sum of the volume of a mole of vacancies at rest, ΔV_f, and the extra volume increase, ΔV_m, caused by a mole of vacancies moving to the activated state. The results of the above-mentioned pressure-diffusion work showed that ΔV is approximately 50 to 60 percent of the molar volume V_M. This implies appreciable relaxation around a vacancy provided that ΔV_m is not negative. Since no experimental values of ΔV_m are available for those metals investigated by Nachtrieb, Lawson et al., the value of ΔV_m for these metals cannot be determined uniquely.

For face-centered cubic metals, particularly for noble metals and especially for gold, other physical properties of vacancies such as E_f and E_m are known. In addition, the value of ΔV_m was determined for gold by Emrick (13) and that of ΔV_f was estimated by Huebener and Homan (14). Their results indicate that for gold ΔV_f and ΔV_m are approximately 55 percent and 15 percent, respectively, of the molar volume. In order to determine the self-consistency of the values of ΔV_f and ΔV_m for gold, the effect of pressure on the self-diffusion rate in gold single crystals was experimentally measured. The details of the experiment were published elsewhere (15). It is sufficient to note here that this experiment required the operation of a gas pressure system up to 9 kbar at temperatures in the neighborhood of 1000°C for the prolonged period of time required for measurable diffusion penetration.

The value of the activation volume ΔV for gold was found to be 7.2 ±0.4 cm^3/mole which corresponds to 71 percent of the molar volume. The direct determination of ΔV_f and ΔV_m yield an estimated value of ΔV as 70 percent of V_M. Considering the errors involved in these three independent experiments the agreement here seems somewhat fortuitous.

Many detailed theoretical calculations for volume relaxation about a vacancy in gold and in copper are found in the literature (16) and their values vary widely between 50 to 90 percent of V_M for ΔV_f. The theoretical value of ΔV_f is sensitive to the choice of repulsive potentials. The present experiment cannot serve to determine the correct choice of the potential, since the variety in the theoretical approximations employed introduces just as much variation in the theoretical value of ΔV_f. Theoretical values of ΔV_m are subject to a considerable error (17).

The accuracy with which one can determine ΔV is limited in part by the various difficulties inherent in the high-temperature, high-pressure experiments. Inaccuracies also arise from the fact that the effect of pressure on the diffusion coefficient, D, in metals is relatively small in the hydrostatic pressure range of 0-30 kbar. It is therefore impossible to obtain an accuracy comparable to that of the activation energy determination for diffusion where the diffusion coefficients may be varied by at least a factor of 10^3 and sometimes as much as 10^{14} (carbon in iron). The variation of D with pressure in the present experiment is less than a factor of 2. A slight error in temperature determination introduces a considerable variation in the outcome of high-pressure diffusion experiments.

It is clear then that the efforts of pressure-diffusion experiments should not be directed toward an area in which the significance of an experiment lies primarily in obtaining the absolute value of ΔV to a high precision. Diffusion experiments at high pressures can be most valuable in the area where the relative magnitude of ΔV or a mere sign of ΔV can lead to a better understanding of the fundamental processes involved.

With the above-mentioned view in mind, the effect of pressure on the diffusion rates of impurities in silver single crystals was studied to test the validity of the Lazarus-LeClaire theory of impurity diffusion (18,19). For this purpose the effect of hydrostatic pressure on the diffusion rates of indium and of antimony in silver single crystals was measured. For the purpose of comparison, the effect of pressure on self-diffusion in silver was also determined. The results obtained are

$$\Delta V_{self} = 9.6 \pm 0.1 \text{ cm}^3/\text{mole},$$

$$\Delta V_{In} = 9.1 \pm 0.6 \text{ cm}^3/\text{mole}$$

and

$$\Delta V_{Sb} = 8.5 \pm 0.5 \text{ cm}^3/\text{mole}.$$

The screening model by Lazarus and LeClaire describes the energy difference between self-diffusion and the impurity diffusion in the same matrix in terms of the interaction between a vacancy and the screened charge of a solute ion. This interaction is calculated in the Thomas-Fermi approximation. Also, the saddle-point configuration is described as a split point charge of $-\frac{1}{2}e$ on each side of the saddle point. LeClaire's calculation yields

$$\Delta H = - \frac{N_0 Ze^2 \delta}{(11/16)r} \exp\left(- \frac{11qr}{16}\right) ,$$

where ΔH is the difference in activation enthalpies between self and impurity diffusion, N_0 is Avogadro's number, Ze is the difference in charge between the solute and solvent ions, δ is a constant which depends on Z only, r is the interatomic distance and q is the screening parameter. q is given by

$$q^2 = \frac{16\pi^2 me^2}{h^2} \left(\frac{3n_o}{\pi}\right)^{1/3} ,$$

where m, e and n_o are the mass, charge and number per unit volume of electrons, respectively. Differentiation of ΔH with respect to pressure at constant temperature yields

$$\left(\frac{\partial \Delta H}{\partial p}\right)_T = -\frac{N_0 Ze^2 \delta K_T}{44}\left(\frac{16}{r}+\frac{11q}{2}\right)\exp\left(-\frac{11}{16}qr\right),$$

where K_T is the isothermal compressibility. The calculation of ΔH by Corless and March (20) differs from LeClaire's expression only in that the factor δ is absent. Thus it is possible to calculate the difference in activation volumes Δ(ΔV) between the self- and impurity diffusion by either theory (Lazarus and LeClaire or Corless and March; L&L and C&M, respectively) through a thermodynamic relation

$$(\partial \Delta H/\partial p)_T = \Delta\Delta V(1 - \alpha T),$$

where α is the volume expansivity. Experimental and theoretical values of ΔΔV are listed in Table I.

Table I

Comparison between $\Delta\Delta V_{exp}$ and $\Delta\Delta V_{cal}$

	$\Delta\Delta V_{exp}$ (cm^3/mole)	$\Delta\Delta V_{cal}$ (L&L) (cm^3/mole)	$\Delta\Delta V_{cal}$ (C&M) (cm^3/mole)
In	0.5	0.3	0.5
Sb	1.1	0.5	1.1

The slight increase in activation enthalpy with pressure in all cases is of the order of 0.2 kcal/(mole kbar). The effect of pressure is greater on the interatomic distance than on the screening radius (1/q) by a factor 2rq which is approximately 10 for silver. The impurity ion, therefore, finds itself further inside the screening potential at high pressures thus requiring less energy to form a vacancy. However, due to the decrease in the interatomic distance around a saddle-point configuration at high pressures, the energy to move a vacancy increases significantly to offset the change in the formation energy.

The authors wish to thank Rodney C. Lowell for the operation of the gas pressure system.

References

1. R. O. Simmons and R. W. Balluffi, Phys. Rev. 117, 52 (1960).
2. R. O. Simmons and R. W. Balluffi, Phys. Rev. 119, 600 (1960).
3. R. O. Simmons and R. W. Balluffi, Phys. Rev. 125, 862 (1962).
4. J. E. Bauerle and J. S. Koehler, Phys. Rev. 107, 1493 (1957).
5. M. Doyama and J. S. Koehler, Phys. Rev. 127, 21 (1962).
6. R. M. Emrick, Phys. Rev. Letters 12, 724 (1964).
7. J. W. Kauffman and M. Meshii, in Lattice Defects in Quenched Metals (Cotterill, Doyama, Jackson and Meshii, Editors, Academic Press, New York, 1965) p. 77.
8. J. Takamura, Acta Met. 9, 547 (1961).
9. N. H. Nachtrieb, J. A. Weil, E. Catalano and A. W. Lawson, J. Chem. Phys. 20, 1189 (1952).
10. N. H. Nachtrieb and A. W. Lawson, J. Chem. Phys. 23, 1193 (1955).
11. N. H. Nachtrieb, H. A. Resing and S. A. Rice, J. Chem. Phys. 31, 135 (1959).
12. C. J. Coston and N. H. Nachtrieb, to be published.

13. R. M. Emrick, Phys. Rev. 122, 1720 (1961).
14. R. P. Huebener and C. G. Homan, Phys. Rev. 129, 1162 (1963).
15. R. H. Dickerson, R. C. Lowell and C. T. Tomizuka, Phys. Rev. 137, A613 (1965).
16. A. Seeger and D. Schumacher, in Lattice Defects in Quenched Metals (Cotterill, Doyama, Jackson and Meshii, Editors, Academic Press, New York, 1965) p. 32.
17. G. Schottky, A. Seeger and G. Schmid, Phys. Stat. Solidi 4, 419 (1964).
18. D. Lazarus, Phys. Rev. 93, 973 (1954).
19. A. D. LeClaire, Phil. Mag. 7, 141 (1962).
20. G. K. Corless and N. H. March, Phil. Mag. 6, 1285 (1961).

SELF-DIFFUSION IN TIN AT HIGH PRESSURE

Norman H. Nachtrieb and Charles Coston
Institute for the Study of Metals
The University of Chicago
Chicago, Illinois

Measurements of the effect of hydrostatic pressure on the diffusive rates of atoms in cubic metals have added considerably to an understanding of the elementary process. Out of the half dozen or so studies that have been made, a clear experimental conclusion has begun to emerge: the activation volume for the overall diffusion process is less than the gram atomic volume and ranges from 25% of this quantity in the case of bcc metals to about 90% for fcc metals. This has led to the conclusion that, although lattice vacancies are the responsible defects, there must be considerable relaxation of the neighboring atoms around them (1). Theoretical estimates have ranged widely, from essentially zero to approximately 50% of the gram atomic volume.

Tin was selected for high pressure diffusion measurements for several reasons. One might a priori expect an anisotropy to appear in the activation volume, corresponding to differences in the rates of atom jumps along the two principal axes of its tetragonal structure. Meakin and Klockholm (2) had observed appreciable anisotropy in the diffusion rates in tin at atmospheric pressure, with D_a approximately twice as large as D_c. In addition, the void volume in white tin is a large fraction (46.5%) of the gram atomic volume, as compared with 32.0% in bcc and only 26.0% in fcc metals. Interstitial diffusion, if it plays any role in the self-diffusion mechanism in pure metals, would seem to be a possibility in white tin. Significantly, no instances of a negative activation volume have been reported for self-diffusion in pure metals ... a finding that would lead us to infer an interstitial mechanism.

EXPERIMENTAL

Single crystals of tin were grown by the Bridgman method from 99.999% pure metal and their crystal perfection was checked by a large beam area back-reflection Laue technique. Mosaic structure was readily revealed by irregularities in the shapes of the Laue reflection spots, and such crystals were rejected. Crystals grown from tin of 99.9% purity invariably showed a cellular network structure on etching, presumably due to impurity segregation, and were not used. The crystals of higher purity showing high structural perfection were cut into cubes of about 1 cm. edge length with a Servomet spark cutter after preliminary orientation by back-reflection Laue techniques in a special goniometer crystal holder. The attempts to cut crystals by means of milling cutters and jeweler's saws were abandoned in favor of this method when it was found that they produced "warping" of crystal planes, even though the crystals remained "single". The final preparation of the surfaces that were to carry the Sn-113 tracer was accomplished by optically aligning the crystal in its goniometer holder in a microtome and removing sections about 2 microns in thickness. A 48-hour vacuum anneal at 200° removed the superficial cold work caused by the microtome blade and restored the symmetry of the wide beam Laue reflections. Sn-113, from which the accompanying Sb-125 impurity had been chemically removed, was vapor-deposited on adjoining (100) and (001) faces of the crystal, to permit the simultaneous measurement of D_a and D_c.

High pressures for the diffusion anneals were generated with a liquid hydraulic system of the Bridgman type, with Dow-Corning 200 silicone fluid. A manganin gage, calibrated against the freezing pressure of mercury at 0°C., was used for the measurement of pressures up to 10 kilobars. Temperatures were measured by means of a calibrated chromel-alumel thermocouple inside the high pressure vessel at a point close to the crystal. Runs were made at atmospheric pressure from 160° to 225° and at temperatures of 192°, 209°, and 226° for pressures of approximately 2,000, 4,000, 6,000, 8,000, and 10,000 bars. The temperatures and pressures in individual runs deviated from these nominal values and were normalized to them by simple correction procedures for the construction of isotherms and isobars.

Sectioning of the crystals along the planes normal to the principal axes was carried out with the microtome after the common edge between the active faces was beveled. Thin sections were counted with a methane flow proportional counter and weighed on a microbalance for the construction of penetration profiles. In this manner D_a and D_c were usually obtained in pairs, whose ratio was less subject to errors arising from fluctuations in temperature and pressure than would have been the case if a- and c- axis runs were separate. Corrections for warm-up time during the diffusion anneal were made, although they were generally less than 2%. Likewise, corrections were made for the thermal expansion and compressibility of tin, even though they seldom exceeded 1% in their effect upon D. The principal uncertainty was in the measurement of temperature; although it could be measured to within $\pm 0.25^{\circ}$, variable gradients within the high pressure vessel and fluid limited the accuracy to about 2°, and the relative error in D is estimated to be between 5% and 7%.

RESULTS

Table 1 lists all of the D values obtained and Figs. 1 and 2 show the corrected isobars (normalized) for a- and c- axis diffusion, respectively. For both axes there is an increase of both D_o and ΔH with pressure, as summarized in Table 2. The activation energy is not determined with high accuracy in this study, however, both because the temperature interval is narrow and generally only three or four temperatures were chosen for each of the six pressures. It is clear, nevertheless, that $(\partial \Delta H/\partial P)_T$ is positive and that $(d\,D_o/dP)$ is also greater than zero. These insure that ΔV is positive, since quite generally:

$$\left(\frac{\partial \Delta H}{\partial P}\right)_T = T\left(\frac{\partial \Delta S}{\partial P}\right)_T + \Delta V \qquad (1)$$

From Table 2, average values for $(\partial \Delta H/\partial P)$ and $(\partial \Delta S/\partial P)$ are 11.5 $cm^3 mole^{-1}$ and 1.31×10^{-2} $cm^3 deg^{-1}$, respectively. For an average temperature

Table 1. Diffusion Data for Tin

P $kg.cm^{-2}$	T, °C.	$D_a \times 10^{11}$ $cm^2 sec^{-1}$	$D_c \times 10^{11}$ $cm^2 sec^{-1}$	D_a/D_c
0	160.4	0.211	...	...
0	162.6	0.249	...	...
0	160.8	...	0.099	...
0	176.3	0.557	0.255	2.18
0	191.7	1.40	0.563	2.49
0	207.7	3.58	1.56	2.30
0	225.8	10.1	4.82	2.09
2158	189.6	0.92	0.47	1.95
1640	207.9	2.89	1.36	2.13
1992	225,6	7.36	3.40	2.16
1932	228.8	7.89	4.38	1.80
4116	190.1	0.80	...	...
3988	192.0	...	0.360	...
3907	211.3	2.75	1.29	2.13
4007	225.1	5.12	2.88	1.78
3977	235.9	10.0	4.87	2.06
6089	190.0	0.550	...	...
6035	192.6	0.679	0.272	2.50
6017	208.1	1.56	0.70	2.22
6104	224.6	4.88	2.30	2.12
6020	239.8	10.8	4.20	2.57
7999	191.9	0.400	0.183	2.18
7909	207.8	...	0.471	...
7959	208.8	1.39	0.664	2.09
7970	224.8	3.64	1.51	2.42
8079	239.8	7.69	3.10	2.45
10090	192.0	0.347	0.159	2.18
10120	207.6	0.984	0.422	2.33
10105	225.3	2.91	1.37	2.13
10134	244.8	7.77	2.73	2.85

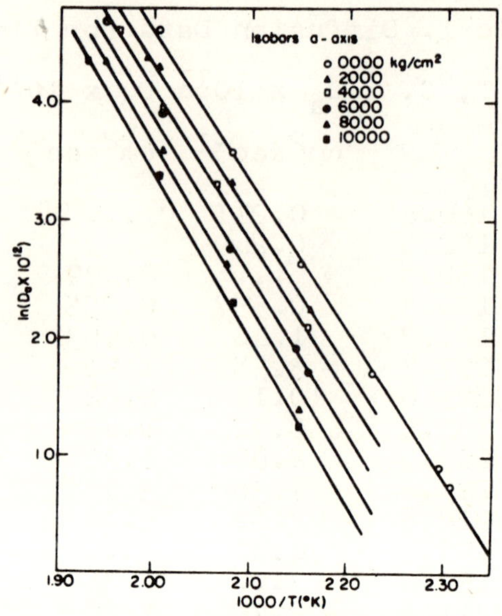

Fig.1 a- Axis Diffusion Isobars

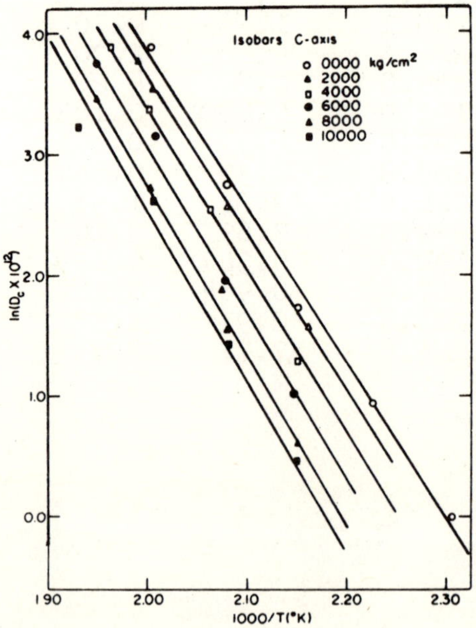

Fig.2 c- Axis Diffusion Isobars

Table 2. Activation Energies and D_o Values

	Pressure (kg.cm^{-2})					
	0	2000	4000	6000	8000	10100
ΔH_a,kcal/mole	25.1	25.6	25.6	26.3	27.2	27.9
ΔH_c,kcal/mole	25.6	25.8	26.7	27.5	28.1	28.5
D_o^a,cm^{2}sec^{-1}	9.1	10.7	9.4	15.2	29.9	42.1
D_o^c,cm^{2}sec^{-1}	6.2	7.2	13.4	21.7	28.8	35.2

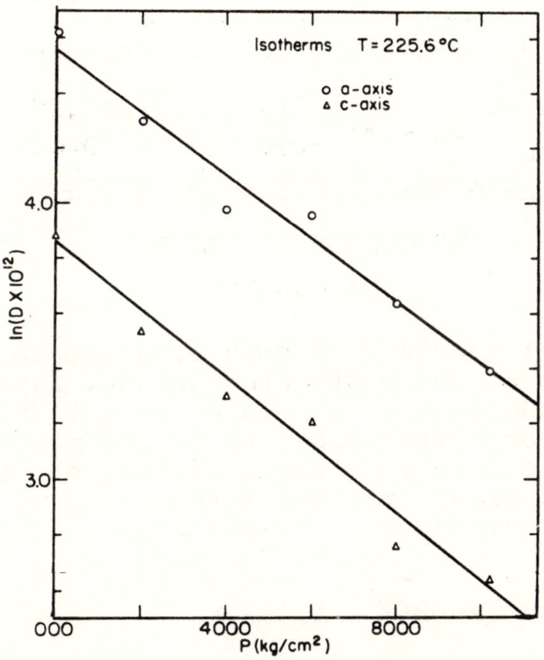

Fig. 3. a- and c- Axis Isotherms.

of $500^\circ K.$, these lead to a mean value of ΔV of 10.9 $cm^3 mole^{-1}$.

A more direct method for evaluating ΔV comes from the isotherms, since it is defined on the basis of the absolute reaction rate theory by

$$\Delta V = - RT\left(\frac{\partial \ln D/\gamma a^2 \nu}{\partial P}\right)_T . \qquad (2)$$

Figure 3 shows one of the isotherms (T = 225.6°C.) and Table 3 lists the values obtained for ΔV_a and ΔV_c for each of the three temperatures. ΔV_c is evidently a function of neither temperature nor pressure. ΔV_a is independent of pressure to 10 kilobars and, although it appears to decrease with increasing temperature, the trend is barely outside the expected error limits. The anisotropy in ΔV is not large and for purposes of later discussion we shall take its average measured value to be 5.3 ± 0.5 $cm^3 mole^{-1}$.

Table 3. Activation Volumes for a-and c-Axes

T, °C.	225.6°	208.8°	192.0°
ΔV_a, $cm^3 mole^{-1}$	4.96±.38	5.20±.25	5.91±.50
ΔV_c, $cm^3 mole^{-1}$	5.31±.38	5.23±.30	5.36±.30

DISCUSSION

Several models have been proposed that lead to estimates of the activation volume. Keyes (3) has used continuum theory to relate the activated state parameters for lattice defects to the macroscopic thermodynamic functions for a variety of processes. Using the strain-energy model and assuming the Grüneisen approximation, he has proposed two relations:

$$\Delta V = 2(\gamma - 1/3)\beta \Delta G \qquad (3)$$

$$\Delta S = 2(\gamma - 1/3)\alpha \Delta G \qquad (4)$$

where β is the isothermal compressibility, α is the thermal expansion coefficient, and γ is now the Grüneisen constant. To a first approximation, $\Delta G \cong \Delta H$, and ΔS from Eq.(4) is approximately

4.2 e.u.. Its substitution into $\Delta G = \Delta H - T\Delta S$ gives a second approximation for ΔG at 500°K (ΔG = 23.0 k.cal.) and further iteration does not sensibly alter either ΔG or ΔS. For this value of ΔG, Eq.(3) yields $\Delta V = 5.6$ cm^{3}mole^{-1}, in good agreement with experiment. However, the value of ΔS (4.2 e.u.) is much lower than those obtained from D_o^a or D_o^c when the usual D_o theory is assumed.

On the basis of Rice's dynamical theory of diffusion (4) and experiments on the pressure-dependence of diffusion in cubic metals (5,6), a correspondence between diffusion and melting has been proposed in the form:

$$\frac{\partial \ln D/\gamma a^2 \nu}{\partial (T_m/T)} = \text{const.} = -\frac{\Delta H}{RT_m} . \qquad (5)$$

This must obviously be true at atmospheric pressure, where T_m is a constant. However, T_m is a function of pressure and dT_m/dP is not generally a constant. Pressure may be transformed into T_m as the variable and Babb's data (7) to 10 kilobars for tin have been used for this purpose. Now both T_m and T or T_m/T are the variables in our experiments.

Figure 4 shows the remarkable experimental fact that for both the a- and c- axes in tin, Eq.(5) is valid even when T_m is varied. It follows that:

$$\Delta H = bRT_m . \qquad (6)$$

To within the experimental error, this suggests that the elementary diffusion process is a cooperative phenomenon that affects the energies of a sufficient number of atoms surrounding a vacancy to show a correspondence with the macroscopic melting parameters. Table 4 shows the slopes, (-b), to be nearly the same for both the a- and c- axes. It also lists the values of ΔH as a function of pressure, based upon Eq.(6) and Babb's values for T_m. Values of $d\Delta H/dP$ evaluated in this way and used in Eqs. (1) and (4) lead to $\Delta V_a = 6.1$ cm^3 mole^{-1} and $\Delta V_c = 6.1$ cm^3 mole^{-1}, in fairly good agreement with experiment.

The concept of this corresponding states principle, relating diffusion to melting, has been stated in another form (5,6):

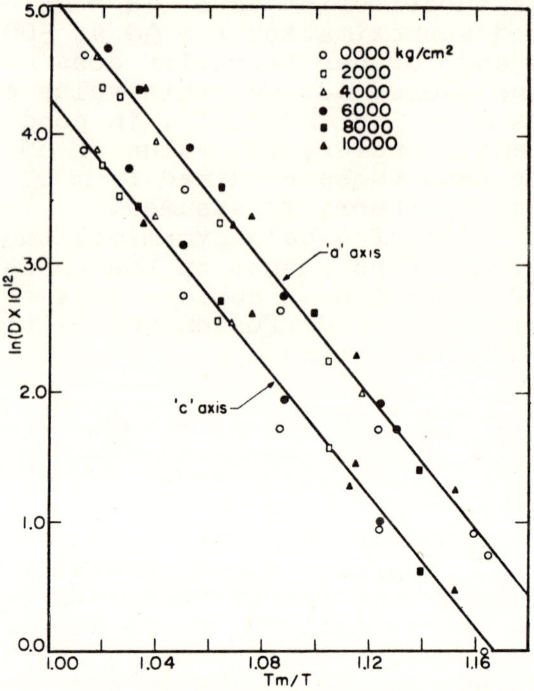

Fig. 4. Reduced Temperature Plot of a- and c- Axis Diffusion Coefficients.

Table 4. Quantities Derived from Corresponding States Test

Quantity	a-Axis	c-Axis
Slope (-b)	-25.69	-25.41
Standard Dev'n.	0.14	0.15
ΔH (P = 0)	25,780	25,500
ΔH (P = 5000)	26,570	26,290
ΔH (P = 10000)	27,350	27,060
ΔV (cm^3 $mole^{-1}$)	6.1	6.1

$$\Delta V = \frac{\Delta H}{\Delta H_f} \cdot \Delta V_f \qquad (7)$$

where ΔV_f and ΔH_f are the volume change and latent heat of fusion. Equation (7) has given remarkably close agreement with experiment for a number of cubic metals including lithium (8), sodium (9), lead (5), and gold (10). With ΔH_f = 1690 cal., ΔV_f = 0.454 cm^3, and ΔH = 25.1 kcal. for tin, however, Eq.(7) yields ΔV = 6.75 cm^{3}mole^{-1}. This is not only in more serious disagreement with experiment, but implies an inconsistency with the other version of the corresponding states calculation that led to ΔV = 6.1 cm^{3}mole^{-1}. The possibility of an error in either the latent heat or volume change on fusion is remote.

The cause of the discrepancy lies in the neglect of the effect of pressure on the entropy of activation in the derivation (5) of Eq.(7). As a rule, this is justified, since D_o has not shown much variation with pressure in cubic metals. In tin, however, both D_o^a and D_o^c increase sharply with pressure (Table 2) and the quantity $Td\Delta S/dP$ (equal to 0.65 cm^3 at 500°K.) must be subtracted from Eq.(7). The important conclusion is that Eq.(7) is only approximate and will be in error to the extent that the entropy of activation is strongly pressure-dependent.

DIFFUSION MECHANISM IN TIN

Table 1 and Fig. 4 both show that diffusion along the a- and c- axes differs by a constant factor, D_a/D_c = 2.20 ± 0.15, over the entire range of pressures and temperatures. Moreover, although ΔH and D_o increase with pressure for both axes, the coefficients $d\Delta H/dP$ and dD_o/dP are the same for both to within experimental error. This, together with the close similarity of ΔH_a and ΔH_c, suggests that only one mechanism of diffusion operates. Since ΔV is about one-third of the atomic volume, furthermore, it is presumably a vacancy mechanism. The diffusion anisotropy might be thought to be due only to differences in the unit cell parameters and the a priori jump factors (γ_a and γ_c) for the two axes. Figure 5 shows the atomic positions in the tin unit cell. Atoms are located at (000),($\frac{1}{2}\frac{1}{2}\frac{1}{2}$),($\frac{1}{2}0\frac{1}{4}$), and ($0\frac{1}{2}\frac{3}{4}$), and the unit cell

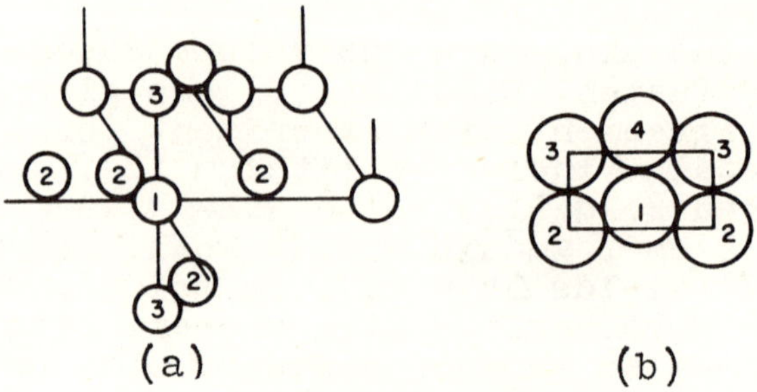

Fig. 5 Atomic Positions in White Tin
(a) Isometric view, (b) a-c Plane

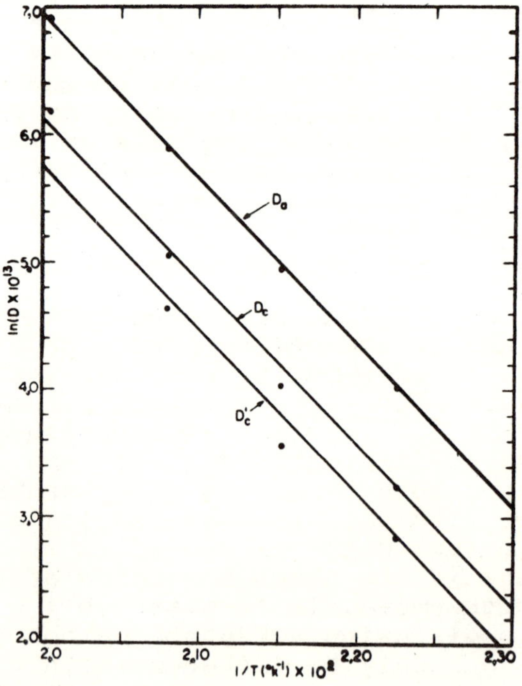

Fig.6 Contributions of a- and c-type Jumps

dimensions are a = 5.820 Å and c = 3.175 Å at 25°C. The axial ratio is c/a = 0.5456. Atom sites labeled 2 in Fig. 5a are all nearest-neighbors to site 1, and site 3 is a next-nearest-neighbor to site 1 along the c- axis. If we consider only nearest neighbor atom/vacancy exchanges, only sites of types 1 and 2 are involved. Such exchanges always advance a vacancy by an amount $\pm \frac{1}{4}c$ and make a contribution of $\pm \frac{1}{2}a$ to only one of the two a-axis motions. We form the ratio:

$$\frac{D_c}{D_a} = \frac{\gamma_c c^2 \nu_c \exp(\Delta S_c/R)\exp(-\Delta H_c/RT)}{\gamma_a a^2 \nu_a \exp(\Delta S_a/R)\exp(-\Delta H_a/RT)} \qquad (8)$$

which simplifies to

$$\frac{D_c}{D_a} = \frac{\gamma_c c^2}{\gamma_a a^2} \qquad (9)$$

if only nearest-neighbor exchanges (a-type jumps) occur. It is readily seen that $\gamma_c = 1/8$ and $\gamma_a = \frac{1}{4}$ under these assumptions. Since $D_c/D_a = 1/2.20 = 0.454$, while $(1/8)c^2/(1/4)a^2 = 0.150$, Eq.(9) cannot be correct. The observed D_c is about three times too large to be explained by means of nearest neighbor atom/vacancy exchanges alone.

The excess rate of diffusion along the c-axis may be accounted for by assuming that next-nearest atom/vacancy exchanges may also occur along the c direction with a coefficient, D_c'. Next-nearest neighbors are separated by 3.175 Å along the c-axis ... only slightly more than the 3.016 Å distance of near neighbors. From the foregoing it is evident that $D_c' = D_c - 0.15D_a$, because an a-type jump makes a contribution of $0.15D_a$ to c-axis diffusion. This is similar to the procedure used by Meakin and Klockholm (2) and the result of correcting the observed D_c values to obtain the D_c' values corresponding to c-type jumps is shown in Fig.6. Since ΔH_a and ΔH_c are equal to within experimental error, it follows that $\Delta H_c'$ must have the same value, approximately 25 kcal. $D_o^{c\prime}$ is less than D_o^c, of course, and has a value of 4.6 cm^2 sec^{-1} at zero pressure.

Since the isobars in Fig.6 are essentially

parallel, it is possible to determine the relative jump frequencies for a-type and c-type exchanges, defining these frequencies as:

$$\Gamma_i = \nu_i \exp(\Delta S_i/R)\exp(-\Delta H_i/RT) \;. \qquad (10)$$

Substitution of Eq. (10) into Eq. (8), together with $\gamma_a = \frac{1}{4}$ and $\gamma_{c'} = 1$ leads to:

$$\frac{D_{c'}}{D_a} = \frac{D_c}{D_a} - 0.150 = 0.304 = \frac{c^2\Gamma_{c'}}{\frac{1}{4}a^2\Gamma_a}$$

$$= 1.19\,\frac{\Gamma_{c'}}{\Gamma_a} \quad , \text{ or } \quad \Gamma_{c'} = 0.257\,\Gamma_a .$$

In other words, the next-nearest neighbor vacancy exchanges along the c-axis occur with about one-fourth the frequency of nearest-neighbor jumps.

REFERENCES

(1) N.H. Nachtrieb and G.S. Handler, Acta Met. 2, 797 (1954).
(2) J.D. Meakin and E. Klockholm, Trans. Met.Soc. AIME 218, 463 (1960).
(3) R.W. Keyes, "Solids under Pressure", W. Paul and D.M. Warschauer, Ed., McGraw-Hill, Inc., New York (1963), p. 71.
(4) S.A. Rice, Phys. Rev. 112, 804 (1958).
(5) N.H. Nachtrieb, H.A. Resing, and S.A. Rice, J. Chem. Phys. 31, 135 (1959).
(6) S.A. Rice and N.H. Nachtrieb, J.Chem.Phys. 31, 139 (1959).
(7) S.E. Babb, J.Chem. Phys. 37, 922 (1962).
(8) R.A. Hultsch, R.D. Engardt, and R.G. Barnes, Bull. Am. Phys. Coc.II, 5, 420 (1960).
(9) N.H. Nachtrieb, J.A. Weil, E. Catalano, and A.W. Lawson, J.Chem. Phys. 20, 1189 (1952).
(10) R.H. Dickerson, R.C. Lowell, and C.T.Tomizuka, Phys. Rev. 137, A613 (1965).

ON THE INFLUENCE OF ELEVATED HYDROSTATIC PRESSURES ON SILVER AND URANIUM SELF DIFFUSION

by
MM. M. BEYELER and Y. ADDA
COMMISSARIAT A L'ENERGIE ATOMIQUE
CENTRE D'ETUDES NUCLEAIRES DE SACLAY
FRANCE

Introduction

The understanding of the mechanisms by which the atoms can move is the major problem in solid state diffusion. Such mechanisms can generally be better understood with the help of the activation volume for diffusion. Its theoretical evaluation is usually difficult because of the uncertainty concerning the relaxation of the atoms around the defects. Therefore, its experimental determination is of considerable interest as it will allow some refinement of the theoretical models.

Measurements of self diffusion in solids under large hydrostatic pressure yield a very accurate value (1) for the total activation volume for diffusion (i.e., the sum of the variation of volume due to the formation and the migration of the defect involved in the process of diffusion). This value is often more accurate than the sum of the volumes determined separately with other techniques. For instance in quenched metals it is sometimes difficult to ascertain which part of the measured volumes must be assigned to divacancies or other defects.

Body centered cubic γ Uranium was chosen for this investigation. This metal has an activation energy which is "anomalously" low ; the determination of the activation volume would thus be of considerable theoretical interest. In order to try to understand the experimental results on polycrystalline uranium, we have been led to investigate the self diffusion of silver under pressure, with single crystals and polycrystals.

Theory of the activation volume

Point defects are generally involved in diffusion processes. At high temperature and, if the solids have been damaged by plastic working or irradiation, also at low temperature, the diffusion occurs by mechanisms involving vacancies or interstitials. It would be extremely interesting on a theoretical point of view to know the actual volume of the point defect in its equilibrium position since it may allow the specification of the mechanism involved in the diffusion ; whence the importance of the activation volume introduced by Nachtrieb (2).

Let us consider a cubic crystal. The diffusion coefficient can be written :

$$D = \gamma a^2 \nu \quad \exp\left(- \frac{\Delta G}{kT}\right), \qquad (1)$$

where the free energy variation ΔG is the isothermal and isobaric work done in displacing an atom from one site to the next. γ is a numerical factor depending on the structure, a is the lattice parameter and ν is the vibration frequency generally taken as equal to the Debye frequency.

If one uses the classical relationship :

$$\left(\frac{\delta \Delta G}{\delta P}\right)_T = \Delta V, \qquad (2)$$

the activation quantity ΔV represents the variation of the volume of crystal during the diffusion jump. In the case of a mechanism involving crystalline defects, this variation may be decomposed into two terms, one ΔV_F corresponding to the formation of the defect, the other ΔV_M corresponding to its migration. Using equation (1) the activation volume is given by

$$V = - RT\left(\frac{\delta \log D}{\delta P}\right)_T + RT\left(\frac{\delta \log \gamma a^2 \nu}{\delta P}\right)_T. \qquad (3)$$

Thus the activation volume may be determined experimentally from equation (3). Experimentally, this involves measuring the diffusion coefficient at a given temperature as a function of pressure if the terms

$\left(\frac{\delta \log a^2}{\delta P}\right)_T$ and $\left(\frac{\delta \log \nu}{\delta P}\right)_T$ are small and can be neglected.

$\left(\frac{\delta \log a^2}{\delta P}\right)_T$ may be evaluated if the compressibility is known at the pressures being used. For a cubic structure the compressibility may be written in the form

$$\chi = -\left(\frac{\delta \log a^3}{\delta P}\right)_T = -3\left(\frac{\delta \log a}{\delta P}\right)_T$$

thus

$$\left(\frac{\delta \log a^2}{\delta P}\right)_T = -\frac{2}{3}\chi$$

For silver χ is $1{,}6 \times 10^{-12}$ cm^2/dynes at 900°C (3) . The correction on ΔV is thus of the order of $-0{,}1$ cm^3/gm atom.

The quantity $\left(\frac{\delta \log \nu}{\delta P}\right)_T$ may be evaluated using the Gruneisen's relationship $\left(\frac{\delta \log \nu}{\delta P}\right)_T = \beta\chi$ where β is the Gruneisen constant which is 3,3 for silver at 900°C (4) , and the correction is 0,5 cm^3/g.atom. Therefore the total correction due to the two quantities $\left(\frac{\delta \log a^2}{\delta P}\right)_T$ and $\left(\frac{\delta \log \nu}{P}\right)_T$ for silver is 0,4 cm^3/g.atom.

For uranium χ is known only for the α phase and this value cannot be extended to γ uranium which has a different structure. If one considers the b.c.c. metals investigated by Bridgman (5) such as iron and tungsten one might expect a maximum compressibility at 900°C and 10 kbars of the order of 1% of the molar volume ; on the other hand with sodium and potassium it could be as high as 10 % of the molar volume. However γ uranium is probably closer to iron than sodium. The term

$\left(\frac{\delta \log \nu}{\delta P}\right)_T$ may be evaluated with Lindemann's relation :

$$\theta_D = C\sqrt{\frac{T_m}{AV^{2/3}}}$$

where Θ_D is the Debye temperature, C a constant, T_m the melting point, A the atomic weight and V the atomic volume. The variation of the melting temperature with pressure must then be known. There are no experimental data on uranium. However, Strong and Bundy (6) have investigated the effect of pressure on the melting point of iron (b.c.c. as uranium). For 10 kbars the melting temperature is raised by 2% at most ; since the Debye frequency is proportional to the square root of the melting temperature the term $\left(\frac{\delta \log \nu}{\delta P}\right)_T$ will be very small.

Experimental techniques

I - High pressure - high temperature apparatus

The following characteristics are necessary .

1. The pressure must be purely hydrostatic to avoid the introduction of defects by plastic deformation .

2. The fluid must not introduce impurities into the diffusion specimens, as small quantities of impurities may have a considerable influence on the diffusion coefficient. The fluid must be chemically stable up to about 1000°C; in particular there is no suitable oil up to this temperature.

3. The heating cannot be done from the outside since no metal would resist high pressures at these temperatures.

These conditions lead to the choice of a Bridgman type apparatus with internal heating and of gas for the fluid. The equipment used (Figs. 1,2) built by "Basset", allows 10000 bars up to 1200°C. There are two compression stages ; the initial pressure (100 to 150 bars from a tank) is raised to 1000 bars in the first stage then to 10000 in the second. The compression is obtained by an oil pump which pushes a piston into the compression chamber. The pressure is measured with a manganin gauge, its resistance being measured with a Wheatstone bridge and potentiometer. It is thus possible to have a continuous measure of the pressure. Pressure control is very important to compensate the slight decreases which may occur because of the compression of material during the long time treatments. The pressure in the

Fig. 1. Pressure apparatus to 10 kbar.

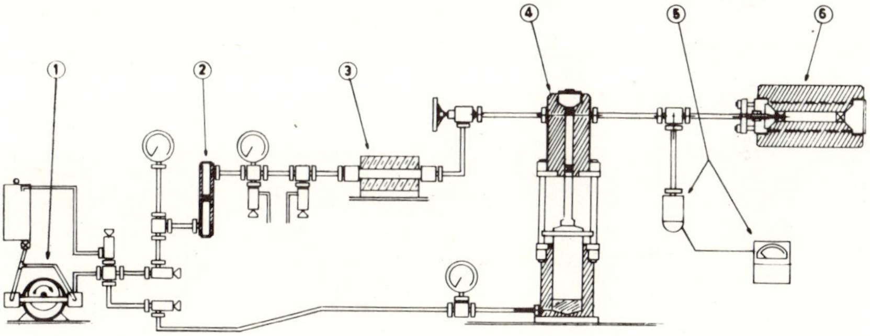

Fig. 2. Pressure apparatus to 10 kbar.

(1) Pump
(2) Separator
(3) Purification furnace
(4) Intensifier
(5) Manganin gauge
(6) High pressure chamber

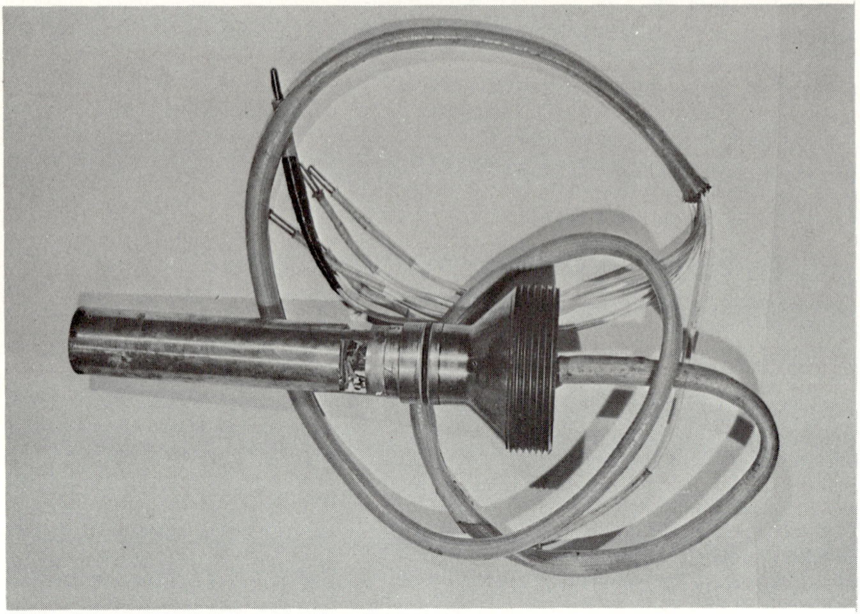

Fig. 3. Closure and furnace (1000°C, 10 kbar).

system varies little; the maximum amplitude is ± 25 bars.

The gas used is pure Argon (quality U) with only a few ppm oxygen. It must however be purified because under high pressure the quantity of oxygen in contact with uranium could be important. Purification is done at 1000 bars ; the gas flows into a nimonic tube filled with copper chips ; the tube is heated externally with a resistance furnace. The specimens, as mentioned above, must be heated from inside the apparatus. This considerably decreases the available space. Without the furnace the pressure chamber is 170 mm long and 36 mm in diameter. When equipped with the furnace 150 mm long mounted on a closure (Fig. 3) the space available for the specimens is only 80 mm long and 10 mm in diameter. The furnace is resistance wound with Pt-22% Rh, 0,5 mm in dia. and insulated with Alumina. The total length of the insulating element is 120 mm while the heating part is only 80 mm long. The current used is the AC. 220 V regulated to ± 0,5 % with an electronic stabilizer ; voltages of 0 to 70 volts at 7 to 15 amperes can be obtained with a transformer and rheostat. A complete range of power steps is indeed necessary since the temperature drops with constant power input when the pressure is raised. This could be attributed to an increase in the thermal conductivity of the gas. However direct measurements have shown that the thermal conductivity varied little with pressure especially at high temperature. Vereschagin and Kalachnikov (7) have published an investigation on the problem of internal heating in the case of vertical and horizontal installations. They report, mainly in the case of a vertical installation, the existence of severe temperature oscillations which they attribute to convection currents in turbulent flow. Indeed in a chamber under pressure, a furnace is surrounded by much colder zones, e.g. gas intake and closure. The gas coming out of the furnace is cooled against the walls. This may lead to variation of pressure. In order to reduce these currents, our chamber is equipped with a horizontal furnace and the space between the furnace and the walls is kept to a minimum. In addition the free space on each side of the specimens inside the furnace is filled with alumina blocks, except for the double barreled insulators for the thermocouple leads.

The temperature measurements are made with Pt-Pt 10 % Rh couples ; they have a good mechanical resistance to 10000 bars and 1200°C. The thermocouple wires 0,2 mm in diameter are guided by 1,5 mm double-bore insulators ; they penetrate 30 mm into the heating zone and 7 mm into 10 mm long specimens. Temperature measurements on silver specimens which are good heat conductors indicate a gradient of the order of 2 to 3°C/cm for diffusion zones of 2 to 4×10^{-2} cm. With a uranium specimen (poorer conductor) the gradients are of the order of 5 to 6°C/cm for diffusion zones of 0,5 to 2×10^{-1} cm.

The reading given by a thermocouple at a given temperature depends on the pressure. This effect has been known for a long time ; Bridgman (8) has attempted a systematic investigation, Birch (9) has given an order of magnitude of the corrections up to pressure of 4000 bars, then Bundy (10) has extended it to pressure of 100.000 bars. If a linear variation is assumed the correction would be 0,5°C to 1°C per 1000 bars depending on the authors.

II- Preparation of diffusion couples

1) Case of Uranium

Natural uranium and uranium enriched 20 % with ^{235}U were used. Spectrographic analysis of the enriched uranium indicates also an enrichment in ^{234}U. (The activity of the enriched uranium is 13 times that of natural uranium.) Cylinders 8,5 mm in diameter and 10 mm high are machined off each metal. A small hole for thermocouple introduction is drilled along the axis of the cylinder to 3 mm from the diffusion interface. The specimens are annealed and welded in the α phase under vacuum; the technique has been described elsewhere (11).

2) Case of Silver

The specimens, single crystals or polycrystals, are also cylinders of the same dimensions. The single crystals are obtained by the Czochralski technique. Drilling the hole for the thermocouple is particularly difficult with the single crystal and has been done with a spark-cutter which gives less plastic deformation.

The single crystals are then chemically etched

and annealed at 900°C. After annealing, Laue diagrams are taken to make sure of their single crystalline nature. Silver 110 obtained by neutron irradiation (10^{12} nvt) of chips machined off the same lot of silver is evaporated onto the end surface of the cylinders. This corresponds to the boundary conditions of the semi-infinite cylinder.

III - Methods of Measurements

1) uranium

After the diffusion treatment under pressure the couples are sectioned on a precision lathe (12). The ^{234}U concentration is determined as a function of penetration by measurements of α emission in an ionization chamber on the surface of the specimens after sectioning. Fig. 4 is an example of a diffusion curve. The solution of Fick's equation is obviously of the form

$$C/C_o = \frac{1}{2}\left(1 - \operatorname{erf}\frac{x}{2\sqrt{Dt}}\right)$$

Use of this solution is fully justified by the examination of the curves C/C_o vs $\frac{x}{t}$ in gausso-arithmetic coordinates. (Fig. 5)

2) silver

After diffusion treatment the specimen is mounted on the precision lathe in a way such that the sectioning is rigorously parallel to the original surface. Before sectioning, approximately 1 mm is machined off the lateral surface of the specimen to eliminate possible surface diffusion effects. The sections are of the order of 15 to 20 microns ; the activity of the turnings is then measured with a scintillation counter. The thickness of the sections is determined by weighing.

Experimental Results

1. Self diffusion of polycrystalline γ uranium under hydrostatic pressure

The self diffusion coefficient has been determined as described above from specimens treated from 4 to 48 hours at 900°C and under pressures varying from zero to 10.000 bars. The values obtained are tabulated in Table 1. Two main features can be noticed :

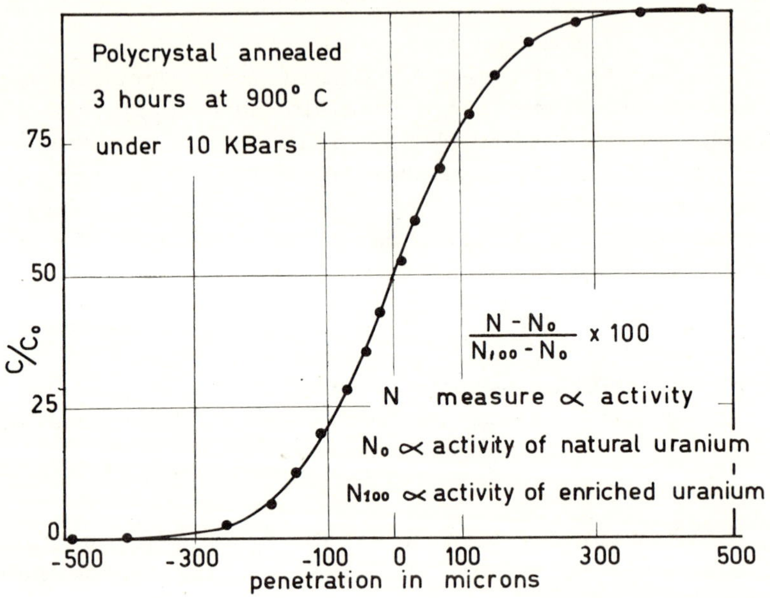

Fig. 4. Penetration curve for γ uranium self diffusion under a hydrostatic pressure of argon.

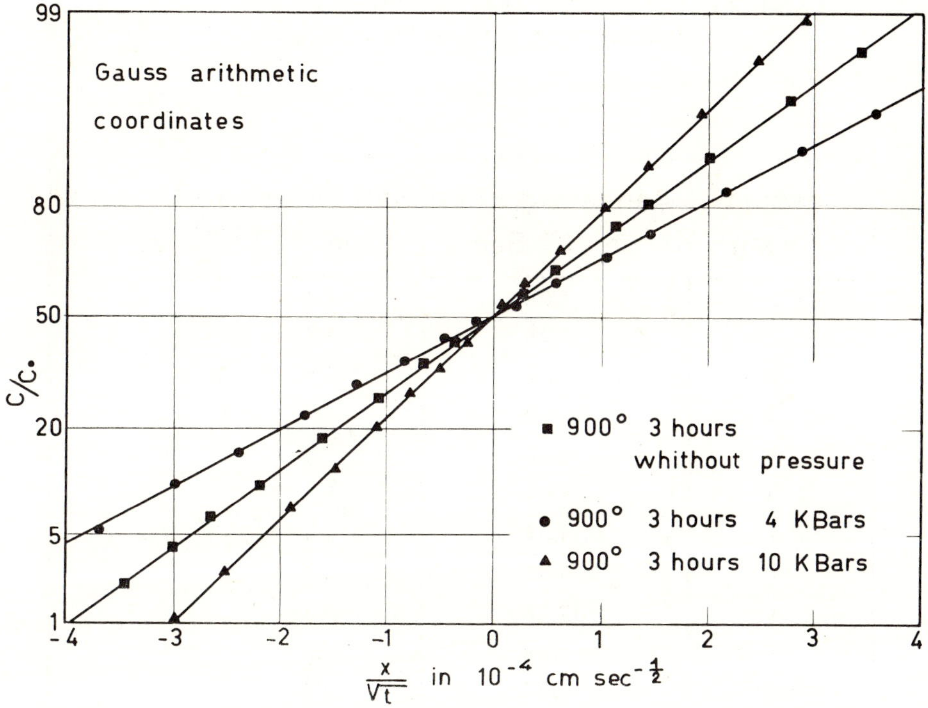

Fig. 5. γ uranium self diffusion under a hydrostatic pressure of argon.

Table 1. γ uranium self diffusion at 900°C under a hydrostatic pressure of argon.

Pressure in Kbars	Annealing time in 10^4 Sec	Self-diffusion coefficient in 10^{-8} cm²/sec
0	1,08	1,4
2	1,08	1,7
2	4,32	1,5
3	1,08	1,9 2.3
3	4,32	1,2
3	8,64	1
4	1,08	2,4 3 2,6
4	4,32	1,8 1,3
4	17,28	0,8
5	1,08	1,6
6	1,08	1,5
6	4,32	1
6	8,64	0,6
8	1,08	1,1 1.3
8	4,32	0,8
10	1,08	0,8

1) The effect of pressure on the self diffusion of uranium is different from the behavior of other metals. Instead of the decrease which was expected, the self diffusion coefficient increases with pressure for some pressures and diffusion times (Fig. 6)

2) At a given pressure, the coefficients decrease with the diffusion times (Fig.7)

These results could be either specific of uranium or related to the polycrystalline nature of the specimens. In order to decide between these possibilities we have been led to investigate the effect of pressure on the self diffusion of silver using single and polycrystalline specimens.

2. Self diffusion of silver under hydrostatic pressure

The specimens were treated 4 hours at 900°C under pressures of 0 to 8.000 bars. The data are presented in Table 2.

1) With single crystals, variation of the self diffusion coefficient with pressure is "normal". Plotting the logarithm of D vs pressure yields a straight line (Fig. 8) the slope of which is related to the activation volume.

We find $\Delta V = 9.2 cm^3/g.atom$, taking into account the evaluation of the second term of equation (3). This represents 88 % of the volume of one gramatom of silver at this temperature. This result is in good agreement with Tomizuka's figure which was 93% (13).

2) With polycrystals, the variation is "anomalous" and similar to that of γ uranium (Fig. 9).

Discussion and conclusion

We have observed an "anomalous" variation of the self diffusion coefficients of silver and γ uranium when using polycrystalline specimens.

In the case of polycrystals of iron and of iron-titanium alloys (bcc)Gertsriken and Pryanishnikov (14, 15) have already pointed out a similar anomaly. The measured values of D(P) show an increase with hydrostatic pressure. There is a difference however between the present work and these authors' ; we observed rather small increases of the self diffusion coefficient with high pressure whereas they have found a large increase in D (the coef-

Table 2. Silver self diffusion at 900°C under a hydrostatic pressure of argon.

PRESSURE IN KBARS	SELF DIFFUSION COEFFICIENT in 10^{-9} cm²/sec			
	Polycrystals		Single crystals	
	Serie 1	Serie 2	Unstrained	Strained
0	2,5	2,5	2,5	
0,5		2,8		
2	2,95	2,5	2,1	
3	2,4		1,9	
4	1,9	2,1	1,7	2,4
6	1,55	1,5	1,45	
8	1,25	1,2	1,2	

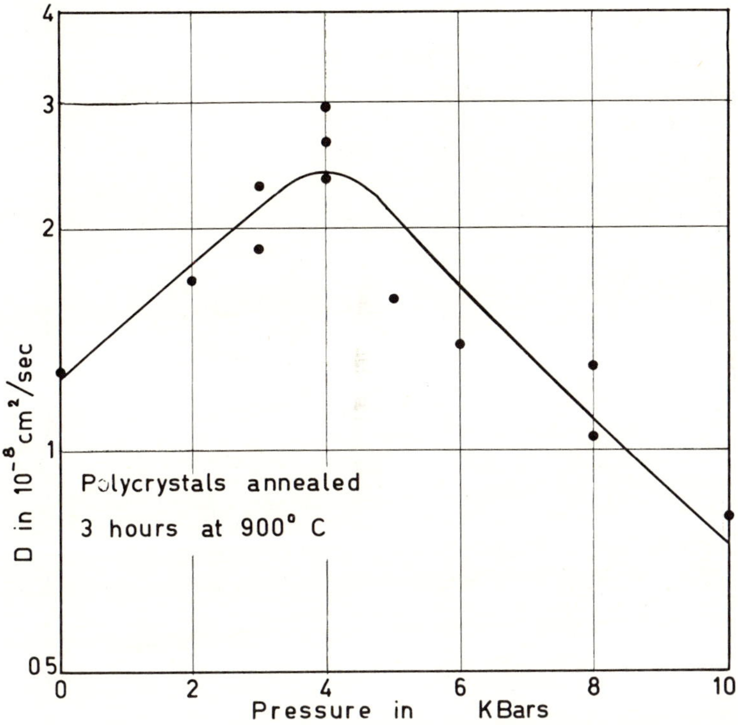

Fig. 6. γ uranium self diffusion under a hydrostatic pressure of argon.

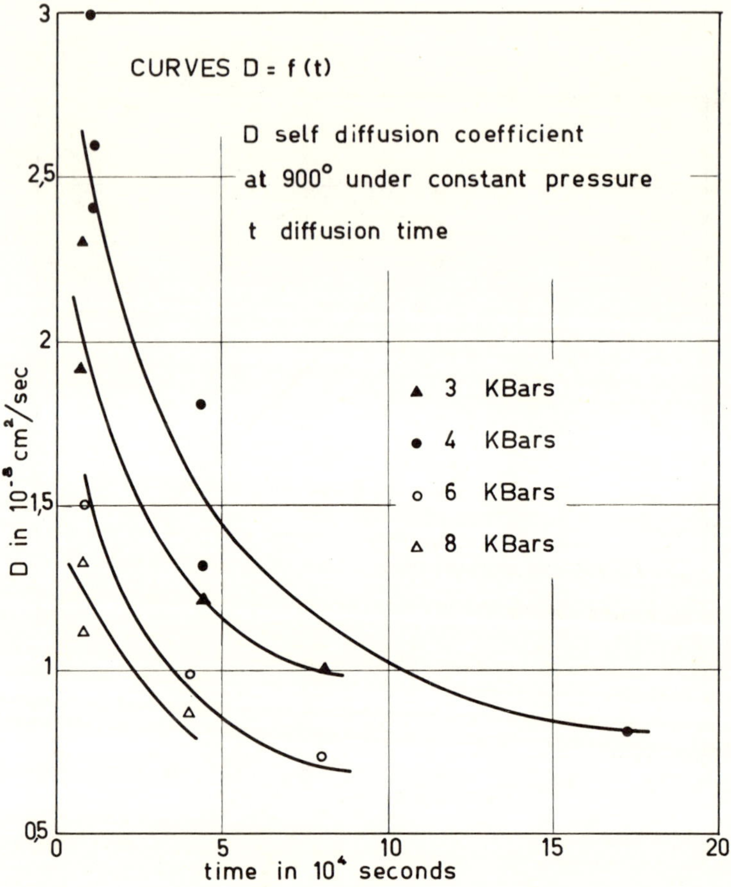

Fig. 7. γ uranium self diffusion under a hydrostatic pressure of argon.

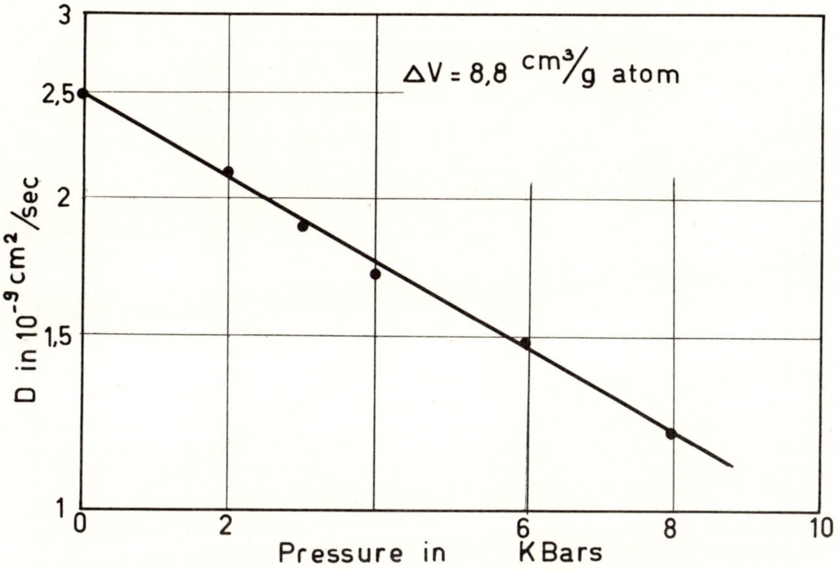

Fig. 8. Activation volume for silver single crystals under a hydrostatic pressure of argon at 900°C.

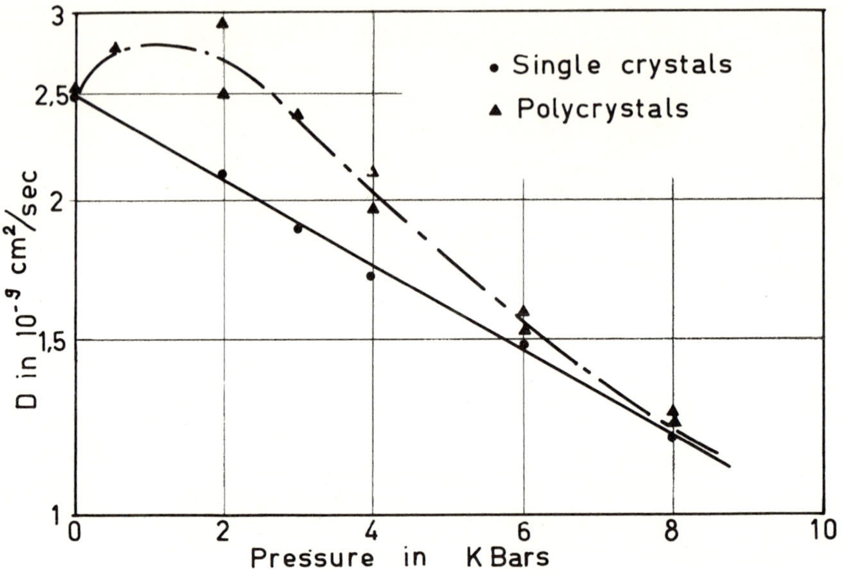

Fig. 9. Silver self diffusion at 900°C under a hydrostatic pressure of argon.

ficients may vary by a factor up to 10) for small pressures (100 to 200 bars).

Several possible interpretations have been considered. Since we have results on polycrystalline and single crystalline silver and since the latter do not show any anomalous variation we shall consider first the case of silver polycrystals.

Grain boundary diffusion

One may at first think of interpreting our results by the contribution of grain boundary diffusion, since the single crystals behave as expected. When grain boundary diffusion takes place it is usually noticeable on autoradiographs or by its modification of the diffusion curves. In our work neither the diffusion curves nor autoradiographs have shown any evidence of grain boundary diffusion. However, the experimental conditions are not the most suitable for the autoradiographic technique (16). On the other hand, the shape of the diffusion curves is not a definitive criterion. Grain boundary diffusion has been detected by autoradiography on specimens showing normal volume diffusion curves. (17)

This is easily explained when the diffusion distances ($2\sqrt{Dt}$) are much larger than the average grain size but it is not the case in our experiments. One may observe an increase in the diffusion coefficient without any change in the curves when the grain boundaries move during the treatment as was shown by Ruoff and Balluffi (18). Moreover the increase in the diffusion coefficient is larger than with static grain boundaries. The maximum value of this increase is given by the relation. $\frac{D'}{D} - 1 = C_b \frac{D_j}{D}$

when D' is the apparent diffusion coefficient, D the volume coefficient, D_j the grain boundary coefficient and C_b the number of sites in the boundaries. A rough computation using values of D_j for silver extrapolated to 900°C from the results of Hoffman and Turnbull (19) gives an increase of the diffusion coefficient of the order of 5 %. This figure is smaller than the experimental observations (20 to 40 %). The diffusion coefficient in a deformed single crystal which recrystallized during treatment was similar to the polycrystal value which qualitatively confirms the importance of the role of boundaries. The difference between the computed (5%) and the observed (20 to 40 %) increases could be explained either by inaccuracies in the calculation or by an enhancement of grain boundary diffusion under pressure.

Recrystallization therefore might explain our results; the question now is whether annealed specimens can recrystallize during a treatment under pressure. This seems difficult to explain.

Contamination of the specimens could play a determining role in this phenomenon. Therefore we have initiated a program to analyze systematically for elements that could come in contact with the specimens. Activation analysis has shown the presence of argon in the vicinity of the surface of the specimens (contents of the order of 10^{-5} to 10^{-4}). These results are insufficient yet to support our assumption.

Plastic deformation

One might assume that under certain experimental conditions, e.g. under non hydrostatic pressure, a plastic deformation of the polycrystals is possible. Since plastic deformation is known to introduce point defects, the diffusion could be speeded up (20, 21, 22). For single crystals, the self diffusion coefficient $D_m \approx n_1 D_1$ where n_1 is the equilibrium atomic fraction of vacancies and D_1 their diffusion coefficient. If the increase of the self diffusion coefficient for the polycrystal with respect to the single crystal value $\Delta D = D_p - D_m$ is attributed to the formation of an excess fraction of vacancies n_1, we can write

$$D_p - D_m = \Delta D \approx D_1 n'_1$$

The energy of formation of a vacancy in silver is known (23) and at 900°C $n_1 \approx 2 \times 10^{-5}$. From the values of the self diffusion coefficient at 900°C and 2 kbars n'_1 should be 10^{-5}. If the vacancies are created by plastic deformations $\frac{dn'_1}{dt} = \dot{\varepsilon} p - \frac{n'_1}{\tau}$

where $\dot{\varepsilon}$ is the deformation rate, p the atomic fraction of vacancies created per unit strain and τ the mean life time of a vacancy.

At equilibrium $n'_1 = \dot{\varepsilon} p \tau$. τ might be evaluated by making the following assumptions.

1. The grain boundaries are the only sources or sinks

of vacancies.

2. The vacancies are created and annihilated at the jogs on dislocations.

The mean life of a vacancy would be then between 5 and 10^{-2} second. If an upper limit of 10^{-3} is taken for p (24) the deformation rates that would be necessary are prohibitively high, between 10^{-4} and 10^{-2} per second.

Diffusion in dislocations

Finally, a third possibility based on diffusion in dislocations has been considered. Some authors (25,26) have observed anomalously high diffusion coefficients with single crystals at low temperatures although the kinetics correspond to a volume diffusion. According to Hart (27) diffusion in the dislocation of the Frank network could explain these phenomena if the average distance between dislocations were much smaller than the diffusion distances. Thin foils of silver treated under pressure seem to indicate by electron microscopy a larger density of dislocations than those without. However the scatter of the results is such that no clear conclusion can be drawn. The increase in dislocation density under hydrostatic pressure could come about from inclusions or micro heterogeneities (28,29). For instance some internal friction measurments (29) on single and polycrystalline aluminium under pressure have been interpreted in terms of nucleation of dislocations induced by microporosities at the grain boundaries. At the temperatures of our experiments such a phenomenon could only take place during the initial stage.

In the case of γ uranium the same ideas could be put forward to explain the anomalies of the self diffusion coefficient altough the two phase changes under pressure could introduce additional complications.

In conclusion, none of the assumptions we have considered is successful in explaining the diffusion anomalies observed. We think that the effect of pressure on grain boundary diffusion and on recrystallization should be looked into very carefully. These phenomena seem to play an important part in polycrystalline diffusion under pressure. In silver single crystals self diffusion agrees well with the theory of the activation volume (2). The measured volume is the sum of ΔV_f and ΔV_m, variation of volume due to formation and migration of the defect respectively. The two values are known for fcc metals only for gold (30) where ΔV_m is only 15 % of the atomic volume. If a similar figure is taken for silver, the volume variation due to the formation of the defect will be 73% of the atomic volume gram atom, indeed very close to Tomizuka's(13) . The values obtained for gold (31, 32, 33) and silver indicate an important relaxation of the atoms around the defect. These results are in good agreement with the theoretical evaluations(34, 35, 36, 37) and with the semi-empirical relation of Keyes (38) and Lawson (39)

References

1. D. Lazarus, N.H. Nachtrieb, Solid under pressure (McGraw-Hill Book Company, 1963), p. 43.
2. N.H. Nachtrieb, T.A. Weil, E. Catalano, A.W. Lawson, J. Chem. Phys. 20, 1189 (1952)
3. I.E. Leksina, S.I. Novikova, Soviet Phys. Solid. State 5, 798 (1963)
4. Y. Pantamo, Ann. Acad. Sci. Fennicae A VI 129 (1963)
5. P.W. Bridgman, The Physics of high pressure (G. Bell and Sons, Londres, 1952)
6. H.M. Strong, F.B. Bundy, Phys. Rev. 115, 278 (1959)
7. L.F. Vereschagin, Kalachnikov, J. Teckn. Fiz. 25 1508, (1955)
8. P.W. Bridgman, Proc. Amer. Acad. 53, 269 (1918)
9. Birch , Rev. Sc. Inst. 10 , 137 (1939)
10. Bundy, Progress in very high pressure research edited by F.P. Bundy, W.R. Hibbard, H.M. Strong (John Wiley and Sons Inc. New-York, 1961) p.256
11. J. Philibert, Y. Adda, C. R. Acad. Sc. France , 245 2507, (1957)
12. Y. Adda, A. Kirianenko, J. Mat. Nucl. 1, 120 (1959)
13. C. Tomizuka, Progress in very high pressure research, edited by F.P. Bundy, W.R. Hibbard, H.M. Strong (John Wiley and Sons Inc. New-York 1961) p.266
14. S.D. Gertsriken, M.P. Pryanishnikov, Ukr. Fiz. Zhur , 3, 651 (1958)
15. S.D. Gertsriken, M.P. Pryanishnikov, Fiz. metal. metalloved, 2, 297 (1960)

16. J. Cabané Thesis Paris 1962
17. G. Seibel Thesis Paris 1963
18. A.L. Ruoff, R.W. Balluffi, J. Appl. Phys. 34 1848 (1963)
19. Hoffman and Turnbull, Acta Met. 2, 419 (1954)
20. C.H. Lee, R. Maddin, J. Appl. Phys. 32, 1846 (1961)
21. J.B. Darby, C.T. Tomizuka, R.W. Balluffi, J. Appl. Phys. 32, 840 (1961)
22. K. Hirano, M. Cohen, B.L.Averbach, N. Ujiiye, Trans. AIME, 227, 950 (1963)
23. Y. Quere Thesis Paris 1962
24. F. Seitz, Advances Phys. 1, 43 (1962)
25. Lidiard, Tharmalingam, Disc. Farad. Soc., 28, 64 (1959)
26. L.G. Harrison, Trans. Farad. Soc., 57, 1191 (1961)
27. Hart, Acta Met. 5, 597 (1957)
28. S.V. Radcliff, H. Warlimont, Phys. Status Solidi 7, K 67 (1964)
29. J.L. Seely, G.S. Baker, P. Gibbs, J. Appl. Phys. 33, 2458 (1962)
30. R. Emrick, D. Lazarus, Bull. Amer. Phys. Soc. Serie II, 5, 181 (1960)
31. W. De Sorbo, Phys. Rev. 117, 444 (1960)
32. R.O. Simmons, R.W. Balluffi, Phys. Rev., 125, 862, (1962)
33. R.P. Huebener, C.G. Homan, Phys. Rev., 129, 1162 (1963)
34. R.A. Johnson, E. Brown, Phys. Rev. 127, 446 (1962)
35. L. Tewordt, Phys. Rev., 109, 61 (1958)
36. K.H. Bennemann, Z. Physik, 165, 445 (1961)
37. A. Seeger, E. Mann, J. Phys. Chem. Solids, 12, 326 (1960)
38. Keyes, J. Chem. Phys. 29, 467 (1958)
39. Lawson, J. Phys. Chem. Solids, 3, 250 (1957)

DISCUSSION

By R. E. Hanneman

I would like to comment on the paper of Drs. Adda and Beyeler and to discuss briefly some recent related results on chemical interdiffusion at high pressures. The experimental diffusion results on single crystal uranium are of considerable interest since they represent the first direct measurements of self-diffusion in non-alkali b.c.c. metals. The value of $V^*/V^M \sim 0.45$ for b.c.c. uranium is quite comparable to values that we have obtained for activation volumes of chemical interdiffusion under high pressure in b.c.c. α Fe-V alloys. Our data were obtained from metal-alloy couples that were diffused at pressures up to 40 kbars and 1300°C. Since the details of this work will appear in the literature (1, 2), I will only briefly present some of the key results. The experimental values of V^*/V^M of the b.c.c. phase lie in the range 0.45-0.6 over the concentration range investigated (~0 to 30 at % vanadium). Also interdiffusion data in the f.c.c. Fe-V and Fe-Ni phases lead to activation volumes of $\sim 0.75 < V^*/V^M < 0.9$. Previous self-diffusion data in f.c.c. metals including the present silver data lie in the range

$$\frac{V^*}{V^M} \sim 0.75 \pm 0.2$$ and are, therefore, comparable with

our f.c.c. interdiffusion results. One of the important conclusions suggested by the above data is that there is significantly more inward relaxation around a vacancy and/or saddle point configuration in normal b.c.c. metals or alloys than in the corresponding f.c.c. case. By means of Kirkendall marker shifts, interdiffusion results also shed light on the effect of pressure on the interaction of a vacancy with different adjacent atoms. In particular the ratio of intrinsic diffusion coefficients $(D_V/D_{Fe})_\alpha$ were found in our study to decrease from ~2 to ~1 as pressure was raised from 1 atm to 40 kbars.

One final point that I would like to stress is that it is often quite important to apply corrections for the pressure dependence of thermocouples in many different types of high pressure studies, including diffusion. Absolute ΔT corrections can amount to ~+50°C at 1200°C and 50 kbar for Pt/Pt10Rh thermocouples as shown by work in our Laboratory (3-5). Fortunately, extrapolation of the thermocouple corrections of Bundy (4) to higher temperatures by Adda and Beyeler for the silver and uranium data leads to only small temperature errors at low pressures. Generally such extrapolations are not justified to high temperatures and pressures (3,5).

References

1. R. E. Hanneman, R. E. Ogilvie and H. C. Gatos, Trans. AIME, 233, 691 (1965).
2. J. I. Goldstein, R. E. Hanneman, and R. E. Ogilvie, Trans. AIME, 233, 812 (1965).
3. R. E. Hanneman and H. M. Strong, J. Appl. Phys., 36, 523 (1965).
4. F. P. Bundy, J. Appl. Phys., 32, 483 (1965).
5. H. M. Strong, R. E. Hanneman and F. P. Bundy to be published.

THE DIFFUSIVITY OF CARBON IN ALPHA IRON AT 1000°K AND 5 KBARS

C. G. Homan and J. F. Cox
Watervliet Arsenal
Watervliet, New York

Introduction:

It has been established that the room pressure diffusivity of carbon in alpha iron exhibits non-Arrhenius type behavior.(1,2,3) Measurements of the diffusivity have been made over the temperature range from -30°C to 865°C. It was shown in reference (1) that the data could be fitted with the expression:

$$D_{exp} = 0.008 \exp(-19{,}800/RT) + 2.2 \exp(-29{,}300/RT) \tag{1}$$

The first term on the right hand side of (1) represents the internal friction diffusivity data due to Wert (4,5). The significance of the second term was explained on the basis of a model involving carbon atom-lattice vacancy interaction (1). This model yielded the expression:

$$D_{exp} = D_{oc} \exp(-E_{mc}/RT) + D_{op} \exp -\left(\frac{E_{fv} + E_{mp} - B}{RT}\right) \tag{2}$$

D_{oc} and E_{mc} represent the pre-exponential and the motional energy respectively for motion of a carbon atom in the "normal" configuration (i.e., when all lattice sites involved in the diffusion jump are occupied by iron atoms). This will be very nearly the case for carbon diffusion as measured by anelastic techniques. Any residual pairs retained in quenched specimens would probably not contribute to the Snoek peak resulting from motion of carbon atoms in the iron lattice.

D_{op} and E_{mp} represent the pre-exponential factor and motional energy, respectively, for motion of a carbon atom in the configuration where a nearest neighbor lattice site is vacant. B and E_{fv} represent the carbon atom-vacancy binding energy and the lattice vacancy formation energy, respectively.

At temperatures below 200 C the second process does not contribute significantly to the diffusivity. Previous measurements (6, 7) of carbon diffusion in alpha iron at low temperatures and at pressures up to 4 kbar imply that the first process is pressure insensitive. With the additional assumption that D_{op} is pressure independent, the following expression for the pressure dependence of D is obtained.

$$D_P = D_{oc} \exp\,(-E_{mc}/RT) + D_{op} \exp -\left(\frac{E_{fv} + E_{mp} - B + P\Delta V}{RT}\right) \qquad (3)$$

Experimental Procedure:

The diffusion couple was prepared from Battelle iron. Details of sample preparation are given in reference (1). Uniformity of C-14 distribution in the doped portion of the diffusion couple and the distribution of C-14 in the couple after the diffusion anneal was determined with a windowless gas-flow proportional counter. The detection chamber was designed to accept cylindrical samples so that the activity on the end of the sample could be measured under conditions of constant geometry. Sectioning was accomplished by surface grinding.

After each section was removed the ground surfact was counted for approximately 10,000 counts. The probability plot of the data gave a good straight line.

The diffusion anneal was conducted in an externally heated high pressure cell using Argon as the pressure medium, the cell hot zone extended over a distance of two inches with temperature gradients of less than 0.5°C/cm. During heat up the specimen was held in a cold zone which remained at or below 100°C. When the pressure cell reached the desired temperature the specimen was dropped into the hot zone from this cool zone. The entry of the specimen into the hot zone caused a sharp drop in the pressure. This pressure drop and the associated pressure recovery time were used to measure the specimen heat-up time, which was about one minute. After the specimen reached temperature, the pressure variation was less than ± 50 bars and the temperature variation was less than ± 1.5°C.

After the anneal was completed the specimen was removed from the hot zone and returned to the cold zone.

Experimental Results:

The diffusivity of carbon in alpha iron at 1000°K and 4850 bars was measured as 7.5×10^{-7} cm^2/sec. The activation volume for diffusion of carbon in alpha iron as calculated from equations 1 and 3 is 14.3 $cm^3/mole$. This value appears larger than would reasonably be expected. It turns out that the volume is very sensitive to the parameters in equation 1. For example, if one uses a recent least squares fit to the anelastic data below 350°C (8) and the value of Dexp at room pressure (1, 2) of 12×10^{-7} cm^2/sec, the value of the activation volume is calculated to be 11.3 $cm^3/mole$.

Discussion:

The data reported here are preliminary and additional measurements of the pressure dependence of the diffusivity are in progress.

However, this measurement does indicate a large pressure effect in the high temperature region which was expected from the two mechanism model proposed.

References:

1. Homan, C. G., Acta Met. 12, 1071 (1964)
2. Smith, R. P., Trans. AIME 224, 105 (1962)
3. Stanley, J. K., Trans. AIME 185, 752 (1949)
4. Wert, C. A., Phys. Rev. 79, 601 (1950)
5. Wert, C. A., and Zener, C., Phys. Rev. 76, 1169 (1949)
6. Bosman, A. J., Brommer, P. E., and Rathenau, G. W., Physica 26, 533 (1960)
7. Bass, J., and Lazarus, D., J. Phys. Chem. Solids 23, 1820 (1962)
8. Lord, A. E., Jr., "Megacycle Damping and the Elastic Stiffness Coefficients of Single Crystal and Polycrystalline Armco Iron from Room Temperature to 400°C", Doctoral Thesis, Columbia University (1964).

ACTIVATION VOLUMES FOR CREEP

A. L. Ruoff
Cornell University, Ithaca, New York

Abstract

High temperature creep is a diffusion controlled process. By using the change of slope method we have obtained the activation volumes for creep (and hence, by implication, for diffusion) for the following materials with cubic crystalline symmetry (and hence isotropic diffusion coefficients): Na, K, Pb, and Al.

Introduction

The rate-controlling step in steady-state high temperature creep is self-diffusion (1). It involves the diffusion of vacancies from a source, at which a local equilibrium concentration exists, through the lattice to sinks at which a (smaller) local equilibrium concentration exists (2,3). Its functional dependence can be written in the form (4)

$$\dot{\epsilon} = \alpha \frac{D}{a^{0.5}} \frac{\sigma^{4.5}}{M^{0.5} G^{3.5} kT} . \qquad (1)$$

Here α is a geometrical constant, D is the self-diffusion coefficient, a is the lattice parameter, G is the effective shear modulus, M is the concentration of dislocation sources, σ is the applied stress, k is Boltzmann's constant. and T is the absolute temperature.

By carrying out a steady-state creep experiment at constant stress and temperature but at two different pressure levels, it is possible to obtain the activation volume for self-diffusion

$$V_d = -RT\frac{d\ \ln D}{dP} = -RT\frac{d\ \ln[\dot{\epsilon}a^{0.5}M^{0.5}G^{3.5}kT/\alpha D\sigma^{4.5}]}{dP}. \quad (2)$$

To a first approximation, if we are cycling between two nearby pressures P_1 and P_2 and obtain respectively strain rates $\dot{\epsilon}_1$ and $\dot{\epsilon}_2$, we have

$$V_d = \frac{RT}{P_1-P_2}\ln\frac{\dot{\epsilon}_2}{\dot{\epsilon}_1}. \quad (3)$$

This usually gives a value too large by 2-5%. The next approximation is

$$V_d = \frac{RT}{P_1-P_2}\left[\ln\frac{\dot{\epsilon}_2}{\dot{\epsilon}_1} - x(P_1-P_2)\right] \quad (4)$$

where

$$x = \left[-0.5\frac{d\ \ln a}{dP} - 3.5\frac{d\ \ln G}{dP}\right]. \quad (5)$$

Similar expressions can be written for the temperature dependence at constant pressure. We have

$$E_d = \frac{RT_1T_2}{T_2-T_1}\left[\ln\frac{\dot{\epsilon}_2}{\dot{\epsilon}_1} - y(T_2-T_1)\right] \quad (6)$$

where

$$y = \left[-0.5\frac{d\ \ln a}{dT} - 3.5\frac{d\ \ln G}{dT} - \frac{1}{T}\right]. \quad (7)$$

If we put $y = 0$ as a first approximation, we obtain a value of E_d which is within 1% of the value obtained by Eq. (6).

Experimental

The methods whereby tensile creep tests were carried out are described elsewhere (5,6). Very briefly, the entire tension creep machine is immersed in the hydrostatic fluid of a pressure vessel and strain is measured potentiometrically.

Results

Table I shows the results for the activation energies and volumes obtained by the creep method using Eqs. (4) and (6). Earlier activation volume results which were analyzed with x = 0 have been corrected. Also shown are values obtained by tracer and NMR methods. The room temperature molar volume is also given for comparison. The activation volumes are about 1/2 atomic volume for sodium and potassium, about 3/4 atomic volume for lead whereas the value is about 4/3 atomic volume for aluminum. Aluminum also differs from other metals when considered on the basis of the pseudopotential calculations. Just as with the alkalis, aluminum shows long range oscillations of these potentials in both the liquid state (21) and the solid state (22). In the alkalis both the nearest neighbors and next nearest neighbors are located in the first region of negative potential while in solid aluminum the second nearest neighbors are located in the region of the second negative potential. Thus an expansion of the lattice could be expected when a vacancy is formed in aluminum.

Acknowledgments

The author wishes to acknowledge the support of the U. S. Atomic Energy Commission and the Advanced Research Projects Agency.

References

1. O. D. Sherby, Acta Met. 10, 135 (1962).
2. J. Weertman, J. Appl. Phys. 28, 362 (1957).
3. H. Hutto and A. L. Ruoff, Cornell Materials Science Center Report #300, April 2, 1965.
4. This is the form given in reference 2. It also is fairly consistent with the empirical result given in reference 1.
5. B. M. Butcher and A. L. Ruoff, J. Appl. Phys. 32, 2036 (1961).
6. B. M. Butcher, R. H. Cornish, and A. L. Ruoff, ASME Publication #64-WA/PT-30, Oct. 13, 1964.

7. N. H. Nachtrieb, E. Catalano, J. A. Weil, J. Chem. Phys. 20, 1185 (1952).
8. D. F. Holcomb and R. E. Norberg, Phys. Rev. 98, 1074 (1955).
9. R. A. Hultsch and R. G. Barnes, Phys. Rev. 125, 1832 (1962).
10. N. H. Nachtrieb, J. A. Weil, E. Catalano, and A. W. Lawson, J. Chem. Phys. 20, 1189 (1952).
11. C. R. Kohler and A. L. Ruoff, Cornell Materials Science Center Report #326, Ithaca, N.Y., March 5, 1965.
12. C. D. Wiseman, O. D. Sherby, and J. E. Dorn, J. Metals 9, 58 (1957).
13. N. H. Nachtrieb and G. S. Handler, J. Chem. Phys. 23, 1569 (1955).
14. B. Okkerse, Phys. Rev. 103, 1246 (1956).
15. K. L. DeVries, G. S. Baker, and P. Gibbs, J. Appl. Phys. 34, 2254 (1963).
16. N. H. Nachtrieb, H. A. Resing, and S. A. Rice, J. Chem. Phys. 31, 135 (1959).
17. J. B. Hudson, Progress in Very High Pressure Physics, ed. by F. P. Bundy, W. R. Hibbard, Jr., and H. M. Strong, John Wiley and Sons, Inc., New York, 1961, p. 269.
18. B. M. Butcher, H. Hutto, and A. L. Ruoff, Manuscript in preparation.
19. T. S. Lundy and J. F. Murdock, J. Metals 13, 676 (1961).
20. J. J. Spokas and C. P. Slichter, Phys. Rev. 113, 1462 (1959).
21. M. D. Johnson, P. Hutchinson, and N. H. March, Proc. Roy. Soc. 282A, 283 (1964).
22. W. Harrison, Phys. Rev. 136, A1107 (1964).

Table I

Activation Volumes for Metals with Isotropic Diffusion Coefficients

	E_c kcal/mole°K	E_d kcal/mole°K	V_c cm^3/mole	V_d cm^3/mole	Ω cm^3/mole
Na	10.45[a]	10.45[b,c,d]	9.8[e]	9.8[f,g]	23.78
K	9.7[h]	--	25.0[h]	--	45.96
Pb	27.2[i] 26.0[j]	24.2[k] 26.0[l]	12.3[m] 13.9[i]	13.0-15.9[n] 11.6[o]	18.27
Al	33.5[p] 34.5[q]	33.6[r] 32.0[s]	13.5[t]	--	9.99

[a] See reference 3.
[b] See reference 7 (Tracer).
[c] See reference 8 (NMR).
[d] See reference 9 (NMR).
[e] Unpublished results of the author.
[f] See reference 9.
[g] See reference 10 (Tracer); see also comments on analysis in reference 11.
[h] See reference 11.
[i] See reference 5.
[j] See reference 12.
[k] See reference 13 (Tracer).
[l] See reference 14 (Tracer).
[m] See reference 15.
[n] See reference 16 (Tracer).
[o] See reference 17 (Tracer at non hydrostatic pressures).
[p] See reference 18.
[q] See reference 12.
[r] See reference 19 (Tracer).
[s] See reference 20 (NMR).
[t] See reference 18.

STATIC AND DYNAMIC DISTORTION AND COLOR CENTER SPECTRA*

D. B. Fitchen†

Department of Physics and
Laboratory of Atomic and Solid State Physics
Cornell University
Ithaca, New York

The study of the properties and behavior of "color centers" in alkali halides is one of the oldest areas of activity in solid state physics. The trapped-electron centers were recognized to be among the most elementary and, at the same time, most readily observable of point defects, and for many years the apparent simplicity was emphasized in the vacancy models.

A correct description of the centers takes into account the discrete neighboring ions and the interactions with the modes of vibration which broaden the spectral lines into bands. Such theories have been presented and they contain fairly explicit assumptions about the nature of the wave functions and the coupling to the vibrational modes (1). Yet as long as only the broad bands are observed it is difficult to extract enough information from the optical spectra to determine the important parameters with the precision required to critically test the theories.

Recent careful analyses of the moments of the broad bands have helped. However, a new approach

* Work supported in part by the U. S. Atomic Energy Commission and the Advanced Research Projects Agency.

† Alfred P. Sloan Fellow

is available which makes it possible to measure directly much more detail related to the electron lattice coupling at color centers. This is the possibility of using the narrow "zero-phonon" lines and discrete vibrational structure associated with a few of the "F-aggregate" bands at low temperatures (2,3), as shown for LiF in Fig. 1.

Zero-phonon lines occur in the optical absorption spectra of defects in solids in much the same way that narrow, "recoil-free" Mössbauer lines occur in the γ-ray spectra of solids.(4) In the optical case, there is a "displacement recoil" associated with an electronic transition as the neighboring ions relax to new equilibrium positions. Such a "displacement recoil" involves the excitation of several vibrational modes of the crystal, and this extra energy must be supplied by the optical photon, resulting in a shift and broadening of the absorption peak. However, since the states of excitation are quantized, there is a finite probability that no modes will be excited. This case contributes a single "zero-phonon" line which is neither displaced nor broadened by energy transfer to the lattice modes.

One usually considers only the two extremes; either strongly coupled defects such as F centers for which virtually all the intensity is in the broad multi-phonon band, or transition and rare-earth impurities whose coupling is so weak that practically all the intensity is in the zero-phonon line. Actually, in the general case, electronic transitions contribute both a broad multiphonon component and a narrow zero-phonon line to the optical absorption spectrum with relative intensities determined by the strength of the electron-lattice coupling.

The comprehensive theories of the broad structureless absorption bands in solids usually include the zero-phonon transition implicity, but fail to recognize its unique status. They allow all discrete transitions to be smeared together on the basis of dispersion of the normal mode frequencies and consider only the envelope of

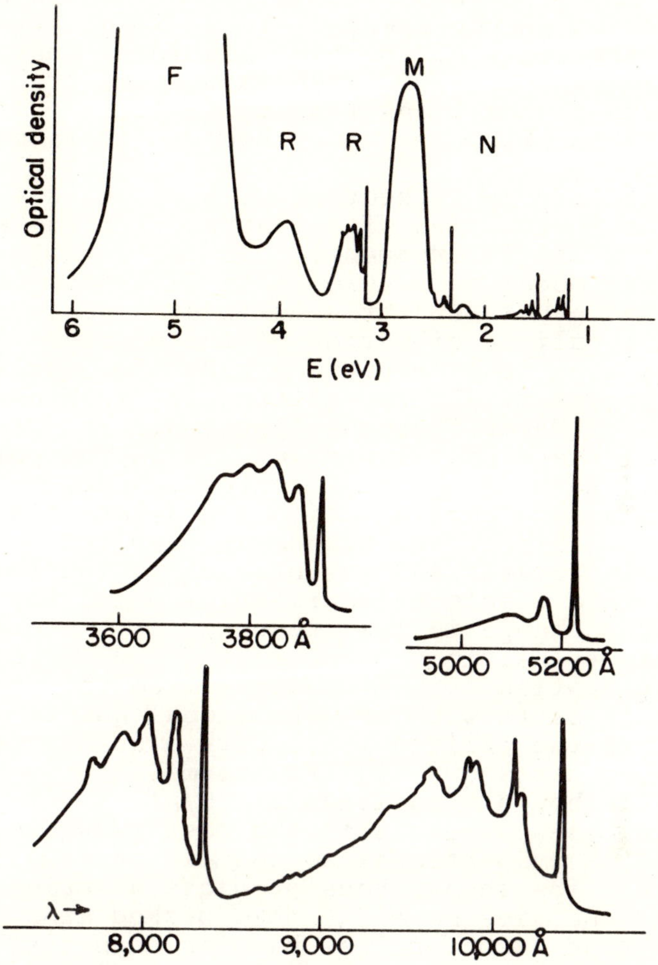

Fig. 1. Absorption Spectra of Irradiated LiF at 78°K.

these lines. In fact, though, the zero-phonon line is not broadened by this mechanism since it exchanges no phonon energy. In an ideal crystal at 0°K, the width of the line should be just lifetime determined. Thus, even though the probability for a zero-phonon transition is small, the line may still be observable just because it is strongly peaked.

The principal theories of broad band spectra treat the interaction of a trapped electron with modes of a single frequency. In the configuration coordinate model as developed by Williams, Klick, Dexter and others (5), the dominant mode is taken to be a localized vibration of the nearest neighbor ions, the "breathing mode" for the symmetric F center. In the alternative weak-coupling approach of Pekar (6), the longitudinal modes of the lattice are most important, as in the polaron problem. Both can be represented schematically with the simple configuration coordinate diagram in Fig. 2. Linear and quadratic coupling between electronic states and a single vibrational mode results in a displaced equilibrium point for the mode and a change in its frequency as the electron undergoes transitions.

Absorption at 0°K proceeds from the ground electronic state with no vibrational quanta present to the excited state with n quanta present. As Markham has shown (7), the normalized probability for the n-phonon transition is $S^n/n! \cdot e^{-S}$ for both cases, where S in the most probable value of n and is proportional to the linear coupling strength. The theory thus predicts a series of lines at energies $E_n = E_o + n\hbar\omega_e$ peaked at n = S. The observed structureless band is then just the envelope of these lines, a Poisson distribution which approaches a gaussian for large S and has a zero-point half-width, H(0), proportional to $\sqrt{S} \cdot \hbar\omega_e$.

At finite temperatures, averaging over occupied initial vibrational states predicts that the band will broaden with temperature as $H^2(T) = H^2(0) \coth \hbar\omega/2kT$. Similarly, this predicts a decrease in the net number of phonons excited and a consequent shift of the absorption peak to lower energies, but this effect is complicated by a shift due to thermal expansion.

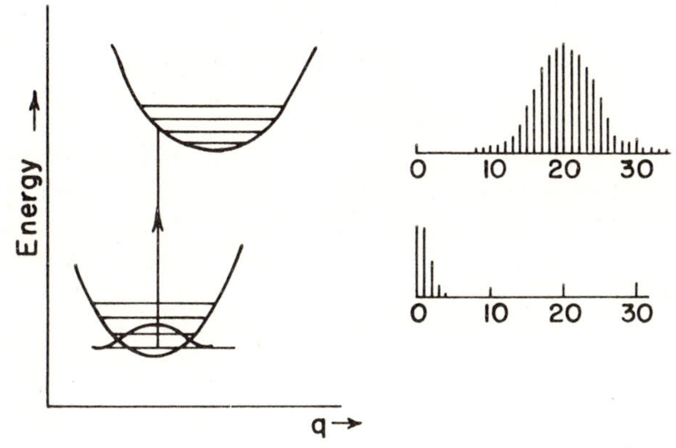

Fig. 2. Schematic Configurational Coordinate Diagram for a Localized Center. The System Energy E is Plotted as a Function of a Coordinate q which Collectively Describes the Positions of the Nearest Ions. The Lower Parabola Represents the Ground Electronic State of the System, and the Upper Parabola Represents the Excited Electronic State. At Low Temperature, only the Lowest Vibrational Level of the Ground Electronic State is Occupied, Corresponding to the Zero-Point Energy of the System. The Arrow Indicates the Most Probable Absorption Transition in this Case, to the Vibrational Level n = S in the Excited State. The Spectra for S = 20 and S = 1 are shown.

A similar argument predicts that for a two level system, emission processes will occur at energies $E'_n = E_o - n'\hbar\omega_g$, causing a broad emission band with a Stokes' shift $S\hbar\omega_e + S'\hbar\omega_g$. Since the temperature dependence of H_{abs} an H_{em} determine both S and ω_e and ω_g, it is customary to use the Stokes' shift for a consistency check. The results are not particularly good. Furthermore, neither the band shape nor the temperature dependence are much affected if ω is actually the mean of many frequencies (7).

In contrast, we obtain considerably more information from bands with S of order unity where the zero-phonon line and discrete vibrational structure are observed. The shape and spacing of the vibrational side bands is directly related to the spectral density of modes, while their intensity is related to the strength of the coupling (8).

In at least seven of the alkali halides, either the R center (F_3 center) or higher F-aggregate centers have intermediate coupling strength so that the zero-phonon line and the broad band are comparable in magnitude. We have studied several of these as a function of hydrostatic pressure, temperature, and isotopic composition to compare the coupling to static and dynamic local strain.

Since the zero-phonon lines and phonon structure can only be observed at low temperatures, the pressure measurements must be done at liquid nitrogen or liquid helium temperatures. The pressure cell with sapphire windows shown in Fig. 3 is similar to one used previously by us (9) in studies of color center formation under pressure and spectra of color centers in high pressure phases. Hydrostatic pressures up to 8 kbar are applied to the sample with compressed helium through small-diameter stainless steel tubing. The optical cryostat fits directly into the sample compartment of a Cary 14R spectrophotometer.

In LiF, the high Debye temperature makes it possible to observe the lines at 78°K with the helium still gaseous, so our pressure data are more complete. On the other hand, the strain achieved is small and no quadratic strain effects were discernable, as indicated in Fig. 4.

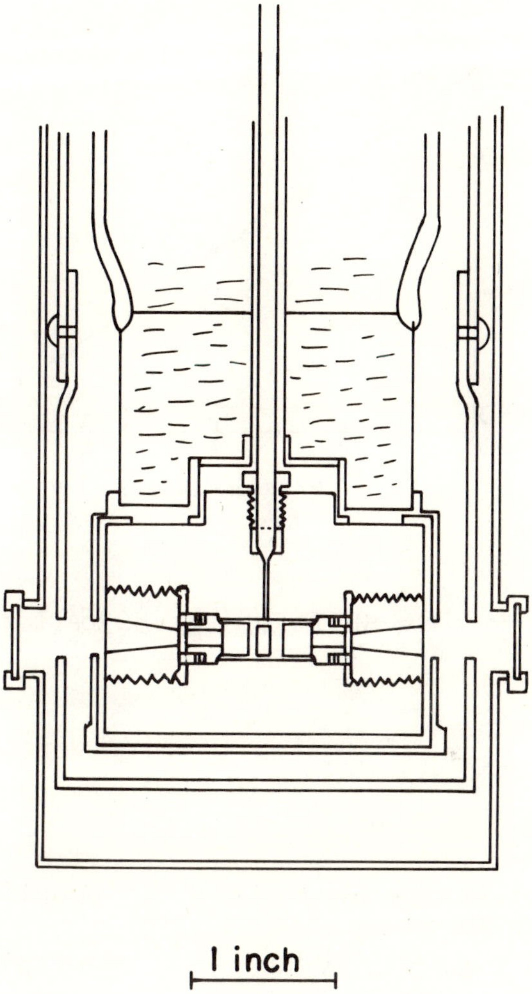

Fig. 3. High Pressure Cell for Optical Studies at Low Temperatures.

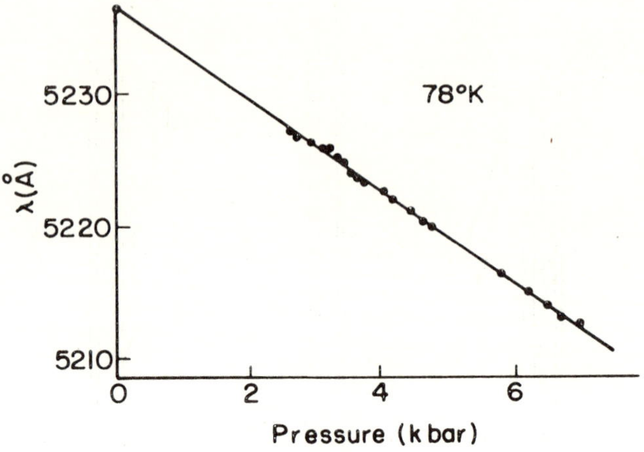

Fig. 4. Pressure Shift of the 5236Å Line in LiF.

Table I. Pressure Shifts and the Coupling Parameter S.

Line	$(\partial E/\partial p)_T$ (mV/kbar)	$(\partial E/\partial p)_T^2$	S_{exp}
3910	2.38	5.66	3.9
5236	1.63	2.66	1.0
8333	1.2	1.4	3.5
10406	0.1	.01	3.4
F Band	~ 6	~ 36	~ 40

In a rough approximation, the square of the linear pressure coefficient should be proportional to the coupling parameter S. Table I indicates the extent of agreement for several lines and the F band, with S determined from the relative intensity of the zero-phonon line and the broad-band width, respectively. Apparently, the coupling is predominatly to cubic modes at the defects.

The important modes in the dynamic perturbation of the defects are revealed in the spectral shape of the one-phonon sidebands, obtained after subtraction of the background due to higher phonon processes. Comparision with the theoretical phonon spectrum for LiF calculated by Karo and Hardy (10) indicates that these centers couple mainly to acoustic modes. Pierce (3) has shown that there is generally very close agreement between the peaks in the side band spectra and phonon critical points at the zone-boundary in the acoustic and/or transverse optical branches.

The frequency dependence of the coupling to these phonons can be derived in a very crude approximation by expressing the local strain at the defect in terms of phonon coordinates. Using the continuum approximation, and adding the Debye approximation, $\omega = ck$, the coupling parameter S_i to the ith mode can be taken as So/ω_i. Thus the absorption constant in the one-phonon sideband should be proportional to $N(\omega)/\omega$, where $N(\omega)$ is the spectral density of coupled modes. Although strictly valid only for phonons at long wavelength, these assumptions probably give reasonable values for coupling to short-wavelength phonons for localized defects. This is, incidentally, the same coupling as in the Mössbauer effect, so that theories of lattice coupling for the two effects may be compared.

The relative integrated absorption in the zero-phonon line at 0°K is e^{-S} for the single mode model and $e^{-\sum_i S_i}$ for many modes. The magnitude of the zero-phonon line decreases with temperature according to a Debye-Waller factor as the probability for stimulated emission of phonons increases with thermal excitation of the lattice. Fig. 6 shows the experimental temperature-induced variation of the zero-phonon

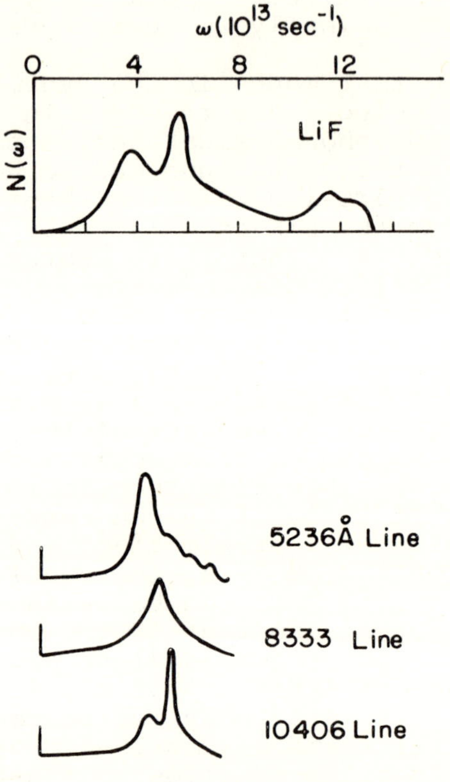

Fig. 5. Theoretical Phonon Spectral Density in LiF (10) Compared with Optical Density in the One-Phonon Side bands of Several Lines.

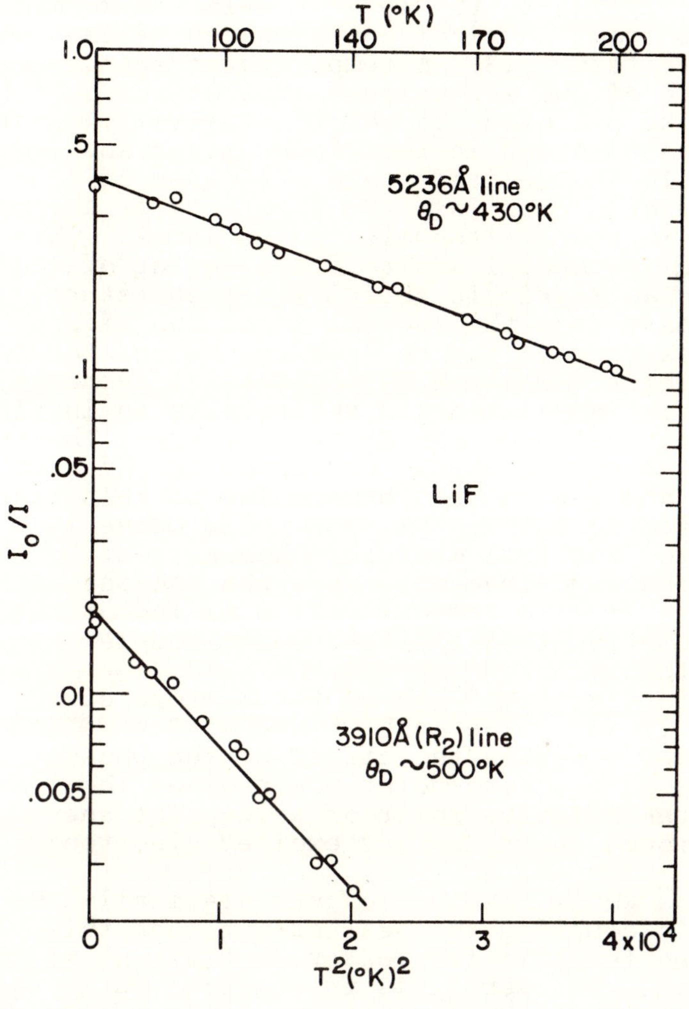

Fig.6. The Integrated Absorption in Zero-Phonon Lines Relative to the Total Absorption in the Bands, as a function of Temperature

line strength and a fitted curve assuming coupling to a Debye spectrum of acoustic modes.

In addition to these effects due to the linear coupling to the local strain, there are also important physical consequences of the quadratic term (11). A temperature dependence of the width of the zero-phonon line arises from Raman-like processes in which a thermally excited phonon is absorbed in the ground state and re-emitted in the excited state. Because of the quadratic coupling, the frequency of any mode differs in the excited electronic state. This energy difference leads to a broadening of the line and an intrinsic shift with temperature, quite apart from the thermal expansion effect. The pressure data can be used to correct for this volume shift. For the 5236Å line for instance, the thermal expansion red shift masks an intrinsic blue shift with increasing temperature. The data for the intrinsic shift, corrected to constant volume, and the thermal broadening of this line are presented in Fig. 7. The solid curve is simply the integral over the phonon spectrum (determined experimentally from the one-phonon sideband) times a thermal weighting factor, on the assumption that the same modes couple quadratically.

Finally, Fig. 8 shows the isotope effect for the 10,406Å line in Li^6F and Li^7F at 2°K. This shows directly the effect of the phonon spectrum on the zero-point position on the line, due to the relative shift of zero-point energies of the modes in ground and excited electronic states.

One would like to measure statically the quadratic coupling to local strain, and this should be feasible for softer crystals at higher pressures. However, this is complicated by the different local compressibility at these vacancy defects. More important are the effects of non-cubic components of strain. Silsbee(12) has shown the uniaxial stress studies to be a particularly fruitful approach in understanding the R center spectra in KCl.

What can be done in connection with physics at high pressures?

One potential application seems intriguing. This is the possibility of using the zero phonon

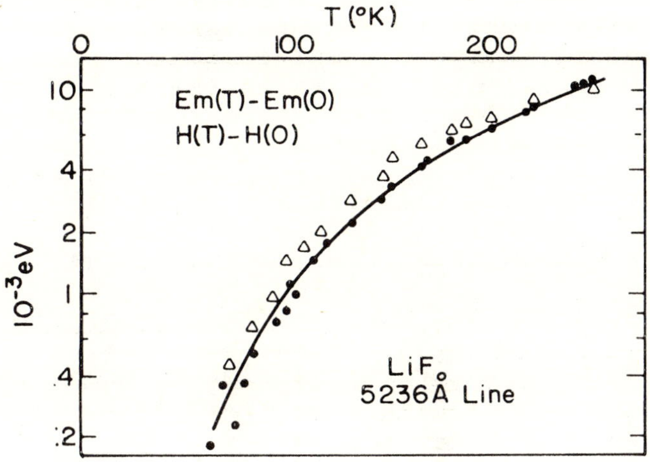

Fig. 7. The Intrinsic Temperature Shift and the Temperature Dependent Half-Width of the 5236Å Line in LiF.

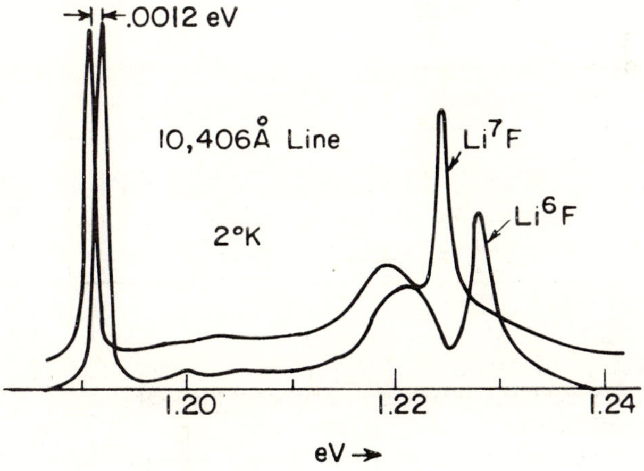

Fig. 8. The Isotope Shift in the 10,406Å Line and One-Phonon Sideband for Li^6F and Li^7F.

lines and vibrational spectra as a probe to study lattice dynamics and phonon-defect interactions at high pressures. In particular, the vibrational frequencies should change in softer crystals, and these shifts might provide one way to distinguish experimentally between quasi-localized modes and normal phonons. Also shifts in the phonon spectra near pressure-induced phase transformations should be observable in this way.

Finally the simplicity of this optical method of investigating lattice modes makes it attractive, especially in high pressure studies, since the window techniques already exist.

1. J.H. Schulman and W.D. Compton, Color Centers in Solids, (Macmillan, New York, 1962).
2. C.J. Delbecq and P.Pringsheim, J. Chem. Phys. 21, 794(1953); A.A. Kaplianskii, Optics and Spectroscopy 6, 267, (1959); D.B. Fitchen, R.H. Silsbee, T.A. Fulton and E.L. Wolf, Phys. Rev. Letters 11, 275 (1963); W. von der Osten and W. Waidelich, Z.Physik 178, 244 (1964).
3. C.B. Pierce, Phys. Rev. 135, A83 (1964).
4. See for example:
   E.D. Trifonov, Soviet Physics--Doklady 7, 1105 (1963);
   K.K. Rebane and V.V. Khizhnyakov, Optics and Spectroscopy 14, 262 (1963);
   R. Engleman and P. Levi, J. Math. Phys. 4, 105 (1963);
   R.H. Silsbee and D.B. Fitchen, Revs. Mod. Physics 36,432 (1964).
   F.E. Williams, J. Chem. Phys. 19, 457 (1951);
5. M. Lax, J. Chem. Phys. 20, 1752 (1952); C.C. Klick, Phys. Rev. 85, 154 (1952); D.L. Dexter, in Solid State Physics,(Academic Press, Inc., New York, 1958) Vol. 6, p.335.
6. T. Muto , Progr. Theor. Phys. (Kyoto) 4, 181 (1949);K. Huang and A. Rhys. Proc. Roy. Soc. (London) A204,406 (1950); S.I. Pekar J. Exptl. Theor. Phys. (USSR) 20, 510 (1952)
7. J.J. Markham, Revs. Mod. Physics 31, 956 (1959).
8. See for instance: E.O. Kane, Phys. Rev. 119, 40(1960); D.E. McCumber, Phys. Rev. 135, A1676 (1964).

9. D.B. Fitchen, Rev. Sci. Instr. 34, 673 (1963).
10. A.M. Karo and J.R. Hardy, Phys. Rev. 129, 2024 (1963).
11. See for instance:
G.F. Imbusch, W.M. Yen, A.L. Schawlow, D.E. McCumber and M.D. Sturge, Phys. Rev. 133, A1029 (1964);
R.H. Silsbee, Phys. Rev. 128, 1726 (1962);
G.F. Imbusch, W.M. Yen, A.L. Schawlow, G.E. Devlin and J.P. Remeika, Phys. Rev. 136, A481 (1964).
12. R.H. Silsbee, Phys. Rev. 138, A180 (1965).

Effect of Pressure on the Luminescence of Zinc Sulfide Phosphors

S. Minomura, S. Shionoya, and T. Koda
Institute for Solid State Physics
University of Tokyo, Tokyo, Japan

Introduction

The effect of pressure on the luminescence of six zinc sulfide phosphors (ZnS:Cl, ZnS:Cu,I, ZnS:Cu,Al, ZnS:Ag,Al, ZnS:Mn and ZnS:Pb) has been measured up to 52 kbars at room temperature.

The luminescence of zinc sulfide phosphors is caused not only by the various impurity centers incorporated with activators of Ib elements (Cu, Ag, Au) and/or coactivators of VIIb elements (Cl, Br, I) or IIIb elements (Al, Ga, In) but also by the native lattice defects of the host crystal. The atomic structure of these centers has not been decided. However, the following four models for the characteristic electronic transition have been presented as illustrated in Fig. 1: Schön-Klasens model for the recombination of free electrons in the conduction band with the acceptor level of Ib elements (1,2); Lambe-Klick model for the recombination of bound electrons in the donor level of VIIb or IIIb elements with free holes in the filled band (3); and two Williams models for the electronic transition between two bound energy levels of inter-ions (4,5) or intra-ions (6). The configuration-coordinate parameters and the related constants have been calculated from the pressure dependence of absorption and emission peaks in alkali halide phosphors by Williams et al. (7) and Alers et al. (8,9).

Experimental

The six zinc sulfide powder phosphors were prepared by the usual method. A single crystal of

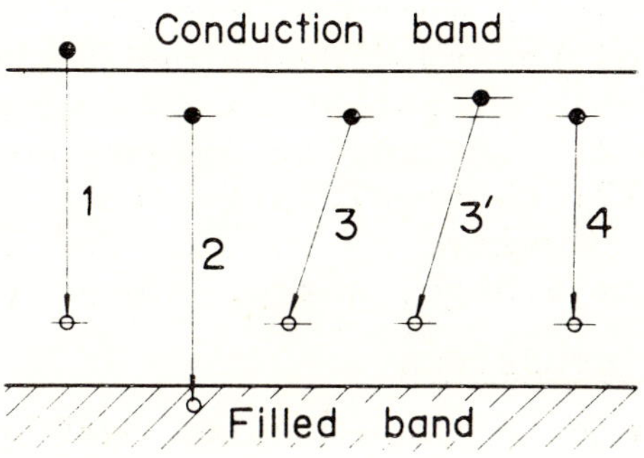

Fig. 1. Models for luminescence transition of zinc sulfide phosphors.

1. Schön-Klasens model
2. Lambe-Klick model
3. Williams model for inter-ion transition
4. Williams model for intra-ion transition

ZnS:Cl obtained from Dr. A. Addamiano of G. E. Co., Cleveland, was also used. The crystal structure, the characteristic luminescence and the firing conditions are summarized in Table 1.

High pressure measurements were made using an optical bomb of the type described by Drickamer (10) which was capable of pressures up to 54 kbar and a Bausch & Lomb 250 mm grating monochromator equipped with an RCA 1P21 photomultiplier tube. The exciting light was provided by the 365 and 315 mμ lines of mercury from a high pressure mercury lamp.

Results and Discussions

The change in the emission peak with pressure is remarkable except for ZnS:Pb as shown in Fig. 2. A red shift is observed for ZnS:Mn and a blue shift for the other phosphors. In every case the data could be fitted to an equation with constants A and B,

$$\Delta E_{ems} = AP - BP^2 . \qquad (1)$$

Using Bridgman's P-V data of ZnS, the change in the emission peak with the lattice contraction ($\Delta a/a_0$) is shown in Fig. 3. The data could be fitted to an equation with a constant C,

$$\Delta E_{ems} = C(\Delta a/a_0). \qquad (2)$$

It has been established that the luminescence of ZnS:Cl takes place within the localized center consisting of the zinc ion vacancy and the coactivator ion located at the nearest neighbor site. The emission-band shape is nearly Gaussian. The temperature dependence of the half-width follows the hyperbolic constant law (6)

$$W_{ems}(T) = W_0[\coth(h\nu_e/2kT)]^{\frac{1}{2}} , \qquad (3)$$

where ν_e is the vibrational frequency of the excited state. Fig. 4 shows a parabolic configuration coordinate diagram for ZnS:Cl. In the notation used here, E_0 is the energy separation between the minima

Table 1. The characteristic luminescence of zinc sulfide phosphors.

Phosphor	Crystal structure	Characteristic luminescence	Firing condition
ZnS:Cl	Cubic	Blue Model 3	950°C in air
ZnS:Cu,I	Cubic	Blue Model 1 or 2	950°C in wet H S
ZnS:Cu,Al	Hex.	Green Model 1 or 2	1150°C in wet H S
ZnS:Ag,Al	Hex.	Blue Model 1 or 2	950°C in wet H S
ZnS:Mn	Cubic	Orange Model 4	950°C in air with 2% NaCl
ZnS:Pb	Hex.	Green Model 4	1100°C in air with 8% NaCl

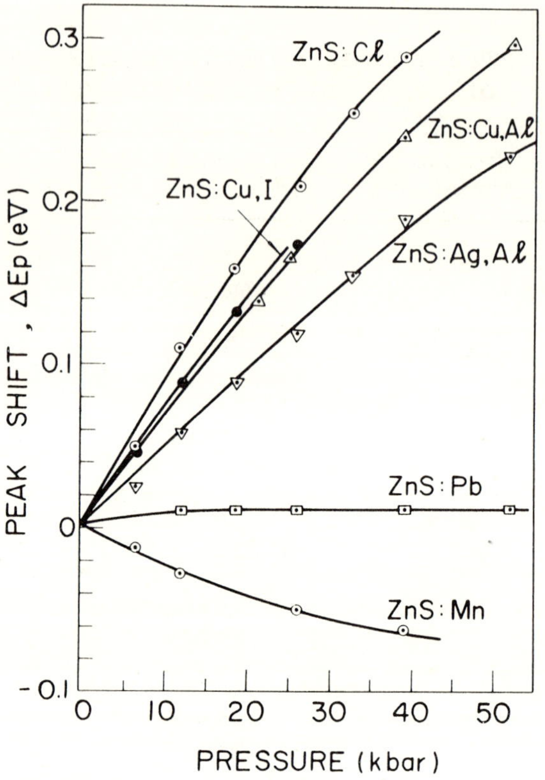

Fig. 2. The change in emission peak with pressure.

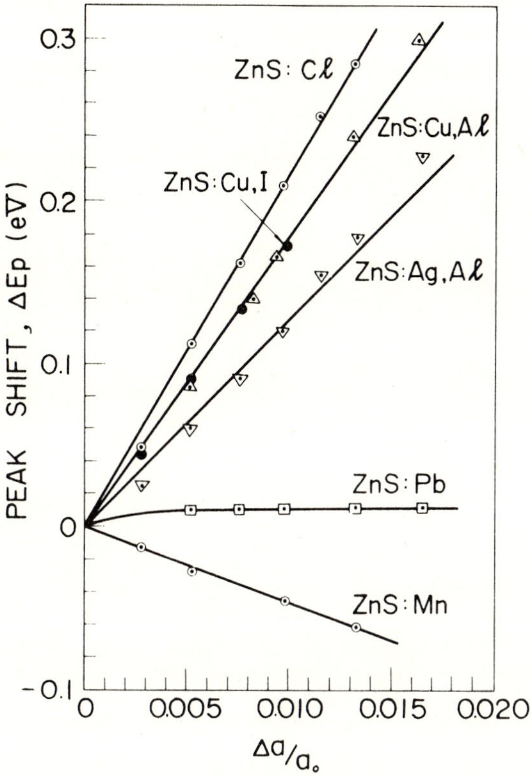

Fig. 3. The change in emission peak with the lattice contraction.

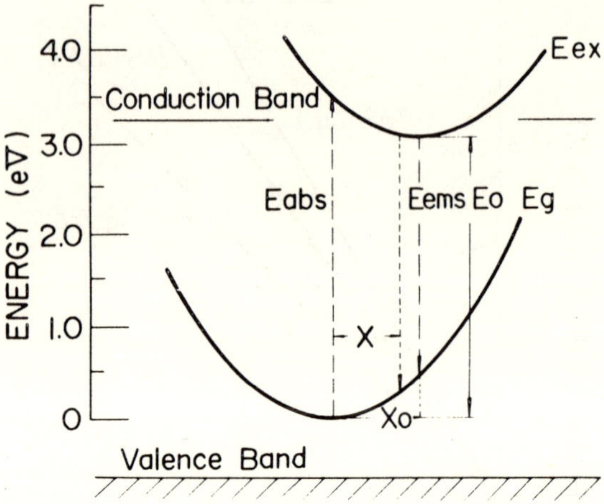

Fig. 4. The configuration-coordinate diagram for ZnS:Cl.

of the ground- and excited-state curves which are characterized by the force constants K_g and K_e and X_o is the coordinate separation. E_{abs} determined from excitation spectra at low temperature is 3.54 eV and E_{ems} determined from absorption spectra is 2.60 eV;

$$E_{abs} = E_o + \tfrac{1}{2}K_e X_o^2 \,, \qquad E_{ems} = E_o - \tfrac{1}{2}K_g X_o^2 \tag{4}$$

Using Hooke's law for pressure application, we obtained the relation between K_e and K_g from the pressure dependence of the emission peak. We assumed that the minimum of the excited-state curve shifted along the zero-pressure curve to a smaller value of the configuration coordinate with increasing stress (σ) of the center,

$$X = X_o - \sigma\alpha/K_e. \tag{5}$$

Here, α is the effective area of the center (assumed to be constant) and σ is defined by the strain ΔS of the center and the elastic constant c as

$$\sigma = c\Delta S. \tag{6}$$

The fractional change in strain of the center is considered to be proportional to that in the lattice constant of the host crystal for moderate pressure,

$$(\Delta S/S_o) = \gamma(\Delta a/a_o). \tag{7}$$

Eqs. (6) and (7) lead to

$$\sigma = \gamma c S_o(\Delta a/a_o). \tag{8}$$

With this assumption, the shift of the emission peak can be expressed by the following equation in terms of σ,

$$\Delta E_{ems}(\sigma) = \frac{K_g}{K_e}X_o\alpha\sigma - \tfrac{1}{2}\left(\frac{1}{K_e} - \frac{1}{K_g}\right)\frac{K_g}{K_e}\alpha^2\sigma^2. \tag{9}$$

Eqs. (8) and (9) lead to an equation in terms of $(\Delta a/a_0)$,

$$\Delta E_{ems}(\Delta a/a_0) = C(\Delta a/a_0) - D(\Delta a/a_0)^2, \qquad (10)$$

where $$C = \frac{K_g}{K_e} X_0 \alpha \gamma c S_0,$$

and $$D = \tfrac{1}{2}\left(\frac{1}{K_e} - \frac{1}{K_g}\right) \frac{K_g}{K_e} (\alpha \gamma c S_0)^2.$$

From Eqs. (2) and (10), we see that

$$K_e = K_g \qquad (11)$$

The coordinate separation (X_0) should be positive to give a blue shift of the emission peak with compression.

Using Eq. (11), the temperature dependence of the half-width and the peak energy of the emission band in ZnS:Cl has been predicted and compared with the experimental values. The temperature dependence of the half-width has been derived from the fact that the contribution from the zero-point vibration becomes appreciable at very low temperatures (11),

$$E_{ems} = (2K_gX_0)(1.386k/K_e)^{\frac{1}{2}}(h\nu_e/2k)^{\frac{1}{2}} \times [\coth(h\nu_e/2kT)]^{\frac{1}{2}}. \qquad (12)$$

Eqs. (3) and (12) lead to

$$W_0 = 1.071 \times 10^{-7} K_g X_0 (\nu_e/K_e)^{\frac{1}{2}} = 0.177 \text{ eV}, \qquad (13)$$

where $\nu_e = 2.9 \times 10^{-7} \text{sec}^{-1}$ and $\frac{1}{2}K_gX_0^2 = (E_{abs} - E_{ems}) = 0.47$ eV. This is in fair agreement with the experimental value, 0.17 eV.

The temperature dependence of the peak energy is expected to be attributable to the change of the occupation probability in the vibrational levels with the thermal excitation, assuming no change of the configuration-coordinate curve with temperature. It can be expressed by the following equation (12),

$$\left(\frac{\partial E}{\partial T}\right)_P = k\left[\frac{K_g - K_e}{K_e} + \frac{8\,K_g^2 E_{ems}}{K_e(K_g + K_e)(E_{abs} - E_{ems})}\right]$$

$$= 1.61 \times 10^{-4}\,\mathrm{eV/deg.} \qquad (14)$$

This is in excellent agreement with the experimental value.

The luminescence of ZnS:Cu,I, ZnS:Cu,Al and ZnS:Ag,Al has been interpreted by the Schön-Klasens model or the Lambe-Klick model as being due to the recombination between free and bound cariers, as illustrated by model 1 or 2 in Fig. 1. The temperature dependence is negative in contrast with that of ZnS:Cl. This coincides with the behavior of the band gap of the host crystal. The pressure dependence of the band gap of zinc sulfide has been observed by Drickamer et al. to be 5.7×10^{-6} eV/kg cm^{-2} for the zincblende structure (13). Piper et al. find the band gap to be 9×10^{-6} eV/kg cm^{-2} for the wurtzite structure of zinc sulfide at low pressure (14). The change in the emission peak with pressure given by the initial slope in this experiment (Fig. 2) is 6×10^{-6} eV/bar for ZnS:Cu,I and ZnS:Cu,Al and 4.5×10^{-6} eV/bar for ZnS:Ag,Al.

The luminescence of ZnS:Mn and ZnS:Pb has been interpreted in terms of the electronic transition of Mn^{2+} and Pb^{2+} ions as illustrated by model 4 in Fig. 1. The emission band could be identified as $^4T_1(G) \rightarrow {}^6A_1(S)$ for the d-d transition of Mn^{2+} and $^3P_1 \rightarrow {}^1S_0$ for the internal transition of Pb^{2+} (15). The red shift of the emission peak with pressure could be predicted by the theory of crystal field spectra by Tanabe and Sugano (16). The experiment for the effect of pressure on the absorption spectra in ZnS:Mn is in progress. The $^3P_1$ band of ZnS:Pb is practically independent of both temperature (15) and pressure as observed in this work.

References:

1. M. Schön, Z. Phys. 119, 463 (1942); Ann. Phys. 3, 333 (1948).
2. H. A. Klasens, J. Electrochem. Soc. 100, 72 (1953).
3. J. Lambe and C. C. Klick, Phys. Rev. 98, 909 (1955).
4. J. S. Prener and F. E. Williams, J. Electrochem. Soc. 103, 342 (1956).
5. E. F. Apple and F. E. Williams, J. Electrochem. Soc. 106, 224 (1959).
6. T. Koda and S. Shionoya, Phys. Rev. 136, A451 (1964).
7. F. E. Williams and P. D. Johnson, Phys. Rev. 113, 97 (1959).
8. P. B. Alers and R. L. Dolecek, J. Chem. Phys. 38, 1046 (1963).
9. P. B. Alers and P. E. V. Shannon, J. Chem. Phys. 41, 1675 (1964).
10. R. A. Fitch, T. E. Slykhouse and H. G. Drickamer, J. Opt. Soc. Am. 47, 1015 (1957).
11. C. C. Klick and J. H. Shulman, Solid State Physics, ed. by Seitz and Turnbull, (Academic Press, 1957), Vol. 5, p. 110.
12. S. Shionoya, T. Koda, K. Era, and H. Fujiwara, J. Phys. Soc. Japan 19, 1157 (1964).
13. A. L. Edwards, T. E. Slykhouse, and H. G. Drickamer, J. Phys. Chem. Solids 11, 140 (1959).
14. W. W. Piper, D. T. F. Marple, and P. D. Johnson, Phys. Rev. 110, 323 (1958).
15. D. S. McClure, Solid State Physics, ed. by Seitz and Turnbull, (Academic Press, 1959) Vol. 9, p. 399.
16. Y. Tanabe and S. Sugano, J. Phys. Soc. Japan 9, 753, 766 (1954).

Effect of Hydrostatic Pressure on Hyperfine Structure of S State Wavefunctions in Ionic Crystals: H and D in CaF_2, F Centers in LiF*

Haywood Blum

Brookhaven National Laboratory, Upton, New York and Massachusetts Institute of Technology, Cambridge, Massachusetts

Abstract

Electron spin resonance studies at room temperature have been performed on H and D atoms interstitial in CaF_2 (1,2) and on F centers in LiF at hydrostatic pressures up to 10 k bars. The hyperfine interactions of the H and D are well resolved and in the notation of Ref. 1, the results, in units of 10^{-6} bar^{-1}, are: $T_{\parallel}^{-1} \partial T_{\parallel}/\partial P|_P = 1.25 \pm 0.01$, $T_{\perp}^{-1} \partial T_{\perp}/\partial P|_P = 1.25 \pm 0.23$, $\Delta^{-1} \partial\Delta/\partial P|_D = 1.25 \pm 0.08$, $C^{-1} \partial C/\partial P = 0.195 \pm 0.013$, $B^{-1} \partial B/\partial P = 0.173 \pm 0.008$. The resolution of the F center lines with this technique is not good enough to give more than an average value $a^{-1}\partial a/\partial P = 3.7 \pm 1.0 \times 10^{-6}$ bar^{-1}. The results will be discussed in terms of the available simple ionic models.

* Work supported in part by the U.S. Atomic Energy Commission and the Advanced Research Project Agency.

[1] J. L. Hall and R.T. Schumacher, Phys. Rev. 127, 1892 (1962).

[2] H. Blum and G. B. Benedek, Bull. Am. Phys. Soc. II, 8, 619 (1963).

Introduction

Since the work of Walsh (1) describing the effects of hydrostatic pressure on paramagnetic ions in ionic crystals appeared, several other workers have also reported (2-4) investigations of the effects of high pressure on the paramagnetic resonance spectra of iron group ions in ionic lattices.

The results of these experiments provide further checks on crystal field theory. However, it is well-known that calculations from first principle of the crystal field parameters are disappointingly unsuccessful (5). It becomes of interest, then, to simplify the problem as much as possible. With this in mind, the hyperfine structure of hydrogen and deuterium atoms substitutionally located in CaF_2 was studied as a function of volume compression.

The F center in LiF was also studied. This system is similar to the $H:CaF_2$ but unfortunately one does not have complete resolution of the hyperfine structure so that a complete investigation will require an ENDOR experiment. Since the techniques used for the F center experiment are more standard and this work will be described in detail elsewhere (6) we shall refer mainly to the CaF_2 work.

Hall and Schumacher (HS) (7) showed in their excellent and careful paper that the paramagnetic resonance spectrum of $H:CaF_2$ can be described by the spin Hamiltonian:

$$H = g\beta H \cdot S + BI \cdot S + \sum_{\alpha} I_{\alpha} \cdot T_{\alpha} \cdot S \qquad (1)$$

where the sum α is over the various fluorine ion neighbors surrounding the hydrogen atom. HS found that T was symmetric about the axis joining the origin (proton site) to the fluorine ion so that

$$T = \begin{pmatrix} T_\perp & & \\ & T_\perp & \\ & & T_\parallel \end{pmatrix} = \left(\frac{T_\parallel + 2T_\perp}{3}\right)\begin{pmatrix} 1 & & \\ & 1 & \\ & & 1 \end{pmatrix} + \left(\frac{T_\parallel - T_\perp}{3}\right)\begin{pmatrix} -1 & & \\ & -1 & \\ & & 2 \end{pmatrix} \qquad (2)$$

The interpretation of the parameters of the spin Hamiltonian is given below

$$g\cdot H = \langle \Psi | g_0 S\cdot H + L\cdot H | \Psi \rangle$$

$$B = (8/3)\gamma_P \beta \ \langle \Psi | \delta(r) | \Psi \rangle$$

$$a = \frac{T_\parallel + 2T_\perp}{3} = \frac{8}{3}\gamma_F \beta \ \langle \Psi | \delta(r - R_\alpha) | \Psi \rangle$$

$$b = \frac{T_\parallel - T_\perp}{3} = \frac{1}{2\pi}\gamma_F \beta \ \langle \Psi | \ \frac{3\cos^2\theta_\alpha - 1}{r_\alpha^3} \ | \Psi \rangle \qquad (3)$$

where Ψ is the eigenfunction of the total Hamiltonian which includes the Hamiltonians of the isolated ions plus the interactions between ions (atoms).

Experimental Details

A sensitive x-band spectrometer with superhetrodyne detection was employed. A block diagram shows the main features (Fig. 1). The high pressure coupling and the microwave cavity are essentially the same as those of Walsh and Bloembergen (1). The pressure was generated with a press from Bridgman's laboratory.

The experiment consists of the measurement, at a given pressure, of the klystron frequency and the magnetic field for resonance of the various hyperfine lines. The magnetic field is locked to the center of each of the lines and is measured with a proton magnetometer. The proton sample is placed on the same magnetic field contour line as the CaF_2 sample. All frequency measurements are referred to a single

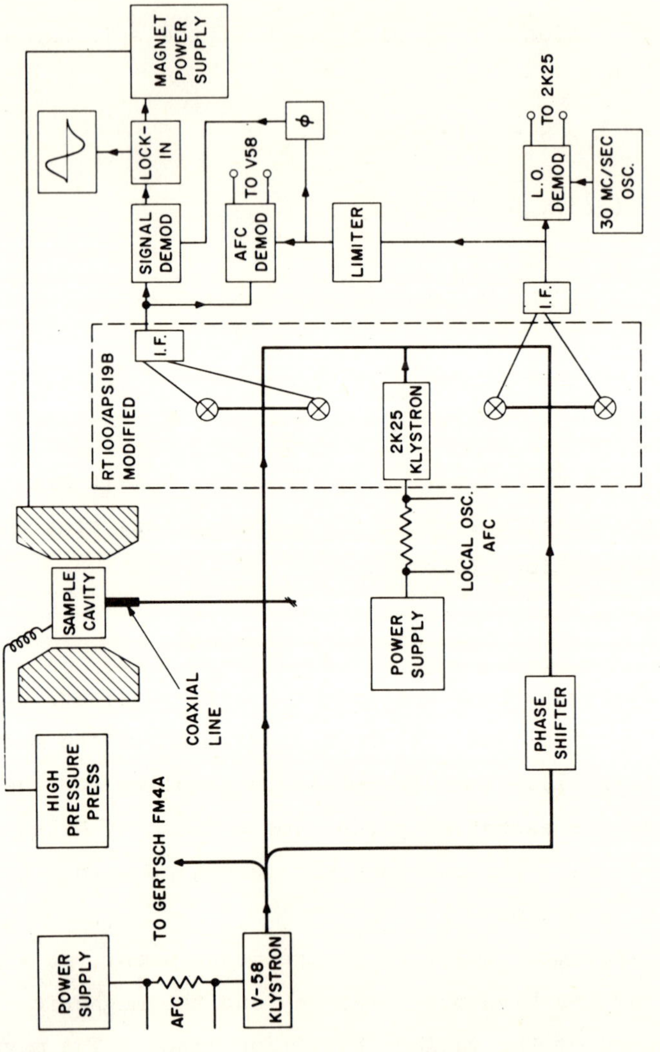

Fig. 1 Block diagram of the electron spin resonance spectrometer.

HP 5245L cycle counter.

Our hydrogen sample was given to us by R. T. Schumacher. The deuterated sample was also made by Schumacher but came to us from J. Castle, Jr. Both were cut into the shape of a torus by the A.D. Jones Optical Company, Cambridge, Mass.

The variation with hydrostatic pressure of B and Δ are shown in Figs. 2 and 3. By measuring Δ at two angles it is possible to determine a and b.

The compressibility of CaF_2 has been determined by Bridgman (8) and confirmed by elastic constants measurements (9). It is used to convert the measured pressure dependence to volume dependence. In using this conversion factor we are assuming that the local compressibility around the hydrogen atom is equal to the compressibility of pure CaF_2.

The results are collected in Table I.

Discussion

In order to understand both the atmospheric pressure values of the parameters of the spin Hamiltonian and their pressure dependences, we write the correct first order wavefunction in the Heitler-London approximation and then calculate to 2nd order in perturbation theory the effects of covalency, spin-orbit coupling and electrostatic field.

Until a much more elaborate but necessary Hartree - Fock treatment is performed we cannot consider these results to be more than an indication of the important interactions in this system.

A. Wavefunctions. The Heitler-London model assumes that the wavefunction of our many electron system can be found by constructing a Slater-determinant with the free ion or free atom wavefunctions. In this approximation we can easily show that

$$\Psi = N \{\Phi - \sum_{\alpha} (\Phi|\chi_{\alpha})\chi_{\alpha}\} \qquad (4)$$

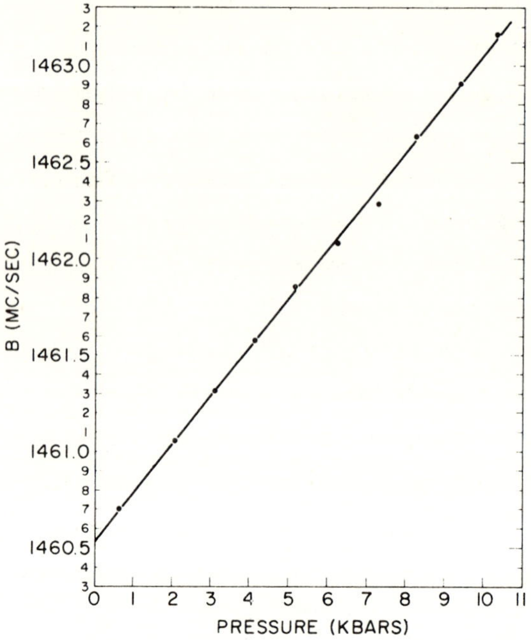

(a)

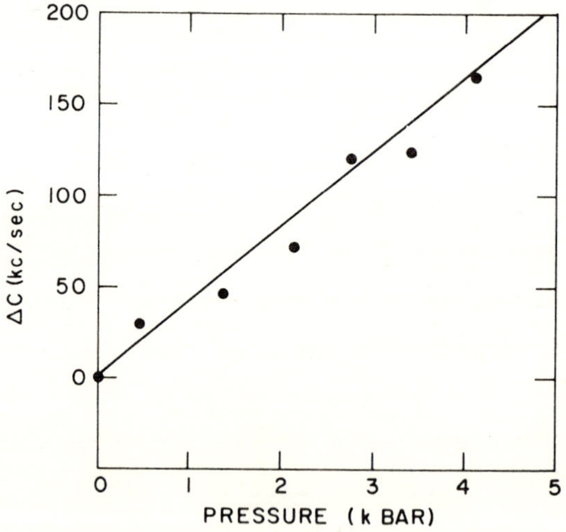

(b)

Fig. 2 The hyperfine constant of (a) hydrogen atoms and (b) deuterium versus pressure.

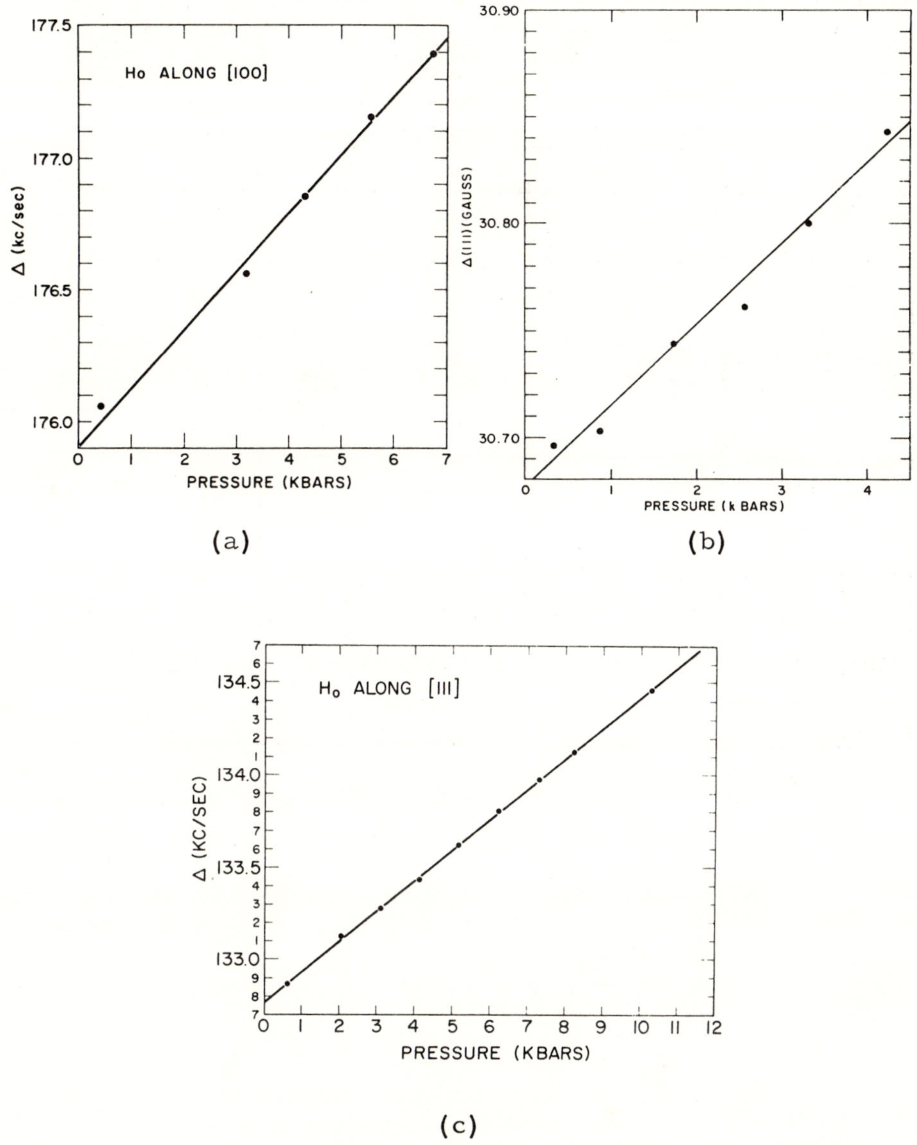

Fig. 3 The fluorine hyperfine interaction versus pressure for various directions of external magnetic field for (a) and (b) hydrogen atoms and (c) deuterium.

Table 1. Parameters of the Spin Hamiltonian

$$H = g\beta H\cdot S + BI_P\cdot S + aI_F\cdot S + b\,(3\cos^2\theta - 1)\,I_F\cdot S$$

	First Order		Second Order	Experiment
	$\gamma = 0$	$\gamma=0.03$		
B	1660	1566	1472	1460.3 mc/sec
$B - B_o$	240	106	52	39.9 mc/sec
a	129	129	122	103.96 mc/sec
b	33	44	35	34.93 mc/sec
$\Delta g/g$	0	0	$+10^{-4}(\gamma=0.03)$ $-2 \times 10^{-3}(\gamma=0)$	10^{-4}
$\partial\ln B/\partial\ln R$	-1.0	-1.0	-0.67	-0.445 ± 0.019
$\partial\ln C/\partial\ln R$				-0.500 ± 0.033
$\partial\ln a/\partial\ln R$	-5.2	-5.2	-3.9	-3.20 ± 0.1
$\partial\ln b/\partial\ln R$	-5.6	-5.6	-4.2	-3.20 ± 0.6
$\partial\ln\Delta g/\partial\ln R$			$-4(\gamma=0.03)$ $+7(\gamma=0)$	-6 ± 4

where

$$N = [1 - \sum_{\alpha} (\Phi|\chi_{\alpha})^2]^{-\frac{1}{2}} \quad \text{and}$$

χ_{α} is one of the fluorine ion wave functions and Φ is the hydrogen atom wavefunction.

To introduce covalency we replace the overlap integral $(\Phi|\chi_{\alpha}) = S_{\alpha}$ by λ_{α}

$$\Psi = N\{\Phi - \sum_{\alpha} \lambda_{\alpha} \chi_{\alpha}\} \qquad N = (1 - 2\sum_{\alpha} \lambda_{\alpha} S_{\alpha} + \sum \lambda_{\alpha}^2)^{-\frac{1}{2}} \tag{5}$$

The ligand wavefunctions are altered also

$$\Psi_L^{\alpha} = M (\chi_{\alpha} + \gamma_{\alpha} \Phi) \tag{6}$$

where

$$\gamma_{\alpha} \sim (\lambda_{\alpha} - S_{\alpha})$$

To keep the calculation under control we arbitrarily assume that there is no covalency with the fluorine 1s and 2s electrons $\gamma_{1s} = \gamma_{2s} = 0$. Furthermore since $(\Phi|2P_{\pi}) = S_{\pi} = 0$, we assume $\gamma_{\pi} = \gamma_{\pi'} = 0$.

B. g value. The fact that the g shift, $g - g_{free\ H}$, is positive is somewhat surprising. However, the phenomenon is good proof of the existence of covalent bonding.

To second order the g shifts are

$$\Delta g_{\gamma} = 0$$

$$\Delta g_{\pi} = \frac{4}{3} \zeta N^2 \lambda_{\sigma} \left\{ \frac{\gamma_{\sigma}}{E-E_{\pi}} - \frac{\lambda_{\sigma}}{E_{ex}-E} \right\} \tag{7}$$

for each fluorine. For eight nearest neighbors we have

$$\Delta g = \frac{16}{3} \Delta g_{\pi} \ . \tag{8}$$

We then chose $(E-E_{\pi}) = 1.28$ eV to fit the experiment and $\gamma_{\sigma} = 0.03$ and $(E_{ex} - E) = 8$ eV. For $\gamma = 0$, $\Delta g_{\pi} = 10^{-3}$.

We hope to show the validity of Eq. (7) elsewhere (6).

The pressure dependence of the g shift can be computed. It comes mainly from the increase in the overlap integrals. If we assume that γ increases proportionally to S_σ then $\frac{\partial \ln \Delta g}{\partial P} = 3 \times 10^{-6}$ bar^{-1}. This is quite small. It is difficult to measure the g shift accurately enough to make a definite comment on this prediction. For the [111] direction we measure $\partial \ln \Delta g / \partial P = (5 \pm 3) \times 10^{-6}$ bar^{-1}.

C. Hyperfine Constants. It is easy to calculate the hyperfine constants (the parameters of the spin Hamiltonian given in Eq. (1)) to first order with $\gamma = 0$ (purely ionic model) and $\gamma \neq 0$ since Hall and Schumacher (7) have evaluated the matrix elements of interest. The answers are collected in Table 1 and are from HS.

In Fig. 4 we have shown the results for $\gamma = 0$ compared with the experimental points at atmospheric pressure of HS. We see that the hydrogen wavefunction falls off too fast with distance from the origin. This defect can be repaired by the introduction of an effective dielectric constant. The hydrogen wavefunction is modified in the region beyond the nearest neighbors. This modification has no appreciable effect on the wavefunction near the origin or the overlap integrals. We would expect the effective dielectric constant to be nearly equal to the optical dielectric constant for CaF_2 which is 2.04. In fact we find $K_{eff} = 1.48$.

In the vicinity of the nearest neighbors it is necessary to consider in detail the interaction between the unpaired electron and the closed shells. We consider now the effect to second order on the hyperfine energy of the crystalline field. This is done in two steps equivalent to second order perturbation theory but somewhat more revealing. First we find the change in the starting wavefunction due to the perturbating crystalline field and then we calculate the hyper-

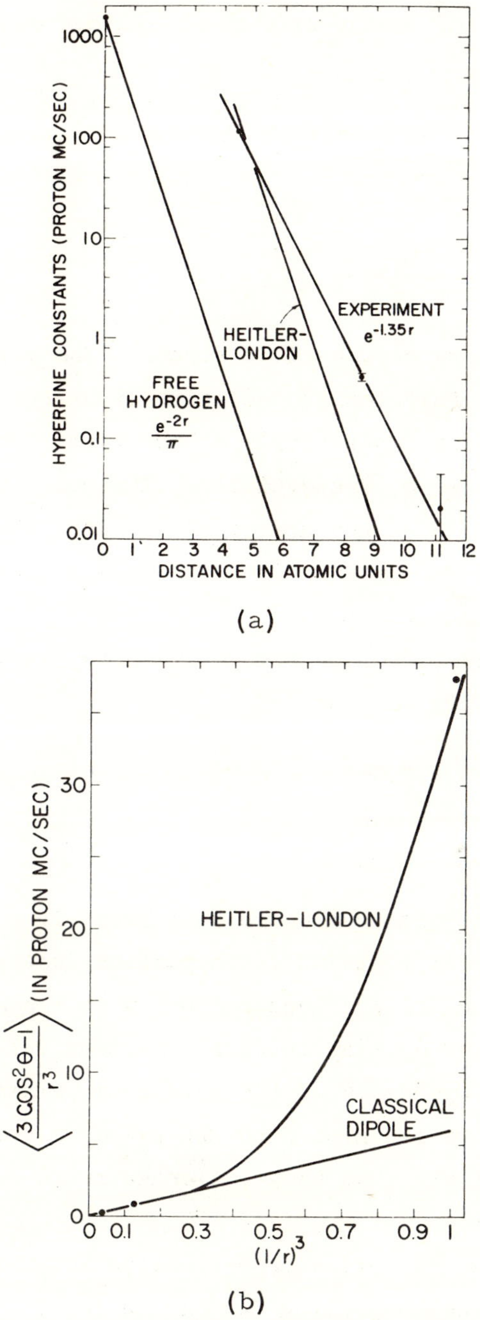

Fig. 4 Hyperfine constants predicted by the Heitler-London model. (a) isotropic (b) dipole-dipole. The experimental points are from Hall and Schumacher.

fine energy to first order with the perturbed wavefunction.

$$\Delta\Psi = N\{\Delta\Phi - \Sigma\lambda_i \Delta\chi_i\} \tag{9}$$

where

$$\Delta\Phi = -\Sigma_j' \frac{\langle\Phi_j|V_L|\Phi\rangle}{E_j - E} \tag{10}$$

$$\Delta\chi_i = -\Sigma_j' \frac{\langle\chi_j|V_M|\chi_i\rangle}{E_j - E_i}\chi_j \tag{11}$$

The sums j are over excited states. The prime on the summation indicates that the ground state is omitted. V_L and V_M are the electrostatic potentials of the eight fluorine ions and the hydrogen atom, respectively. The point here is that V_L causes excited hydrogen states to mix into the hydrogen ground state and V_M mixes excited fluorine states into the fluorine ground state.

Without going into further detail here, we find that the hyperfine constants are reduced

$$\Delta B/B = -0.06$$
$$\Delta b/b = -0.02$$
$$\Delta a/a = -0.05 \tag{12}$$

These results are also collected into Table 1.

The pressure effects on the hyperfine constants can be calculated in the Heitler-London model with the aid of the derivatives of the overlap integrals already determined by HS. If we assume a dependence of the covalency parameter γ on overlap, say $\gamma \propto S$, we can also find the pressure derivatives with covalency. In Table 1 we have collected these estimates.

D. Deuterium

We find that the pressure derivative for C is about 10% larger than the corresponding derivative for B. This indicates that the region of the deuterium site is somewhat more compressible than the region around the hydrogen atom. One can understand this result qualitatively in terms of a local vibration mode (10) of the hydrogen. This is indeed a highly localized mode since the pressure derivatives of a and b in deuterated and hydrogenated CaF_2 are identical.

E. F Center. The F center spin Hamiltonian is similar to Eq . (1) with B = 0 since the F electron is centered on a vacancy. We find an average value for the variation of the hyperfine constants with pressure

$$\Delta^{-1}(\partial\Delta/\partial P) = (3.7 \pm 1.0) \times 10^{-6} \text{ bar}^{-1} \tag{13}$$

which can be compared with the predictions of the Gourary and Adrian (11) point-ion model.

$$(\partial \ln_a/\partial \ln R)_{\text{theory}} = -3 \tag{14}$$

Combining Eq s. (13) and (14) with the bulk compressibility of LiF measured by Bridgman (12) we can calculate the ratio, f, of the effective local compressibility to the bulk compressibility. We find

$$F_{ESR} = 2.5 \tag{15}$$

This can be compared with the semi-empirical results of optical studies for F centers in alkali halides which is

$$F_{\text{optical}} = 2. \tag{16}$$

Conclusions

We find that the various interactions we invoked have brought the cal culated hyperfine constants into suspiciously close agreement with experiment although the pressure derivatives are still much further off. The system CaF_2: H remains an excellent one to use as a test of a first principles calculation.

References

1. W. M. Walsh and N. Bloembergen, Phys. Rev. 107, 904 (1957); W. M. Walsh, Phys. Rev. 114, 1473 (1959); ibid, 114, 1485 (1959); ibid 122, 762 (1961).
2. D. B. Larson, Thesis, Brigham Young University, 1963.
3. D. F. Wait, Phys. Rev. 132, 601 (1963).
4. L. Rimai, T. Deutsch and B.D. Silverman, Phys. Rev. 133, A1123 (1964).
5. A bibliography for the recent literature can be found in R. E. Watson and A. J. Freeman, Phys. Rev. 134, A1526 (1964).
6. H. Blum, in preparation.
7. J. L. Hall and R. T. Schumacher, Phys. Rev. 127, 1892 (1962).
8. P. W. Bridgman, Proc. Amer. Acad. 77, 187 (1949).
9. H. B. Huntington in Advances in Solid State Physics, edited by F. Seitz and D. Turnbull (Academic Press, Inc., New York, 1958), Vol. 7, p. 214.
10. D. W. Feldman, J. G. Castle, Jr. and J. Murphy, Phys. Rev. (to be published).
11. B. S. Gourary and F. J. Adrian, in Solid State Physics, edited by F. Seitz and D. Turnbull (Academic Press, Inc., New York, 1960), Vol. 10, p. 127.
12. P. W. Bridgman, Proc. Am. Acad. 70, 285 (1935).

SESSION E

PHASE TRANSFORMATIONS

Chairman: H. B. HUNTINGTON

THE ELECTRONIC AND THERMODYNAMIC NATURE OF DENSE VAPORS*

J. A. Krumhansl

Laboratory of Atomic and Solid State Physics
Cornell University
Ithaca, New York

Abstract

The purpose of this work is to review theoretical and materials science considerations relevant to dense vapor systems. Particular attention is given to "metallic" vapors. In single species metallic vapors one may expect an insulator to conductor transition as density increases. In mixed vapor systems the situation may be analogous to doped semi-conductors. In addition, the change in electronic character with increasing density suggests that a phase transition may frequently, but not necessarily, accompany the onset of metallic conductivity. A qualitative thermodynamic treatment is given. The behavior of sodium-ammonia systems is discussed as a model.

1. INTRODUCTION. This is not a report of completed work on a well developed subject. It cannot be, for the subject is only in a preliminary state. Thus, instead of providing definite answers, we only examine the nature of some questions which should be asked. The writer believes this to be a most interesting topic about which all too little is known; the experiments are difficult, but the reward can be the deeper understanding of the physics of matter under most unusual circumstances.

Our purpose is to examine the probable behavior of dense vapors or low-density liquids (i.e., matter in a form with a density of perhaps 1 to 10 percent of normal solid or liquid density). We

---

*This work has received support from the Advanced Research Projects Agency, and was carried out in part at General Atomic under support by NASA Lewis Research Center.

restrict consideration to temperatures less than about 2500°K, above which direct thermal ionization effects will occur. Well above that temperature one makes a transition to a high temperature plasma (10^4 - 10^5°K) where the electronic considerations of crucial importance in the present context no longer play a role.

The immediate interest in this subject derived from the objective of examining the electrical conductivity of dense vapors at high pressures and intermediate temperatures to determine their possible application to MHD systems (1). An early appreciation of the possibilities was due to Miller (2), who employed the phrase "mercury doped vacuum" to suggest a semiconductor analogy.

As a basic scientific problem, dense vapors have not received much attention, particularly for elements which are metallic in their condensed phase. The reasons are apparent with regard to both experimental and theoretical aspects. Experimentally, high pressures and difficult materials problems arise. Theoretically, neither to constitution nor transport properties can we apply the time-honored and fortuitous self-deception embodied in the one-electron approximation and the Boltzmann transport equation.

Recently, interest in the theoretical aspects of this problem has increased through the introduction of proper many-electron descriptions of the state of such matter (3-6) as well as generalized techniques for describing hopping and tunneling transport in such systems when the Boltzmann equation is not applicable (7-10). We have attempted to incorporate the essential concepts of these into the discussion to follow, but it must be emphasized that quantitative application of a theory still is beyond our ability.

2. ELECTRONIC BEHAVIOR OF PURE SINGLE-SPECIES MATTER AT LOW AND HIGH DENSITIES. There is a fundamental qualitative difference between the properties of low-density potentially-metallic species (e.g., atomic hydrogen, monovalent alkali metals, etc.) and those of its high density form (3). (Substances which form insulators, with filled bands, at normal solid densities are neither of concern nor interest in the present discussion.) This essential and conceptual difference is as follows: If the one electron (Hartree-Fock)

approximation were valid, we could, in principle, describe the valence electrons in a dilute alkali-metal gas by Bloch-like extended orbitals, half filling a "band" or continuum of states. Though the width of this band would decrease with decreasing density, the lowest lying excited states above the ground state would still, in principle, lie infinitesimally close and the behavior would be metallic in nature in all densities.

This one-electron view is substantially wrong, in essence, as Mott (3), Kohn (4), and Hubbard (5) have pointed out. Correlation effects, which become significant in the low density limit when electrons are placed in the same atom, fundamentally modify the description. The appropriate wave functions at large atomic separations really localize electrons in disconnected regions of configuration space, and the ground state will be detached by a finite energy from a continuum of excited states. It is not a matter of degree, but it is essential that the configuration should be that of an insulator. As the density is increased, a distinct transition to a quite different configuration is expected -- namely, the metallic state described by the extended orbitals referred to above.

Additional physical feeling for the situation can be obtained from Mott's article (3), whose essential points we restate. Consider an alkali metal gas (assumed to be held on a lattice) which we will compress through a range of densities, which might take us continuously up to solid or liquid densities. At low densities we expect the atoms to be described by atomic orbitals which place exactly one valence electron on each atom. However, this complete localization cannot be precisely correct and we must examine the nature of electron transfer and exchange. There are two competing factors: (a) When two atoms are brought together, and assuming an "atomic orbital" (AO) basis, there is possible an electron transition from the m-th to n-th atom which is quantum-mechanically specified by an overlap matrix element, v_{mn} for brevity. It is this interaction matrix element which leads to molecular binding, or relates to the band widths in a tight binding description of solid-state electron energy levels. In these cases, its magnitude is several electron volts at normal molecular and atomic densities. As the density decreases, overlap falls off rapidly, usually exponentially; this overlap matrix element between atomic orbitals can produce delocalization of the electrons by interatomic transfer. (b) When an electron is envisioned as transferring from the m-th to n-th atom, without allowing the other electrons on the n-th atom to "make way" by transferring in turn, a substantial amount of "correlation" energy must be provided. In the very low density limit, a positive and negative ion are thus produced, requiring an amount of energy I-A where I is the ionization energy and A is the electron affinity for formation of a negative ion. This quantity (I-A) is typically several eV in pure systems, and may depend also on the magnetic state. In any case, this effect opposes the preceding factor (a) and tends to localize electrons in a quota of just one per atom and also leads to a ground state detached by finite energy from the excited state.

The formalization of this state of affairs may be embodied in a Hamiltonian (4,5) written in second quantized form with an AO or Wannier function basis. A reasonable approximation is to include only the correlation correction for electrons on the same atom. Since we will not use the equations here references 4 and 5 may be consulted for details. Models may differ, but there is no disagreement on the general behavior to be expected. At low densities, factor (b) dominates and the system is truly insulating. At higher densities, many-electron (transfer) effects become important and can relieve the electron excess on any one atom by compensating transfer. Mott's criterion is that this takes place when v_{mn} becomes comparable to I-A, whereas a dramatic change to metallic character takes place. Mott's criterion allows us to draw conclusions about the likely conditions for achieving the metallic transition; namely, pure vapors would have to be very dense indeed (an estimate is 10 percent of solid densities) before transition to the metallic state occurs. This follows from requiring $v_{mn} \simeq I - A \simeq 1$ eV.

However, it must also be noted that we have no way of estimating how density fluctuations, which could be significant in the present situation, will affect multiple transfer. This, of course, is the nastiest part of the problem, about which we can say nothing further. Young (6) has estimated the critical densities for the metallic transition in the alkali metals to be about 20 percent of solid densities.

From these general considerations, it follows that computations of conductivity which do not specifically take electron correlation effects into account (1) will be quite inapplicable to vapors of single species at low densities, i.e., densities less than, say, 10 percent of solid and lower densities. We comment further on this matter in Section 4.

3. MIXED OR "SEMI-CONDUCTING" VAPOR SYSTEMS. In a "pure" vapor, very high densities will be required for a transition from the insulating vapor state to a metallic vapor state. It is worth reconsidering an alternative possibility for achieving electronically conducting vapors. A natural analog to a solid state system would be the doped insulator or "semi-conducting" vapor. By semi-conducting vapor we mean that though a majority species is the principal component of the vapor it also includes a minority component whose function is either to donate or accept an electron from the majority component (11,12,13).

The problem of describing electron transfer at low densities is much simpler here. Presuming a low concentration of donors or acceptors, electron transfer in donor systems takes place by interchange of an electron from one majority atom to another and merely amounts to changing the identity of the negative ion (or positive ion in the case of acceptor systems). Thus, there is no net change in correlation energy; the intra-atomic correlation has been accounted for in forming the ions in the first place and does not alter when the identity of the ion is simply interchanged. Thus, there is conceptual justification in using one-electron models for conductivity. For these there exist two limiting cases: (a) The band model for a system with a small carrier concentration (of either electrons or holes), including impurity bands in compensated semiconductors; or (b) An atomic model with transport by hopping from one site to another.

An application of the former approach is contained in Reference 1, and we believe it to be appropriate to doped vapors. The latter view is developed in various discussions of hopping and tunneling conductivity in dilute or narrow-band semi-conductors (14-16). The density dependence of the overlap potential matrix elements which determine electron transfer, can be estimated with some order of magnitude confidence in some impurity band semiconductors, and in transition metal oxides.

Since this may be an interesting area for high pressure research it is worth giving criteria for choosing materials: (a) In a solid there is not too much a priori difference between donor and acceptor systems; this is not true for vapors. Electron transfer is desired at low densities by interaction transition due to overlap potentials. Negative ions are much larger than positive ions, so that for a given density, electron transfer between a negative ion and the same-species neutral is far more likely than in the positive ion case. Thus, donor systems are preferable. (b) A minimum energy should be required to ionize the donor to transfer an electron to the majority species. The majority species should be capable of forming stable negative ions. Thus, we wish to minimize $(I_d - A_m)$, where I_d is the donor ionization energy, and A_m is the majority electron affinity. (c) Chemical compatibility is required between majority and donor species. Electron affinities and ionization potentials are obtainable from the literature (17,18). Vapor pressure data may similarly be calculated from thermochemical data; a particularly useful compilation for solid and liquid elements is given by Honig (19).

One combination which might be studied as the prototype of the "donor" system is an "insulating" Hg vapor doped with Cs; that is, Hg is the majority species. The following data are relevant. At 643°K, Cs has a vapor pressure of 10 torr, while Hg has a vapor pressure of 10^3 torr; this represents a relative donor concentration of 1 percent, though at a relatively low pressure. At 980°K, Cs has a vapor pressure of 10^3 torr (1.32 atm), while Hg has a pressure (1) of approximately 60 atm. With regard to electronic properties, the ionization potential of Cs is 3.87 eV, and the electron affinity of Hg to form Hg^- is 1.79 eV; the difference $I_d - A_m = 2.08$ eV is significantly less than any single-species ionization potential. Chemically, it is likely that Cs and Hg will be compatible.

4. ELECTRON TRANSPORT IN DENSE VAPOR SYSTEMS. An attempt to analyze electrical conductivity in systems such as those discussed herein is a most interesting experience. It serves to illustrate the specialized nature of the basic premises on which the great bulk of conventional transport theory is based.

The conventional method of describing electron transport is by means of a statistical theory. It is based on a one-particle distribution function (reduced from an N-particle), referring to a basis description (free particles or Bloch functions in a crystal) whose states are eigenstates of the electron current operator, and considers scattering to be a non-essential part of the transport process, except insofar as free transport is limited thereby. The distribution function itself must be obtained from a defining integro-differential equation (Boltzmann) which has validity only for a continuum of spatial and momentum states. With the historical origin of transport theory in gas kinetic problems, and with the variety of plasma and electronic systems for which the idealization to these conditions is not a troublesome approximation, this is not surprising.

By contrast, electron transfer in vapor systems illustrates the most opposite situation. Except at the very highest densities, the appropriate basis states are atomic orbitals which carry no current. The basis states do not form any simple continuum. In "pure" single-species systems, electron correlation must play an essential role. Thus, no one-particle distribution function obtained from a Boltzmann equation is applicable. Furthermore, scattering from one atomic state to another is an essential part of the transport process itself. Finally, the spatial distribution and correlation of the atoms which carry the electrons must play an essential role.

Fortunately, during the past few years there have been many developments in transport theory which may be used with greater generality. The description of transport is based on correlation functions (7) and Green's-function methods for finding these correlation functions. They are general enough to handle either the limiting case of nearly free particles (8), (9), (20), (21), or the opposite case of tightly bound electrons (14). It should be stated immediately that we have not applied these methods yet in detail to the vapor systems under discussion. Rather, these remarks are intended to call attention to the fact that transport calculations based on free-particle or Bloch-function states can only be valid in the limit of essentially solid densities and, just as in discussing the electronic constitution, must also fail in an essential qualitative manner for vapors. This situation has been encountered in narrow band solids, and in some semi-conducting systems when conduction takes place by hopping from site to site (15), (16), (22), (23). (In the case of semi-conductors, the transfer matrix elements are usually not the simple potential overlap terms, but also contain phonon contributions.)

To complete this survey we summarize the principal formal steps in the theory. First, according to Kubo's method (7), we must compute an appropriate transform of the current-current correlation function; this may be understood qualitatively by recognizing that an applied electric field couples energy into the system by acting on moving charges, and we subsequently also observe the effect by measuring the current associated with this charge distribution. Thus, (for an isotropic medium) the static conductivity is

$$\sigma = \lim_{s=0^+} \int_0^\infty dt \int_0^\beta d\lambda \, \exp(-st) \, [J(-ih\lambda), J(t)] \tag{2}$$

where the meaning of the symbols is standard, and the review article by Chester (10) or the paper by Matsubara and Toyozawa (14) should be consulted for details. To use this equation, we need the equations of motion for the current operators J(t), which are in the Heisenberg representation.

To determine these J, the method of Green's function may be employed (9). The appropriate basis representation for the vapor case would be atomic orbitals or Wannier functions. The significant features will then be contained in a Hamiltonian, written in second quantized form using creation or annihilation at the ith or j^{th} site:

$$H = \sum_{i,j} (t_{ij} + v_{ij}) a^+_{i,\sigma} a_{j,\sigma} + \sum_{i,\sigma,\sigma'} E_{\sigma\sigma'} n_{i,\sigma} n_{i,\sigma'} \tag{3}$$

where $n_{i,\sigma}$ is the number operation; $a^+_{i,\sigma}$ is the creation operator; $a_{j,\sigma}$ is the annihilation operator; and σ's are spin indices. The first term contains both the intra-atomic and interatomic one-electron kinetic-energy and potential-energy matrix elements (and thus would determine the usual one-electron energy bands and related parameters in the crystalline case). The second term is an approximation mentioned before; it is the correlation energy which results from multiple occupation of the same i^{th} site and omits correlations between different sites. This type of Hamiltonian is considered by Kohn (4) and Hubbard (5).

The electron-current operator has zero expectation for a single atomic orbital. But the solutions to the interacting system defined by Eq. (3) may definitely have non-zero value. The complete current operator would contain all the information about the average "intra-cell" current as well as the "inter-site" current where the electron is transferred through the system. It is the latter which is of interest, and Matsubara and Toyozawa (14) give an appropriate expression which simply localizes the electron of the nth nucleus as long as it is on the nth atom. The current is then

$$J = (ie/h) \left[H, \sum_m R_m \, a^+_{m,\sigma} \, a_{m,\sigma} \right] \tag{4}$$

where the brackets are the usual commutator giving the time rate of change of this approximate electron-position distribution.

In the "pure" vapor at low densities, the electrons referred to in Eq. (3) are about one per atom and the correlation term in the equation dominates so that the ground state is an essential insulator (discussed in Section 2). To compute the conductivity properly in this limit, it would be quite essential that the Hamiltonian, Eq. (3), be treated completely.

By contrast, in doped vapors, a Hamiltonian similar to Eq. (3) could be written; however, the correlation term would be modified. The computation in this case is far more tractable. Specifically, Matsubara and Toyozawa (14) have carried out a development of just this model in connection with compensated impurity-band conduction in a randomly-doped semi-conductor (partially compensated). The atomic orbitals are assumed to be hydrogen-like. They compute expressions for both the level density and conductivity in the limiting case where the Hamiltonian Eq. (3) is approximated by only the interatomic matrix elements of potential v_{ij}. The Green's function and the equation of motion, Eq. (4), of the current operator J have been obtained by averaging over various random configurations of the impurity atoms.

The purpose of this section was to survey the nature of the electron-transport problem in the dense vapor state. When conductivity data are available for mixed systems, there is some hope of applying existing methods to analyze the data.

5. THERMODYNAMIC CONSIDERATIONS IN PHASE TRANSITIONS IN THE DENSE VAPOR. The preceding sections emphasize the differences in electronic constitution of two alternative configurations of an array of atoms. Of primary interest has been the electronic transition from insulating to metallic state.

In this section we compare electron behavior on the one hand, and thermodynamic constitution on the other hand. It is not *a priori* true that there is a relationship between them, insofar as the insulator to metal transition is concerned. However, we are led to the conclusion that although a thermodynamic phase transition does *not necessarily* accompany the insulator to metal change, the conditions are highly conducive for such. We can thus usually expect to get first order transitions between a low density insulating phase and a high density liquid (or solid) metallic phase as we increase the vapor density and pressure.

To discuss thermodynamic behavior assume that the molar free energy of the system is composed of two parts: $F = F_e + F_t$ where F_e is the electronic free energy and F_t the configurational and translational free energy. Unlike the insulator-metal transition in semiconductors, the species here are assumed to be free to move as a gas or liquid, allowing phase transitions to occur freely as required by thermodynamic considerations.

Only the simplest model is envisaged for F_t, such as a free lattice gas model or a free particle model at low densities, and an Einstein oscillator model at high density. To include spatial correlations and clustering would be needed for complete description, but these lie quite beyond our scope. Accordingly it is assumed that

$$F_t = - kT \log(v/\lambda_T^3) \quad \text{low densities}$$

$$F_t = - kT \log(kT/h\omega_e) \quad \text{high densities}$$

where the symbols have the usual meaning, v is volume per particle, λ_T is the thermal de Broglie wavelength and ω_e is an oscillator frequency which increases as the density increases. The main physical consideration is that F_t becomes monotonically more negative with increasing v or T.

The unique consideration in the present case is that unlike conventional interactions in statistical mechanics, the electronic interaction between atoms changes sharply in an essential qualitative way as the insulator to metallic quantum transition occurs. Perhaps a jump in F_e might be expected at this condition, but examination of either Kohn's or Hubbard's models suggests that this will not occur; and this is not postulated here. On the other hand one will expect an essential change in the derivative of F_e with respect to density; this is proposed i.e. that there is a change from van der Waal's interaction to metallic cohesion.

Therefore (Figure 1) the curve of F_e as a function of volume is expected to have a pronounced change in slope in the vicinity of the metallic transition; the greater the cohesive energy of the metallic phase the more pronounced this effect will be.

In the simplified model presented here, a first order phase transition can be expected (Figure 1) if at the temperature in question F versus v has two points of common tangency. It then is clear that if the break in slope at the metallic transition is very pronounced it will very likely separate two points of common tangency.

However, the break in slope will not be infinitely sharp, and for sufficiently high

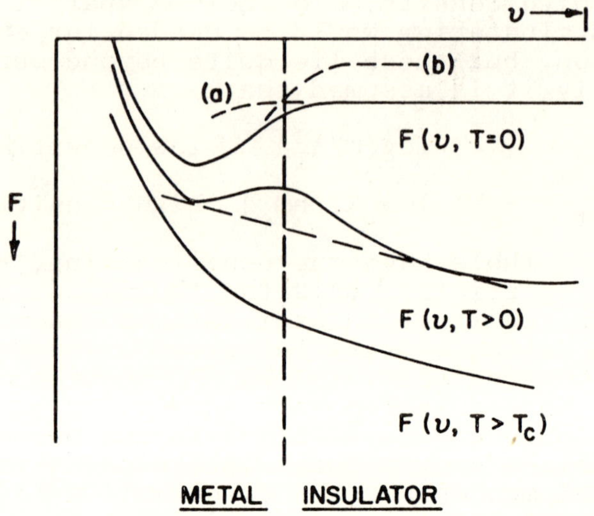

Fig. 1. Free energy versus atomic volume. For T = 0 the dotted curve (a) depicts van der Waal's cohesion and (b) depicts the Hartree-Fock energy of the metallic configuration, uncorrected for correlation.

temperatures it is possible to reach a critical temperature for which

$$\frac{\partial^2 F_t}{\partial v^2} > 0 > \left| \frac{\partial^2 F_e}{\partial v^2} (< 0) \right|$$

and the two phases will coalesce in the usual manner. Critical temperature and cohesive energy data have been reviewed in Gmelin (24). It is interesting to note how little experimental work has been done; most data are computed thermochemical extrapolations. The following comparison is interesting: cohesive energy vs. critical temperature -- Li, 36.5 Kcal.mol.$^{-1}$, $T_c \backsim 2500°C$; Na, 26.1 Kcal.mol.$^{-1}$, $T_c \backsim 2000°C$; Hg, 15 kcal. mol.$^{-1}$, $T_c = 1460°C$. The trend with cohesive energy is as suggested above.

An attempt was made for sodium to compute F_e using a Slater-Kirkwood van der Waal's energy (25) at large atomic separation, and using the Wigner-Seitz metallic model (26) with various exchange and correlation corrections [Brooks (27)] at smaller atomic separations. A polarizability of $8.8 \times 10^{-24} cm^3$ and van der Waals cohesive energy of $0.0532 (r_{so}/r_s)^6$ Ryd. were computed. Though the questionable validity of the methods at the crossover point was recognized a reasonable attempt was made to compute in some detail. The principal result was that in fact a pronounced flattening of the electronic energy per atom occurs at $r_s \simeq 2r_{so}$ i.e. 12.5% of the normal solid density (Figure 2). The balancing of spurious coulomb energy by correlation energy corrections was notable. The results are far too coarse to compute meaningful second derivatives.

Of course, we do not know what effect clustering will have on these arguments but must eventually settle such questions experimentally, or by extensive theoretical analysis (31,32).

This view may be summarized pictorially as in Figure 3, where a "constitution" diagram is plotted on the T, $\rho = v^{-1}$ plane. It indicates an electronic constitution change at some density ρ_{trans}; it also places a thermodynamic phase transition in the same density range around ρ_{trans}, with a critical point indicated at T_c and $\rho_c = \rho_{trans}$. This diagram is oversimplified certainly; there may be further phase changes in the liquid

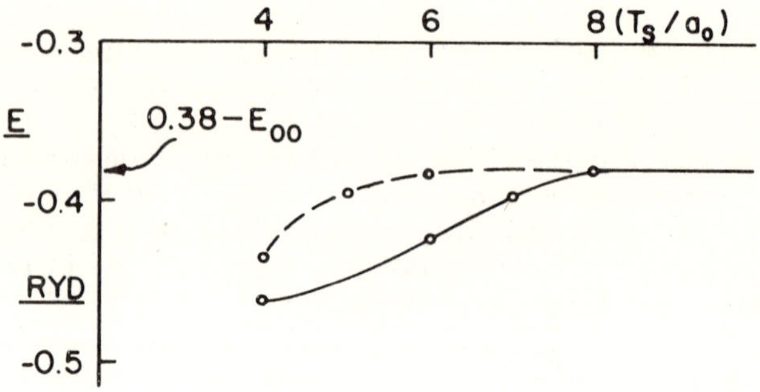

Fig. 2. Van der Waal's cohesive energy (dashed), and metallic cohesive energy (solid) in Rydbergs per atom, versus unit cell radius (in Bohr units). $E = E_o + (0.221/r_s^2) + (0.284/r_s) + .031 \log r_s - .115$.

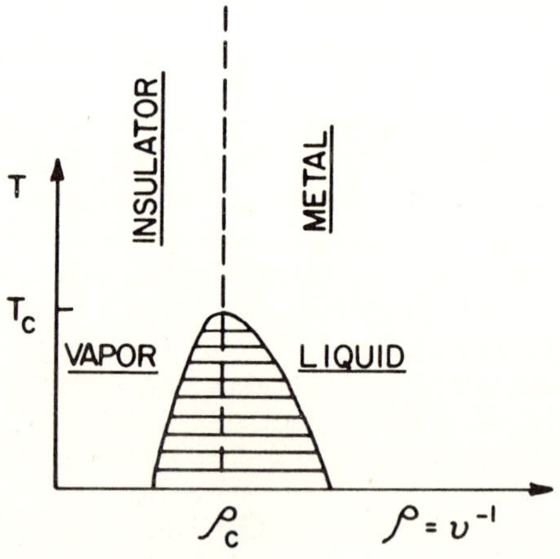

Fig. 3. Schematic constitution diagram plotted in temperature-density plane, T and ρ, respectively.

or solid region as density increases.

Turning to the experimental situation, it has been found on searching the literature that very little data pertinent to this type of transition has been taken. Experiments have been carried out by F. Birch (28) on Hg at temperatures and pressures sufficiently high to apparently see both the critical point and the onset of "metallic-vapor" conduction -- but barely, and the data are limited. Otherwise no directly relevant data have been found.

Accordingly we looked for model systems and have thereby been led to metal-ammonia solutions. These have been examined over many years by chemists, and during recent years many electrical, magnetic and optical methods have been employed to characterize their properties. An excellent summary is to be found in the reviews by G. Lepoutre and M. J. Sienko (29) and by T. P. Das (30). I am particularly grateful to M. J. Sienko for discussions and suggestions in this context. The literature of this subject is too vast to review here. The principal viewpoints relevant to the present discussion are: (1) Apparently the ammonia may act as a dielectric (liquid) into which relatively large amounts of sodium may be dissolved; (2) For very dilute solutions the sodium dissociates into sodium ions and solvated electrons, but for concentrations higher than a few tenths of a mole percent most of the electrons are associated with the sodium; (3) It is then found that there are two regimes -- a dilute low conductivity region at low to intermediate concentrations, going over into a metallic behavior as concentration is increased; (4) There is also, for sodium solutions below - 41.6°C, a phase separation into two liquids -- one being essentially metallic in properties and the other having very high resistivity. The critical concentration is 4.2 At.% Na.

Kraus and Lucasse have determined the phase diagram and some conductances. **(See Lepoutre and** Sienko (loc. cit.)). Much more recently S. Naiditch (loc. cit.) has studied the conductivity over a wide range of concentrations and temperatures. The data of Naiditch may be analyzed in a straightforward way to identify the concentration at which the conductivity changes character sharply to a metallic behavior. The phase diagrams of

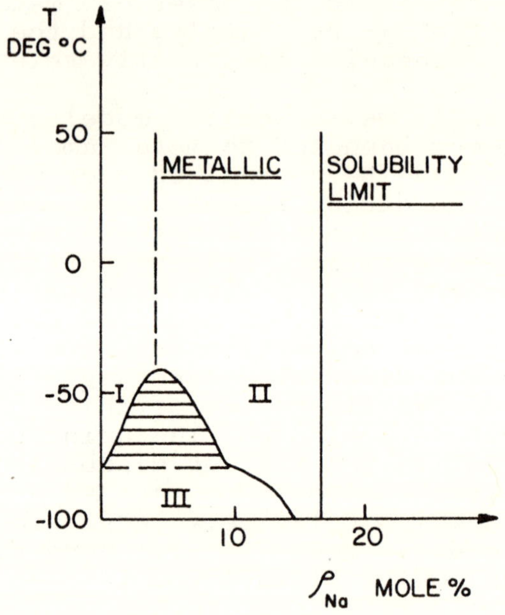

Fig. 4. Phase diagram for sodium-ammonia solutions, after Kraus et al. Region I is low conductivity liquid, region II is metallic like and liquid. Region III comprises solid NH_3 + liquid II. The conductivity boundary is found from S. Naiditch's data.

Kraus et al may then be combined to give the constitution plot shown in Figure 4. The relation of the indicated section to Figure 3 should be apparent.

Thus, we suggest that the metal ammonia system is a model system for the insulator-metal transition. In sodium-ammonia solutions, below -41.6°C there is a phase separation into an "insulating vapor" and a "metallic liquid metal," with the ammonia being simply a liquid dielectric "vacuum." Above this critical temperature the electronic transition occurs without a thermodynamic transition, but bears out the hypothesis that the density at which this occurs is essentially that at which the thermodynamic phase change occured below T_c.

It is a pleasure to acknowledge discussions with J. F. Colwell, S. W. Kurnick, and the hospitality of General Atomic during the summer of 1964.

REFERENCES:

1. Colwell, J. F., et al., The Effect of High Pressure on the Conductivity of Vapors, General Atomic report GA-5253, May 15, 1964.
2. Miller, P. H., Jr., J. H. Taylor, Bull Am. Phys. Soc. 1, 120.
3. Mott, N. F., Phil. Mag. 6, 287 (1961).
4. Kohn, W., Phys. Rev. 133, A171 (1964).
5. Hubbard, J., Proc. Roy. Soc. A276, 238 (1963); A281, 401 (1964).
6. Young, W. H., (private communication).
7. Kubo, R., "Some Aspects of the Statistical-Mechanical Theory of Irreversible Processes," Lectures in Theoretical Physics, Summer Institute for Theoretical Physics, University of Colorado, Boulder, 1958. W. E. Brittin and L. G. Dunham, eds. (Interscience, New York 1959), Vol. 1, pp. 120-203.
8. Kohn, W., and Luttinger, J. M., Phys. Rev. 108, 590 (1957).
9. Edwards, S. F., Phil. Mag. 3, 1020 (1958).
10. Chester, G. V., Reports Prog. Phys. 26, 411 (1963).
11. The concern of the present paper is with the constitution of the vapor considered as a "material." The thermoelectric properties of these vapors may be of potential interest in energy conversion systems; the following references (12), (13) are suggested.

12. Rosa, R. J., Phys. Fluids 4, 182 (1961).
13. Lin, S. C., Resler, E. L., and Kantrowitz, A., J. Appl. Phys. 26, 95 (1955).
14. Matsubara, T., and Toyozawa, Y., Progr. Theoret. Phys. (Japan) 26, 739 (1961).
15. Yamashita, J., and Kurosawa, T. J., J. Phys. Soc. Japan 15, 802 (1960).
16. Appel, J., "On the Small Polaron Motion in NiO" (to be published).
17. Loeb, L. B., Fundamental Processes of Electrical Discharges in Gases, (John Wiley and Sons, New York, 1939), p. 297, Table 19.
18. Branscomb, L. M., "Negative Ions," in Advances in Electronics and Electron Physics, L. Marton, ed. (Academic Press, New York, 1957) Vol. 9, p. 46, Fig. 1.
19. Honig, R. E., RCA Review 23, 567 (1962).
20. Chester, G. V., and Thellung, A., Proc. Phys. Soc. (London) 77, 1005 (1961).
21. Langer, J. S., Phys. Rev. 127, 5 (1962); ibid. 128, 110 (1962).
22. Herring, C., "The Current State of Transport Theory," in Proc. Int. Conf. Semi-Conductor Phys., 4th Prague (1960), (Academic Press, New York, 1961), p. 60.
23. Holstein, T., Ann. Phys. (N.Y.) 8, 325 (1959).
24. "Gmelin's Handbuch d. anorganischen Chemie," Verlag Chemie, G.m.b. H Berlin, Volumes 20 (Li), 21 (Na), 34 (Hg).
25. Slater, J. C., and Kirkwood, J., Phys. Rev. 37, 682 (1931).
26. Wigner, E., and Seitz, F., Phys. Rev. 43, 804 (1933); 46, 509 (1934).
27. Brooks, H., Nuovo Cimento (Suppl. ser. 10) 7, 165 (1958).
28. Birch, F., Phys. Rev. 41, 641 (1935).
29. "Solutions Metal-Ammoniac," Memoires et Travaux Faculte Catholique Lille, G. Lepoutre and M. J. Sienko, ed., W. A. Benjamin (1964).
30. Das, T. P., Advances in Chemical Physics IV, Interscience Publishers (1962).

31. Preliminary results have been reported by T. Arai, Bull. Am. Phys. Soc. 10, 302 (1965) who also suggests the phase transition in a manner similar to that proposed here.

32. Sundfors, R. K., and Holcomb, D. F., Phys. Rev. 136, A810 (1964); using resonance techniques the apparent onset of the metallic transition for acceptor states has been seen in heavily doped Si:P. This method has the advantage of being potentially capable of studying spatial distributions, though the "metallic" species is fixed in the lattice in doped solids.

Structural Transitions in Solids at High Pressures

John C. Jamieson
University of Chicago
Department of the Geophysical Sciences
Chicago 37, Illinois

Abstract

The boron annulus technique for X-ray diffraction studies in the 150 kilobar range has been modified to allow temperature variation of the experiments. Using liquid nitrogen as a coolant, temperatures at least as low as -100°C may be reached on pressurized samples. Both internal and external heating elements have been added in a second version. The external heating permits controlled temperatures to at least 300°C. The internal heating arrangement shows promise of a much greater temperature range. Its capabilities are under study.

The transition of the B1(NaCl type) to B2 (CsCl type) with increasing pressure is of particular theoretical interest. Using the heating version of the high pressure X-ray apparatus, an intensive study is under way to verify this transition in the sodium halides as reported by other workers to occur at quite low pressures. No verification has been achieved as yet. A study of a 60% Na-40% KCl mix-crystal shows its transition to the B2 type to occur at about 110 kbar, much higher than could be reasonably expected on thermodynamic grounds if both of the pure end members NaCl and KCl transform circa 20 kbar.

Introduction

This article has two purposes. One is to bring up to date earlier reports (1) (2) (3) on apparatus for X-ray diffraction studies on materials under pressures over 150 kilobars, in particular the addition of temperature as a variable. The second purpose is to present preliminary results on the alkali halides NaF and a NaCl-KCl mixed crystal series. These results suggest very strongly that the <u>equilibrium</u> transition in NaF and NaCl to the CsCl type structure occurs actually at quite high pressures (above 200 kilobar) rather than the much lower figures of other recent workers.

Techniques

The technique used at present in these high pressure studies is based on that developed in conjunction with A. W. Lawson. (1) It consists of a .015" diameter sample about .010" high surrounded with "amorphous" boron trapped between carbide flats. A collimated X-ray beam is passed parallel to the flats and the diffracted beams are recorded on film. The pressure distribution within the approximately 1/8" diameter boron annulus is peaked at the central sample. Pressures somewhat greater than 150 kbars may be reached on most materials and their X-ray powder diffraction pattern obtained. Filtered Mo target radiation is used and exposure times are comparable to those of ordinary powder cameras. Since the diffraction pattern of a material determines its volume, pressures are obtained by consulting known compressibility data or by mixing a substance of known compressibility with the sample. The overall load on the sample is generated by a small, low tonnage ram adjusted to give a maximum 50 kbar average load across an 1/8" diameter face. The camera diameter is 114.6 mm. Values of 2θ as high as 60° may be measured corresponding to d-spacings of 0.7A. Typical diffraction patterns consist of ten to twenty lines which may be used to solve the new structures of many materials. Further detailed description of this technique has been published elsewhere (2) (3).

This previous work has all been performed at room temperature. However the great advantages of being able to vary the temperature while maintaining such high pressures are obvious and over a period of time several quasi-abortive attempts have been made in this direction, both to reach high temperatures and low temperatures. Initially, it was felt that either heating or cooling the sample assembly should be done by heat conduction so as to avoid possible interference with the X-ray diffraction patterns. For low temperature work, a liquid N_2 system consisting of an insulated can and various other insulators and heat conducting paths was easily assembled. Unfortunately, due to the large thermal mass of the apparatus, temperatures of only $-40^{o}C$ were reached. In addition, there was the formation of ice on the boron annulus which interfered grossly with the transmission of the X-rays. This latter could, of course, be prevented in a number of ways but the ineffectual temperature achieved made the whole system unattractive. Similar (but converse) difficulties appeared when heating the system was tried. The direct method which consisted of wrapping a small furnace around one carboloy piston did reach $300^{o}C$. The comparative lowness of this temperature was again due to the thermal mass of the apparatus.

The actual solutions to both temperature problems were essentially the same. If the sample itself can be heated or cooled directly, the effect of the surrounding apparatus is minimized. With this in mind, actually comparatively minor alterations in the apparatus permit experiments at greater than 100,000 atmospheres pressure to be conducted at both quite low and quite high temperatures. As to low temperatures, it has been found possible to re-design the upper piston assembly so that liquid N_2 may be circulated within 3/4 inch of the sample. This device is shown in Figure 1. A hollow collar has been made containing a cavity and guide ring for the upper piston. Liquid N_2 is forced by tank gas pressure to circulate through the assembly. Even with this arrangement the sample is 3/4 inch ($\frac{1}{2}$ inch carbide + $\frac{1}{4}$ inch stainless steel) away from the coolant. This affects the ultimate temperatures achieved. Calibration experiments with a thermocouple at the position of the sample showed that about -100°C was the lower limit of the current one-sided cooling. At this time of writing a second such collar is being machined for the lower piston assembly. It is anticipated that the combination of two such collars with the sample at the mid-point should allow temperatures very close to those of liquid N_2 to be achieved. The ice formation problem will be avoided by a stream of dry gaseous cold N_2 fluxing the exterior of the boron annulus, an already known technique in low temperature X-ray spectroscopy.

With the now added interest in temperature variation, techniques for measurement of sample temperature had to be evolved. A thermocouple was placed at the bottom of a 1/16 inch diameter well drilled in the Carboloy piston to a depth of approximately $\frac{1}{4}$ inch from its working face. Such wells affected in no way the stress behavior of the Carboloy. In such runs with a second thermocouple at the sample position, the reading of the two couples were compared for a calibration. Such a support (for the camera) and lower piston assembly with piston and guide ring removed is shown also in Fig. 1. Although any thermocouple materials can be used, the current general practice is to use Chromel-Alumel P due to its flexibility for temperatures both above and below room temperature.

The drilling of holes in carbide proved fairly expensive. In casting around for alternate solutions, a better arrangement was discovered which not only solved the drilling problem but effected a five-fold financial savings in the cost of routine pistons. The discovery was that composite pistons functioned just as well as ones made completely of carbide. Such an arrangement is illustrated in Figure 2. The body of the piston is now stainless steel drilled for a

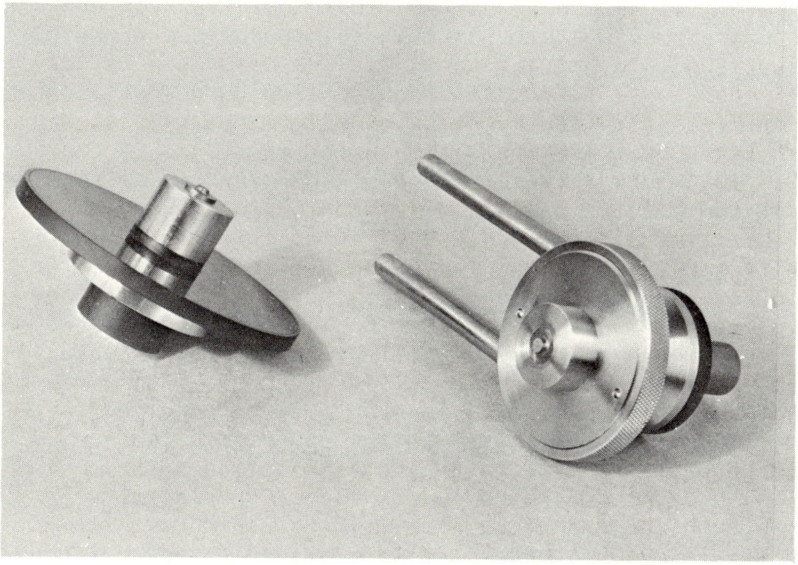

Figure 1 - Low Temperature Adaptions to High Pressure X-ray Camera.

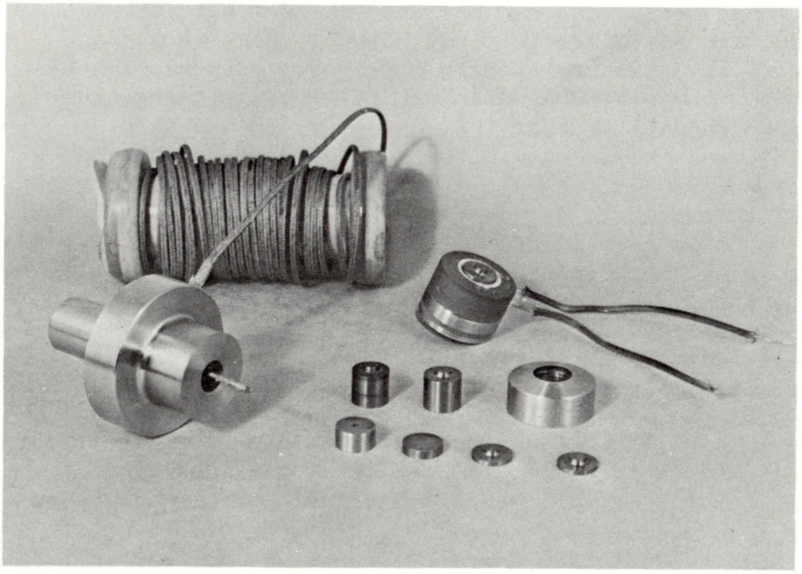

Figure 2 - Furnace, Composite Pistons and Thermocouple Arrangement.

thermocouple. It is capped by two 1/8 inch thick x ½ inch diameter ground "disposable inserts" (4) on the upper one of which a working face has been ground. The upper cap is destroyed in each run. However this destruction occurs only on release of pressure. Series of exposures on the same sample have been taken over a month's time using this type of cap.

Also in Figure 2 is shown one of the nichrome wire furnace piston holders capable of operation to 300°C. When in use, the film holder is cooled by a small fan to protect the film from over-heating. Two other different methods for high temperature-high pressure studies have been evolved and proto-typed experiments performed. Both methods depend on internal heating in an opposed-anvil geometry; a technique which has very recently been described by Bundy. (5) In one method the .015" sample is encased in a thin walled cylinder of pyrolytic graphite, in the other powdered pyrolytic graphite is mixed with the sample (and boron). In both cases, heat is generated by passage of an electrical current through the graphite. The major break-through came when it was discovered that in neither case did the addition of the graphite have a deleterious effect on the diffraction patterns, either at one atmosphere or at high pressures. The graphite sleeves are electrically more satisfactory and reproducible than the admixed powders, but they also are much more tedious and time consuming to manufacture on a jewelers lathe. The admixed powders were easy to form but somewhat irreproducible as to electric resistance. For the purpose of testing, several samples were made and pressurized until one of a few ohms (at pressure) was obtained. To test the temperature capabilities of the apparatus, the phase diagram for Sn was chosen due to the control afforded by the work of J. D. Barnett, et al (6). A drilled carboloy piston served as thermocouple well. Temperatures of at least 350°C were recorded. There were confirmed by "after" pictures of the Sn which showed it to be coarsely crystalline almost certainly indicative of melting. No pictures at high pressures and temperatures were taken in this particular experiment due to excessive heating of the camera and press assembly. This can be avoided by cooling coils in later work.

To summarize this section briefly, there now seems to be no insurmountable physical objection to the addition of the full capabilities of internal heating in a high pressure system. This means that studies may be made at coexisting temperatures above 1000°C and pressure above 100 kbars. Possible chemical objections such as reaction of the sample with carbon, boron, or carboloy or each with each other have still to be evaluated. Tests have been made which show that the "amorphous" boron annulus may be replaced with carbon thus eliminating one variable. It may be necessary to replace carbide anvils with a ceramic.

Studies on Alkali Halides 1. NaF

Few studies have been made of the volume behavior of NaF under compression. Bridgman (7) studied its compressibility to 12 kilobars while much more recently Pistorious and Admiraal (8) report a sluggish phase transition at room temperature and 16.8 kbar which they suggest is the appearance of the B2-CsCl type. At room pressure NaF is the B1-NaCl type. Bridgman (9) had earlier searched to 50 kbar at 25°C, 125°C, 150°C and 175°C for this type of transition and came to the conclusion that no such large volume change transition existed in his range.

Due to its low absorption coefficient, conditions were ideal for X-ray studies of NaF. To obtain maximum clarity on the films of diffraction patterns, it was desirable to take pictures of undiluted NaF in which pressures were determined by a knowledge of the compressibility of NaF. This was determined using NaCl as an internal pressure standard. Mixtures of NaF "Baker's Analysed" grade and NaCl, Mallinckrodt's "Analytical Reagent" were made into samples and exposed to varying pressures. The films thus obtained were excellent, in general hkl spacings to 420 were measurable for both phases. The cubic a spacing was calculated for each line (discarding the 200 spacing) and averaged. The camera was calibrated to the one atmosphere values of a_o = 5.6402 Å for NaCl (9) and a_o = 4.6344 Å for NaF (10). The results are presented in Figure 3 together with a straight line representing Bridgman's (7) volume data converted to the lattice parameter a for each salt. The conversion to the volume of NaF as a function of pressure may be made easily by consulting any of several sources (11) (12) (13) for the compression of NaCl. When this is done the maximum pressure of the data in Figure 3 is about 120 kbar, the minimum is about 33 kbar. The gap between the current data and that of Bridgman precludes any statement one way or the other as to the existence of a discontinuity in the plot corresponding to a phase transition there. The sample for one segment of this experiment was pressurized (at room temperature) for four weeks at varying pressures of 50 kbar to 120 kbar with no alteration in its pattern. For all pressures the NaCl-type cubic indexing for each phase sufficed to calculate a with no gross deviations of its value from line to line.

One long term experiment on undiluted NaF was conducted at room temperature. A sample was held at 60 kbar with occasional film exposures for two weeks. No alteration could

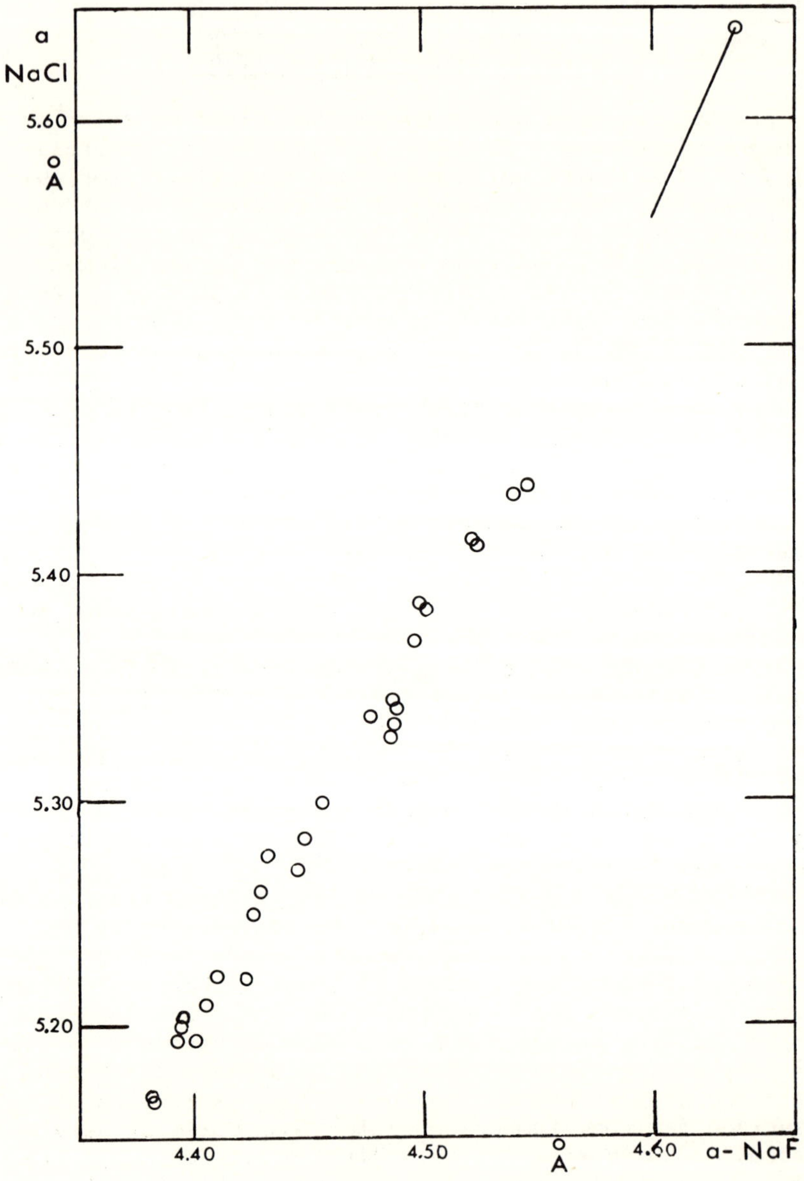

Figure 3 - Coexisting Cubic Lattice Parameter a of NaF and NaCl.

be perceived in its diffraction pattern. Without release of pressure, the same sample was held for an additional 53 days. A film at the end of that period showed only normal NaF at a pressure of 30 kbar. The lower pressure was due to a slight oil leak in the ram over a period of time.

Further experiments were conducted at several different elevated temperatures. Pressures were typically about 100 kbar. Alternate exposures were made at room temperature and at temperatures of about 140°C and about 185°C using a furnace around the upper piston. Temperature exposures were six hours in length. No change occurred in either the high temperature or "after" patterns. Yet this is precisely the temperature region (157.5°C) in which Pistorious and Admiraal report that the transition is clear but exceedingly sluggish and incomplete at 45 kbar.

Studies on Alkali Halides 2. NaCl

The case for the transition B1-NaCl type to B2-CsCl type occurring at low (16 kbar range) pressures is much stronger for NaCl than for NaF. Not only has a transition been reported as occurring in piston cylinder apparatus (14) but in X-ray diffraction studies, Evdokimova and Vereshchagin (15) had earlier assigned the B2-CsCl structure to the high pressure form. Although on the basis of piston-cylinder experiments an assignment of crystal structure may be regarded only as a suggestion, it is difficult to ignore the X-ray evidence. Unfortunately several other workers (16) (17) (13) have failed to confirm the transition using their own diffraction techniques. The writer (3) (18) has used NaCl as an internal pressure standard for several years without observing any transitional effects at least to 160-170 kbar. As to other techniques, there is a suggestion in the shock wave studies by Christian (11) which has been amplified by Alder (19) that the CsCl structure occurs in NaCl at pressures above at least 250 kbar. The theoretical approach has been thoroughly reviewed recently by Tosi (20). Although various attempts to calculate the B1-B2 equilibrium pressures for the alkali halides have been made in the past, the quantitative aspects of the calculations have been disappointing. Probably the best that can be said is that the experimental trend of increasing pressure of transition in going from rubidium halides to potassium halides is qualitatively justified by any of the theoretical approaches and hence one should find the transition in sodium halides at higher pressures than the potassium halide transitions. It is of interest to compare energetically a "low pressure" transition in NaCl at say 18 kbar and $\Delta V/V_0 = -14\%$ with a "high pressure" transition at say 250 kbar and $\Delta V/V_0 = -5\%$. Neglecting compressibility differences, the former leads to the NaCl type being stable by 1.6 kcal at one atmosphere, the latter (in which a distinct effort to underestimate has been made) makes the NaCl type stable by 8.1 kcal. Considering that lattice energies of NaCl (as NaCl type) are the order of 182 kcal, a "high pressure" transition in NaCl seems much more amenable to theoretical analysis.

An initial effort to settle this question was made in analogous fashion to the studies on NaF above. Indeed the compressibility data presented on that salt also serves as a long term (four weeks) study on NaCl in which no changes occurred. Even more drastic heat treatment was accorded a mixture of NaCl and graphite pressurized to 110 kbar. Pictures taken first for $5\frac{1}{2}$ hours while 48 watts were dissipated at the sample showed no extra reflections. In a further study of the same sample 120 watts were fed into the system. An exposure of $5\frac{1}{2}$ hours duration during this process showed no alteration in the NaCl structure as did pictures taken of the sample afterward both at high pressure and one atmosphere. It is not possible at present to know the actual sample temperatures of these experiments, nor can the power input be used to make a meaningful estimate since the distribution of power dissipation along the entire heating path is not known. But the Carboloy pistons were badly corroded in a small area corresponding to the sample position so it is likely that several hundred of degrees were achieved. This approach would have been pursued further had it not occurred to the writer that there was a flaw in the logic of this type of study. The most that a non-observance of a given transition can prove is that the observer failed to achieve it in his apparatus with his technique, which may be sufficiently different from that of the original observer as to suppress the transition. The correct indirect approach thus becomes apparent. It is to start in the systems where the B1-B2 transition is well known and then trace it using mixed crystals toward the sodium halides. KCl forms a complete solid solution with NaCl at elevated temperatures while the B1-B2 transition in KCl is well known and reproducible at high pressure and room temperature. Hence, this system was chosen for this approach.

The actual phase diagram for (Na,K)Cl consists (21) (22) of a liquidus-solvus with minimum at about 660°C and 50% Na and an asymmetrical unmixing curve with maximum at about 500°C and 64% Na. Except for very minor substitution on both sides of the phase diagram these mixed salts are quite unstable and unmix readily in the presence of the ordinary water vapor in the atmosphere. Mixed salts were prepared by weighing mixtures of the pure members at 10% atomic weight intervals (including 5% and 95% Na) then fusing in platinum crucibles. Each sample was annealed at 650°C for two hours and air quenched to room temperature. A portion of each sample was placed in an agate mortar under silicone oil (Dow-Corning DC 200) and ground. The resulting mixtures was smeared onto a glass slide for analysis on a General Electric XRD-5 Diffractometer. Except for minor substitution ($\leq$ 10% Na and $\geq$ 90%) the freshly ground material unmixed in a few hours to almost pure end members. The unground material persisted a day or two under silicone oil. To obtain the lattice parameters of the mix-crystals while they were unmixing, a routine was evolved in which the region 27°-32° in 2θ (filtered Cu radiation) was scanned at 2°/min. This region contains the 200 diffraction peaks of (Na,K)Cl crystals, then the region 66°-75° in 2θ, again at 2°/min to locate the 420 peak (s). When located, a slow scan across a 1° region containing the given 420 peak at 1°/5 min was made for a precision determination of the lattice parameter. A final scan at 2°/min in the 27°-32° region completed the routine. Some two weeks later the same slides were re-scanned in the 27°-32° region. Unmixing was complete except as above for the fringes of the phase diagram. The relative intensities now for the NaCl and KCl 200 peaks gave the necessary correction for the differing scattering power of Na and K. This routine was evolved to correct the lattice parameter of each composition for possible differential decomposition. In the region of main interest, 20%-80% Na, this proved not to occur. When a sample was unmixing the ratio of the intensities of the 200 peaks of NaCl and KCl remained constant while the peaks individually grew and the 200 peak of the mix-crystal declined to zero. The lattice parameters obtained in this fashion were very useful for sample control, but intrinsically lower in absolute accuracy than those reported by Barrett and Wallace (22), hence no report will be made of their values at this time. They were in fair agreement with both (21) (22) of the earlier studies and confirmed completely that the volume increased on forming a mix-crystal from the pure end-members.

In the usual routine for high pressure X-ray studies, a one-atmosphere picture of about six hours duration is used to calibrate the camera and correct for several errors (3). This, unfortunately, provided sufficient time for the freshly ground material to unmix. Several attempts to load unground material or material ground under benzene produced, on application of pressure, only patterns characteristic of NaCl in the NaCl type coexisting with KCl in both the NaCl type and CsCl type. Further increase in the pressure converted all the KCl to the latter type. Even this provided a crumb of information since in bulk mixtures as diverse as 20% NaCl +

80% KCl and 80% NaCl + 20% KCl, the KCl invariably transformed readily while at the same pressures, NaCl did not to the limits of detection (about 5% of a given phase). It is possible that the end members resulting from decomposition prior to pressure application contained a few percent of the alternate cation. This would be impossible to detect under pressure.

When the above difficulties were sufficiently appreciated, success in pressurizing solid solutions of the (Na,K)Cl salts was achieved. An annealed (650°C) sample was air-quenched, ground in air and loaded into a boron annulus in the usual fashion (3) and pressurized immediately. A diffractometer trace of this material showed it to be approximately half the 60% Na-in-KCl salt and the remainder ~ 60% NaCl + ~ 40% KCl. A second sample of this 60% Na-in-KCl was prepared by grinding under DC 200 silicone oil and loading into a polystyrene annulus (23). This material was about 80% of the 60% Na-in-KCl plus end members. Again this was pressurized immediately. In both cases the initial independent camera calibration had to be abandoned and a slightly rougher one used obtained from experience on other samples of the same absorption coefficient. The data collected on both samples are presented in Table 1 where it can be seen that the B1-B2 transition has been achieved and reversed at pressure for each. Compressions are given for each phase, followed in the case of NaCl and KCl by pressures estimated from the volume data of Christian (11) for NaCl and Percy-Albuerne and Drickamer (13) for KCl. A figure of 100 kbar seems reasonable for the transition. Pending more detailed studies a liberal error of ± 20 kbar seems justified at present.

A simple model suffices to test the compatibility of the same B1-B2 transition occurring at 20 kbar in KCl, 100 kbar in $(Na_{.6}K_{.4})Cl$ and 18 kbar in NaCl. Although the unmixing curve for (Na,K)Cl in the B1 structure is known (21) (22) to be asymmetic, its asymmetry is comparatively slight. In addition, the unmixing curve for (Na,K)Cl in the B2 structure is unknown. Hence, it will be assumed that both structure types obey a regular solution theory model (24) with the Gibb's Free Energy of a composition x (mole fraction of NaCl) given by

$$G_i(P,T,x) = xG_{iN}(P,T) + (1-x)\,G_{iK}(P,T) + RT\{x\ln x + (1-x)\ln(1-x)\} + w_i x(1-x) \quad (1)$$

in which i=1 corresponds to the B1 structure type and i=2 to the B2 type. Standard derivations lead to

$$w_i = 2RT_{ci} \quad (2)$$

for the critical temperature (at x = .5 in this model) and bringing in the needed molal volumes

Table 1 - Transitions and Compressions in $(Na_{.6}K_{.4})Cl$

Sample 303 - $\frac{1}{2}(Na_{.6}K_{.4})Cl + \frac{1}{4}$ KCl

Load	60%-B1	NaCl	KCl-B2	60%-B2
3 tons	.78	.80/100 kb	.69/90 kb	Not Obs.
4	.76	.78/115 kb	.67/110 kb	.71
5	.73	.77/120 kb	.66/120 kb	.69
6	.73	.76/125 kb	.66/120 kb	.68
3	.77	.77/120 kb	.67/110 kb	.71
1	.84	.84/62 kb	.72/50 kb	Not Obs.
1	.87	.87/48 kb	.73/50 kb	" "
6	.84	.84/62 kb	.72/60 kb	" "
8	.81	.83/70 kb	.70/80 kb	" "

Sample 304 - $8/10(Na_{.6}K_{.4})Cl$ + 1/10 NaCl + 1/10 KCl

Load	60%-B1	NaCl	KCl-B2	60%-B2
4	.74	Not Obs.	.66/120 kb	Not Obs.
2	.93	" "	.81/20 kb	" "
6	.76	.81/82 kb	.69/90 kb	" "
7	.76	.78/115 kb	.67/110 kb	" "
8	.76	.79/108 kb	.68/100 kb	.71
7	.74	.78/115 kb	.67/110 kb	Not Obs.
6	.77	.80/100 kb	.67/110 kb	" "

Where pressures are quoted for compressions, they are obtained from references (11) and (13).

$$\left(\frac{\partial G_i}{\partial P}\right)_{T,x} = x\, V_{iN}(P,T)+(1-x)V_{iK}(P,T)+ x(1-x) \left(\frac{\partial w_i}{\partial P}\right)_{T,x} \quad (3)$$

now the molal volume of any composition can be written

$$V_i(P,T,x)=xV_{iN}(P,T) + (1-x)V_{iK}(P,T)+x(1-x)\, V_{mi}(P, T, x)\ . \quad (4)$$

From eq. (3) and eq. (4) it follows that

$$(\partial w_i / \partial P)_{T,x} = V_{mi}(P, T) \quad (5)$$

In the term on the right side of eq. (5), V_{mi} has been made independent of x to agree with the regular solution hypothesis that w_i is so. A fit of the volume data of mixing of Barrett and Wallace reveals that this is not a bad approximation. From their data on mixing in the B1 type, the 50% composition deviates .41 cm^3/mole from a linear volume dependence . Hence $V_{mi} \cong 1.6\ cm^3$/mole. From eq. (5) then w_i increases with pressure at a rate of about 40 cal/kbar initially. This favors unmixing in the B1 system at higher pressure yet diffusion rates should be depressed. The experimental fact that the salts unmix readily at one atmosphere and yet apparently not at all at high pressure (cf the data on the 60% Na phase) suggests that diffusion control is dominant. Hence the B1-B2 transition for a composition x will be considered as occurring at unchanged composition and for an equality of Gibb's Free Energy in the two phases. Assuming KCl and NaCl to transform at the same 20 kbar pressure, using eq. (1) for each phase and $\Delta G = G_1 - G_2$,etc.

$$\Delta G(20,T,x) = w(20) \times (1-x) \quad (6)$$

Since an iso-composition B1-B2 transition seems called for, $(\partial\Delta G/\partial P)_{T,x} = \Delta V(P,T,x)$. If P_x is the equilibrium pressure of transition of composition x, then the integrated approximation is

$$-\Delta G(20,T,x) = \Delta V(P,T,x)\ (P_x-20) = \Delta w(20)x(1-x) \quad (7)$$

Setting x = .6, P_x = 100 kbar and using the measured .05 volume decrement (Table 1) combined with Barrett and Wallaces (calculated) molal volume for 60% Na it appears that $.24\Delta w(20)$ $= 1.6\ cm^3 \times 80$ kbar = 12.7 kcal or since $w_1 \cong 3.5$ kcal from eq. (2), $w_2 \cong 16.2$ kcal or the critical temperature for NaCl-KCl unmixing is about 4000°K. A large w such as this usually is caused by a gross difference in size of the atomic radius. This is paradoxical since a gross difference in atomic radius would favor the NaCl type structure over the CsCl. This, therefore, seems highly unreasonable and suggests that the hypothesis $\Delta G_N(20,T)=0$ is wrong.

Now if the B1 and B2 types have comparable values of w_i, then approximately

$$\Delta G(P,T,x) = x\,\Delta G_N(P,T) + (1-x)\Delta G_K(P,T) \quad (8)$$

$$\Delta G(20,T,\ .6) = .6\,\Delta G_N(20,T) \tag{9}$$

and for P_e, the equilibrium pressure of the B1-B2 transition in pure NaCl

$$-\Delta G_N(20) \cong (P_e - 20)\Delta V_N \tag{10}$$

$$\cong 1.6\ \Delta G(20,T,\ .6) \tag{11}$$

$$\cong 1.6 \times 80\ \text{kb} \times 1.6\ \text{cm}^3 \tag{12}$$

If the volume change at transition is set equal to the molal volume of NaCl times a fraction δ relations (10)-(12) give

$$P_e = 20\ \text{kb} + (7.9/\delta)\ \text{kb} \tag{13}$$

For a high pressure say 300 kb of transition in NaCl, $\delta \cong .03$. Considering that similarly δ for KCl is about .11 and δ for $(Na_{.6}K_{.4})Cl$ is .05, this seems quite reasonable.

The above calculations are rough and really serve only to point the way to future research. Even at this stage, it seems to the writer that the 18 kbar transition in NaCl must be regarded as either due to other effects or non-equilibrium phenomena. If the notion of a high pressure for the NaCl transition is accepted, then the reported "transition" in NaF is certainly not of the B1-B2 type.

Acknowledgments

The writer wishes to thank Professor H. E. Drickamer for many helpful communications during this work. Professor R. C. Newton provided stimulating discussion. Mr. H. Zahn was very helpful in construction and design of apparatus while Mr. D. Dandekar was of great assistance in the conduction of experiments.

This Research was supported by The National Science Foundation Grant GP 3551 and The Petroleum Research Fund of the American Chemical Society Grant No. 12 PRF - 1408 - C6.

References

1. J. C. Jamieson and A. W. Lawson, J. Appl. Phys. 33, 776 (1962).
2. J. C. Jamieson, Science 139, 762 (1963).
3. J. C. Jamieson, in "Metallurgy at High Pressures and High Temperatures", M. T. Hepworth, N.Q.D. Pardee, K. Gschneidner. Jr., Eds. (Gordon and Breach, New York, 1964).

4. Commerically available in Grade 999 Carboloy from General Electric Metallurgical Products Department, Detroit, Michigan.
5. F. P. Bundy, Science 146, 1673 (1964).
6. J. D. Barnett, R. B. Bennion and H. T. Hall, Science 141, 1041 (1963).
7. P. W. Bridgman as corrected in Handbook of Physical Constants, Eds. F. Birch, J. F. Schairer and H. C. Spices, Geol. Soc. America Spec. Paper 36(1942).
8. C. W. F. Pistorius and L. J. Admiraal, Nature 201, 1321 (1964).
9. H. E. Swanson and R. K. Fuyat, Natl. Bur. Stds. Circ. 539 vol. II (1953).
10. H. E. Swanson and E. Tatge, Nat. Bur. Stds. Circ. 539 vol. I (1953).
11. R. H. Christian, UCRL-4900 University of California Radiation Laboratory, Livermore (1957).
12. D. L. Decker, J. Appl. Phys., 36, 157 (1965).
13. E. A. Percy-Albuerne and H. G. Drickamer, Private Communication (1965).
14. C. W. F. T. Pistorius, J. Phys. Chem. Solids, 25, 1477 (1964).
15. V. V. Evdokimova and L. F. Vereshchagin Soviet Physics J.E.T.P. 16, 855 (1963).
16. G.J. Piermarini and C. E. Weir, Private Communication reported by Pistorius (1964).
17. W. A. Bassett and T. Takahaski, Paper 64-WA/PT-24, Winter Annual Meetings A.S.M.E. New York (1964).
18. J. C. Jamieson, p 389 Panel Discussion in "High-Pressure Measurement 1962" Ed. A. A. Giardini and E. C. Lloyd, Butterworths, Washington (1963).
19. B. J. Alder, in "Solids Under Pressure" eds. W. Paul and D. M. Warschauer, McGraw-Hill Book Company Inc., New York (1963).
20. M. P. Tosi in Solid State Physics vol. 16, Eds. F. Seitz and D. Tumbull, Academic Press, New York (1965).
21. A. J. H. Bunk and G. W. Tichelaar, Konink. Ned. Ahad. van Weten ser. B. vol. 56, 375 (1953).
22. W. T. Barrett and W. E. Wallace J. Amer. Chem. Soc. 76, 366 (1954).
23. J. C. Jamieson, Science 145, 572 (1964).
24. E. A. Guggenheim, "Mixtures" Oxford University Press London, (1952).

DISCUSSION

By D. B. Larson

Recently I have observed shock-induced phase-transformations in bismuth, KCl, KBr, and NaCl. In these experiments I have used a quartz pressure gauge* to record the multiple wave structure produced in the sample by the phase transformation. The following transformation pressures have been obtained from the data:

Bismuth	(cast)	26.8 ± 1.1 kbs at 25°C
KCl	(single crystal)	20.0 ± 0.8 kbs at 25°C
KBr	(single crystal)	18.5 ± 0.8 kbs at 25°C
NaCl	(single crystal)	29 kbs
NaCl	(pressed powder)	23 kbs

*R. A. Graham, F. W. Neilson, W. B. Benedick, "Piezoelectric Current from Shock-Loaded Quartz - A Submicrosecond Stress Gage", J. Appl. Phys., to be published.

X-RAY STUDIES OF SOLIDS AT HIGH PRESSURE

L. F. Vereschagin
Institute of High Pressure Physics
Moscow, USSR

In order to understand solids under high pressure, we at our Institute feel that it is necessary to have as much information as possible. First of all, we are very much interested in the influence of pressure on the shape of the Fermi surface. Recently at our laboratories in Moscow we have made such measurements on graphite, zinc, and cadmium. We have also investigated several questions on the behavior of the mechanical properties of metals, as well as electrical conductivity, magnetoresistance and optical properties at pressures up to 10 kbar.

Any measurement of the physical properties of solids at high pressure requires first a knowledge of the compressibility as a function of pressure. In general we have an increase or decrease of some physical property which depends on the decrease of the distance between atoms. In several cases, for example in the case of metallic cerium, we have the appearance of a new crystalline structure at high pressure. We have approximated what was first established by Drs. Lawson and Ting-Yuan Tang (1) by x-ray methods. In each case we must know how the interatomic distance decreases with pressure in both the old and new crystal lattices.

In our laboratories at the Institute of High Pressure Physics in Moscow, we worked out several techniques and used as well several very interesting techniques which were developed by Dr. Jamieson and his colleagues. In the main we used two techniques. One used is a well-known method which was worked out by Dr. Lawson and the second is the new technique which was developed by Dr. Jamieson. The first technique requires the use of metallic beryllium as a high pressure vessel. This vessel, which we supported by two heavy steel rings, now reaches a pressure of about 20 kbar.

We have some measurements of electrical conductivity at high pressure which may give useful information about new polymorphic transitions. Such a pressure induced transition between electron shells is indicated by a break in the smooth curve of resistivity versus pressure. Fig. 1 shows the resistivity of lanthanum of different purities (2). The pressure scale is that of Bridgman, not of Kennedy and La Mori. For lanthanum, the break in the curve is seen to occur at a pressure of about 110 kbar. Fig. 2 shows the resistivity data for neodymium (2), indicating a transition near 90 kbar. Fig. 3 for cerium (2) shows the abnormal behavior of electrical resistivity versus pressure.

It was also very interesting to investigate the velocity of sound as a function of pressure. We developed special techniques for carrying out studies up to 10 kbar. The sample was placed in a steel vessel. Two quartz transducers were used; one as a vibrator and the other as a receiver. Fig. 4 shows that up to 10 kbar we have an increase in the velocity of sound for beryllium, molybdenum and iron.

It was interesting to note that in contrast with other metals which we investigated, cerium exhibited regions of normal behavior as well as very strange behavior of several elastic moduli. Fig. 5 shows the pressure dependence of several elastic moduli of cerium both above and below the transition pressure. It appears that the atoms have some sense of the approaching polymorphic transition with increasing pressure. Before the transition point is reached the compressibility behaves abnormally; the behavior becomes normal after the polymorphic transition. It was also observed that for increasing pressure, the Debye temperature passes through a broad maximum and then decreases to the transition point. Thereafter, the Debye temperature increased normally as for other metals. Fig. 5, curve E, shows the behavior of the Young's modulus of cerium under high pressure. We may conclude that at the approach of the polymorphic transition there is a weakening of the bonds in the low-pressure crystalline form and a strengthening of the bonds in the high pressure form. However, in the case

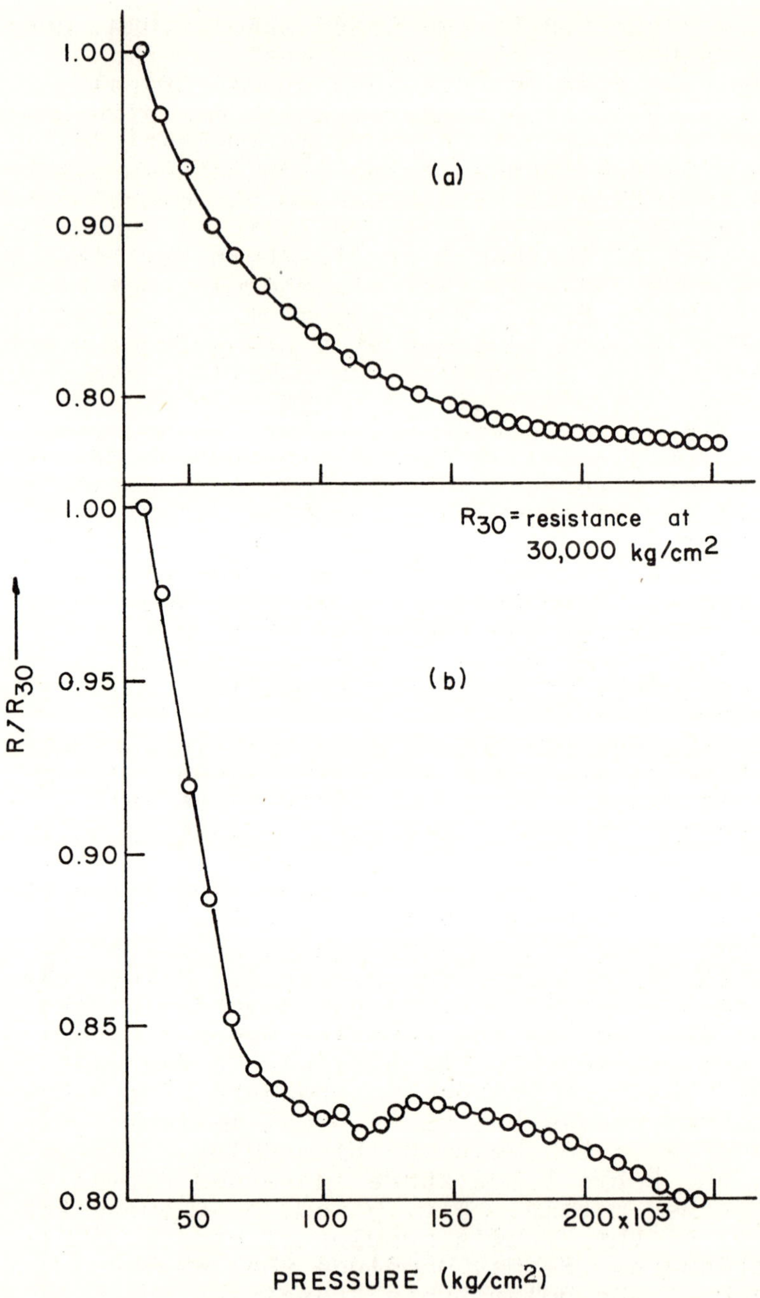

Fig. 1. Resistivity of Lanthanum as a Function of Pressure: a) Low Purity b) Highest Purity

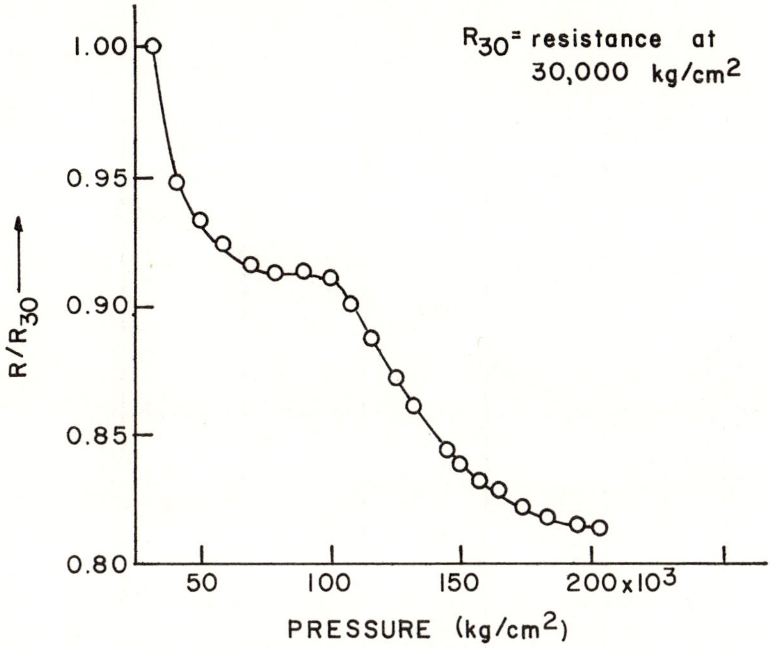

Fig. 2. Resistivity of Neodymium as a Function of Pressure.

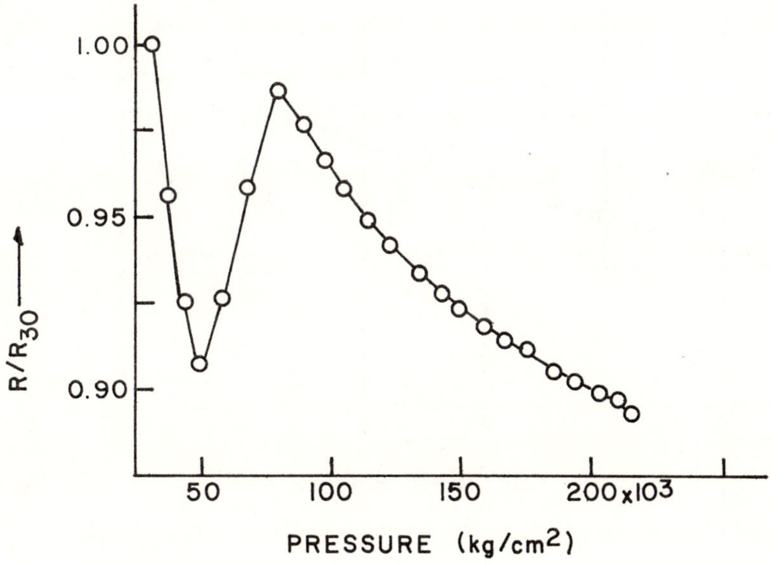

Fig. 3. Resistivity of Cerium as a Function of Pressure.

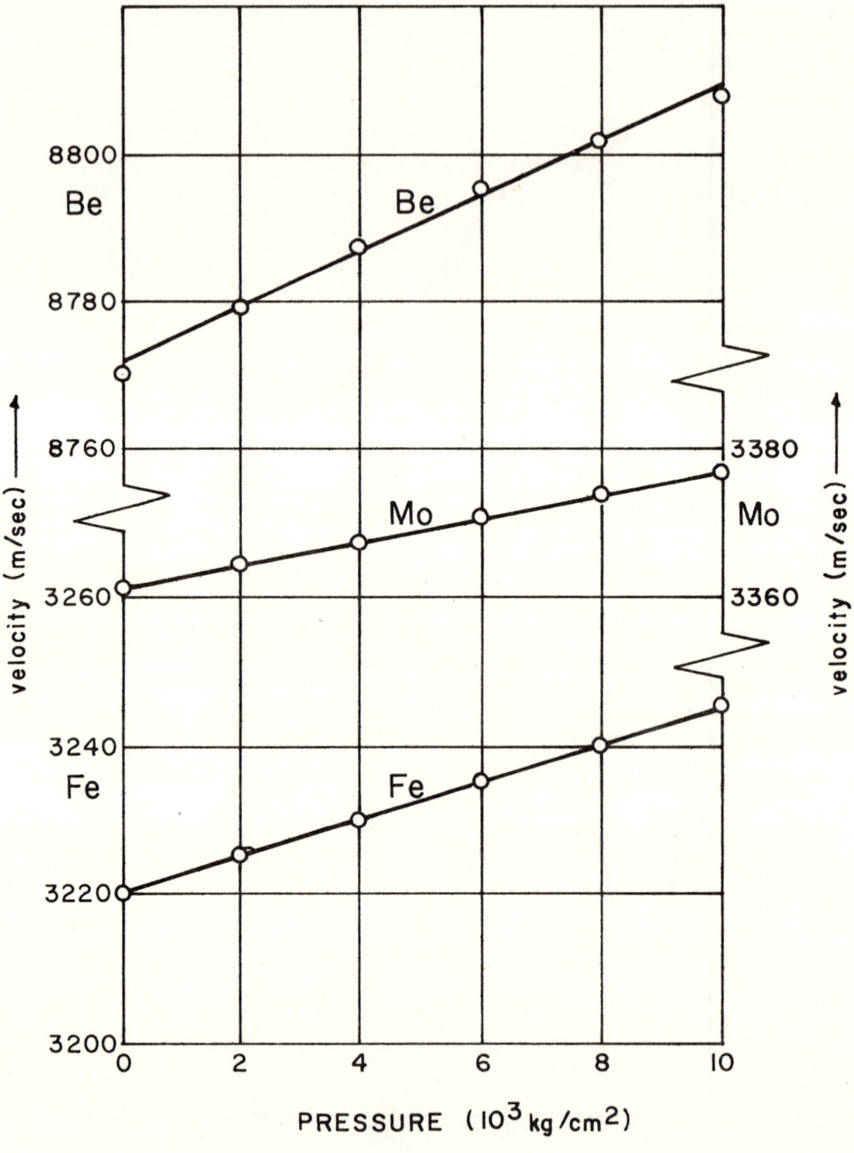

Fig. 4. Velocity of Sound as a Function of Pressure for Beryllium, Molybdenum and Iron.

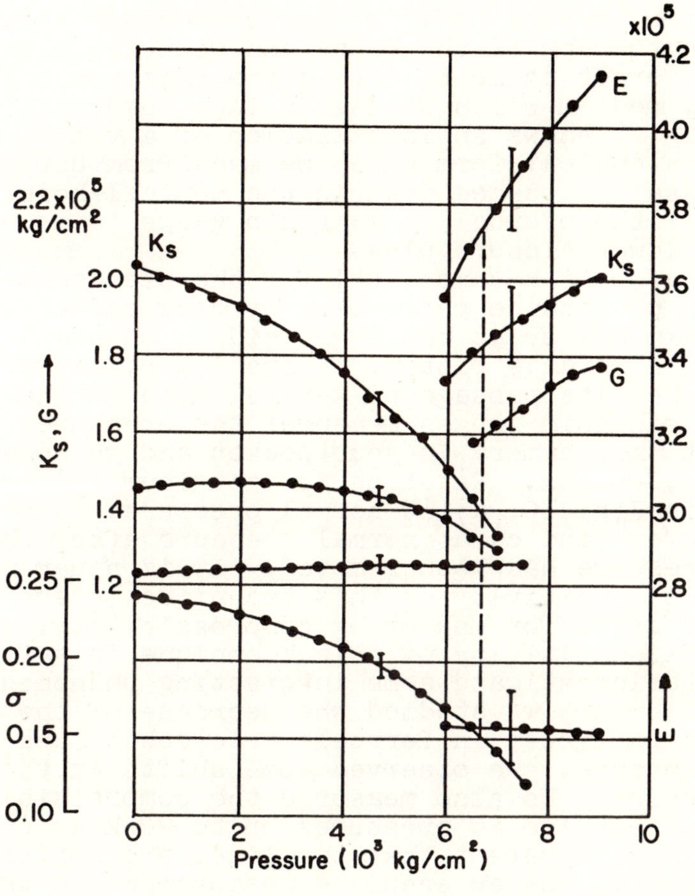

Fig. 5. Pressure Dependence of the Elastic Constants of Cerium.
K_S - Adiabatic Bulk Modulus
E - Young's Modulus G - Shear Modulus
σ - Poisson's Ratio

of cerium, as was pointed out by Dr. Lawson, a new crystal structure is not formed but rather an electronic transition between the shells of atoms occurs.

As I have mentioned, we can investigate solids up to pressures of 20 kbar using a beryllium vessel. Fig. 6 is a drawing of such an apparatus. Of particular interest is the very small manganin gauge for precise pressure measurements. Fig. 7 is a photograph of the gauge and mounting plug.

I forgot to mention that the pressure transmitting medium which we use is light hydrocarbon.

Fig. 8 shows an installation of a vessel (3) having a conical form which we made from beryllium. The vessel is pushed down in the conical hole by means of the piston. A manganin gauge is mounted on the lower closure plug.

Fig. 9 is a camera (4) for the precise measurement of lattice parameters by back reflection on a film holder at the right side. On the left side is a movable piston. A small beryllium window permits passage of x-rays. One of the miniature manganin gauges measures the pressure. You can put this camera in your pocket and go home with 10,000 atm. Patterns for aluminum (4) are shown in Fig. 10; a) at normal pressure, b) at 4400 kg/cm^2 and c) at normal pressure after the high pressure has been removed. It is apparent that the sensitivity of this technique is sufficiently great for measuring compressibility.

The sensitivity of this technique is sufficient to investigate some interesting phenomena. In particular, we studied the decrease of the lattice parameter in ferroelectric substances at high pressure. We observed some shifts in the Curie point. We also measured the compressibility of boron nitride at pressures up to 20 kbar (Fig. 11). It is apparent that the lines are shifted sufficiently for an accurate measurement so that we could calculate the compressibility. We have calculated the compressibility of graphite, boron nitride in the hexagonal form, calcium, beryllium and several others. Fig. 12 shows the dependence of the compressibility of graphite (3) on pressure. It is apparent that the value of $\Delta c/c$ increases very rapidly with pressure. Fig. 13 shows that the compressibility of boron nitride is much more than the compressibility of graphite. We may

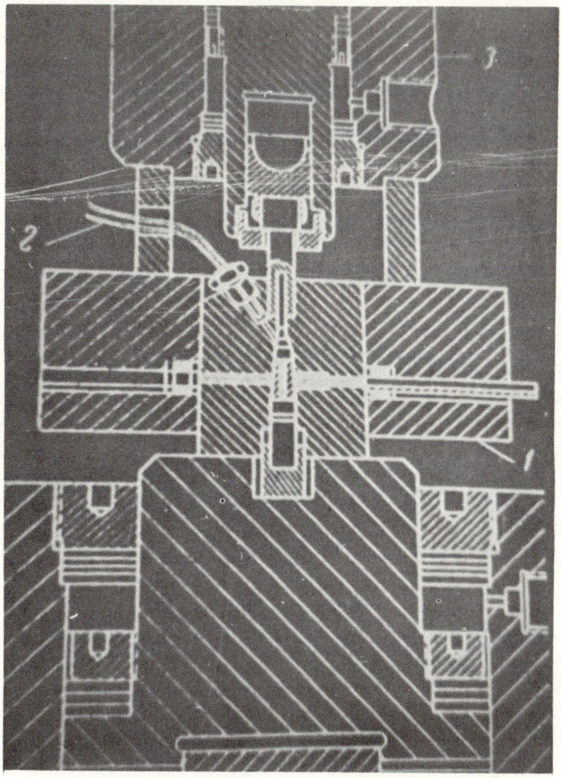

Fig. 6. Apparatus for Producing 20 kbar in a Beryllium Vessel.
1. Steel Ring 2. Manganin Gauge Leads
3. Press

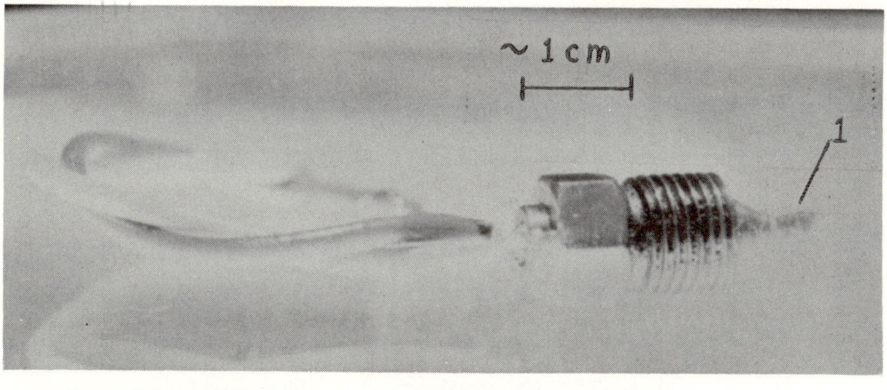

Fig. 7. Photograph of Manganin Gauge and Mounting Plug.
1. Manganin Coil.

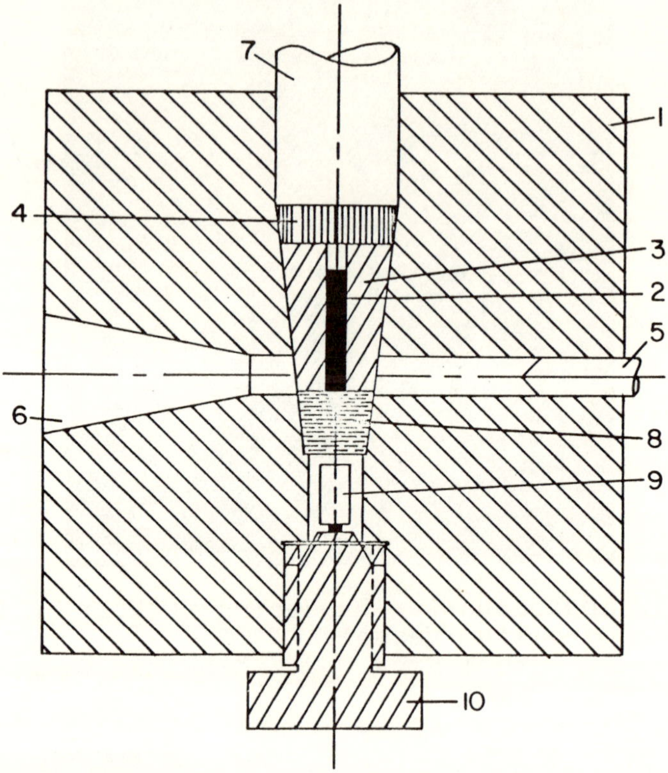

Fig. 8. Diagram of High Pressure X-Ray Camera to 15,000 kg/cm^2.

1. Steel block
2. Specimen (0.4-0.5mm dia.)
3. Beryllium cone
4. Steel stopper
5. X-ray entrance diaphram
6. X-ray exit
7. Steel piston
8. Pressure fluid
9. Manganin gauge
10. Electrical lead plug

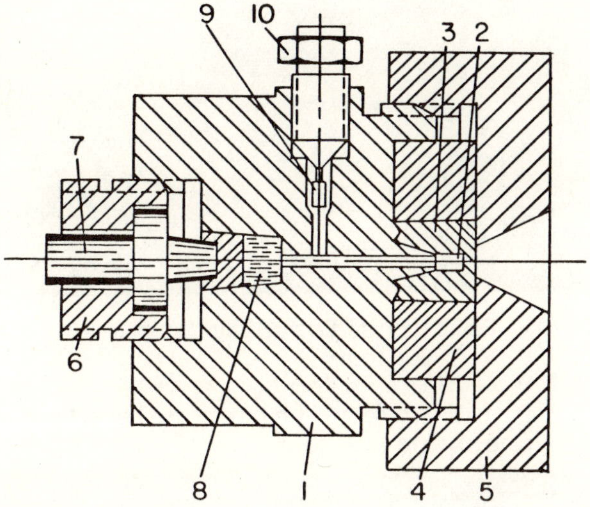

Fig. 9. Diagram of High-Pressure Back-Reflection X-Ray Camera.
1. Steel body. Sample, 2, is placed in the recess of a beryllium piece, 3, pressed into a collar, 4. The nut, 5, with a conical opening for x-rays holds the collar tight. Pressure is created by piston, 7, held at desired position by nut, 6. Pressure fluid, 8, is light hydrocarbon. 9 is a manganin gauge connected through an electrical lead plug, 10.

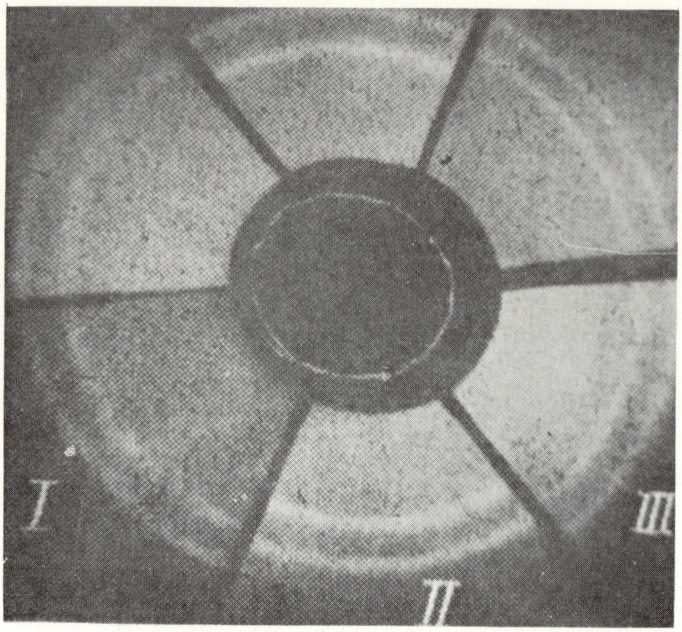

Fig. 10. X-Ray Patterns for Aluminum.
I. P = 1 kg/cm^2.
II. P = 4400 kg/cm^2.
III. P = 1 kg/cm^2 (after releasing pressure).

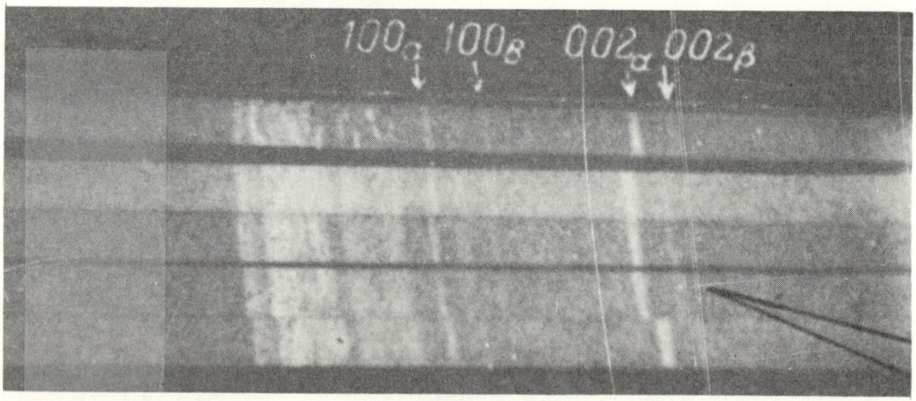

Fig. 11. Five X-Ray Photographs of Boron Nitride Seen on One Film:
a) P = 1 kg/cm^2 (before applying pressure).
b) P = 11,800 kg/cm^2.
c) P = 10,800 kg/cm^2.
d) P = 12,400 kg/cm^2.
e) P = 1 kg/cm^2 (after removing pressure).

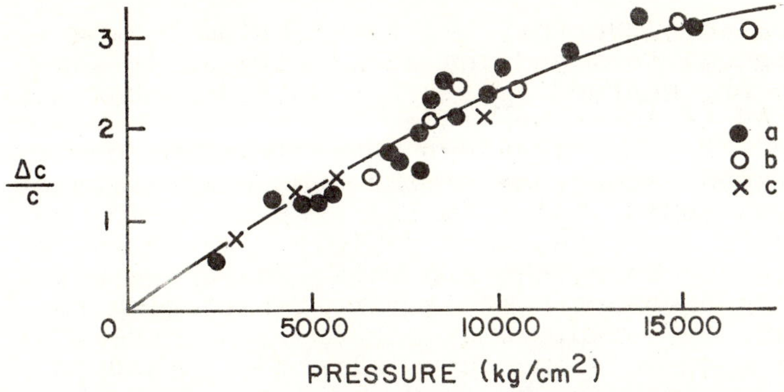

Fig. 12. Solid Line Gives $\Delta c/c$ as a Function of P according to $\Delta c/c = 28 \times 10^{-7}$ P $- 45 \times 10^{-12}$. a) Artificial graphite in camera 1, b) Artificial graphite in camera 2, c) Ceylon graphite in camera 2.

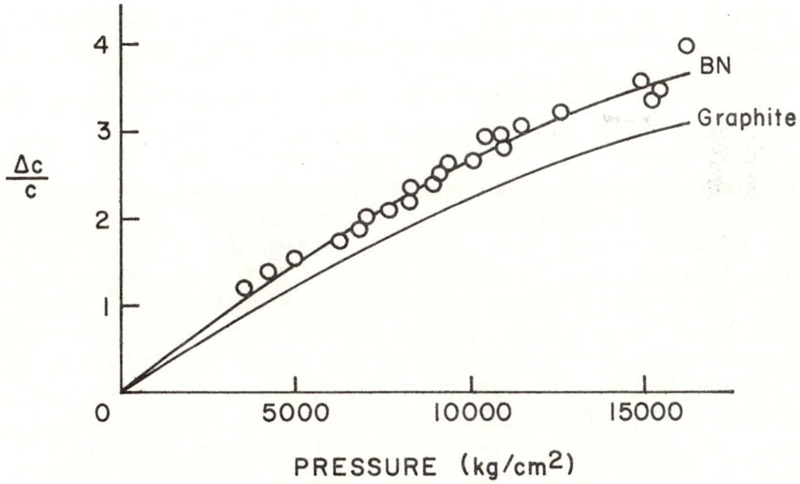

Fig. 13. Variation of $\Delta c/c$ with Pressure for Boron Nitride and Graphite.

therefore conclude that the equilibrium between the hexagonal form and the cubic form as it was found by Dr. Wentorf (5) in the United States may be shifted to lower pressure.

Fig. 14 summarizes our investigations of the pressure dependence of the compressibility of $BaTiO_3$ and $PbTiO_3$ (4). In the case of $BaTiO_3$ we find that c decreases more rapidly with pressure than does a. Thus, the c/a ratio decreases with increasing pressure. It is very interesting to note that for $PbTiO_3$, c decreases with increasing pressure whereas a increases slightly. Therefore it is very dangerous to draw conclusions about all parameters from experiments in which there is only uniaxial stress. We then found that for $PbTiO_3$ the c/a ratio crosses the pressure axis at the point where we have a transition from ferroelectric to paraelectric properties.

We have made many compressibility measurements using both x-ray and moving piston methods. It is interesting to see (Fig. 15), as first shown by Richardson, that there is an oscillation in the plot of compressibility versus Z (6), where Z is the atomic number. The amplitude of these oscillations decreases with increasing pressure. Some elements even change their relative positions. For example, cerium makes a sudden drop as the pressure increases.

We have also investigated polymorphic transitions. For such investigations we use the x-ray technique which was worked out by Dr. Bridgman and found it to be very useful and very interesting. This was also true of a technique developed by Dr. Lawson. We found that there is a polymorphic transition in tellurium (7) at a pressure between 15 and 20 kbar. In such a technique we cannot use a manganin gauge so we must make a more rough estimate of the pressure. The transition goes reversibly without a volume change to a hexagonal cell with parameters a = 4.208 Å and c = 12.036 Å. The transition is masked by the scatter of the $\Delta a/a$ and $\Delta c/c$ plots shown in Fig. 16. It is detectable however when we plot the c/a ratio as in Fig. 17. There is also a transition at 42-45 kbar to a quite new structure which we have not yet determined. There were too few lines to decide the structure. However, there appears to be no further change in structure in the range 45-90 kbar.

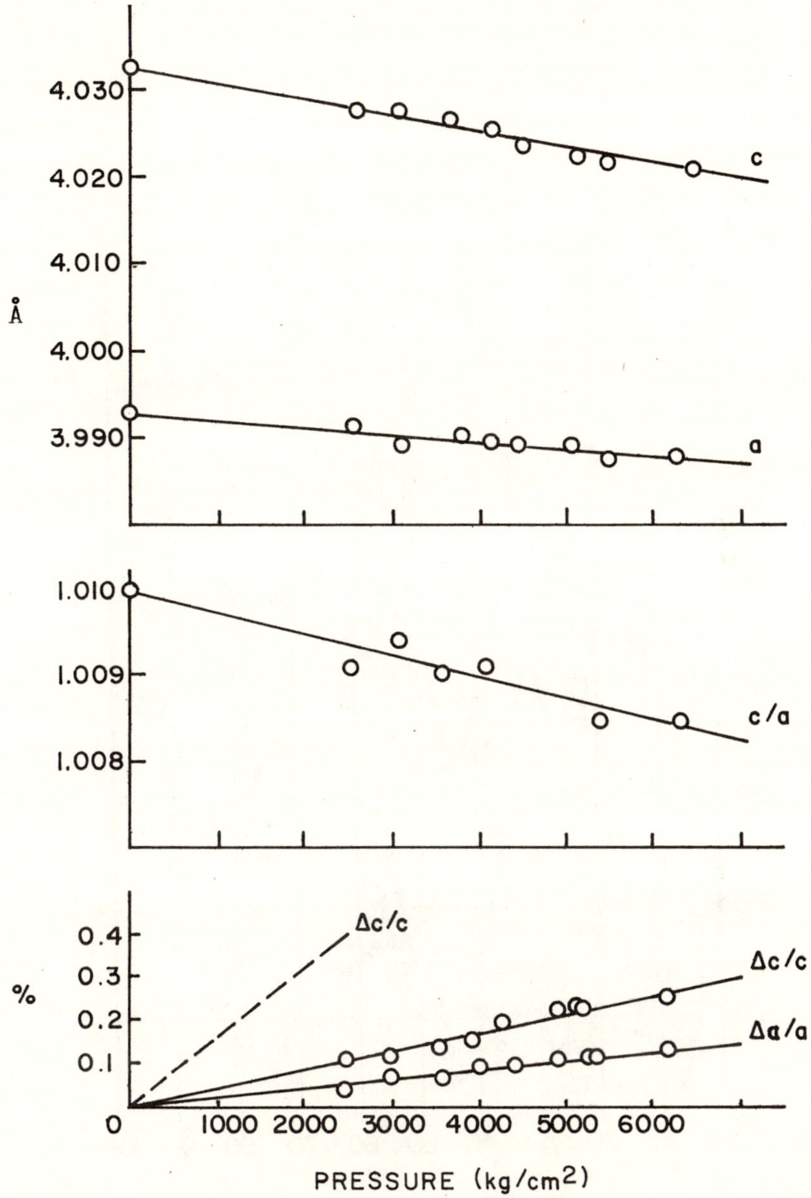

Fig. 14. Variation of a, c, c/a, Δa/a and Δc/c with Pressure for $BaTiO_3$. Variation of Δc/c with Pressure for $PbTiO_3$ is shown by the dotted line.

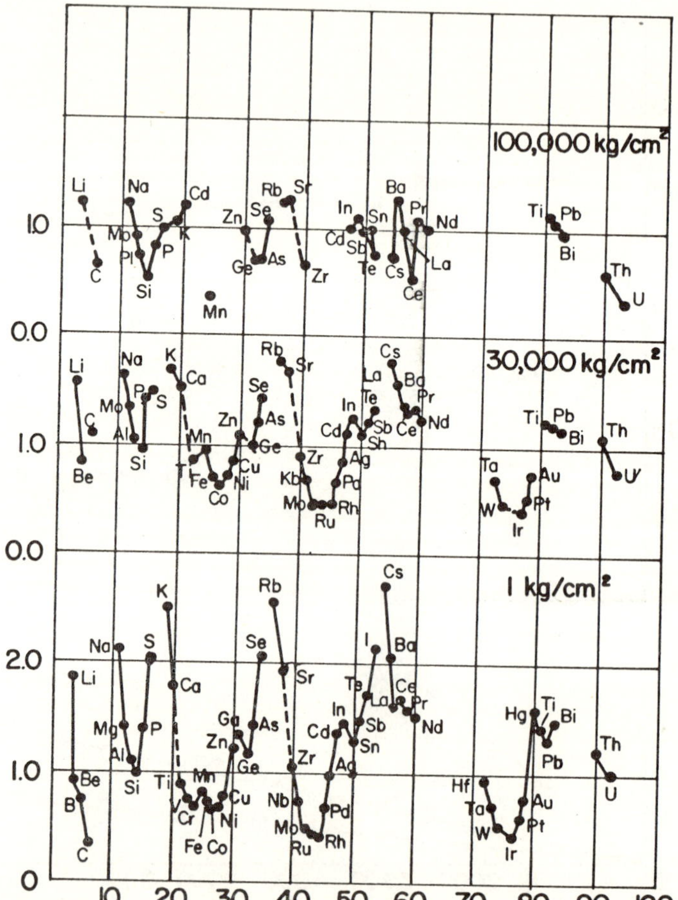

Fig. 15. Compressibility as a Function of Atomic Number Z for Various Pressures.

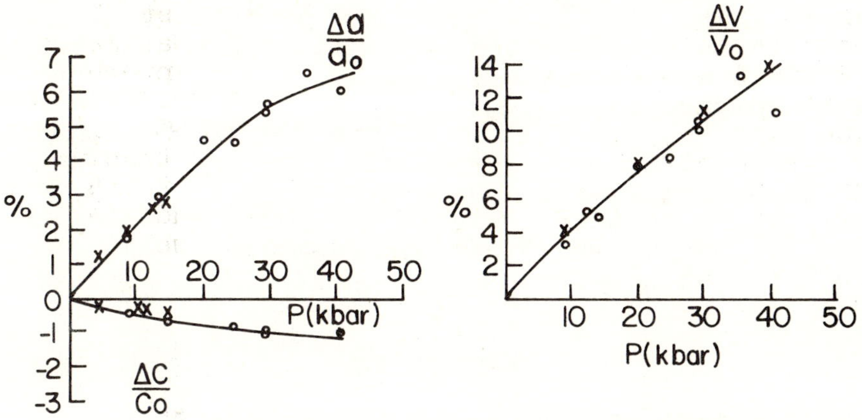

Fig. 16. a) Dependence of Linear Compressibility of Tellurium on Pressure
b) Dependence of Volume Compressibility of Tellurium on Pressure (Circles - x-ray data; crosses - Bridgman's data.

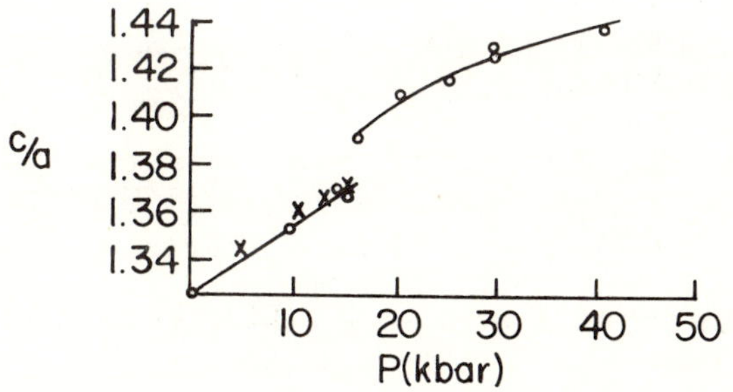

Fig. 17. Dependence of c/a on Pressure for Tellurium. Circles denote x-ray diffraction data (for the A7 phase at P = 15 kbar c/2a was used). Crosses denote values of c/a calculated from Bridgman's data.

For antimony (8) we found that at a pressure near 70 kbar there is a quite new crystal structure. It is simple cubic with a lattice parameter of 2.986 Å and with $-\Delta V/V_0$ = 12%. It is interesting to compare this result with Bridgman's. He found that the decrease in volume was 12.4%. Thus the agreement is good. We found a second transition in antimony in the pressure range 75 to 85 kbar. It is hexagonal with a = 3.346 Å and c = 5.296 Å. $-\Delta V/V_0$ was 3.6% which compares well with Bridgman's data.

REFERENCES

(Subjects were referenced only when they could be reasonably well identified from the context of Dr. Vereschagin's talk. - Ed.)

1. A. W. Lawson and Ting-Yuan Tang, Phys. Rev. 76, 301 (1945).

2. L. F. Vereschagin, A. A. Semerchan and S. V. Popova, DAN 138, 1059 (1961) [Soviet Physics - Doklady 6, 488 (1961)].

3. S. S. Kabalkina and L. F. Vereschagin, DAN 131, 300 (1960) [Soviet Physics - Doklady 5, 373 (1960)].

4. S. S. Kabalkina, L. F. Vereschagin and B. M. Schulenin, DAN 144, 1019 (1962) [Soviet Physics - Doklady 7, 527 (1962)].

5. R. H. Wentorf, Jr., Progress in Very High Pressure Research (John Wiley and Sons, New York, 1961), p. 82-89.

6. L. F. Vereschagin and A. J. Likhter, DAN 86, 745 (1952).

7. S. S. Kabalkina, L. F. Vereschagin and B. M. Schulenin, JETP (USSR) 45, 2073 (1963) [Soviet Physics - JETP 18, 1422 (1964)].

8. L. F. Vereschagin and S. S. Kabalkina, JETP (USSR) 47, 414 (1964) [Soviet Physics - JETP 20, 274 (1965)].

NOTE: Some of the figures in Professor Vereschagin's article were taken from photographs of slides shown at the Conference. Although care was taken to compensate for the distortion, they should be considered as only a qualitative representation of the phenomena. The references should be consulted for figures once-removed from the original drawing (twice-removed, of course, if the English translation is used).

RARE EARTH METALS AT HIGH PRESSURES AND TEMPERATURES

A. Jayaraman
Bell Telephone Laboratories, Murray Hill, N. J.

INTRODUCTION

In the metallic state rare earth elements are all trivalent with the exception of Eu and Yb. The latter are divalent and exhibit a discontinuous jump in atomic volume, melting point and other properties in the otherwise smoothly varying physical behavior (decreasing A.V., increasing M.P.) shown by the trivalent members of the series with atomic number. Cerium undergoes a first order phase transformation under the influence of pressure (1-5) or at low temperatures (6-8) (<100°K at atmospheric pressure) in which the valence state changes due to electron promotion involving 4f → 5d (9-10) states indicating that these states are energetically quite close.

The energy differences between the 4f and 5d states in the rare earth metals are probably not large and pressure and temperature might be expected to influence the normally found electronic distribution and hence the valence state of the atoms in the metal. Encouraged by the example of Ce, an extensive study of the fusion behavior and solid-solid transformations of other rare earth metals was undertaken. Several interesting facts concerning high temperature, high pressure properties of the rare earth metals have emerged from these studies and these will be discussed in this paper.

EFFECT OF PRESSURE ON THE CRYSTAL STRUCTURE

The rare earth metals crystallize in close-packed structures (11) with the exception of Eu which has a body-centered cubic (bcc) arrangement. Normally, La, Pr and Nd have the double hexagonal close-packed (dhcp), Sm a rhombohedral, Ce and Yb the face-centered cubic (fcc) and Gd to Lu the hexagonal close-packed (hcp) structure. These structures represent different stacking arrangements of the atomic layers; hcp - ABAB..., Sm-type - ABABCBCACABAB..., dhcp - ABACABAC... and fcc - ABCABC.... Further, most of the rare earth metals undergo a temperature-induced transformation to the bcc structure (12-13) at high temperatures before melting, with the possible exception of Er, Tm and Lu.

Bridgman reported resistance anomalies in a number of R.E. metals; resistance as well as volume discontinuities indicative of a phase transition were noted in Ce (1), Yb (14), La (15) and Gd (16) (rather sluggish transformations in the latter two). The transformation in Ce occurred at about 7 kbars at room temperature. The same transformation is encountered when γ-Ce is cooled below 100°K (8). The high pressure X-ray study by Lawson and Tang (9) and low temperature structural investigation of Schuch and Sturdivant (8) have established that there is no symmetry change accompanying the so-called γ to α-Ce transition (fcc-Ce remains fcc) but only a decrease in the lattice constant giving rise to a macroscopic volume diminution of about 8% at room temperature. A neutron diffraction study on Ce (10) has confirmed that the γ to α-Ce transition involves 4f-5d electron promotion, earlier suggested by Pauling, and independently also by Zacharisen (9). Hall et al (17, 18) have shown that the transition in Yb at 40 kbars at room temperature is to the bcc structure; the latter has a higher density relative to fcc. Lanthanum transforms at about 25 kbar and room temp. to the fcc (19). Piermarini and Weir (20) have reported a phase change in Pr and Nd to the fcc structure at pressures above 40 and 50 kbars, respectively.

Gadolinium, when subjected to pressures of about 30 kbars and about 400°C, transforms to the Sm-type structure (21) which is metastably retained at atmospheric pressure. Samarium under similar treatment goes over to the dhcp structure (22). The sluggish transformation in Gd noted by Bridgman (16) at about 25 kbars must be the hcp to the Sm-type structure. McWhan and Stevens (23) have presented evidence based on magnetic and X-ray studies under pressure that a phase change analogous to Gd (to the Sm-type) takes place in Tb, Dy and Ho. Thus, there appears to be a well defined sequence of pressure-induced structural transformations in the R.E. metals from hcp → Sm-type → dhcp → fcc. This sequence is in the order of increasing cubicity. In the Sm-type 1/3 and dhcp, 1/2 of the layers have the environment of cubic close-packing and the rest hexagonal close-packing. It has been suggested (21) that in Gd the transformation to Sm-type structure might be connected with an increase in c/a. The latter ratio (11) is found to be distinctly higher for the lighter rare earths compared to Gd and Lu ($c/a \approx 1.58$). McWhan and Stevens (23) find that in Gd and also in Tb, Dy and Ho the c/a ratio increases with pressure. Rare earth alloys crystallizing in Sm-type structure have c/a ratios very close to 1.604 ± 4. These facts suggest that a rare earth metal with hcp structure would transform to the Sm-type structure when this value (~ 1.6) is realized for c/a, under pressure.

Hall et al (17) have suggested that the atomic radius observed in the bcc Yb at 40 kbar reflects a change in the valence state of Yb from 2 to 3 involving the promotion of a 4f electron to the 5d state. However, the observed atomic radii of bcc-Yb and bcc-Sr could be accounted for when corrected for the change of coordination number (24). When the volume of 33.5 cubic angstroms per atom observed for the fcc-Yb at 39.5 kbar (18) is used to calculate the lattice constant of the bcc-Yb at that pressure, one obtains 4.06 Å while the observed value is 4.02 Å (18). This difference almost exactly accounts for the 3% macroscopic volume change (18) observed for the transformation at 40 kbar. The bcc phase is found to be denser than fcc (25, 26) or hcp in several metals, e.g., the bcc phase is denser than hcp in Zr, Ti (27) Tl (28) and Gd, denser than fcc in Ce, Yb and Sr. The general notion that the bcc arrangement represents a less efficient packing relative to fcc or hcp is based on a hard sphere model. In the bcc arrangement each atom has 8 nearest neighbors and the 6 next nearest neighbors are only 15.5% farther away. If the second neighbor interactions are strong an effective coordination higher than 12 and therefore a higher packing efficiency could be realized in the bcc. The shape, rigidity and the nature of the binding interactions would play a role in this and apparently in the above examples all these factors favor a bcc phase with a higher density. The seemingly conflicting circumstance could thus be rationalized.

PHASE DIAGRAMS OF R.E. METALS

Cerium

The phase diagram of Ce is presented in Fig. 1. The fusion curve as well as the fcc-bcc boundary have negative slopes, -4.7°/kbar and -1.4°/kbar respectively and meet at a triple point near 26 kbar and 674°C. Beyond the triple point the fusion curve continues with the same trend and gradually reverses slope, exhibiting a rather broad minimum at about 33 kbar and 662°C.

The boundary between γ to α-Ce was delineated using the resistance discontinuity method. In these experiments temperature was held constant while the pressure on the sample was varied by slow compression. Transition pressures were read at the commencement of the discontinuous resistance drop. The data are plotted in Fig. 1. It may be noted that the γ to α-Ce boundary when extrapolated intersects the melting curve at the minimum. The resistance vs pressure isotherms are shown in Fig. 2. These curves indicate that the magnitude of the vertical drop in resistance progressively diminish with temperature. The isotherm for 598°K does not show any discontinuous resistance drop, and the resistivity of the sample smoothly varies with pressure. This behavior is reminescent of the changes in the volume discontinuities noted by Beecroft and Swenson (29) at corresponding temperatures.

Europium and Ytterbium

The phase diagrams of Eu and Yb are reproduced in Figs. 3 and 4. The fusion curve of Eu (Fig. 3) exhibits very definitely a maximum, located at about 35 kbar and 995°C. It begins with a steep initial positive slope, rapidly flattening in the region of 25 to 35 kbar and takes on a negative slope at higher pressures. The fusion curve Yb (Fig. 4) has a steep positive initial slope, rapidly changing at higher pressures. Probably it has a maximum at higher pressures.

The fcc-bcc transformation temperatures in Yb

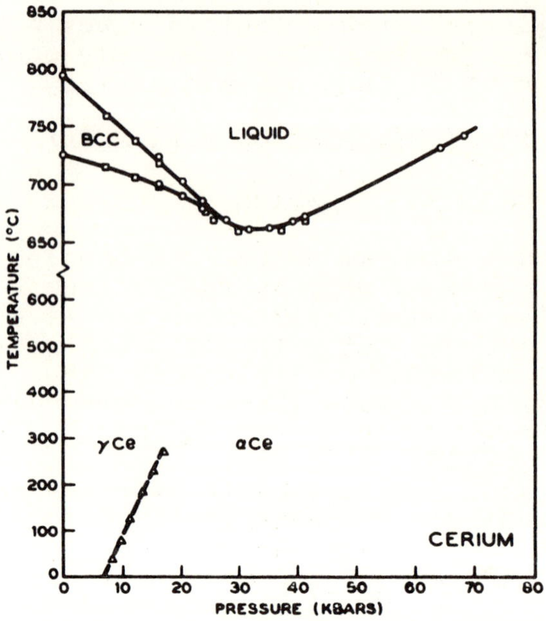

Fig. 1. Phase diagram of cerium

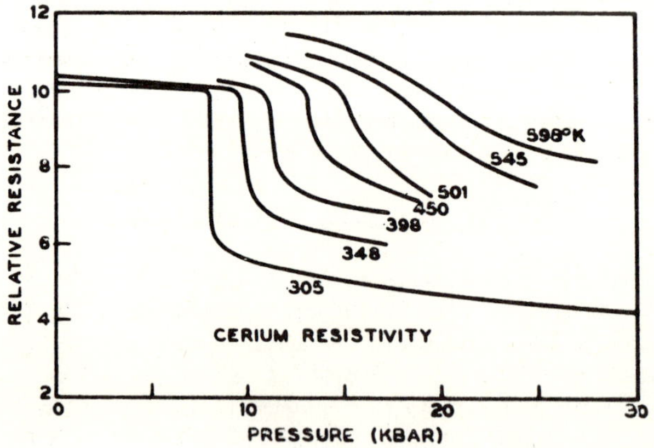

Fig. 2. Pressure-resistance isotherms of fcc-cerium

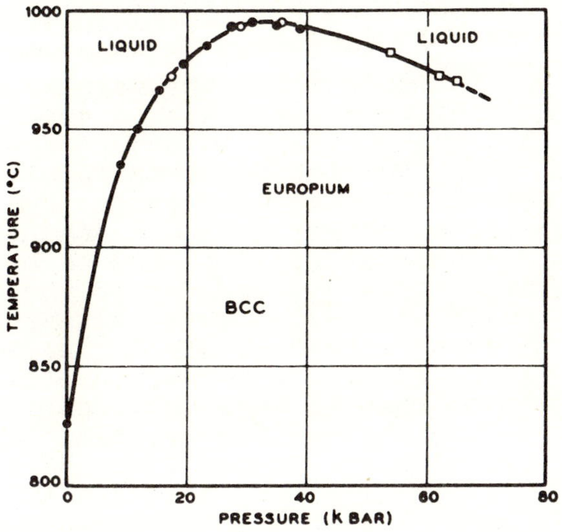

Fig. 3. Phase diagram of europium

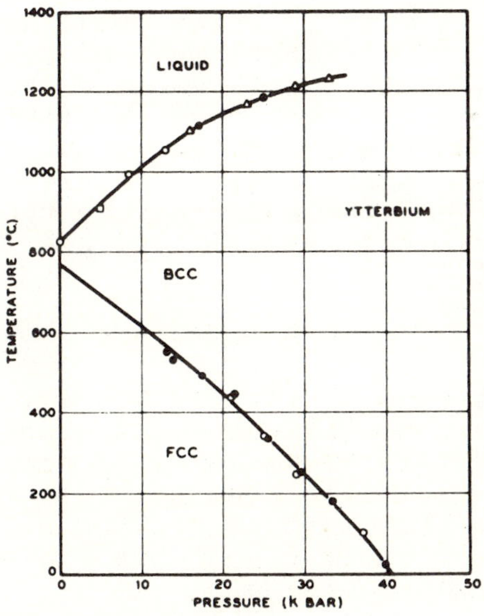

Fig. 4. Phase diagram of ytterbium

were determined from the resistance discontinuity noticed upon heating; the data are plotted in Fig. 4. As in the case of Sr (30) the character of the resistance discontinuity accompanying the transformation varies with the pressure; to about 24 kbar, it is an increase in resistance, the magnitude progressively diminishing with increasing pressure. Above 24 kbar the discontinuity is an abrupt decrease in resistance. Also, the fcc-bcc transformation has associated with it certain temperature hysteresis and sluggishness which are pressure dependent. The transformation temperatures obtained on heating and cooling exhibit little hysteresis and the transition is sharp to about 18 kbar. Above this pressure the hysteresis interval progressively increases and the transition is spread over a temperature range. At pressures greater than 21 kbar the fcc-bcc transition is very sluggish, and the reverse transformation does not take place on cooling. Reversion could only be accomplished by reducing the pressure to about 14 kbar. Figure 5 shows isobars at 35 and 17.5 kbar, typifying the nature of the resistance discontinuity following the fcc-bcc transition.

At pressures above 30 kbar bcc-Yb shows an anomaly in resistivity at about 470°C (see in Fig. 5 the isobar at 35 kbar). After transformation to bcc, the temperature coefficient of resistance is steeply rising until a well defined break occurs at 470°C. The latter temperature is practically insensitive to pressure and is presumably connected with some kind of a transition, the nature of which is not clear. A resistance anomaly in bcc-Yb was first observed by Hall and Merrill (18) but they report a temperature of 625°C.

Other R.E. Metals

Phase diagrams of La, Pr, Nd, Sm, Gd and Tb have been determined (31) in some cases up to 65 kbar but generally to 45 kbar. Fusion data at about 8 kbar have also been obtained for Dy, Ho, Er, Tm and Lu. The initial melting slopes derived from these are collected in Table I along with some pertinent thermodynamic data. In calculating the initial slopes corrections have been made on the basis of Hanneman and Strong's (32) findings for the pressure effect on the thermal emf of Pt - Pt 10% Rh thermocouples which were used throughout.

Table I. Thermodynamic Data of Interest

R.E. Metal	MP °C	dT/dP°/kbar	ΔS/mol fusion	ΔV/mole fusion
La	920	2.5	1.34	0.14
Ce	797	-4.7	1.157	-0.22
Pr	935	0.1	1.365	0.005
Nd	1024	3.5	1.314	0.19
Sm	1072	11.5	1.53	0.72
Eu	826	15	2.23	1.39
Gd	1312	6.5	1.53	0.41
Tb	1356	9	1.56	0.61
Dy	1407	~12	2.04	0.93
Ho	1461	~15	2.25	1.4
Er	1497	~15	2.67	1.66
Tm	1545	~12	2.31	1.25
Yb	824	19	1.64	1.26
Lu	1652	~8.5	1.71	0.64

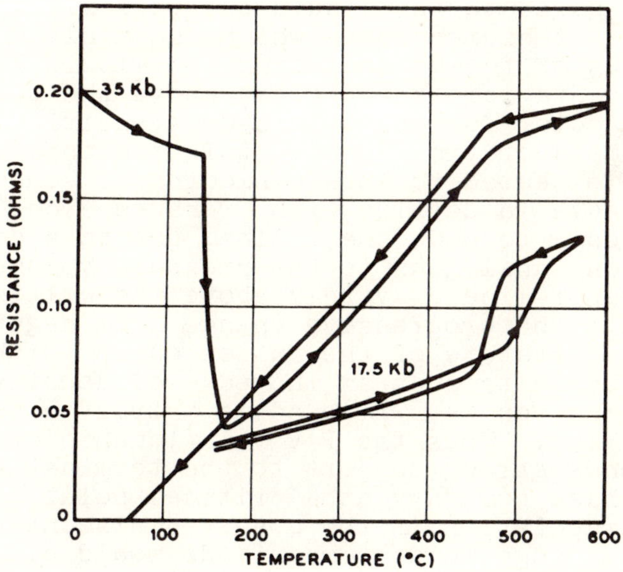

Fig. 5. Resistance-temperature isobars of ytterbium

DISCUSSION

Fusion Behavior and Critical Phenomena in Cerium

The trajectory of a fusion curve in the P-T plane would be determined by the sign and magnitude of the volume change ΔV accompanying fusion as well as by changes in the entropy of fusion. Conversely from the melting curve, using the Clapyeron equation ($dT/dP = \Delta V/\Delta S$), it is possible to obtain qualitative information regarding the changes in the density of the solid or the liquid as the case may be, provided ΔS does not vary appreciably with pressure. Thus a fusion curve maximum indicates a rapid increase in the density of the liquid while a minimum in the fusion curve, a rapid increase in the density of the neighboring solid.

In the case of Ce the thermal arrest signals appear to have the same magnitude in the pressure range investigated, indicating that the latent heat of fusion and hence ΔS does not change notably with pressure. Assuming that there is no rapid density change in liquid Ce with pressure, we interpret the fusion curve of Ce in terms of density variation in the neighboring fcc-Ce. Initially the solid is less dense compared to the liquid but this situation reverses at pressures above which the minimum occurs. The region of fcc-Ce bordering the liquid lies in the phase diagram (see Fig. 1) above the postulated critical point for the γ to α-Ce transition. If there were a critical point, this region of solid would be expected to show a smooth but somewhat rapid increase in density with pressure along the extrapolated γ to α-Ce phase line, due to a continuous transition analogous to the pressure-volume behavior of vapor-liquid systems above the critical temperature. The progressive change from negative to positive curvature of the fusion curve with a broad minimum reflects such an increase of density of the neighboring fcc-Ce, which seems a truly supercritical behavior. Thus the P-V relationship of this solid lends strong support to the termination of the γ-α-Ce phase boundary at a critical point.

Extrapolation of the observed discontinuous resistance change to zero magnitude would place the critical point for the γ to α-Ce transition at about 550°K and 17.5 kbars. It would be seen from Fig. 2 that the isotherm for 545°K hardly shows any abrupt resistance drop. Beecroft and Swenson (29) show that the discontinuous volume change upon transition noted as a function of temperature would extrapolate to zero magnitude at about 630°K and 20 kbar. But these authors have commented that the extrapolation may not be very accurate due to uncertainties in locating the transition pressure at higher temperatures. Ponyatovskii (3) found that at P = 18 kbar and T = 553°K the heat of transition vanished. The presently derived value for the critical point is in close agreement with the latter.

It has been pointed out (33) that a critical point can exist only for phases between which the difference is of a purely quantitative nature and which do not differ in internal symmetry; the different crystalline modifications of a substance or a crystalline solid and liquid are ruled out. The γ to α-Ce transformation, being isostructural and differing in the valence state of Ce atoms ($4f^1$, $5d^1$, $6S^2$ and $4f^0$, $5d^2$, $6S^2$, respectively), apparently satisfies these conditions. Beecroft and Swenson (29) have speculated that the degrees of 4f character of the wave functions of the two states approach each other along the equilibrium line with pressure and temperature until they merge for the two phases at the critical point. This would imply that the size of the Ce ion varies continuously depending on the extent of 4f-5d character. In support of this, they have quoted studies on alloys of Ce (34, 35) at zero pressure which seem to indicate a continuous variation of the size of the Ce-ion with composition and temperature, supposedly due to a continuous change in the valence state of Ce. According to Jamieson's (36) model the critical point for the γ to α-Ce transition would occur when the degree of ionization ($4f \rightarrow 5d$), which is a function of pressure and temperature, is 50%. Beecroft and Swenson (29) have suggested that the influence of the γ to α-Ce transformation on the equation of state would be felt along the extrapolated transition line well beyond the critical point. The equation of state of fcc-Ce deduced from the fusion curve is in full accord with this suggestion.

The negative slopes of the fcc-bcc and the bcc-liquid phase boundaries of Ce indicate that the densities of the three phases are in the sequence fcc $<$ bcc $<$ liquid. To account for the higher density of liquid Ce relative to the fcc and bcc phases, the Ce atoms in the liquid must have either a higher effective coordination (probably > 13) or a smaller radius. Although the attainment of such high coordination in liquids has been discussed, (37) we believe the latter circumstance as more probable and suggest that the atoms in liquid Ce have collapsed to a smaller size because of $4f \rightarrow 5d$ electron promotion, a phenomenon known to take place in fcc-Ce (10).

Fusion Behavior of Other R.E. Metals

Most of the rare earth metals undergo a transformation to the bcc structure before melting and thus offer a series whose fusion properties can be compared. In Fig. 6, the initial melting slopes of the members have been plotted against atomic number. Although this plot has some common features to the plot of atomic radius versus atomic number, it shows some striking deviations. The atomic radius smoothly decreases with atomic number except at Eu and Yb, the anomalous properties of which are attributed to their divalent state. From Fig. 7 it can be seen that La, Nd, Gd and Lu fall on a line designated trivalent and Eu and Yb on another designated divalent. While all the heavier rare earths and Sm fall above the line labelled trivalent, Ce and Pr are below the latter. Except for Ce the melting slopes of all the rare earth elements are positive.

The P-T behavior of Ce (25) suggests that in interpreting high pressure-high temperature behavior of the rare earth metals, the valence state of the atoms in the liquid as well as in the solid is an important factor, as this would influence the ΔV term. The atomic size for the rare earth elements depends on the valency and increases in the sequence tetravalent < trivalent < divalent. If the atoms in the liquid have a lower valency compared to the atoms in the solid, ΔV of fusion would be positive and larger than usual. In the opposite situation, ΔV of fusion would be negative and consequently the melting slope also negative (example of Ce). Another factor to be reckoned with is the magnetic contribution to the entropy, especially important in the case of heavier rare earths, but at high temperature this is probably negligibly small.

For the rare earth series, it is experimentally established that the divalent metals Eu and Yb (26)

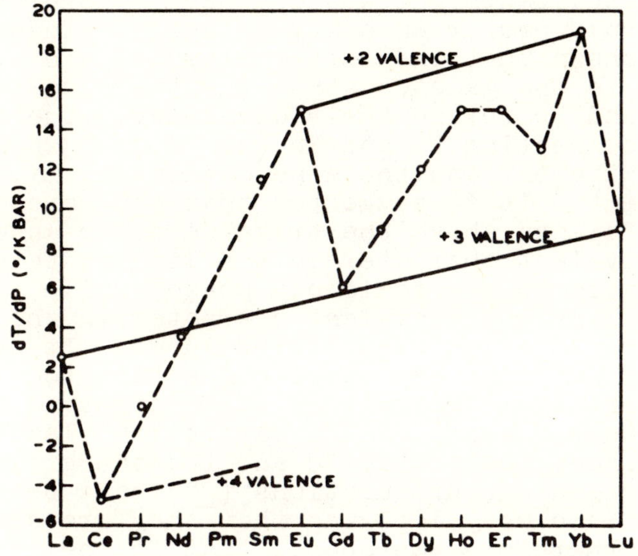

Fig. 6. Initial melting slopes of R.E. metals as a function of atomic number

have the highest dT/dP of fusion (see Table I and Fig. 6). In these, both the solid and the liquid would be expected to maintain divalency, for the half filled and completely filled 4f states are preferred. For the same reason Gd and Lu would maintain trivalency both in the solid as well as in the liquid state. Lanthanum would behave accordingly, for it would prefer an empty 4f shell. Indeed, it has been shown that the latter elements are trivalent even in the gaseous state (neutral atoms) (38), while Eu and Yb retain their divalency. Therefore the line connecting Eu and Yb (see Fig. 6) has been labelled divalent and the line connecting La, Gd and Lu trivalent. It is suggested that the rare earth elements falling above the trivalent line assume partial divalent character in the liquid state and possibly also in the solid bcc phase, while those below the line a tetravalent character in the liquid, the latter situation exemplified in Ce. It should be pointed out that the above suggestion receives some support from the valence state observed for the gaseous neutral atoms of the rare earth elements. With the exception of La, Gd and Lu, the ground state electronic configurations (38) have been found to be $4f^{1-14}5d^{0}6s^{2}$, implying that they are divalent. It is not, therefore, improbable that in the liquid the atoms assume partial divalent character. On the basis of this picture we discuss the results.

The fusion behavior of Ce has already been discussed. For the very small positive initial slope of Pr indicative of a vanishingly small ΔV of fusion, it is suggested that there is a decrease in the atomic size in the liquid due to partial excitation of the 4f electron into 5d. This would compensate for the usual expansion on melting. Neodymium falls on the trivalent line and apparently adheres to trivalency in both the solid and in the liquid. A melting slope of 7°/kbar is predicted for Pm. Samarium has a dT/dP well above the trivalent line and may therefore be expected to have a large divalent character. A divalent oxidation state is easily formed in the case of Sm and this indicates that the above situation is a plausible one. The sudden drop in dT/dP in passing from Eu to Gd and from Yb to Lu is because of the strict adherence of Gd and Lu to trivalency irrespective of the state of aggregation, temperature and pressure condition. The partial divalent character postulated for those members which lie above the trivalent line apparently increases in going from Nd to Sm and from Gd to Ho or Er. We have no explanation to offer for this tendency in the heavier rare earths although in the case of Sm there is some support. We have also not explained why there is a reversal of trend commencing at Er. It may be noted that Er and Tm do not in all probability undergo a transformation to the bcc structure before melting.

SOME GENERAL REMARKS ON FUSION BEHAVIOR AT HIGH PRESSURES

The occurrence of a fusion curve minimum in Ce and a maximum in the case of Cs, (39) Ba (40) and Eu (26) reveal that the melting behavior of simple metallic elements could be complex. It appears that while a fusion curve maximum is due to a progressive increase in the density of the liquid presumably due to an increase in the coordination in the liquid relative to the solid, the minimum is related to an increase of density in the solid phase. When the solid is close-packed as is Ce, such a density increase would seem to require a progressive change in the size of the atom. The mechanism that provides this in the case of Ce is definitely related to the 4f → 5d electronic promotion. The Cs transition at 40 kbar (39) is regarded as electronic in nature (6s → 5d) (41) but the fusion curve does not show the kind of minimum observed in the case of Ce. In this case, however, the solid-solid boundary does not end before intersecting the melting curve. It seems therefore that the presence of an electronic transformation which terminates at a critical point is the requirement for the occurrence of a fusion curve minimum. If this is the case, Ce might be a unique example.

It may be pointed out that the rapidly increasing density of liquid Cs in the region of 30 to 45 kbars (39) is presumably connected with the progressive change in the size of Cs atoms in the liquid due to 6s-5d electron promotion. Since CsII appears to be fcc (42) this view would be in line with the proposed explanation for the higher density of liquid Ce compared to fcc Ce.

SOME HIGH PRESSURE PROPERTIES OF INTEREST

The similarities in physical properties between Eu (43) and Yb on the one hand and Ba and Sr on the other have been pointed out. High pressure experiments have brought out further close correlations that exist amongst them. Both Eu and Ba (40) exhibit fusion curve maxima and have compressibilities which are about the same (43). Likewise, the melting behavior and compressibilities of Yb and Sr are very similar. The fusion curves of Sr (30) and Yb have initial slopes of 20°/kbar and 19°/kbar, respectively and rapidly flatten in the 35 kbar region in either case. The most striking parallelism in the behavior of Yb and Sr is to be found in the fcc-bcc transformation and resistivity of these metals under pressure. The resistivity of both fcc Yb and Sr rises rapidly with pressure and over a certain region (around 30 kbars) they exhibit a negative temperature coefficient of resistance, a characteristic of an intrinsic semiconductor. The semiconducting fcc-phase is terminated in either case through a sluggish phase transformation to the bcc structure. The nature of the resistance discontinuity accompanying the fcc-bcc transformation and the temperature and pressure hysteresis associated with this transformation are very similar in the two metals. The resistivity of bcc Yb decreases very slowly with pressure in contrast to the fcc phase which has a very large positive pressure coefficient of resistance. The differences in the resistance discontinuity accompanying the fcc-bcc transformation referred to earlier owe their origin to this circumstance. At low pressures the bcc phase has the higher resistance compared to fcc, but the situation reverses at high pressures because of the rapid increase in resistivity of the fcc phase with pressure. The fcc-bcc phase line has a negative slope in either case.

The similarities to the alkaline earths appear to be compelling evidence for the divalent character of metallic Eu and Yb. Further, their metallic nature evidently results from overlapping first and second Brillouin zones as is found for the alkaline earths (44); the overlaps are conceivably in the same directions in k space. To account for the large pressure coefficient of resistivity in Ca and Sr, Mott (45) proposed that a decrease in the zone overlaps occurs with pressure (46). It is believed that a similar mechanism would account for the resistivity of Yb under pressure. The recent findings of a semiconducting region under pressure in Ca (47), Sr (30) and Yb (48) lend support to Mott's proposal. It may be noted that a large positive pressure coefficient of resistance and the existence of a semiconducting region under pressure seem to be characteristic of fcc metals with two electrons per atom.

The magnetic behavior of the metastably retained Sm-form of Gd and dhcp-Sm has been studied. The bulk magnetic susceptibility of the Sm-form of Gd is strikingly lower compared to ferromagnetic hcp Gd and it seems probable that Gd in the Sm-form is antiferromagnetic. The difficulty in establishing the antiferromagnetic ordering in metastably retained samples arises from the rapid reversion of Sm-type Gd to the hcp form during cooling, which affects the susceptibility measurements. However McWhan and Stevens (23) in their A.C. susceptibility measurements on Sm-type Gd under pressure obtain signals which are indicative of antiferromagnetic ordering; they see the same kind of signals for Tb, Dy and Ho which are known to order antiferromagnetically. The Néel temperature of dhcp Sm- is 27°K compared to 14°K for the normal rhombohedral form. This shift would seem to represent the effect of volume on the ordering temperature, for the high pressure phase has a smaller volume relative to the normal form. From this it may be expected that the Néel temperature of Sm would increase with increasing pressure.

I wish to thank Dr. D. B. McWhan for helpful comments and R. G. Maines for technical assistance.

REFERENCES

1. P. W. Bridgman, Proc. Am. Acad. Arts Sci. 62, 207 (1927); 79, 164 (1951).
2. A. I. Likhter, Yu. N. Ryabinin and L. F. Vereschagin, Sov. Phys. J. Exp. Theor. Phys. 6, 469 (1958).
3. E. Ponyatovskii, Dokl. Akad. Nauk SSSR 120, 1021 (1958); Soviet Physics, Doklady 3, 498 (1958).
4. E. S. Itskevich, J. Exptl. Theor. Phys. (USSR) 42, 1173 (1962).
5. M. G. Gonikberg, G. D. Shakovskii and V. P. Butuzov, Zh. fiz. Khim. 3, 350 (1957).

6. C. J. McHargue and H. L. Yakel, Jr., Acta. Met. 8,637 (1960).
7. K. Gschneidner, Jr., R. O. Elliott and R. R. McDonald, J. Phys. Chem. Solids 23, 555, (1962).
8. A. F. Schuch and J. H. Sturdivant, J. Chem. Phys. 18, 145 (1950).
9. A. W. Lawson and T. Y. Tang, Phys. Rev. 76, 301 (1949).
10. M. K. Wilkinson, H. R. Child, C. J. McHargue, W. C. Koeller and E. D. Wollan, Phys. Rev. 122, 1409 (1961).
11. "The Rare Earths", Editors, F. H. Spedding and A. H. Daane, John Wiley and Sons, Inc., p. 185, (1961).
12. F. H. Spedding, J. J. Hanak and A. H. Daane, J. Less Common. Metals 3, 110 (1961).
13. A. E. Miller and A. H. Daane, Trans. A.I.M.E. 230, 568 (1964).
14. P. W. Bridgman, Proc. Am. Acad. Arts Sci. 83, 1-21 (1954).
15. P. W. Bridgman, Proc. Amer. Acad. Arts and Sci. 81, 165 (1952).
16. P. W. Bridgman, Proc. Amer. Acad. Arts and Sci. 82, 95 (1953).
17. H. T. Hall, J. D. Barnett and L. Merrill, Science 139, 111-112 (1963).
18. H. T. Hall and L. Merrill, Inorg. Chem. 2, 618 (1963).
19. D. B. McWhan and W. L. Bond, Rev. of Sci. Instr. 35, 626, (1964).
20. G. J. Piermarini and C. E. Weir, Science 144, 69 (1964).
21. A. Jayaraman and R. C. Sherwood, Phys. Rev. Letters 12, 22 (1964).
22. A. Jayaraman and R. C. Sherwood, Phys. Rev. 134A, 691 (1964).
23. D. B. McWhan and A. L. Stevens (to be published).
24. D. B. McWhan and A. Jayaraman, Appl. Phys. Letters 3, 129 (1963).
25. A. Jayaraman, Phys. Rev. 137A, 179 (1965).
26. A. Jayaraman, Phys. Rev. 135A, 1056 (1964).
27. A. Jayaraman, W. Klement and G. C. Kennedy, Phys. Rev. 131, 644 (1963).
28. A. Jayaraman, W. Klement, R. C. Newton and G. C. Kennedy, Jour. Phys. Chem. Solid 24, 7-18 (1963).
29. R. I. Beecroft and C. A. Swenson, J. Phys.

Chem. Solids 15, 234 (1960).
30. A. Jayaraman, W. Klement and G. C. Kennedy, Phys. Rev. 132, 1620 (1963).
31. A. Jayaraman (under publication).
32. R. E. Hanneman and H. M. Strong, J. Appl. Phys. 36, 523 (1965).
33. L. D. Landau and E. M. Lifshitz, "Statistical Physics", Pergamon Press, Ltd., Addison-Wesley Publishing Co., Inc. 1958, p. 260.
34. L. F. Bates and M. H. Newmann, Proc. Phys. Soc. (London) 72, 345 (1958).
35. J. H. Wernick and S. Geller, Acta. Cryst. Cambridge 12, 662 (1959).
36. J. C. Jamieson, Fourth Annual Report to ONR on High Pressure Research, NONR-2121 (03), (1960).
37. J. D. Bernal, Zeit. fur Krist. 112, 4 (1959).
38. A. Y. Cabezas, I. Lindgren and R. Marrus, Phys. Rev. 122, 1796 (1961).
39. G. C. Kennedy, A. Jayaraman and R. C. Newton, Phys. Rev. 126, 1363 (1962).
40. A. Jayaraman, W. Klement, Jr., and G. C. Kennedy, Phys. Rev. Letters 10, 387 (1963).
41. R. Sternheimer, Phys. Rev. 78, 235 (1950).
42. H. T. Hall, L. Merrill and J. D. Barnett, Science 146, 1297 (1964).
43. F. H. Spedding. J. J. Hanak and A. H. Daane, Trans. A.I.M.E. 212, 379 (1958).
44. M. F. Manning and H. M. Krutter, Phys. Rev. 51, 761 (1934).
45. N. F. Mott, Proc. Phys. Soc. (London) 46, 680 (1934).
46. S. L. Altmann and A. P. Cracknell, Proc. Phys. Soc. 84, 761 (1964).
47. R. A. Stager and H. G. Drickamer, Phys. Rev. 131, 2524 (1963).
48. P. C. Souers and G. Jura, Science 140, 481-483 (1963).

ELASTIC CONSTANTS, PRESSURE AND THE ALKALI HALIDES

Charles S. Smith and D. E. Schuele, Physics Department,
Case Institute of Technology, Cleveland, Ohio
and W. B. Daniels, Solid State and Materials Programme,
Princeton University, Princeton, New Jersey

Abstract

The variation with pressure of the single crystal elastic constants of the alkali halides LiF, NaF, NaCl, KCl and RbI has been measured by the ultrasonic pulse-echo method. The results can be expressed directly in terms of mode gamma's, $\gamma_i = -d\ln\omega_i/d\ln V$, by the relation $\gamma_i = -B_T d\ln t_i/dP$, where t_i is the observed transit time of the i-th acoustic mode. The values of γ_i can be used to compute exactly values, $\overline{\gamma}_L$, applicable in the T^3 region, of the Gruneisen parameter $\gamma = \beta V B_s/C_p$, and with some approximation, values, $\overline{\gamma}_H$, applicable to high temperatures. The $\overline{\gamma}_L$ values agree well with γ_0, obtained by White from direct thermal expansion measurements in the T^3 region. In the alkali halide sequence given, $\overline{\gamma}_L$ falls monotonically from a 'normal' value of 1.85 for LiF to approximately zero for RbI. This behavior is intimately connected with the stiffness C_{44} and its γ_i, both of which also fall monotonically in the sequence, γ_{44} becoming negative for KCl and RbI. These facts can be understood qualitatively from the simplest Born model for the crystals, as Blackman had outlined before the data were available. The cation-anion repulsive interaction contributes negatively to C_{44}, and increasingly so to γ_{44} as the exponent, n, of the interaction increases in the sequence.

INTRODUCTION

It is well known that the measurement of the elastic constants of a solid leads to a value of the Debye temperature that describes very well the low temperature lattice specific heat of the solid. The pressure variation of the elastic constants leads similarly to a value of the Gruneisen constant that describes closely that part of the low temperature thermal expansion property that arises in lattice vibrations.

It is the purpose of this paper to summarize measurements of the latter type that have been made in recent years on a selected sequence of five alkali halides at Case and at

Princeton. The detailed data will be found in papers by Miller and Smith (1) (LiF and NaF), Bartels and Schuele (2) (NaCl and KCl), and Daniels (3) (RbI). This research on the pressure variation of the elastic constants of single crystals was stimulated by the pioneering work of Lazarus (4) on the same subject. Our interest in the connection of such measurements with the low temperature Gruneisen constant has been especially great because of the possibility of comparison with the recent direct thermal expansion values of White (5) (6) on most of the same materials.

The thermal expansion of alkali halides with the NaCl structure-type is of special concern because of the possibility of a strong temperature dependence of the Gruneisen constant for certain members of the structure type. There is also the possibility of a wide variation within the family of the low temperature Gruneisen constant. These possibilities were pointed out by Blackman (7) from a theoretical analysis of the Born model for the binding energy long before the data were available. We conclude this paper by analyzing the experimental results, which confirm Blackman's predictions, in terms of the same model, with emphasis on the second point.

EXPERIMENT

The ultrasonic pulse-echo method for determining the elastic constants of a crystal is the technique that has been employed in the pressure investigations to be summarized here. The elastic constants are obtained from the measurement of the transit time, t, of an acoustic pulse in a crystal of length L. The experimental propagation direction in cubic crystals is usually [110], where one has the following relations,

$$C'_{11} = \rho\left(\frac{2L}{t_2}\right)^2, \quad C' = \rho\left(\frac{2L}{t_4}\right)^2 \text{ and } C = \rho\left(\frac{2L}{t_5}\right)^2. \qquad (1)$$

In these equations ρ is the density, $C = C_{44}$, $C' = (C_{11}-C_{12})/2$ and $C'_{11} = B_s + C'/3 + C$ (B_s is the adiabatic bulk modulus). The time t_2 refers to the longitudinal wave, and t_4 and t_5 to the two transverse waves. The shear constants C and C' play a central role in all considerations of the elastic property and it is an important feature of the method that they can be obtained without combining observations.

The pressure measurements consist, quite simply, of observing the change, Δt, of each transit time with pressure. Data for the longitudinal wave in NaCl are shown in Fig. 1. The pressure range of 1.5 kbar was deliberately kept low, but results in a 1 per cent change in transit time. Data for two fixed temperatures, 295°K and 195°K (dry ice) are shown; both curves were obtained under strictly hydrostatic conditions.

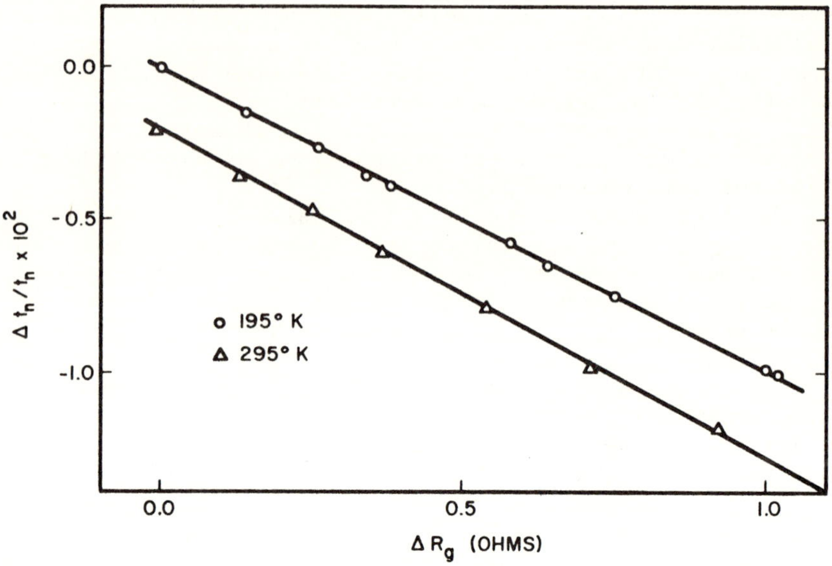

Fig. 1. Fractional change in transit time versus pressure gage reading for the longitudinal wave in NaCl. Pressure range is 1.5 kbar.

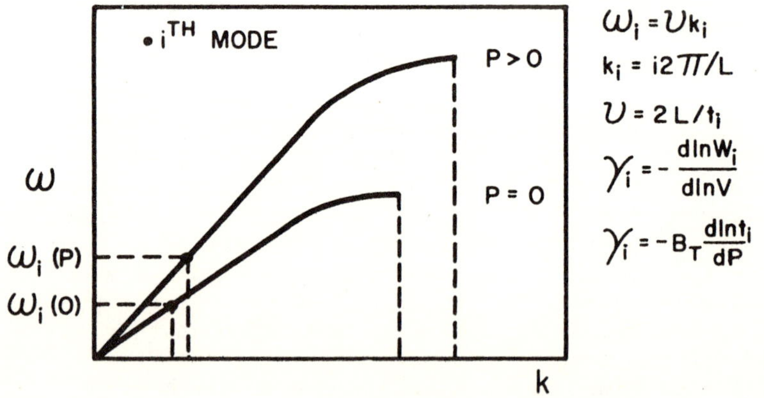

Fig. 2. Schematic dispersion curve for a single branch, in a given direction, of the acoustic spectrum. The effect of pressure on a given mode in the branch is shown.

Bartels and Schuele (2) describe in detail these measurements at two temperatures in NaCl and KCl. We shall be interested in the initial slope of such curves which have slight but detectable non-linearity. There is a typical 7 per cent difference in slope for the two temperatures of Fig. 1; the rule seems to be that 295° data shows the more negative slope.

The slopes of data plots like Fig. 1 are directly related to the mode gamma, $\gamma_i = -d\ln\omega_i/d\ln V$, that describes the anharmonic character of the acoustic vibrational mode of the crystal that has frequency ω_i. The relationship is derived in Fig. 2, which shows, schematically, dispersion curves $\omega(k)$ fastening attention on a particular mode i, using elementary definitions, and making the basic assumption that ω is linear with k. Upon further recognizing that the transit time t_i of modes on a given branch j of the acoustic spectrum is independent of i where the assumption applies, one has

$$\gamma_{ij} = -B_T \frac{d\ln t_j}{dP} , \qquad (2)$$

where B_T is the isothermal bulk modulus. This relation can be put in another useful form by using Eqs. (1), namely,

$$\gamma_{ij} = -\frac{1}{6}\left[1 + \frac{dC_j/d\ln r}{C_j}\right] . \qquad (3)$$

The mode gamma given by these expressions applies to all low frequency modes i on a given branch j.

The gammas obtained in this way for the three [110] acoustic waves in KCl are shown in Fig. 3. The values obtained from Eq. (2) at 295° and at 195° are plotted as a function of temperature and extended to T = 0 only to exhibit more plainly the possible temperature variation in γ_{ij}. This variation is less, generally, than that of $d\ln t/dP$ because of the temperature dependence of B_T and is usually a decrease with temperature. Figure 3 shows that the gammas for the three acoustic branches differ widely, contrary to Gruneisen's famous assumption. In particular, the shear modes C and C' differ especially much in alkali halides with the NaCl structure, γ_C being actually negative for KCl and RbI. This difference has the consequence, and arises in the physics, to be discussed below.

THERMAL EXPANSION

Mode gammas such as those shown in Fig. 3 are connected with the thermal expansion gamma, $\beta V B_S/C_p = \beta V B_T/C_v$, through the relation,

$$\frac{\beta V B_T}{C_v} = \overline{\gamma} = \frac{\Sigma_i \gamma_i c_i}{\Sigma_i c_i} , \qquad (4)$$

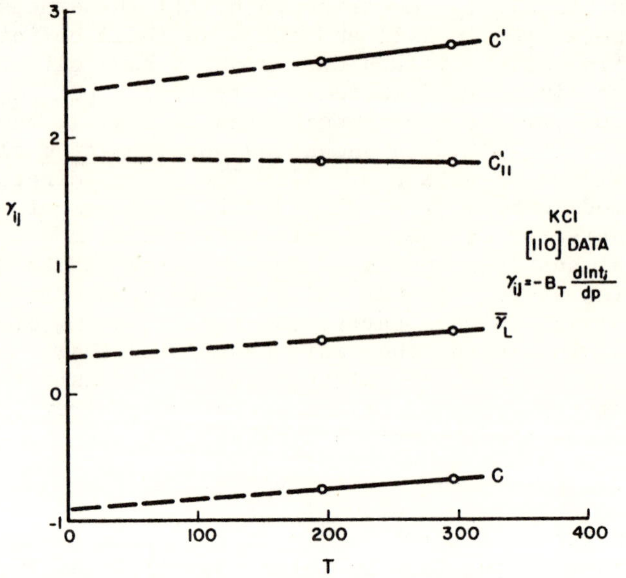

Fig. 3. Mode gammas for the three [110] acoustic waves in KCl. The lines are extended to T = 0 only to suggest more plainly the possible temperature variation.

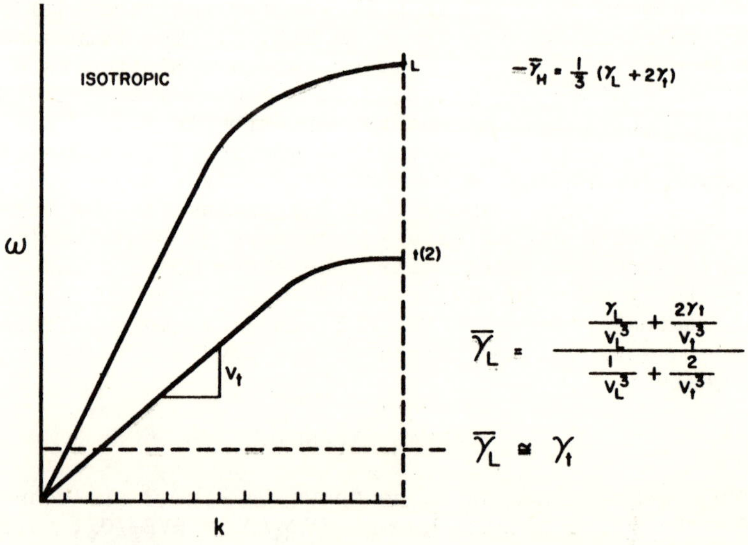

Fig. 4. Longitudinal and transverse dispersion curves for an isotropic solid. These have mode gammas, γ_ℓ and γ_t, respectively. $\bar{\gamma}_L$ denotes the derived, very low temperature, Gruneisen constant.

which follows directly from the Gruneisen equation of state. It is seen that the gamma describing thermal expansion is really a weighted average over all mode gamma's, the weighting factor being the Einstein specific heat, c_i, of the mode, which is essentially a measure of its degree of excitation. Thus $\overline{\gamma}$ may become temperature dependent through the weighting factors if the γ_i differ as they do.

The essential physical point concerning the weighting factors can be illustrated simply in terms of the isotropic model shown in Fig. 4, where there is a branch of longitudinal vibrational modes with velocity v_ℓ and with γ_i denoted by γ_ℓ, and two branches of transverse modes with velocity v_t and with γ_i denoted by γ_t. We suppose the γ's to be independent of index i except for the difference in branch type. At temperatures high enough that all modes are classically excited, Eq. (4) yields a value of the Gruneisen parameter denoted by $\overline{\gamma}_H$ and given by $\overline{\gamma}_H \doteq (\gamma_\ell+2\gamma_t)/3$. At very low temperatures the number of modes that is classically excited is proportional to v^{-3} as in ordinary Debye theory. The specific heat weighting in Eq. (4) may be replaced by a v^{-3} weighting (rigorously) to yield another value of the Gruneisen parameter, $\overline{\gamma}_L$, given by $(v_\ell^{-3}\gamma_\ell+2v_t^{-3}\gamma_t)/(v_\ell^{-3}+2v_t^{-3})$. We note that both $\overline{\gamma}_H$ and $\overline{\gamma}_L$ will be constant in the temperature ranges that are applicable, but that they may be different if <u>both</u> of two conditions hold, i.e. $\gamma_\ell \neq \gamma_t$, and $v_\ell \neq v_t$. The last condition is always met in solids and in isotropic materials one commonly has $v_\ell = 2v_t$, in which case $\overline{\gamma}_L$ becomes approximately γ_t because of the 16:1 weighting of the transverse modes. Thus, longitudinal modes can be ignored in the discussion below, although they have not been neglected in the calculations to be described. The expression given above for $\overline{\gamma}_H$ is approximate as it is based on the assumption that the γ_i are independent of i even into the dispersive range of $\omega(k)$. The expression for $\overline{\gamma}_L$ is rigorous, however, because the mode gammas are independent of i **in** the linear range of a given branch j.

The problem of calculating $\overline{\gamma}_L$ in real crystals from experimental data is enormously complicated by the fact of anisotropy; so much so that the physics of the weighting factors tends to become lost. The appropriate expression is

$$\overline{\gamma}_L = \sum_{j=1}^{3}\int_\Omega \gamma_j C_j^{-3/2} d\Omega \Big/ \sum_{j=1}^{3}\int_\Omega C_j^{-3/2}\, d\Omega, \qquad (5)$$

where $d\Omega$ is an element of solid angle and the v^{-3} weighting of the discussion has been replaced by $C_j^{-3/2}$, where C_j is the stiffness of the acoustic branch, j. We ignore the longitudinal branch for the reason given above and concentrate on the transverse branches, which in a cubic crystal show extremal values of the stiffness, C and C', in the [110] direction. In other directions the stiffnesses are intermediate. It is easy to show that the transverse γ_j will have extremal values in the [110] direction also, with intermediate values elsewhere.

We are led then to examine the mode gammas for C and C' and the weight of C relative to C' which will be given by $(C/C')^{-3/2}$. These quantities are shown in Fig. 5, for five alkali halides, where they are plotted for later reference as a function of n, the exponent in the repulsive term in ordinary Born theory. It is seen that $\gamma_{C'}$ is high and essentially invariant in the sequence, that γ_C is always small but decreases rapidly in the sequence and goes negative for the heavy members, KCl and RbI. Furthermore the relative weight of the C mode increases rapidly in the same sequence from a value less than unity for LiF to a very high value of 8 for RbI. We may then expect the numerous modes governed principally by stiffness C to play an increasingly dominant role in this sequence and to lead to a decreasing value of $\overline{\gamma}_L$ in the sequence, with $\overline{\gamma}_L$ possibly going negative if γ_C becomes negative enough and dominant enough.

Schuele (8) has carried out numerical calculations of $\overline{\gamma}_L$, using Eq. (5), the appropriate experimental data, and a computer program that sampled 184 equally spaced points over the unit triangle of a cubic crystal. The fine scale of the integration is a necessity in highly anisotropic situations like KCl and RbI. We emphasize that this calculation involves no microscopic theory and that there are really no assumptions involved other than that $\omega \sim k$ for acoustic modes of low enough frequency. The result, $\overline{\gamma}_L$, applies therefore in the T^3 region of temperature with just the same validity as does the Debye temperature, θ_D, computed from the elastic constants of the crystal.

The numerical results, $\overline{\gamma}_L$, are plotted in Fig. 6 for the five alkali halides. Also shown for comparison are values, $\gamma_0 = \beta V B_T / C_V$ obtained by White (5) from thermal expansion measurements in the T^3 region. As expected, $\overline{\gamma}_L$ falls rapidly in the sequence plotted because of the fall of γ_C and its rising weight in the same sequence. The 'elastic' $\overline{\gamma}_L$ also agree reasonably well with the 'thermal expansion' γ_0 in the sequence. There is no theoretical reason why they should not agree exactly; the residual discrepancies may be removed in the future as noted below.

The values given in Fig. 6 for NaCl and KCl were obtained (2) by carrying through the calculation using 295°K and 195°K pressure data and then making a linear volume extrapolation to T = 0. (In all cases weighting factors $C_j^{-3/2}$ characteristic of T = 0 were used.) This process subtracted 0.12 from the $\overline{\gamma}_L$ computed from 295° data for both NaCl and KCl. Only 295°K pressure data are available for LiF and RbI but one may plausibly expect for these the same small temperature dependence of the experimental gammas shown in Fig. 3 for

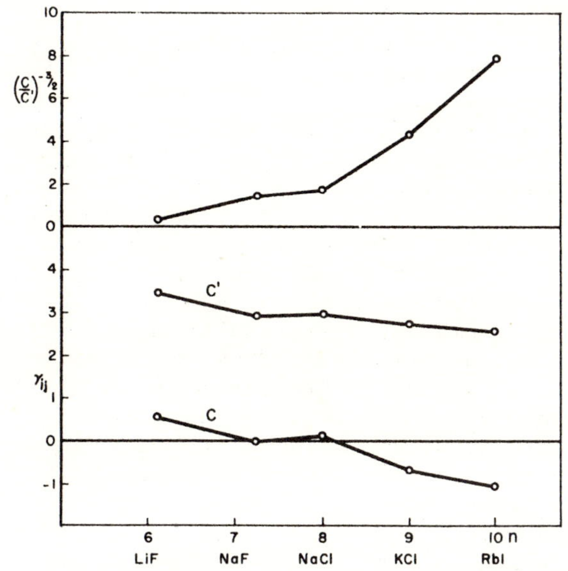

Fig. 5. Mode gammas for the two transverse modes in the [110] direction, C and C', and their relative weight $(C/C')^{-3/2}$ for five alkali halides. The two transverse modes in other directions have γ's that are intermediate and a relative weight nearer unity.

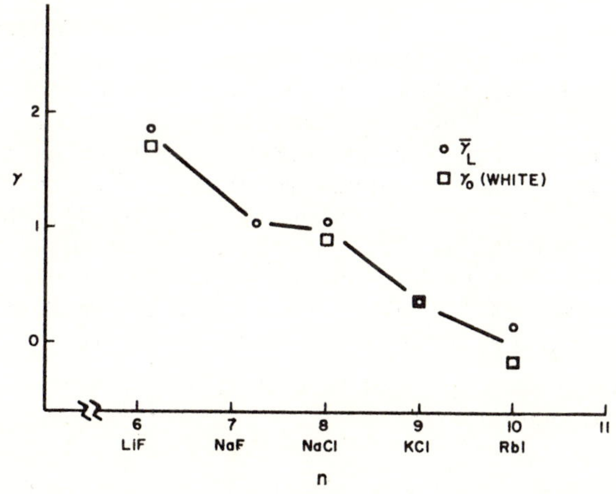

Fig. 6. The Gruneisen constant describing thermal expansion in the T^3 region for five alkali halides. $\bar{\gamma}_L$ is the 'elastic value', γ_0 the 'thermal expansion' value due to White.

KCl. Subtraction of 0.12 from the values plotted would give excellent agreement with γ_0 for LiF and materially improve that for RbI.

Both $\overline{\gamma}_L$ and γ_0 involve primary data, elastic gammas on the one hand and thermal expansion on the other. Both involve auxiliary elastic data, zero temperature $C_j^{-3/2}$ for $\overline{\gamma}_L$ and the zero temperature bulk modulus B for γ_0. We feel, surprisingly enough, that uncertainties in these quantities may be responsible for the residual discrepancies between $\overline{\gamma}_L$ and γ_0 for NaCl and RbI. The elastic constants of NaCl are not well known at T = 0; a change from the values used would affect both $\overline{\gamma}_L$ and γ_0, principally the latter. Low temperature C_j have never been measured for RbI; the values of $\overline{\gamma}_L$ and γ_0 shown are based on a plausible extrapolation. Different elastic constants would not visibly shift γ_0 because it is so small but different values, particularly the critical shear stiffness, C, would change $\overline{\gamma}_L$ as White has suggested (8). The low temperature elastic constants of these two materials are being carefully measured at Case Institute.

Apart from these details, the major feature of Fig. 6 is the steep decline in the sequence of the low temperature Gruneisen parameter which appears to go negative at RbI. This decline is expected from the purely phenomenological considerations of this section. The microscopic considerations underlying the experimental facts are considered in the next section.

BORN MODEL

A very simple model of the NaCl structure accounts for the major portion of the trend of γ_C and of $(C/C')^{-3/2}$, and hence of $\overline{\gamma}_L$, from one alkali halide to another that is shown in Figs. 5 and 6. In the classical Born theory of ionic crystals the binding energy per molecule is written as the sum of a Coulomb energy ϕ^c and a repulsive energy ϕ^r. In the simplest case the latter may be taken as $\phi^r = 6I(d)$, where I(d) is a two-body, central interaction between an anion and its 6 nearest-neighbor cations located at equilibrium distance d_o. It is also convenient to take I(d) in the older form Ad^{-n}. The binding energy of the crystal then becomes,

$$\phi = 6Ad^{-n} - \alpha e^2 d^{-1}. \qquad (6)$$

The equilibrium condition,

$$\frac{1}{6} \alpha \frac{e^2}{d_o} = nI(d_o), \qquad (7)$$

and the expression for the bulk modulus,

$$\Omega B = -\frac{4}{9}\alpha\frac{e^2}{d} + \frac{2}{3}(n+3)nI(d), \tag{8}$$

allow the parameters describing the short range repulsive interaction to be determined from a knowledge of d_o and B_o. (Ω_o is the molecular volume and α the Madelung constant.) In this work room temperature values of d_o and B_o (adiabatic) have been used in order to treat the five alkali halides uniformly. The resulting short range parameters are conveniently expressed in the form nI and n; the repulsive exponent, n, obtained in this way has been used in Fig. 5 and 6 and later ones.

The dependence of ϕ^c and ϕ^r on elastic shear strain S is readily calculated for this model (9) and yields for the shear constants C' and C the following expressions:

$$\Omega C' = -1.34e^2d^{-1} + n^2I, \tag{9}$$

and

$$\Omega C = +1.28e^2d^{-1} - 2nI. \tag{10}$$

The numerical factors arise from Coulomb lattice sums in a manner similar to the Madelung constant. The equilibrium condition has not been used in Eqs. (8), (9) or (10).

The motion under the shears C and C' of nearby ions in the (100) plane with respect to a central anion is shown in Fig. 7, which is incomplete but does permit qualitative features of Eqs. (9) and (10) to be deduced. It can be seen that the nearest-neighbor (N^2) cations will indeed contribute negatively to C but positively to C' because the distance, d, increases for all (two) ions that move in the former case. Because d changes only in the second order of small strains, n appears to the first power in C, but to the second power in C' where there is a first order change of d with strain. Equations (9) and (10) show that the Coulomb contribution to the stiffness also has opposite sign in C and C'.

Before leaving Fig. 7 one may also note the effect of a possible repulsive interaction between next-nearest-neighbor (N^3) anions of which only four of twelve are shown. The relative motions of these are first order in the strain for C and the distance of two of them decreases. The corresponding stiffness contribution to C will therefore be positive and involve the square of an appropriate exponent. The same conclusion is reached for C' when all twelve anions are considered. The important point is that in ΩC the N^3 interaction contributes with sign opposite to that of the N^2 interaction; the same statement holds for the pressure variation of C, expressed as $\Omega dC/d\ln r$. These considerations will be used qualitatively in subsequent discussion, but for the present we revert to the simplest model with only N^2 interactions.

The anisotropy of a cubic crystal is described by the

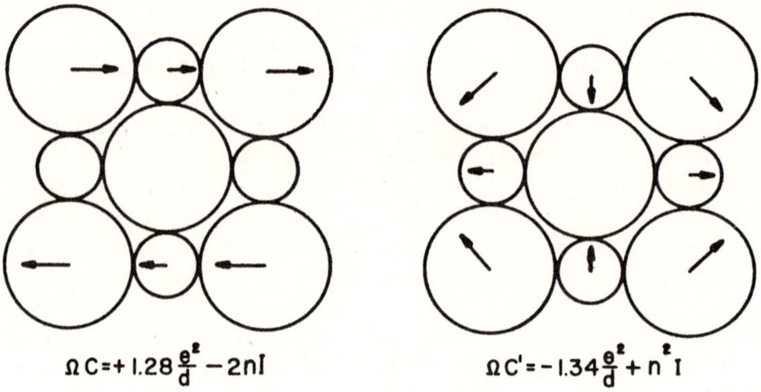

Fig. 7. A (100) plane of the NaCl structure type with the motions, relative to the central anion, of nearby ions under the shear strains C and C'.

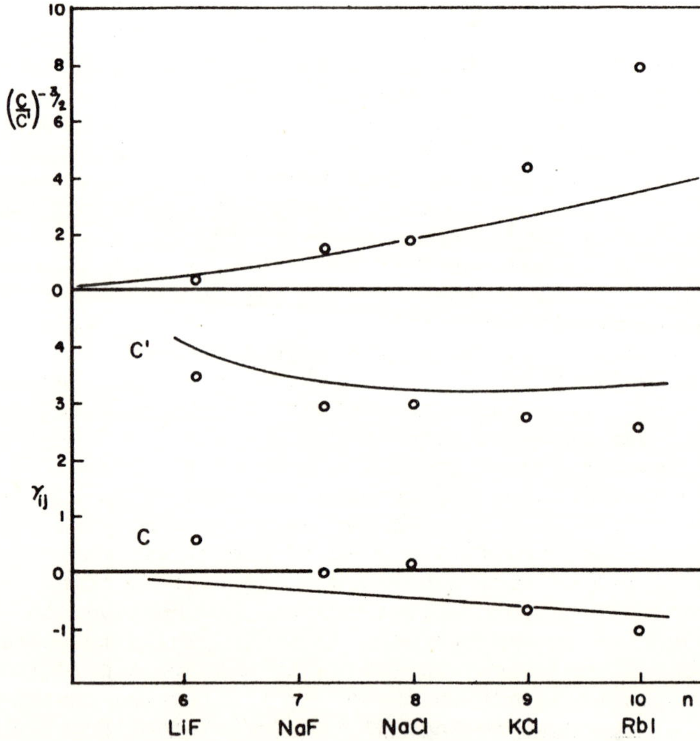

Fig. 8. The mode gammas of transverse waves of stiffness C and C' and their relative weight $(C/C')^{-3/2}$. The abscissa is the exponent describing the repulsive, nearest-neighbor interaction in the Born model, the predictions of which are shown as solid curves.

ratio C/C'. If the equilibrium condition is now inserted in Eqs. (9) and (10), eliminating the term nI, this ratio takes the simple form

$$(C/C')^{-1} = -1.92 + 0.42n \tag{11}$$

a linear function of the repulsive exponent alone. This expression, raised to the three-halves power is just the relative weight of the C acoustic mode that is predicted by the model.

The model also provides another simple, but pertinent, relation. Differentiation of Eq. (10) with respect to d and the use of Eqs. (3) and (7) followed by some rearrangement yields the extraordinarily simple relation,

$$\gamma_C = -\frac{1}{3}\left(\frac{C}{C'}\right)^{-1}. \tag{12}$$

Thus the gamma of the critical acoustic branch is a linear function of n for this model and is always negative.

The model described above is the one for which Blackman (7) made the prediction that $\overline{\gamma}_L$ should go negative for a sufficiently large n. His approximate calculation of $\overline{\gamma}_L$ for the value n = 21, gave $\overline{\gamma}_L = -0.62$. The central physical reason for this prediction lies in the dependence on n of γ_C and its relative weight $(C/C')^{-3/2}$, as set forth above in a straightforward fashion.

The relations (11) and (12) are shown in Fig. 8 with the corresponding experimental data for the five alkali halides. It will be observed that the relative weight $(C/C')^{-3/2}$ for the model increases quite rapidly with n, that the model predicts the observed values well for LiF, NaF and NaCl, but that KCl and RbI exceed the model weighting factor by a large margin. The γ_C for the model is small, is always negative, and decreases with increasing n. The experimental γ_C are low, decline in the same general way, but more rapidly, starting from positive values and winding up below the theoretical value for KCl and RbI. Thus the simplest model predicts the trends of the important quantities correctly and hence accounts for the major portion of the large decline of $\overline{\gamma}_L$ shown in Fig. 6.

It appears that γ_L goes negative at RbI (n = 10) rather than at a much large value of n, as predicted by Blackman, because of effects, not contained in the model, that raise the weighting factor and lower the gamma of the stiffness C. Two microscopic effects that may be considered are the possible presence of N^3 interactions and the certain presence of many-body non-central character in the N^2 interaction.

A detailed comparison of all three elastic constants and their pressure variation with this simplest Born model has been made for all five alkali halides with a view to suggesting which of these possibilities is the more important in accounting for the deviations. Without going into detail, it may be said that for all five materials the model accounts very well for C' (and of course B), and for the pressure variation of C' and B. Large discrepancies occur for C and its pressure variation, however. It is in just these quantities that the N^3 interaction occurs with sign opposite to that of the N^2 interaction, as noted above. One may use this fact to enquire qualitatively if the agreement between a more general central force model and experiment would be better.

For both KCl and RbI, the simple model accounts very well for the pressure variation of C but predicts a stiffness that is too large. Thus the predicted weighting factor (Fig. 8) is low and the predicted γ_C (see Eq. (3)) not negative enough. Consideration of the N^3 interaction would worsen agreement in both respects, contributions attributed to many-body effects being increased. It can therefore be asserted that the N^3 interaction is absent in KCl and RbI and that these materials deviate from the simple model because of the many-body character of the N^2 interaction.

LiF is a different case. Here the anisotropy C/C' is greater than unity and this circumstance is popularly supposed to arise because of the N^3 interaction. Figure 8 shows, however, that the simple model itself is quite capable of giving the right value of the anisotropy in the absence of the N^3 interaction, and in fact, gives $C/C' > 1$ for all values of n less than ~ 7. The model does not, however, give a positive value of γ_C as observed. A detailed analysis of LiF by Miller and Smith (1) shows that the N^3 interaction does improve the agreement of theory with experiment for both C and its pressure variation, but that very little of it is required. They suggest that an N^3 contribution to the equation of equilibrium of about 8 per cent is required, leaving a modest portion of the total elastic picture to be accounted for by many-body effects. However, the major portion of the elastic property of LiF still arises in the N^2 interaction of the simple model.

The detailed analysis of NaF and NaCl is different and is an intermediate case. Here consideration of the N^3 interaction worsens the detailed agreement for C but improves it for the pressure variation of C, leaving the other quantities roughly unchanged. Miller and Smith estimate that only as little as 1 per cent of N^3 interaction in the equation of equilibrium can be tolerated unless many-body effects are to become implausibly large. In these materials, again, N^2 interactions account in the main for the elastic property and hence for their values of $\overline{\gamma}_L$.

This discussion may be summarized as follows:

a) The major portion of the decline of $\overline{\gamma}_L$ with n is implicit in the simplest Born model of the NaCl structure with only nearest-neighbor repulsive interactions.

b) $\overline{\gamma}_L$ for LiF, NaF, NaCl deviate from this model on the high side because of the presence of small amounts of next-nearest-neighbor interaction.

c) $\overline{\gamma}_L$ for KCl and RbI deviate on the low side because of the actual many-body character of the nearest-neighbor interaction.

ACKNOWLEDGEMENTS

The Atomic Energy Commission has supported the preparation of this paper. The authors are grateful to G. K. White for discussions, suggestions and pre-publication data.

REFERENCES

1. R. A. Miller and C. S. Smith, J. Phys. Chem. Solids 25, 1279 (1964).
2. R. A. Bartels and D. E. Schuele, J. Phys. Chem. Solids 26, 537 (1965).
3. W. B. Daniels, 1963 Copenhagen Conference on Lattice Dynamics.
4. D. Lazarus, Phys. Rev. 76, 545 (1949).
5. G. K. White, Proc. Roy. Soc. A (to be published).
6. J. G. Collins and G. K. White, Progress in Low Temperature Physics IV, 450 (1964).
7. M. Blackman, Phil. Mag. 3, 831 (1958).
8. D. E. Schuele and C. S. Smith, J. Phys. Chem. Solids 25, 801 (1964).
9. H. B. Huntington, Solid State Physics 7, 213 (1958).

PRESSURE DERIVATIVES OF THE ELASTIC CONSTANTS IN SOME ALKALI HALIDES

P. J. Reddy and A. L. Ruoff
Cornell University, Ithaca, New York

Abstract

The pressure derivatives of the adiabatic elastic stiffness constants of KBr, KI, RbBr, and CsBr at 300°K and of CsBr at 77°K have been measured by ultrasonic interferometry. The values found are:

	$\frac{dC_{11}}{dP}$	$\frac{dC_{12}}{dP}$	$\frac{dC_{44}}{dP}$
KBr	12.96	1.59	-0.328
KI	13.99	2.42	-0.241
RbBr	13.52	3.05	-0.362
CsBr	5.810	4.37	3.201
CsBr (77°K)	5.681	4.34	3.186

These data have been analyzed in terms of the Born model. Evidence is given which indicates that the fcc → sc transformation is a martensitic (shear) transformation.

Introduction

The original Born model of ionic binding in the alkali halides is a simple first order approximation which accounts surprisingly well for cohesive energies (1). However as would be expected the agreement is less good when attempts are made to calculate such physical quantities as the elastic constants (2) or the pressure of the fcc → sc phase transition (3), although the latter can be approximately calculated if different repulsive parameters are attributed to the ions in the two phases consistent with the usual empirical approach associated with the Born model (4).

The second order elastic constants and third order elastic constants provide the theorist with convenient tests of theories of cohesion. The third order constants can be obtained by measuring pressure derivatives and uniaxial stress derivatives of the second order elastic constants. The availability of these values at low temperatures would be particularly useful for this purpose. It is the goal of this laboratory to carry out such measurements. For several reasons we started out a year and a half ago with the more modest initial task of first measuring second order constants accurately and then moving into the area of pressure measurements using liquid media in the vicinity of room temperature. After these became routine we began using gas as a pressure medium and then at liquid nitrogen temperatures. Such measurements are now routine. There are several obvious extensions which we plan to make. We will very shortly attempt measurements at liquid helium temperatures. We have already tried to measure the effects of uniaxial stresses on the elastic constants of KI but microplasticity ffects have hindered progress thus far even in crystals irradiated with Co-60. However, Chang (5), in Huntington's laboratory, has succeeded in obtaining two combinations of uniaxial derivatives in irradiated NaCl and KCl, so that the experiment is plausible.

The present paper records some of our initial results. It has three facets:

1) It presents pressure derivatives of the second order elastic constants of KI, KBr, RbBr, and CsBr near room temperature and hence complements the earlier data on NaCl and KCl (6) and on LiF and NaF (7).

2) It presents the first set of pressure derivatives of elastic constants at low temperature.

3) It provides evidence to the effect that the fcc → sc transformation is a martensitic transformation.

Experimental

Sample Preparation.

The KBr, KI, and CsBr crystals were obtained from the Harshaw Chemical Company. The RbBr crystals were obtained from Semi-Elements Inc. The as-received crystals were cubes with 1/2" edges and either [100] or [110] orientation (within 2°). The acoustic surfaces were prepared by lapping with Clover Lapping and Grinding Compound 4A. Orientations were found by Laue back reflection. Final orientations were obtained by mounting the crystals with paraffin into lapping blocks and by finish lapping which produces accurately parallel faces. The distance between parallel faces was measured using a Pratt and Whitney supermicrometer and independently a Van Keuren light wave micrometer. These measurements being carried out in a temperature controlled gauge room. The orientations were known to within 0.5° and the thickness to 2×10^{-5} in. The crystals were annealed for one day at $.75T_M$ in a pure argon atmosphere; the anneal was followed by furnace cooling. After this treatment the crystals were handled with exceedingly great care.

Transducers of 0.4" diameter x-cut and y-cut quartz crystals with coaxial chrome and gold coated electrodes and 10Mc resonant frequency were obtained from the Valpey Crystal Co. The electrode assembly was constructed of a brass cylinder, which served as one electrode contact, surrounded by an insulating ring of boron nitride which in turn was surrounded by a brass sleeve which served as the other electrode contact.

Bonding of the transducer to the crystal was achieved with Salol (phenyl salicylate) for the 300°K studies and with Nonaq stopcock grease for the 77°K studies. The Salol was applied to the crystal and melted and the transducer was squeezed onto the crystal in a special arrangement between the jaws of a micrometer, the entire assembling being carried on at 65°C. The maximum thickness of the Salol is estimated to be less than 0.0001cm from the calculated rate of radial flow of liquid Salol. It was found that Salol was a satisfactory bond in pressurized silicone oil. There was a tendency for some reaction of Nonaq at room temperature, but this was not a problem in the low temperature studies.

Sample Holder.

The specimen-transducer assembly could be readily placed in and removed from the jig shown in Fig. 1. The entire jig assembly could conveniently be inserted into or extracted from the pressure vessel.

Ultrasonic Interferometer.

The electronic system for measuring transit

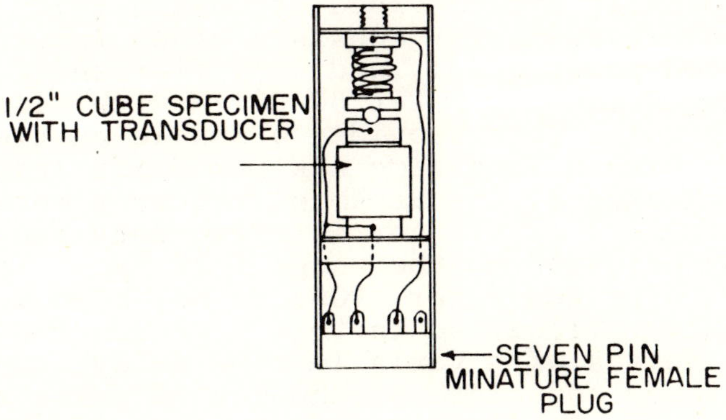

Fig. 1. Jig for Measuring Pressure Effects on Elastic Constants.

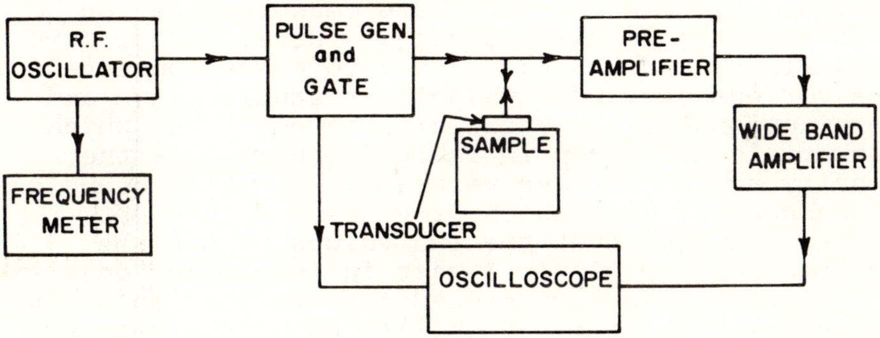

Fig. 2. Ultrasonic Interferometer.

times is illustrated in Fig. 2. The method used was first described in detail by Williams and Lamb (8). Further details are provided by Colvin (9).

The rf-oscillator consists of an internally constructed power supply and amplifier and a surplus U. S. Army Frequency Meter BC-221-N with slight modifications. This provides a stable carrier frequency with a long time maximum drift less than 5cps as checked by the Beckman frequency meter (1 part in 10^7). This carrier frequency can be continuously adjusted on the range 8-12.5Mc. The Arenberg pulse generator (PG650c) has been modified with a gated circuit so that it produces two phase coherent pulses with variable time delay, such that the amplitude and width of the second pulse can be adjusted. The receiver consists of an Arenberg preamplifier (620-SN-209), an Arenberg wide band amplifier (WA-600) and a Techtronics oscilloscope (545A) with a fast rise calibrated preamp plug in unit (Type L).

The procedure used to analyze the data is described in detail in reference 9, where a sample calculation is described in Appendix A.

Velocity - Elastic Constant Relation.

In all, two longitudinal wave velocities and three shear wave velocities are measured for a given type of crystal. All of these measurements can be made on a single [110] crystal. Here, two or more crystals have always been used to check consistency. The various pure wave velocities studied here and combinations of adiabatic elastic constants which they yield are shown in Table I. Here ρ is the density and v the wave velocity. There is hence a redundancy of data which provides a convenient check. We find that the two values of C_{44} are reproducible to within 1 part in 10^4 if and only if very careful handling procedures are used following annealing. The reproducibility of C_{11} between two crystals as measured directly from a [100] longitudinal wave is similarly 1 part in 10^4. We therefore use the directly obtained values of C_{11} and C_{44}. To calculate C_{12} we use the two remaining waves. The expected fractional error in C_{12} will of course be larger inasmuch as it depends upon the difference of two relatively larger numbers. The C_{12} values obtained here are consistent to better than 2 parts in 10^3.

Table I

Wave Type	Wave Direction	Particle Motion	ρv^2
Longitudinal	[100]	[100]	C_{11}
"	[110]	[110]	$\frac{(C_{11}+C_{12}+2C_{44})}{2}$
Shear	[100]	[010]	C_{44}
"	[110]	[001]	C_{44}
"	[110]	[1$\bar{1}$0]	$\frac{C_{11}-C_{12}}{2}$

The Liquid Pressure System.

The general features of the liquid pressure system are shown in Figs. 3 and 4. By rotating the generator piston a pressure close to the desired value could be obtained directly from the press load. Pressure was then measured, after an appropriate wait for temperature to return to normal following a pressure change, by use of two manganin gauges, one in the generator, the other in the test vessel. The manganin gauges were calibrated using the Hg freezing pressure of 7.565 kbars at 0°C as seen by volume change. We have made the assumption that the manganin gauge behaves in a linear fashion. Conical seals with nylon sleeves were used for electrical leads and a seven prong male connector is attached to the top of the piston. The jig assembly of Fig. 1 can be conveniently plugged into this connector.

The Gas Pressure System.

The gas generation system is similar in design to that described elsewhere (10). It is capable of producing pressures of 18 kbars although the present low temperature data (CsBr at liquid N_2 temperatures) involved pressures up to only 3 kbars.

The low temperature pressure vessel is made of Vascomax maraging steel. It is essentially a dead-ended cylinder with a single piston. The piston-cylinder seal is of the Bridgman type with successive washers made of 70-30 brass, tellurium lead, and potassium. The helium gas is brought into the chamber through 5/16" OD, 1/16" ID high pressure

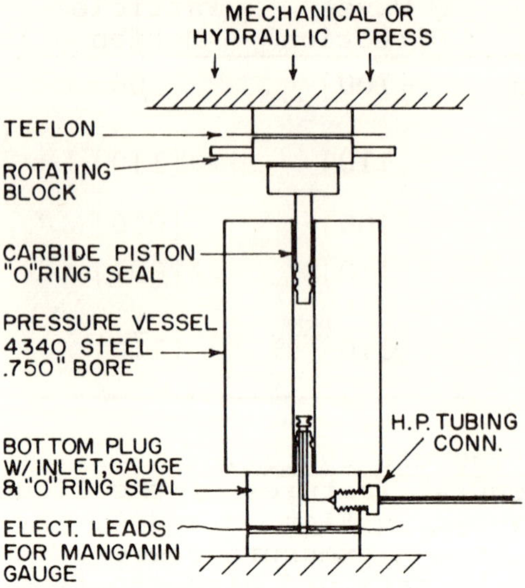

Fig. 3. Pressure Generator for Liquid System.

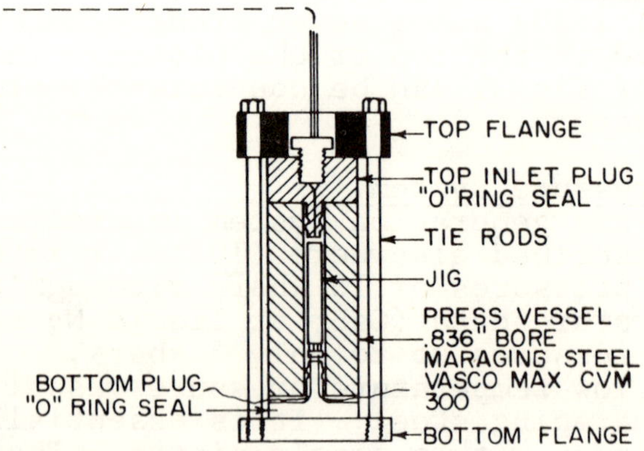

Fig. 4. Pressure Vessel for Liquid System.

stainless steel tubing which is silver soldered into (and passes through) the Bridgman mushroom. The single electrical lead (in addition to the vessel itself) passes through the tubing through a T-connection and into the liquid containing pressure chamber of Fig. 3 where it is connected to a lead passing out of that system. The helium gas is brought into the third port of the T-connector. This process eliminates the need for sealing electrical leads at low temperatures and has proved itself absolutely foolproof. Since it involved a liquid chamber which was already available, no additional work was involved. A similar concept has been used elsewhere (11).

The Nature of the fcc → sc Transformation.

Bridgman's (12) data show the fcc → sc transformation in RbBr beginning at 4450 bars and ending at 5050. We have measured the elastic constants to 4288 bars with no apparent deviation in any of the dC/dP from linear. The elastic constant matrix for the fcc lattice is clearly positive definite so that the transformation does not occur via a macroscopic shear instability.

When single crystal specimens were carried into the transformation region a burst phenomenon occurred near 4730 bars, i.e., at a pressure substantially above the transformation pressure. The shock is so intense (the volume change is 13.4%) that a loud report is associated with the transformation even though it takes place within the liquid pressure media. A similar burst phenomenon occurs in 69-31 Fe-Ni alloys, in this case at a temperature substantially below the transformation temperature characteristic of the polycrystalline alloy (13).

Microscopic examination at atmospheric pressure (before which the reverse transformation has also taken place) showed the surface upheavals characteristic of martensitic transformations. There are hence several features which strongly indicate a martensitic transformation:

(a) The sluggishness of the reaction, i.e., the range of pressure over which it occurs.

(b) The presence of burst phenomena.

(c) The characteristic surface upheavals.

When the kinetics of diffusion processes are considered it seems quite likely that many and perhaps most very high pressure transformations that occur at relatively low temperatures are martensitic in nature. It should be noted that the transformation pressure is the pressure at which the ac phase becomes more stable.

The exact mechanism of the transformations in RbBr is not known but it is likely that the transformation proceeds via a dislocation pole mechanism of the type involved in twinning (14) when the free energy favors the other phase.

Elastic Constant Results

The experimental results for the adiabatic constants are shown in Table II (15). These results were converted to isothermal data and bulk moduli were calculated; from the bulk moduli and the lattice parameter, the constants in the power law form of the Born theory were calculated. From this binding theory the second order elastic constants and their pressure derivatives were calculated using existing formulae (16). These relations account for nearest neighbor repulsive terms only. The results for the fcc crystals are shown in Table III. The agreement is good. It is, in fact, considerably better for the pressure derivatives than for the constants themselves.

The exponential law has also been used. Here the formulae of Nŕanyan have been applied (17,18). It should be noted that reference 17 contains a correction to equations in reference 18. We have used the formula

$$\begin{aligned} C'_{11} &= C_{11} + \eta(2C_{11} + 2C_{12} + 6C_{111} + 4C_{112}) \\ C'_{12} &= C_{12} + \eta(-C_{11} - C_{12} + C_{123} + 4C_{112}) \\ C'_{44} &= C_{44} + \eta(C_{11} + 2C_{12} + C_{44} + \tfrac{1}{2}C_{144} + C_{166}) \\ \text{and}\quad \eta &= -p/(C_{11} + 2C_{12}) \end{aligned} \qquad (1)$$

for calculating dC_{ij}/dP. The notation of the third order constants is that of Birch (19). The results for the fcc crystals are shown in Table IV. Here we note large discrepancies in the experimental and calculated values of dC_{44}/dP. The direction of the error is always the same.

Similar calculations have been carried out for CsBr. These are shown in Table V. Once again

Table II
Experimental Adiabatic Elastic Constants

Material	Constant	C_{ij} (10^{11} dyne·cm^{-2})	$\frac{dC_{ij}}{dP}$	T°K
KBr	C_{11}	3.419	12.96	300
KBr	C_{12}	.520	1.591	"
KBr	C_{44}	.508	-0.328	"
KI	C_{11}	2.677	13.99	"
KI	C_{12}	0.405	2.421	"
KI	C_{44}	0.369	-0.241	"
RbBr	C_{11}	3.155	13.52	"
RbBr	C_{12}	0.493	3.049	"
RbBr	C_{44}	0.3802	-0.362	"
CsBr	C_{11}	3.091	5.810	"
CsBr	C_{12}	.841	4.370	"
CsBr	C_{44}	.747	3.200	"
CsBr	C_{11}	3.343	5.681	77
CsBr	C_{12}	1.019	4.338	"
CsBr	C_{44}	.972	3.186	"

Table III
Isothermal Elastic Constants
(The Elastic Constants have Units of 10^{11}dyne·cm^{-2}.)

Mat'l	Constant	Calculated (Power Law) C_{ij}	dC_{ij}/dP	Experimental C_{ij}	dC_{ij}/dP
KBr	C_{11}	2.900	12.203	3.357	12.91
KBr	C_{12}	.682	1.649	.457	1.542
KBr	C_{44}	.682	-0.360	.508	-0.328
KI	C_{11}	2.28	12.44	2.616	13.66
KI	C_{12}	0.514	1.621	0.345	2.096
KI	C_{44}	0.514	-0.378	0.369	-0.241
RbBr	C_{11}	2.808	12.99	3.098	13.33
RbBr	C_{12}	0.582	1.587	0.436	2.856
RbBr	C_{44}	0.582	-0.414	0.3802	-0.362

Table IV
Isothermal Elastic Constants
(The Elastic Constants have Units of 10^{11}dyne·cm^{-2}.)

Mat'l	Constant	Calculated (Exponential Law) C_{ij}	dC_{ij}/dP	Experimental C_{ij}	dC_{ij}/dP
KBr	C_{11}	2.903	11.01	3.357	12.91
KBr	C_{12}	0.681	2.399	.457	1.542
KBr	C_{44}	0.681	-0.015	.508	-0.328
KI	C_{11}	2.283	14.42	2.616	13.66
KI	C_{12}	0.514	2.51	0.345	2.096
KI	C_{44}	0.514	-0.0105	0.369	-0.241
RbBr	C_{11}	2.802	10.621	3.098	13.33
RbBr	C_{12}	0.582	2.256	0.436	2.856
RbBr	C_{44}	0.582	-0.194	0.3802	-0.362

Table V
Isothermal Constants for CsBr
(The Elastic Constants have Units of 10^{11}dyne·cm^{-2}.)

Constant	T°K	Calculated (Exponential Law) C_{ij}	dC_{ij}/dP	Experimental C_{ij}	dC_{ij}/dP
C_{11}	300	3.413	5.246	2.969	5.559
C_{12}	"	.500	3.794	.719	4.159
C_{44}	"	.500	1.491	.747	3.200
C_{11}	77	3.730	4.118	3.300	5.592
C_{12}	"	0.822	3.102	.978	4.249
C_{44}	"	0.822	1.029	.972	3.186

dC_{44}/dP is algebraically too small.

These large deviations for dC_{44}/dP can be partially attributed to the need of considering second nearest neighbor interactions (20).

Acknowledgment

This work was supported by the Advanced Research Projects Agency.

We would like to thank the following people for their help: G. T. Chevalier and D. E. Testut in design and operation of pressure equipment and R. Martinson in design and construction of the ultrasonic system. We also appreciate discussions with P. B. Ghate and K. M. Koliwad.

References

1. M. Tosi, Preprint of: "Cohesion of Ionic Solids in Born Model" Solid State Physics, Volume 16, edited by F. Seitz and D. Turnbull, Academic Press, Inc., New York, 1965.
2. H. B. Huntington, "The Elastic Constants of Crystals" reprinted from Solid State Physics, Volume 7, edited by F. Seitz and D. Turnbull, Academic Press, Inc., New York, 1958.
3. C. E. Weir and G. J. Piermarini, J. Research National Bureau of Standards.
4. M. P. Tosi and F. G. Fumi, J. Phys. Chem. Solids 23, 359 (1962).
5. Z. P. Chang, Ph.D. Thesis, Rensselaer Polytechnic Institute, Troy, New York, 1964.
6. D. Lazarus, Phys. Rev. 76, 545 (1949).
7. R. A. Miller and C. S. Smith, J. Phys. Chem. of Solids 25, 1279 (1964).
8. J. Williams and J. Lamb, J. Acoust. Soc. Am. 30, 308 (1958).
9. A. D. Colvin, M.S. Thesis, Rensselaer Polytechnic Institute, Troy, New York, 1959.
10. D. Langer and D. M. Warschauer, Rev. Sci. Instr. 32, 32 (1961). The author would like to thank D. M. Warschauer and T. Deutch of the Raytheon Co. for valuable discussions and for drawings of their systems.
11. W. S. Gorel, B. McDowell, and T. A. Scott, Rev. Sci. Instr. 36, 99 (1965).
12. P. W. Bridgman, Z. Krist. 67, 363 (1928).
13. E. S. Machlin and M. Cohen, J. Metals (Trans. AIME) 3, 746 (1951).
14. A. H. Cottrell and B. A. Bilby, Phil. Mag. 42, 573 (1951).
15. The room temperature values of CsBr have also been measured by K. M. Koliwad in this laboratory.
16. W. B. Daniels and C. S. Smith in "The Physics of High Pressures" Gordon and Breach Science Publishers, Inc., New York (1963), p. 50.
17. A. A. Nŕanyan, Soviet Physics-Solid State 5, 129 (1963).
18. Ibid, 5, 1361 (1964).
19. F. Birch, Phys. Rev. 71, 809 (1947).
20. P. B. Ghate, Materials Science Center Report #331, March 31, 1965.

DISCUSSION

By D. Lazarus and A. L Ruoff

D. Lazarus: Are your values of C'_{ij} effective elastic constants or true elastic constants?

A. Ruoff: We have used the $\rho_o v_o{}^2$ to calculate the C_{ij} and ρv^2 to calculate the C'_{ij}; thus the C'_{ij} are effective constants. There are several other ways of writing effective constants (1).

We have given the theoretical expressions for our effective constants and have made a comparison of calculated and experimental values of $(C'_{ij} - C_{ij})/P$; this is perfectly legitimate. However, for other purposes, it is important to convert the effective elastic constants C'_{ij} to the constants C''_{ij}, which are derivatives of internal energy. We have not done this and the reader who wishes to use our results in this manner should convert our data (2).

1. R. N. Thurston and K. Brugger, Phys. Rev. 133, A1604 (1964).
2. R. N. Thurston, J. Acoust. Society Am. 37, 348 (1965).

DISCUSSION

By H. G. Drickamer

We have measured lattice parameters of a number of crystals having the NaCl or CsCl structure to high pressure (NaCl - 260 kbars, KCl - 340 kbars, CsCl - 460 kbars, MgO - 320 kbars, CaO - 230 kbars, CaS - 240 kbars). The alkali halides all showed good agreement with Bridgman's data to 100 kbars and with shock wave data to 250 kbars.

A simple version of the Born-Mayer-Huggins Equation for ionic crystals can be written

$$P = -\left(\frac{\partial W_L}{\partial V}\right)_T + \frac{\alpha}{\beta} T \qquad (1)$$

where $\alpha = + \frac{1}{V}\left(\frac{\partial V}{\partial T}\right)_P$, $\beta = - \frac{1}{V}\left(\frac{\partial V}{\partial P}\right)_T$, and the lattice energy

$$W_L = - \frac{\alpha_m (Ze)^2}{R} - \frac{C}{R^6} - \frac{D}{R^8} + Be^{-R/\rho}. \qquad (2)$$

The terms involved are the Madelung term, van der Waals' terms and repulsion term. α_m, C, and D can be calculated from known data at one atmosphere. B and ρ can be evaluated from the volume at one atmosphere and the initial compressibility. Using these values our data for NaCl, KCl, CsCl and MgO can be fitted to the highest pressures. It is rather remarkable that such a simple, semi-classical equation fits the data over such a range of pressure and volume. (At the highest pressures V/V_0 for CsCl is 0.565.)

For CaO, and more strongly for CaS, the high pressure compressibility is definitely less than is predicted by the above equations. It can be shown that no series of exponentials of the form of the last term in (2) will fit the data. Evidently there is a significant component of covalent binding in these compounds even though they have the NaCl structure.

SESSION F

PHASE TRANSFORMATIONS AND LATTICE PROPERTIES

Chairman: A. B. PIPPARD

The Interaction of Local Modes with Stress Fields

L. A. Girifalco
School of Metallurgical Engineering
and
Laboratory for Research on the Structure of Matter,
University of Pennsylvania

1. Introduction

Many processes of solid-state and metallurgical interest are controlled, or strongly influenced by point defects. Such processes include diffusion, radiation damage recovery, homogenization of solid solutions, phase transformations and creep. Furthermore, all crystals contain defects that have strain fields associated with them, so that kinetic processes in crystals always take place in the presence of internal stresses. Often, the rates of such processes are determined by diffusion to sources of internal stress such as precipitate particles, grain boundaries and dislocations. It is apparent, therefore, that an analysis of the energetics of defect formation and migration in the presence of stress fields is important in determining the extent to which stresses influence solid state reactions.

In particular, the temperature dependence of formation and migration volumes, and of the binding between defects and internal stress sources, is determined by the perturbation of crystal vibration spectra by stress fields. This paper consists of a discussion of the application of the lattice dynamics of defect crystals to the theory of the entropy of defect formation and migration in the presence of stresses.

A considerable amount of work has been done in recent years on the lattice dynamics of defect crystals.[1,2] This work shows that a defect can have two different effects on the frequency spectrum of the perfect crystal. First, the normal mode frequencies can be shifted in value by an amount of the order of magnitude of the spacing between the frequency levels of the crystal. Second, a few of the modes may be displaced completely out of the band of perfect crystal vibrations and give rise to modes in the defect crystal having frequencies higher than the maximum frequency of the perfect crystal. Since the perfect part of the crystal can transmit only waves that have in-band frequencies, such displaced modes will have amplitudes that are highly localized in the vicinity of the defect. The physical origin of the localization of defect modes is to be found in the changes of mass and force constants among atoms when a defect is introduced in the crystal. An extremely light impurity atom, for example, will behave as an Einstein oscillator thereby giving rise to local modes. Also, an interstitial atom will induce local mode formation since the coupling between the interstitial and its neighbors is much stronger than the coupling among other atoms in the crystal. Local mode formation may be thought of as the result of an effective uncoupling of the defect from the rest of the crystal caused by differences in masses and force constants between the defect and perfect crystals.

In the theory of point defects, the localization of vibrational modes and the perturbation of the in-band modes is reflected in the entropies of defect formation and migration. For stress free crystals containing a small number of defects, the local modes are not important thermodynamically except insofar as they determine the values of the entropies. In stressed crystals, however, we are interested in the free energy of interaction of the defect with the stress field and then the local modes may be quite important. To see this, consider a point defect in the field of an internal stress source such as a dislocation. To compute the entropy of interaction of the defect with the stress field, we need to know the effect of the stresses on the vibrational frequencies. The in-band modes are extended throughout the crystal so that they are affected by the entire strain field. The localized modes, however interact with the field in the vicinity of the defect. The extended and localized modes must therefore be treated differently.

The quasi-harmonic model represents the simplest approximation that can be made in treating the interaction of normal modes with stress fields. In this model, the potential energy of the crystal is taken as quadratic in the atomic displacements, but the atomic coupling coefficients, or force constants, are taken to be functions of strain. Within the quasi-harmonic approximation, a rigorous derivation of the effect of stresses on the frequencies would require the determination of the variation of the coupling coefficients with strain, and the solution of the lattice dynamics problem of a defect crystal subject to a generalized stress. Such an approach would meet with such great practical difficulties, that every attempt should be made to develop simpler methods. In this article, a simplified treatment will be presented based on a generalized Gruneisen formulation.

In Section II, a brief presentation of the lattice dynamics of defect crystals is given. We are primarily interested in the defect induced changes in mass and force constants necessary for local mode formation, and will therefore adopt the work of Litzmann[(3)] which is convenient for this purpose.

In Section III, the problem of determining numerical criteria for the formation of local modes in copper will be examined. Consideration will be restricted to the vacancy, the vacancy saddle point configuration, the split-interstitial and the split-interstitial saddle point configuration. A generalized Gruneisen method for describing the strain dependence of the vibration frequency will be developed in Section IV. Finally, the results will be incorporated into the theory of the entropy of defect formation and migration in Section V, and numerical results will be presented for copper.

II. Lattice Dynamics of Defect Crystals

We first consider a perfect, monatomic crystal with a unit cell defined by three vectors $\vec{a}_1$, $\vec{a}_2$, $\vec{a}_3$. The crystal contains N atoms and has the same shape as the unit cell so that it forms a parallelepiped defined by the three vectors $N^{1/3}\vec{a}_1$, $N^{1/3}\vec{a}_2$, $N^{1/3}\vec{a}_3$. In the harmonic approximation,

$$V = 1/2 \sum_{\substack{m,n \\ \alpha,\beta}} \Phi^{\alpha\beta}_{mn} u^m_\alpha u^n_\beta \qquad (1)$$

where u^n_β is the βth component of the displacement of the nth atom from its equilibrium position and the coupling coefficients, or force constants, are defined by

$$\Phi_{mn}^{\alpha\beta} = \left(\frac{\partial^2 V}{\partial u_\alpha^m \, \partial u_\alpha^n} \right)_0 \tag{2}$$

the derivatives being evaluated at zero displacement. α and β take on the values 1, 2, 3; m and n take values from 1 to N, and the energy of the static crystal at equilibrium is taken to be zero.

The equations of motion of the crystal are readily obtained from equation (1) and are given by

$$M \ddot{u}_\alpha^n + \sum_{m,\beta} \Phi_{mn}^{\alpha\beta} u_\beta^m = 0 \tag{3}$$

where M is the atomic mass. It is well known that if we adopt the periodic boundary conditions

$$u_\alpha^{n'} = u_\alpha^{n''} \tag{4}$$

where n' and n'' label corresponding atoms on opposite faces of the crystal, that equation (3) has solutions of the form

$$u_\alpha^n = A_\alpha \, e^{i\vec{k}.\vec{R}_n} \, e^{-i\omega t} \tag{5}$$

where $\vec{A}$ is a constant vector, $\vec{R}_n$ is the lattice vector given by

$$\vec{R}_n = n_1 \vec{a}_1 + n_2 \vec{\alpha}_2 + n_3 \vec{a}_3 \tag{6}$$

(n_1, n_2, n_3) being the integers labeling the mean position of the nth atom, and k is the wave number vector given by

$$\vec{k} = \frac{1}{N} 1/3 \, [n_1 \vec{b}_1 + n_2 \vec{b}_2 + n_3 \vec{b}_3] \tag{7}$$

The $\vec{b}$'s are the reciprocal lattice vectors defined by

$$\vec{a}_i . \vec{b}_j = 2\pi \, \delta_{ij} \tag{8}$$

and the $\vec{k}$ - values are restricted to the first Brillouin zone. The frequencies $\omega = \omega(\vec{k})$ are solutions of the secular determinant

$$| C_{\alpha\beta} - M\omega^2 \delta_{\alpha\beta} | = 0 \tag{9}$$

where

$$C_{\alpha\beta} = \sum_{\ell} \Phi^{\alpha\beta}(\ell)\, e^{i \vec{k}.\, \vec{R}_\ell} \tag{10}$$

The two subscripts (m, n) in the force constants have been replaced by a single index, since $\Phi^{\alpha\beta}_{mn}$ depends on m and n only through the difference ℓ = m-n.
The general solutions of equation (3) can be written as linear combinations of the special solutions (5).

$$u^n_\alpha = \sum_{\vec{k}} A^{\alpha}_{\vec{k}}\, Q_n(\vec{k}) e^{-i\omega(\vec{k})t} \tag{11}$$

where

$$Q_n(\vec{k}) \equiv e^{i\vec{k}\,.\,\vec{R}_n} \tag{12}$$

The $Q_n(\vec{k})$ satisfy the following conditions:

$$\sum_n Q_n(\vec{k})\, Q^*_n(\vec{k}') = N \delta_{\vec{k},\vec{k}'} \tag{13}$$

$$\sum_{\vec{k}} Q_n(\vec{k})\, Q^*_{n'}(\vec{k}) = N \delta_{nn'} \tag{14}$$

Equations (13) and (14) follow from the translational symmetry of the lattice and the restriction of the $\vec{k}$ values to the first Brillouin zone.

Now consider a crystal containing a point defect, but otherwise perfect. We are interested only in defect crystals with atomic masses that are all identical. Separate the crystal into two regions D and P such that the region D contains the defect and all atoms that are sensibly affected by the defect, and P contains the rest of the crystal. The force constants in region P are the same as in the perfect crystal, but in region D, the force constants take different values which we denote by $U^{\alpha\beta}_{mn}$

The frequencies of the local modes can be determined as follows:(3)
First, write the equations of motion in regions P and D.

Then seek trial solutions in the usual way for the defect region

$$u_\alpha^n(t) = V_\alpha^n e^{-i\,{}^*\omega t} \quad \text{(D region)} \qquad (15)$$

but write the amplitudes for the P region as a linear combination of the modes of the perfect crystal. The coefficients of this linear combination can be determined from the equations of motion of the defect crystal when the D region is assumed frozen, using the normalization condition (13). The expression for the coefficients can then be substituted into the equations of motion for region D, and it is found that the frequencies are determined by the determinantal equation

$$|M\,{}^*\omega\,\delta_{\alpha\beta}\delta_{mn} - (B_{mn}^{\alpha\beta} + U_{mn}^{\alpha\beta})| = 0 \qquad (16)$$

where

$$B_{mn}^{\alpha\beta} = \sum_{m',n'\beta'}^{P} A_{m'n'}\,\Phi_{m'n}^{\alpha\beta'}\,\Phi_{mn'}^{\beta'\beta} \qquad (17)$$

$$A_{m'n'} = \frac{1}{N}\sum_{\vec{k}} \frac{Q_{m'}(\vec{k})\,Q_{n'}^*(\vec{k})}{[{}^*\omega^2 - \omega^2(\vec{k})]} \qquad (18)$$

${}^*\omega$ and ω are the frequencies of the defect and perfect crystals respectively. Equation (18) defines the Green's function matrix used in the work of Montroll, Maradudin, et. al.

The computational difficulties arise in the calculation of $B_{mn}^{\alpha\beta}$. For localized modes, ${}^*\omega$ is greater than the maximum frequency of the perfect crystal, and then the Green's function can be expanded in powers of ${}^*\omega^{-2}$. It is then possible to write $B_{mn}^{\alpha\beta}$ in a series containing only ${}^*\omega$ and the force constants so that if the force constants are known, the local mode frequencies can be computed.

II. The Criteria of Local Mode Formation

We are interested in the values that the force constants must have to insure localized mode formation. If ω_m is the maximum frequency of the perfect crystal, localized modes are produced by force constants that give solutions to equation (27) for which

$$\frac{\overset{*}{\omega}}{\omega_m} > 1 \qquad (19)$$

Actual calculations of the criteria for local mode formation have been performed by Litzmann[(3)] for a substitutional impurity in a simple cubic crystal, and by Maradudin for a mass defect in a face-centered cubic crystal.[(2)] In both cases only nearest neighbor forces were considered. Although Litzmann considered changes in both force constant and mass, we will be concerned primarily with defects for which the atomic masses are the same as in the perfect crystal. For this case, Litzmann finds that if

$$\frac{k}{k_P} > 2 \qquad (20)$$

where k and k_P are force constants in the defect and perfect crystal respectively, then a local mode will be produced for which $\overset{*}{\omega}/\omega_m > 1.2$.

It is of interest to compare the Litzmann calculation for a mass defect with Maradudin's work in order to obtain an estimate of the effect of crystal structure on local mode formation. For a mass defect with no change in force constants, Litzmann finds that for

$$\frac{M'}{M} < 0.45$$

A local mode is produced for which

$$\frac{\overset{*}{\omega}}{\omega_m} > 1.2$$

Maradudin's results show that for the mass defect in the F. C. C. Crystal

$$\frac{M'}{M} < 0.40$$

for

$$\frac{\overset{*}{\omega}}{\omega_m} > 1.2$$

We see, therefore, that the criteria for local mode formation do not seem to depend very much on the crystal structure (for three dimensional crystals). In trying to decide on thenumber of local modes to associate with a defect, we will therefore ignore the particular geometry of the defect and apply the Litzmann criterion by assuming that a local mode appears only when (20) is satisfied. Since (20) corresponds to a local mode that is displaced out of the band by at least 20%, such a procedure clearly contains a safety factor. It is probable that when this procedure is in error, it underestimates, rather than overestimates the number of local modes. This is suggested by the work of Land and Goodman[4] who investigated the localized vibrations at a vacancy and found that for a model using Born-Mayer potentials a local mode exists, but it is very close to the band edge ($\omega^*/\omega_m \simeq 1.01$). As will be seen below, using Born-Mayer potentials to compute the force constants and applying the Litzmann criterion gives the result that no local modes exist for the vacancy.

The precise number of local modes to associate with the changes in the force constants around a defect is difficult to determine without a full scale analysis of the dynamical problem. Welch[5] has assumed that one normal mode is formed for every force constant that satisfies (20). There is some justification for this, in that the work of Litzmann shows that this is the case when (20) is satisfied for the coupling of a single impurity atom to its neighbors. Below, we give Welch's results[5] for the vacancy (V), the vacancy saddle point (SPV), the split-interstitial (I) and the split-interstitial saddle point (SPI). The structure of these defects is shown in Figure 1.

For defects in copper, it is common to use the Born-Mayer potential to represent the nearest neighbor interactions. This potential is repulsive, so that crystal stability is maintained by the interactions between the ion cores and the conduction electrons. This is often simulated by surrounding the defect region by an isotropic elastic continuum.

The Born-Mayer interaction for an ion pair is

$$U_{ij} = Ae^{-\alpha(r_{ij}-r_o)/r_o} \tag{21}$$

where r_{ij} is the distance between ions i and j, r_o is the equilibrium nearest neighbor distance, and A and α are

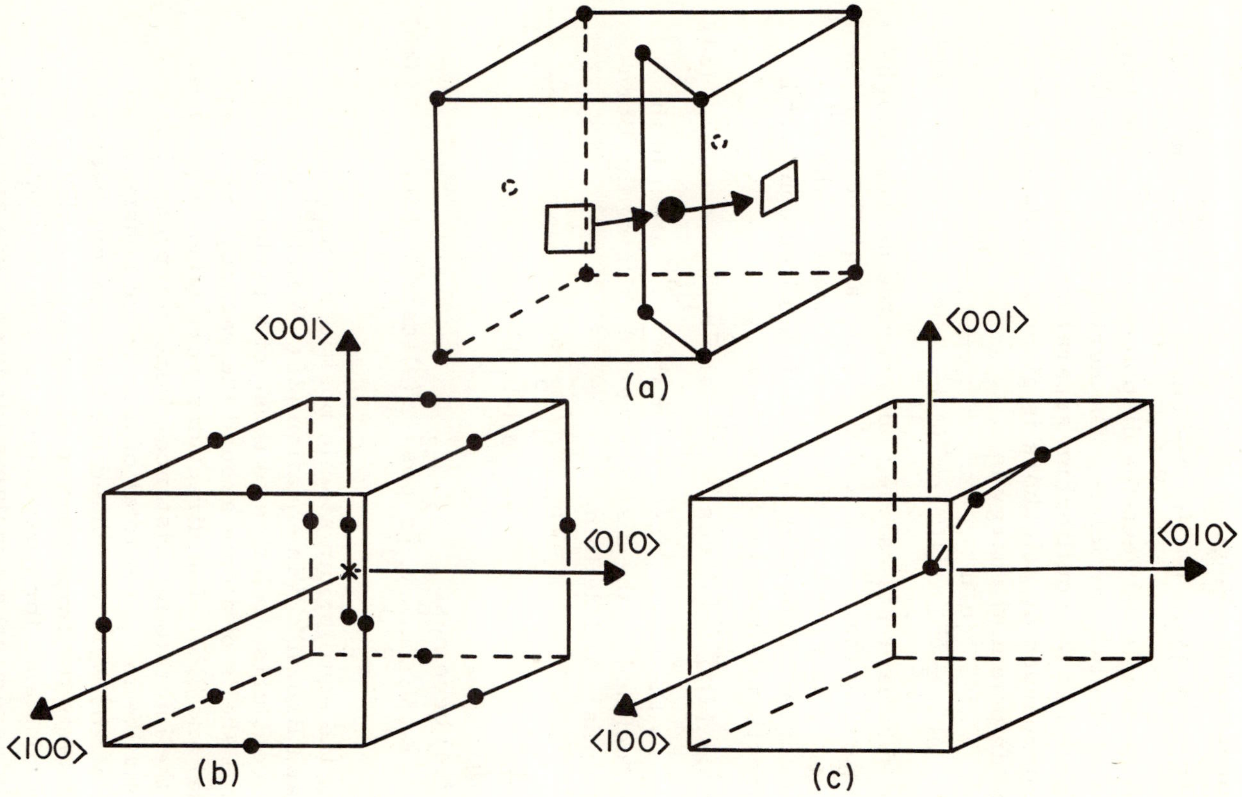

Figure I. Defect Configurations
a. Saddle Point Vacancy, b. Split Interstitial, c. Saddle Point Interstitial

constants. We adopt the values derived by Huntington [(6)] from the elastic constants of copper: A=0.053 e.v. and α=13.9. As shown by Huntington, for a central force interaction between two particles, U(r), if one of the particles is given a displacement $\vec{\delta}$, the change in potential energy is

$$\Delta U = U'(r_o)\delta\cos\theta + U''(r_o)\frac{\delta^2}{2}\cos^2\theta + U'(r_o)\frac{\delta^2}{2r_o}\sin^2\theta \quad (22)$$

where θ is the angle between the displacement δ and the line of centers of the two particles. The derivatives U' and U'' are evaluated at the equilibrium separation. Therefore, if an atom surrounded by neighbors is given a displacement $\vec{\delta}$, the change in potential energy is

$$\Delta U_{ij} = \frac{1}{2}\delta^2 \sum_j^{n.n.} [U''\cos^2\theta_{ij} + \frac{U'}{r_o}\sin^2\theta_{ij}] \quad (23)$$

where the summation is carried out over the nearest neighbors surrounding the ith ion, and θ_{ij} is the angle between the displacement $\vec{\delta}$ and the line of centers connecting the ith and j^{th} ion cores. From (23), the effective force constant k for the direction defined by $\vec{\delta}$, is given by

$$k_i(\ell) = \sum_j^{n.n.} [U''\cos^2\theta_{ij} + \frac{U'}{r_o}\sin^2\theta_{ij}] \quad (24)$$

where $k_i(\ell)$ is the force constant for the ith atom in the direction ℓ defined by the displacement $\vec{\delta}$. Using the Born-Mayer potential, and defining

$$\left.\begin{aligned} b_{ij} &= r_{ij}/r_o \\ y_{ij} &= e^{\alpha(1-b_{ij})} \end{aligned}\right\} \quad (25)$$

equation (24) becomes

$$k_j(\ell) = \frac{A}{r_o^2} \sum_j^{n} y_{ij}[\alpha^2\cos^2\theta_{ij}(\ell) - \alpha\sin^2\theta_{ij}(\ell)] \quad (26)$$

Using the equilibrium atomic positions computed by Johnson and Brown [(7,8)] in equation (26), Welch [(5)] computed the force constants, relative to those of a perfect crystal, for several shells of atoms around the vacancy, SPV, split interstitial and SPI. The directions were taken to be those in which the interatomic distances were compressed or expanded relative to r_o. A total of nine force constants for forty two atoms were computed for the vacancy, twenty one force constants for twenty seven atoms for the SPV, twenty one force constants for forty atoms for the split interstitial, and twenty seven force constants for twenty one atoms for the SPI.

The results of the calculations show that the vacancy very probably does not give rise to local modes. The largest value of k/k_p was 1.16, which is considerably less than 2, so that the assumption of Girifalco and Kuhlmann-Wilsdorf[9] assigning 12 local modes to the vacancy, is quite unjustified.

For the SPV, one value of k/k_p satisfies the Litzman inequality, with $k/k_p = 2.15$. Thus, there is probably one localized mode for the SPV.

For the split interstial there are 14 values of $k/k_p > 2$, ranging from 2.02-2.93. Four of these values were very close to 2. The probable number of localized modes for the interstitial is therefore between 10 and 14.

For the SPI, there are 8 values of $k/k_p > 2$, ranging from 2.33 to 3.58, and 8 values of $k/k_p \approx 2$. The probable number of localized modes in this case is somewhat between 8 and 16.

The number of local modes obtained by the above procedure is certainly open to question, but they can be taken as approximate values until more detailed studies of the lattice dynamics of defect crystals are available.

IV. The Strain Dependence of the Vibration Frequencies

The effect of volume changes on the normal mode frequencies has usually been treated by the Gruneisen method in which it is assumed that

$$\frac{d \ln \nu_j}{d \ln V} = -\gamma \tag{27}$$

or

$$\frac{\Delta \nu_i}{\nu_i} = -\gamma \bar{\epsilon} \tag{28}$$

where γ is a positive constant and $\bar{\epsilon}$ is the average dilation.

The Gruneisen assumption is adopted so that the anharmonic contributions to the potential energy can be at least partly accounted for without sacrificing the convenience of the harmonic approximation. In fact, it can be shown that the Gruneisen constant is related to the third order force constants in the expansion of the potential energy. A major criticism of (28) is that there should be a separate γ for each normal mode. However, for short range forces the number of distinct γ's is small, and it would probably be adequate to define one for longitudinal and one for transverse modes. If localized modes are present, a third γ should be defined for them. However, the Gruneisen assumption adequately describes the thermal expansion at the high temperatures of interest in point defect and diffusion work. Accordingly, we will retain the simplicity derived from using just one γ.

For crystals of low symmetry, the frequencies depend on all components of the strain matrix and it is necessary to define a corresponding γ-matrix

$$\gamma_{ij} = -\frac{1}{\nu}\frac{d\nu}{d\epsilon_{ij}} \tag{29}$$

For cubic crystals, however, it may be shown that $\delta_{ij} = \gamma d_{ij}$. We will therefore take (28) as the representation of the strain dependence of the frequencies of in-band modes.

For the local modes, however, the frequency is determined by the coupling among atoms close to the defect and the defect may seriously alter the local symmetry. To account for this, we adopt an approach proposed by Girifalco and Kuhlmann-Wilsdorf[9] and extended by Welch[5] in which instead of equation (28) we write

$$\frac{\Delta\nu_i}{\nu_i} = -3\gamma\,\frac{\Delta\lambda}{\lambda} \tag{30}$$

where $\Delta\lambda$ is the change in interatomic distance induced locally by a strain matrix ϵ_{ij}, and is related to the strain by

$$\Delta\vec{\lambda} = \vec{\lambda}\,.\,\Psi \tag{31}$$

where Ψ is the strain dyadic. For small $\Delta\lambda$ we have

$$\frac{\Delta\lambda}{\lambda} \approx \frac{\vec{\lambda}\cdot\Delta\vec{\lambda}}{\lambda^2} = \frac{\vec{\lambda}\cdot\Psi\cdot\vec{\lambda}}{\lambda^2} \tag{32}$$

so that the fractional change in interatomic distances between two atoms i and j is

$$\left(\frac{\Delta\lambda}{\lambda}\right)_{ij} = \vec{n}_{ij}\cdot\Psi\cdot\vec{n}_{ij} \tag{33}$$

where $\vec{n}_{ij}$ is a unit vector in the direct ion from the ith to the jth atom.

Now consider a region around the defect small compared to the dimensions of the crystal but large enough to enclose the atoms affected by the defect. The average fractional change in interatomic distance in this region is

$$\left\langle\frac{\Delta\lambda}{\lambda}\right\rangle = \frac{\sum_{i=1}^{n}\sum_{j=1}^{k_i} \vec{m}_{ij}\cdot\Psi\cdot\vec{m}_{ij}}{\sum_{i=1}^{m} k_i} \tag{34}$$

where k_i is the number of nearest neighbors of the ith atom, and N is the number of atoms in the region considered. If this region were perfect, it is easy to show that for a cubic crystal, (34) reduces to

$$\left\langle\frac{\Delta\lambda}{\lambda}\right\rangle \text{perfect} = \frac{1}{3}\epsilon \ . \tag{35}$$

For the vacancy, we obtain

$$\left\langle\frac{\Delta\lambda}{\lambda}\right\rangle_V = \frac{1}{3}\,\epsilon \tag{36}$$

because of the cubic symmetry of the defect.

The other three defects we wish to consider do not have cubic symmetry, so that a specific evaluation of (34) must be made, taking into account the relaxed atomic positions around the defect. This was done by Welch using the displacements of Johnson (8) for copper and he found that

$$\frac{\Delta\lambda}{\lambda}\,\text{SPV} = \frac{1}{3}\left[0.99\epsilon_{11} + 1.01\epsilon_{22} + 1.01\epsilon_{33}\right] + 0.22\epsilon_{23} \tag{37}$$

$$\left\langle\frac{\Delta\lambda}{\lambda}\right\rangle_I = \frac{1}{3}\left[1.04\epsilon_{11} + 1.04\epsilon_{22} + 0.925\epsilon_{33}\right] \tag{38}$$

$$\left\langle\frac{\Delta\lambda}{\lambda}\right\rangle_{SPI} = \frac{1}{3}\left[1.005\epsilon_{11} + 0.995\epsilon_{22} + 0.995\epsilon_{33}\right] + 0.01\epsilon_{23} \tag{39}$$

The region of localization was taken to include only nearest neighbors to the defects. The departure of the above results from those for cubic symmetry is seen to be quite small. This departure would be even smaller if a larger region of localization were considered. We can therefore write, to an excellent approximation, for all of the defects,

$$\left\langle\frac{\Delta\lambda}{\lambda}\right\rangle = \frac{1}{3}\epsilon \qquad (40)$$

or

$$\left(\frac{\Delta\lambda}{\lambda}\right)_{\substack{\text{local}\\ \text{modes}}} = -\gamma\epsilon \qquad (41)$$

V. Entropy of Defect Formation and Migration in Stress Fields

The entropy of formation of a point defect is related to the normal modes by (10)

$$\Delta S_d^f = k\sum_{i=1}^{3N} \ln\frac{\nu_i}{\nu_i^d} \qquad (42)$$

where ν_i and ν_i^d are the normal mode frequencies for the perfect and defect crystals respectively. We are interested in the change in formation entropy when the crystal is strained.

$$\delta S_d^f = \Delta S_d^f(\epsilon)-\Delta S_d^f(0)=k\sum_{i=1}^{3N}\left[\ln\frac{\nu_i(\epsilon)}{\nu_i(0)}-\frac{\nu_i(\epsilon)}{\nu_i^d(0)}\right] \qquad (43)$$

Since all $[\nu(\epsilon) - \nu(0)]$ will be small, we expand the logarithms in series and retain only terms first order in $\Delta\nu(\epsilon) - \nu(0)$. Then (43) becomes

$$\delta S_d^f(\epsilon) = k\sum_{i=1}^{3N}\left[\frac{\Delta\nu_i(\epsilon)}{\nu_i} - \frac{\Delta\nu_i^d(\epsilon)}{\nu_i^d}\right] \qquad (44)$$

The first term in the sum refers to in-band modes of the perfect crystal, so we will adopt the ordinary Gruneisen approximation. The second term contains both in-band and local modes of the defect crystal. The in-band modes can be treated by the ordinary Gruneisen formula, but the local modes must be treated according to equation (41) in which the local dilatation appears. Thus, if there are Z_d local modes in the defect crystal, using (28) for in-band modes and (41) for the local modes, equation (44) becomes

$$\delta S_d^f(\epsilon) = -3Nk\gamma\bar{\epsilon} + (3N-Z_d)k\gamma\bar{\epsilon}' + Z_d k\gamma\epsilon' \qquad (45)$$

where $\bar{\epsilon}'$ is the average dilatation in the defect crystal, $\bar{\epsilon}$ is the average dilatation in the corresponding perfect crystal and ϵ' is the local dilatation in a small region surrounding the defect in the defect crystal.

Eshelby (11) has shown that if a crystal contains a spherical inclusion whose elastic constants differ from those of the matrix, the dilatation induced by a stress in the inclusion is proportional to the dilatation of the crystal without an inclusion.

$$\epsilon' = K_d \epsilon \tag{46}$$

Outside the inclusion, $\epsilon' = \epsilon$. Eshelby's result can be adopted to our problem by assuming that the defect and its nearest neighbors act as an elastic inclusion. Then (45) becomes

$$\delta S_d^f(\epsilon) = 3kn\gamma(K_d - 1)\epsilon + k\gamma Z_d(K_d\epsilon - \overline{\epsilon}) \tag{47}$$

where n is the number of atoms in the localized region defining the elastic inclusion. The constant K_d is related to the compressibilities of the inclusion and the matrix. The change in local compressibility accompanying defect formation was computed for copper (5) using the Born-Mayer potential and the displacements of Johnson (7, 8). If a constant C_d is defined by

$$C_d = n\frac{\Delta\beta}{\beta} \tag{48}$$

where $\Delta\beta$ = the local change in bulk modulus on forming a defect and β is the modulus of the perfect crystal, then

$$K_d = 1 - \frac{2}{3}\frac{C_d}{n} \tag{49}$$

and (47) becomes

$$\delta S_d^f(\epsilon) = -2k\gamma C_d \epsilon + Z_d \gamma k(\epsilon - \overline{\epsilon}) \tag{50}$$

The values of C_d computed by Welch are shown in Table 1.

The calculation for migration entropies proceeds in a similar manner. The jump frequency formula for defect motion is (12)

Table 1. Values of $C_d = N\dfrac{\beta_d - \beta}{\beta}$ computed by Welch

Defect	C_d
Vacancy	-1.34
SPV	-0.594
⟨100⟩ - Split Interstitial	13.6
SPI	13.7

Table 2. Entropies of Interaction of Defects with Stress Fields in Units of k per unit Dilatation

Defect		$\delta S/k\epsilon$ (homogeneous external stress)	$\delta S/k\epsilon$ (Internal stress)
Formation entropies	V	5.36	5.36
	I	-54.4	-30.4
Migration Entropies	SPV	-5	-1
	SPI	-2.4	-.4

$$\gamma = \frac{\prod_{i=1}^{3N} \nu_i^d}{\prod_{i=1}^{3N-1} \nu_i^{SPd}} e^{-\Delta\Phi/kT} \tag{51}$$

where $\Delta\Phi$ is the energy of migration, and ν_i^{SPd} are the normal modes when the defect is in the saddle point configuration. We will define the entropy of migration to be

$$\Delta S_d^m = k \sum_{i=1}^{3N} \ln\nu_i^d - k \sum_{i=1}^{3N-1} \ln\nu_i^{SPd} \tag{52}$$

We adopt this definition for the present work, rather than one in which the activated and normal states have the same number of degrees of freedom, because we are interested in the entire vibrational contribution to the jump frequency.

Proceeding exactly as above, we find that

$$\delta S_d^m \equiv \Delta S_d^m(\epsilon) - \Delta S_d^m(0)$$

$$= -\gamma k[2(C_{SPd} - C_d)\epsilon + \epsilon -(Z_{SPd} - Z_d)(\epsilon - \epsilon)] \tag{53}$$

The values of C_{SPd} are shown in Table 1. The numerical estimates of the entropy of interaction of the defects with stress fields given in Table 2 for copper were computed from equations (50) and (53) taking $\nu = 2$, $Z = 0$, $Z_{SPV} = 1$, and $Z_I = Z_{SPi} = 12$. In using equation (50) and (53), it must be remembered that for externally imposed homogeneous strain $\epsilon = \bar{\epsilon}$, and for internal strain $\bar{\epsilon} = 0$.

It is difficult to estimate the accuracy of the results presented in Table 2, but some qualitative conclusions, at least can be given. In the first place, strains of the order of 1% or more are needed before any significant effect on the entropies can be observed. In the second place, the effect of local modes appears only for non-homogeneous strains or strains arising from internal stress fields. It is evident that the local modes contribute an appreciable amount to the entropy changes in a stress field. For the interstitial in an internal stress field, the local mode contribution amounts to nearly 50% of the total entropy of interaction.

The results of Table 2 show that measurements of defect formation and migration under hydrostatic pressure should give formation and migration volumes that are independent of temperature unless the pressures are high ($\epsilon \geq 1\%$).

An important application of the theory of the dependence of entropy on strain is to the interaction of point defects with dislocations. At a distance of 2b (b=Burgers vector) the dilatation along the slip plane of an edge dislocation is $\geq 2.5\%$. This gives an interaction entropy of a vacancy and an interstitial with the edge dislocation of $\sim 0.1k$ and $\sim -k$ respectively. This corresponds to contributions to the free energy of interaction at the melting point of copper of the order of 0.01 and -0.1 e.v. for the vacancy and interstitial respectively. Estimates of the binding energy of vacancies and interstitials to edge dislocations at r=2b give 0.09 e.v. and 0.64 e.v. respectively. The thermal contribution to the binding energy to dislocations is therefore probably of the order of 10-20% for point defects.

This research was sponsored by the Advanced Research Projects Agency under Contract SD-69.

References

1. Maradudin, A.A., Montroll, E.W. and Weiss, G.A., "Theory of Lattice Dynamics in the Harmonic Approximation" Solid State Physics, Suppl. 3 (1963) Academic Press

2. Maradudin, A.A., "Dynamic Theory of Imperfect Crystals in Astrophysics and the Many-Body Problem" W.A. Benjamin Inc. N.Y. (1963)

3. Litzman, O., Czechosl. Journ. Phys. 8, 521(1958)

4. Land, P.L. and Goodman, B., Lattice Dynamics, Proceedings of the International Conference, Copenhagen, August 5-9, 1963, Edited by R.F. Wallace, Pergamon Press, P. 451

5. Welch, D.O., "A Theoretical Investigation of Point Defects and Diffusion in Metals Containing Elastic Stress Fields." Ph.D. Thesis (1964) School of Metallurgical Engineering, University of Pennsylvania

6. Huntington, H.B., Phys. Rev. 91, 1092 (1953)

7. Johnson, R.A., and Brown, E., Phys. Rev., 127, 446(1962)

8. Johnson, R.A., "Point Defects in Copper" Ph.D. Thesis (1962) Rensselaer Polytechnic Institute

9. Girifalco, L.A. and Kuhlmann-Wilsdorf, D., J. App. Phys. 35, 438 (1964)

10. Vineyard, G. and Dienes, G.J. Phys. Rev., 93, 265(1954)

11. Eshelby, J.D., Progress in Solid Mech. 11, 112 (1961)

12. Vineyard, G.H., "J. Phys. Chem. Solids, 3, 121 (1957)

EQUATION OF STATE OF SIMPLE SOLIDS

C.A. Swenson*

Institute for Atomic Research and Department of Physics, Iowa State University, Ames, Iowa.

The alkali metals and the solidified inert gases form simple solids which are of interest in themselves, but which also are of interest because they can be used as "model" substances to study the qualitative behaviour of solids at high compressions. For instance, the volume reductions in 20 kbars are roughly 30 percent for potassium metal and 1.5 percent for copper. Hence, one would expect to obtain more general information per kbar of pressure for potassium than for copper. Similarly, the Debye temperatures θ_D for solid xenon (55°K) and potassium metal (93°K) are low and the melting points correspond to approximately $3\theta_D$ in each case. Hence, the thermal properties of these solids can be studied over a wide range of relative temperatures and compressions while using temperatures and pressures which are quite moderate in absolute terms.

Our approach to this problem has involved the use of the piston-displacement technique to obtain pressure-volume-temperature data on elementary solids. In these experiments, the sample serves as its own pressure transmitter and the changes in molar volume of a sample with pressure and temperature are reflected directly in the motion with applied force of the two pistons which close the ends of a cylinder containing the sample (1,2,3,). The results are based on the equation of state of indium metal (4). The reliability of the results

---

* Temporarily at the CSIRO National Standards Laboratory, Division of Physics, Sydney, Australia during the tenure of a National Science Foundation Postdoctoral Fellowship. Much of the work described was performed in the Ames Laboratory of the United States Atomic Energy Commission.

depends both on the data for indium and the degree to which the sample has a shear yield stress which is negligible when compared with the pressures used. Fortunately, the "simple" solids are quite plastic even at low temperatures and are quite compressible. This large compressibility reduces considerably the dependence of the results on the properties of indium, and volume changes in 20 kbars can be measured for these solids with an absolute precision of roughly one percent.

The analysis of the equation of state data which follows is based on the separation of the Helmholtz Free Energy F(T,V) into absolute zero and thermal contributions

$$F(V,T) = U_o(V) + F^*(V,T).$$

Here, $U_o(V)$ is the volume-dependent cohesive energy at absolute zero and the thermal term $F^*(V,T)$ must vanish at absolute zero along with its derivatives. Lattice vibrations make the predominant contribution to F^* for the alkali metals and the solidified inert gases, although these effects may be small compared to those due to magnetic or other interactions in more complex materials (5).

The equation of state P(V,T) and the isothermal bulk modulus B_T are given by

$$P(V,T) = -(\partial F/\partial V)_T = P_o(V) + P^*(V,T), \qquad (1a)$$

and $$B_T(V,T) = -(\partial P/\partial \ln V)_T = B_o(V) + B_T^*(V,T). \qquad (1b)$$

Here, experimental values for $P_o(V)$ and $B_o(V)$ provide direct tests of a cohesive energy calculation while P^* (the thermal pressure) can be analyzed in terms of the Mie-Grueneisen equation of state

$$P(V,T) = \gamma(V)\, U^*(V,T)/V. \qquad (2)$$

$U^*(V,T)/V$ is the lattice vibrational energy content per unit volume (and is distinct from the enthalpy), while $\gamma(V)$ is a Grueneisen constant and is given by $\gamma(V) = -d\ln\theta/d\ln V$ if a characteristic temperature $\theta(V)$ can be defined for the solid. In this case, the volume thermal expansion β is related to the specific heat at constant volume C_V by the Grueneisen relationship

$$(\partial P^*/\partial T)_V = \beta B_T = \gamma(V) C_V/V. \qquad (3)$$

The mathematical identity $(\partial P/\partial T)_V = \beta B_T$ has been used in the left-hand side of Eq.(3). In general, a unique characteristic temperature cannot be defined for a solid and the γ's defined by Eqs.(2) and (3) are not identical and show slight temperature dependences.

Hence, a discussion of pressure-volume-temperature data for a solid breaks up naturally into a discussion of the absolute zero and thermal equations of state. The $P_0(V)$ data can be analyzed in terms of theoretical models which will be quite different for the alkali metals and the solidified inert gases. On the other hand, the P* data can be analyzed best in terms of the Mie-Grueneisen equation of state. Differences between solids will be reflected basically in different values of the Grueneisen constant γ and the Debye temperature θ_D.

Recent work on lithium metal illustrates the type of data which can be obtained using the piston-displacement technique (6). Lithium is not an ideal subject for these investigations since it has a considerably greater shear yield stress than the other alkali metals and is relatively less compressible. The basic data take the form of pressure-volume isotherms at various temperatures. The room temperature and extrapolated absolute zero isotherms are shown in Fig. 1 as smooth curves, since the raw data were smoothed graphically and smoothed corrections were applied. To eliminate systematic errors, data were taken using cylinders which differed in area by a factor of two, and which, consequently, had different values of the maximum pressure available.

Fig. 2 gives an isobaric plot of 18 isotherms. The molar volumes for each pressure were read from the individual isotherms, and the size of the "data points" so obtained represents an uncertainty in the shape of the smoothed isotherm (Fig. 1). These uncertainties are roughly ± 0.2 kbar or ± 0.02 cm^3/mole. Fig. 2 shows the extrapolation of the data to absolute zero. Indeed, $20^\circ K$ is effectively absolute zero, but the rapid increase of shear yield stress with decreasing temperature makes the low temperature isotherms somewhat unreliable at low pressures.

Once the extrapolated absolute zero isotherm has been obtained it is possible to obtain values for the thermal pressure P* for each molar volume

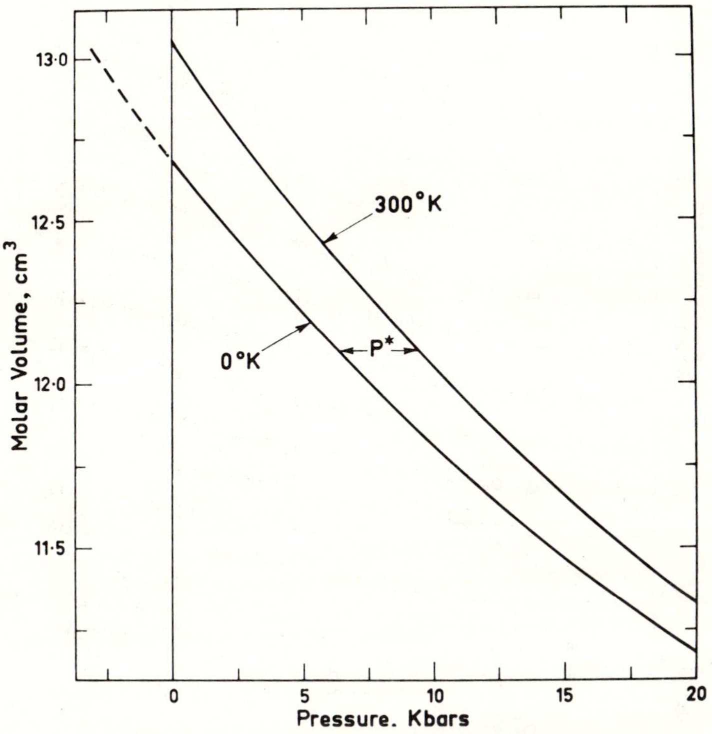

Fig. 1. The room temperature and extrapolated absolute zero isotherms for lithium metal (6).

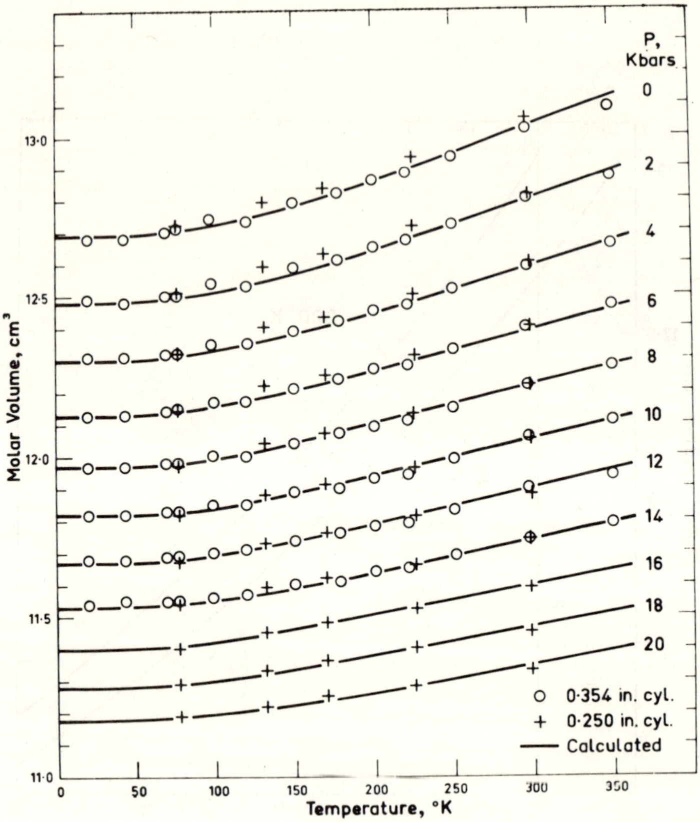

Fig. 2. An isobaric plot of the data from 18 isotherms similar to those in Fig. 1. Sample holders of two different diameters (0.250in. and 0.354in.) were used to eliminate systematic errors. The maximum pressure was greater for the smaller diameter sample holder, with a corresponding lower precision in the low pressure data.

(P^* is defined by Eq.(1a) and is indicated in Fig. 1). P^* is found to be independent of volume for a given temperature, implying that all pressure-volume isotherms have the same slope for a given molar volume. This is equivalent to stating that the isothermal bulk modulus is independent of temperature at constant volume. Or, using Fig. 1 as an example, a translation of the 300°K isotherm parallel to the pressure axis by a pressure $P^* = 3.05$ kbars causes it to coincide with the absolute zero isotherm. Values of P^* which were obtained for all of the isotherms are shown in Fig. 3, where the size of the error bars indicates the range of P^* values which were found for each isotherm. Finally, the extrapolated absolute zero isotherm and the smoothed P^* curve shown in Fig. 3 can be used to produce an equation of state which is given by the solid lines in Fig. 2. The agreement of these curves with the data is in general within the indicated uncertainty of ±0.2 kbar. The zero pressure isobar is obtained by extrapolation from a minimum pressure of one kbar and is hence unreliable. Nevertheless, it is in excellent agreement with X-ray lattice constant data (6).

RESULTS

The Low Temperature Alkali Metal Isotherms

The sodium (1), potassium (3), and lithium (6) data for the extrapolated absolute zero isotherms all obey the simple relationship $(dB/dP) = 3.8(\pm 0.2)$. The zero pressure bulk moduli (at appropriate volumes) and this derivative are in good agreement with room temperature elastic constant data for both sodium and potassium (7,8). Thus, the shapes of the $P_0(V)$ relationships for these three metals are very similar and the actual curves are completely characterized by the values of the molar volume and the bulk modulus at zero pressure and absolute zero. Cesium appears to behave somewhat differently (9), and reliable data do not exist for rubidium. In addition, the product $B_0(V_0).V_0 = 1.63(\pm 0.1) \times 10^5$ Joules/mole for all three metals, and possibly for cesium and rubidium also. Very crudely, the common value of dB/dP implies that all three metals obey a common reduced $U_0(V)$ (or cohesive energy) curve (with the exception of an additive constant) with different values of characteristic energy and range parameters. The second empirical relation-

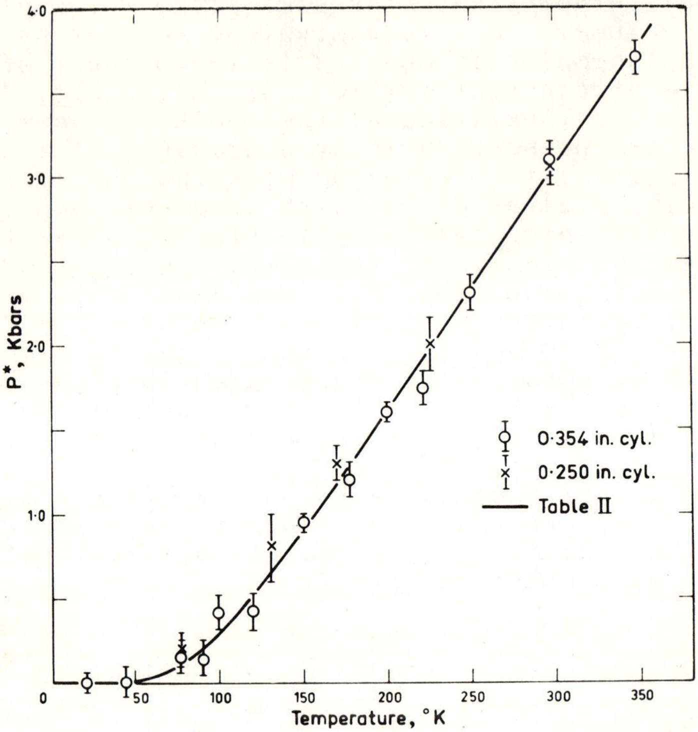

Fig. 3. The thermal pressure P* (Eq.(1a)) for the data summarized in Fig. 2. The length of the error bars is indicative of the range of P* values found for each isotherm.

ship implies that the characteristic energy is the same for all three metals to a good approximation, leaving only the range factor (which could be the diameter of the ionic core) as the distinguishing feature of an individual metal. Indeed, Stokes has shown that the ionic core radii of the alkali metals are a linear function of the nearest neighbor distance at absolute zero (10). Recently, elementary theories of metals (and the alkali metals in particular) have been proposed which consider primarily the electrostatic nature of the electron-electron and electron-ion interaction, the Fermi energy of the conduction electrons and the finite size of the ion core (10,11). The above relationships do not follow explicitly from either of these calculations.

The Inert Gas Solids at Absolute Zero

The interaction between two inert gas atoms is approximated by a Lennard-Jones 6-12 potential function:

$$V(r) = 4\varepsilon[(\sigma/r)^{12} - (\sigma/r)^{6}], \qquad (4)$$

where ε and σ can be determined from non-ideal gas properties. This same potential function also should apply to two-body interactions in the solid since the electronic shells of the atoms are closed and there is no covalent or ionic binding. The theoretical problem for these solids is complicated by the relatively large zero-point motion kinetic energy which effectively "blows up" the solids to a molar volume greater than the classical value. The effect of the zero-point motion is directly related to a quantum parameter $\lambda^2 = \hbar^2/2m\varepsilon\sigma^2$ which is very large for the heliums, quite important for solid neon, but small for solid xenon. Explicit expressions can be obtained for the cohesive energy E_0, the molar volume at absolute zero and zero pressure V_0 and the absolute zero isotherm in terms of ε, σ and λ, providing that λ is not too large (12). The agreement between the theory (with no adjustable constants) and experimental values of E_0 and V_0 is quite good.

A more stringent test of the theory is given by the experimental absolute zero isotherms for xenon (2), argon (13) and neon (14) which are shown as solid curves in Fig. 4. The agreement of these data with the theoretical (dashed) curves is quali-

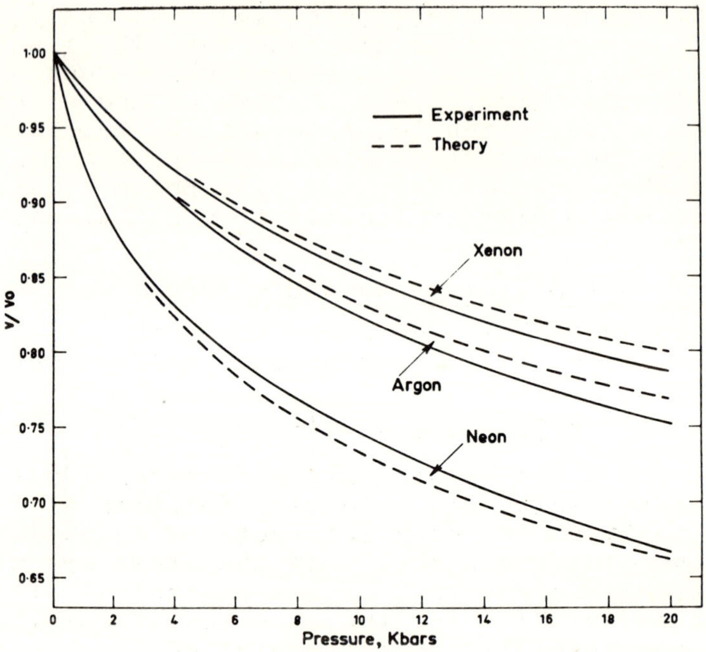

Fig. 4. A comparison with theory of the experimental absolute zero isotherms for solid xenon, argon and neon.

tatively good. In this comparison, the experimental values for E_0 and V_0 were combined with the theoretical expressions to obtain values of ε and σ which are characteristic of the solid state. The differences between these constants and the non-ideal gas values are less than one percent. Both the xenon and the argon data appear to agree with the theoretical curve at pressures of up to approximately 6 kbars, after which the deviation occurs. The neon data are quite old, and the author quotes an experimental uncertainty of up to five percent in $\Delta V/V_0$, mostly due to the extrapolation of the data from a few kbars to zero pressure (14).

These high pressure experiments cover a wide range of interatomic separation, with the nearest-neighbor distance varying from 1.16σ at zero pressure for neon to σ at 20 kbars for each solid. It would be surprising if the 6-12 potential were valid over this wide range. On the other hand, the effects of correlations (and 3-body forces) have not been considered in the theory, and the amplitudes of the zero point vibrations are of the order of 0.07σ for neon even at the highest pressures. A further complication arises in the calculation of thermal properties by Horton and Leech (15) which shows that the experimental heat capacity data for these solids can be understood only by using a potential function which has a stronger repulsive term than the Lennard-Jones 6-12 potential. This is directly opposed to our conclusions although the zero pressure thermal data in general correspond to greater nearest-neighbor distances than the low temperature high pressure data. Clearly, one needs more precise data for these solids.

Temperature-dependent Equation of State (4)

The piston-displacement data for sodium (1), potassium (3) and lithium (6) show a volume-independent thermal pressure P^*, or, equivalently, an isothermal bulk modulus B_T which is independent of temperature at constant volume to within a few percent. The xenon data (2) show a slight decrease in P^* with decreasing volume or a small negative value of B_T^* (Eq.(1b)). For all these data, $(\partial P^*/\partial T)_V$ is constant over a wide range of temperatures [from the melting point to θ_D for xenon (165° to 55°K), sodium (360° to 150°K), and potassium (321° to 90°K), and to $\theta_D/2$ for lithium (350° to 150°K)]. Hence, one is tempted to conclude that for a given solid at a temperature greater than the Debye temperature, P* is approximately a function of the temperature only, and $(\partial P^*/\partial T)_V$ is approximately independent of both temperature and volume. These postulates lead to the conclusion that $B_T^* = [-V(\partial P^*/\partial V)_T]$ and its intrinsic temperature dependence $[(\partial B_T^*/\partial T)_V = (\partial B_T/\partial T)_V]$ both should be small. Since both $(\partial P^*/\partial T)_V$ and C_V are roughly constant above θ_D, it follows from Eq.(3) that the Grueneisen constant will be approximately a linear function of the volume at these temperatures. This gives a valuable guideline for the extrapolation of room temperature thermal data to higher temperatures and pressures. θ_D increases as the volume decreases, however, so the inequality $(T>\theta_D)$ may break down. Quite often phase transitions will occur before this is a serious concern.

B_T^* and its intrinsic temperature dependence $[(\partial B_T^*/\partial T)_V]$ may be obtained also from ultrasonic measurements of elastic constants as a function of temperature and pressure (4). The temperature dependence of elastic constants is most commonly carried out at constant pressure, so the high pressure data must be used to make a correction to constant volume conditions. In addition, the ultrasonic data give most directly the adiabatic bulk modulus and its derivatives, and these must be converted to the equivalent isothermal values which are relevant to our discussion. Such a survey of the ultrasonic data (4) shows that with few exceptions $B_T(V_o,T) \simeq -0.01B_o(V_o)$, and $(\partial B_T^*/\partial T)_V$ is zero within experimental precision. This result is essentially in agreement with the alkali metal and solid xenon results, and implies that the isothermal bulk modulus decreases slightly with increasing temperature at constant volume, and then becomes roughly constant for temperatures greater than θ_D. No such statement can be made about the intrinsic temperature dependence of shear constants which in general shows systematic differences between various solids.

These conclusions can be rationalized through the use of the Mie-Grueneisen equation of state [Eq.(2)]. For temperatures well above θ_D, the lattice vibrational energy is given by

$$U = C_V T - (9/8)\, R\theta_D,$$

where R is the gas constant, and $C_V = 3R$. If it is assumed that γ/V is independent of volume, then

$$B_T^* = -V(\partial P^*/\partial V)_T = V(\gamma/V)[(9/8)R(d\theta/dV)] = -(9/8)\gamma^2 R\theta/V, \quad (5)$$

which is independent of temperature. At lower temperatures, C_V will be a function of volume, and γ/V will most certainly be volume dependent. Qualitatively, B_T would be expected to decrease slightly at constant volume as the temperature is increased from absolute zero, and then to approach a roughly constant value for temperatures greater than the Debye temperature. The values of B_T^* obtained for most solids from Eq. (5) are of the order of (-)0.01 $B_o(V_o)$ and are within a factor of two of those which are obtained from the elastic constant data (4). For solid xenon, B_T is approximately (-)0.03 $B_o(V_o)$, and is in approximate agreement with the observed value of (-)0.06(±0.03)$B_o(V_o)$ (2).

CONCLUSIONS

The objective of the foregoing discussion has been to emphasize the similarities between the behavior of various solids, both at absolute zero and at elevated temperatures. The solidified inert gases, for instance, can be understood quite well in terms of the L-J 6-12 potential when the effects of zero-point motion are included in the theory, although small discrepancies exist. The differences between the alkali metals qualitatively would seem to be due almost entirely to the size of the ionic core. The high temperature behavior of most solids can be understood in terms of the Mie-Grueneisen equation of state and an isothermal bulk modulus which is to a good approximation independent of temperature at constant volume. The interest in the individual solids lies in their deviations from these simple generalizations, and since the differences are small, precise data are necessary. The effects of anharmonicity or defect formation in the high temperature work have not been considered, although these should be important.

REFERENCES

1. R.I. Beecroft and C.A. Swenson, J.Phys.Chem. Solids 18, 329 (1961).
2. J.R. Packard and C.A. Swenson, J.Phys.Chem. Solids 24, 1405 (1963).
3. C.E. Monfort, III and C.A. Swenson, J.Phys.Chem. Solids 26, 291 (1965).
4. C.A. Swenson, Thermal Contributions to the Equation of State of Solids, USAEC Rept. IS-870 (1964).
5. C.E. Monfort III and C.A. Swenson, J.Phys.Chem. Solids 26, 623 (1965).
6. C.A. Swenson, to be published.
7. W.B. Daniels, Phys.Rev. 119, 1246 (1960).
8. P.A. Smith and C.S. Smith, J.Phys.Chem.Solids 26, 279 (1965).
9. C.A. Swenson, unpublished.
10. R.H. Stokes, J.Amer.Chem.Soc. 86, 233 (1964).
11. J.C. Raich and R.H. Good, Jr., J.Phys.Chem. Solids, to be published.
12. N. Bernardes, Phys.Rev. 120, 807 (1960).
13. J.W. Stewart (private communication).
14. J.W. Stewart, J.Phys.Chem.Solids 1, 146 (1956).
15. G.K. Horton and J.W. Leech, Proc.Phys.Soc. A 82 816 (1963).

Remarks on Some Electronic Phase Transitions

W. Kohn
University of California, San Diego
La Jolla, California

I would like to speak on two subjects in the area of electronic phase transitions on which we have been doing some thinking in La Jolla. These are only loosely related to each other, and the preliminary conclusions in the two cases are of quite different character. For neither case do we have definitive answers but, rather, a number of interesting questions.

1. Localized Electronic States in Metals

The dominant concept for understanding the electronic structure of metals is the Bloch wave describing an electron propagating throughout the system. However, there exist also metals in which some of the electrons, at least, seem to be more appropriately described by wave functions localized around individual atoms. Examples are the d- and f- electrons in the transition and rare earth metals respectively, and the localized magnetic states which arise when certain elements are introduced as impurities into a non-magnetic metallic matrix, for example Mn in Cu.

We have been interested for some time in finding a precise criterion for deciding whether, in a given situation, one should or should not talk about localized electronic states. Recently, C. Majumdar and I (1) have been able to give a complete analysis of a model of a single non-magnetic impurity in a non-magnetic matrix. We consider this situation as a function of the strength of the impurity potential, λ, and find that even though, in a one-electron approximation, bound states will suddenly appear at critical values of λ, the physical properties of the system as a whole are completely smooth functions of λ, even at these critical values. As an example, Fig. 1 shows a schematic plot of the electronic density n, at some fixed point near the perturbing potential, as function of λ. λ_1 is a value of λ at which, in an approximation of independent electrons, a bound state appears. For systems to which this model applies, the notion of a localized electron has then no sharp meaning.

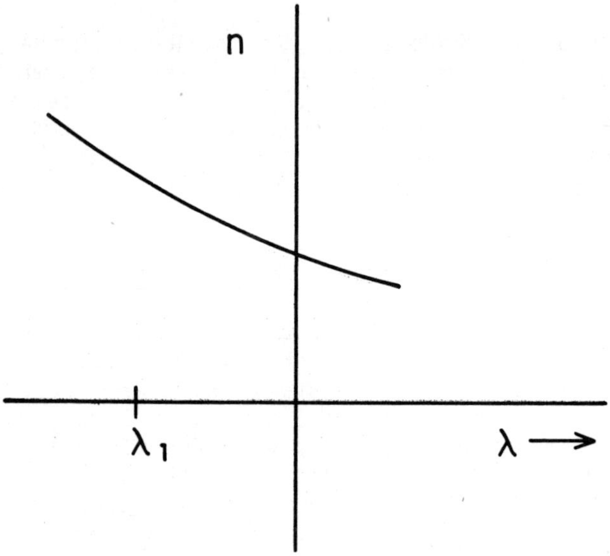

Fig. 1. Schematic plot of density as function of the strength λ of the perturbing potential.

We believe that a similar situation holds also for some systems in which instead of a single perturbing potential there is an entire periodic array. Recently, there has been a report of experimental work on metallic Ce (2). This work shows the possibility of a smooth transition between two phases, one of which is generally regarded as arising from the other by the release of an f-electron from a "bound" state into the "conduction band" (See Fig. 2). This situation is understandable in terms of our theoretical model.

We do however by no means wish to rule out the possibility of a sharp distinction between localized and unlocalized impurity states in some systems, but our considerations indicate that (for a smoothly varying lattice parameter) sharp transitions can come about only when interactions between the electrons play an essential role. Examples of such systems are probably metals with magnetic impurities.

2. Transition between the Metallic and Insulating State

About a year ago, a paper was published (2a) proposing a mathematically sharp distinction between an insulating and conducting wave function of a many-electron solid. In the insulating case, the wave functions consist of a set of disconnected pieces, while, in the metallic case, all parts of the wave function are connected to each other. The disconnectedness property of the insulating wave function was established to all orders of perturbation theory, but, of course, such an argument is not really a mathematical proof since the convergence of the perturbation expansion can be questioned. However, even if this result is accepted, many physical questions still remain to be answered. Among these are the following: Granted that there is a sharp distinction between insulating and metallic wave-functions (as pure states, in the terminology of statistical mechanics), is there also a sharp phase transition between insulating and conducting states (in the thermodynamic sense) at finite temperatures? Is it ever possible to pass from an insulating to a conducting state without crossing a phase-boundary?

These questions are trivially answered for a model of a periodic lattice with two non-interacting electrons per unit cell and energy bands which have an energy gap G passing through zero as function of the lattice parameter d (Fig. 3). In this case, at $T = 0$, the electronic energy is an analytic function for $d > d_0$ ($G > 0$: insulator) and a different analytic function for $d < d_0$ ($G < 0$: metal). As a function of d there is then a non-analyticity or "phase transition" at $d = d_0$. However, for $T > 0$, the number of holes and elec-

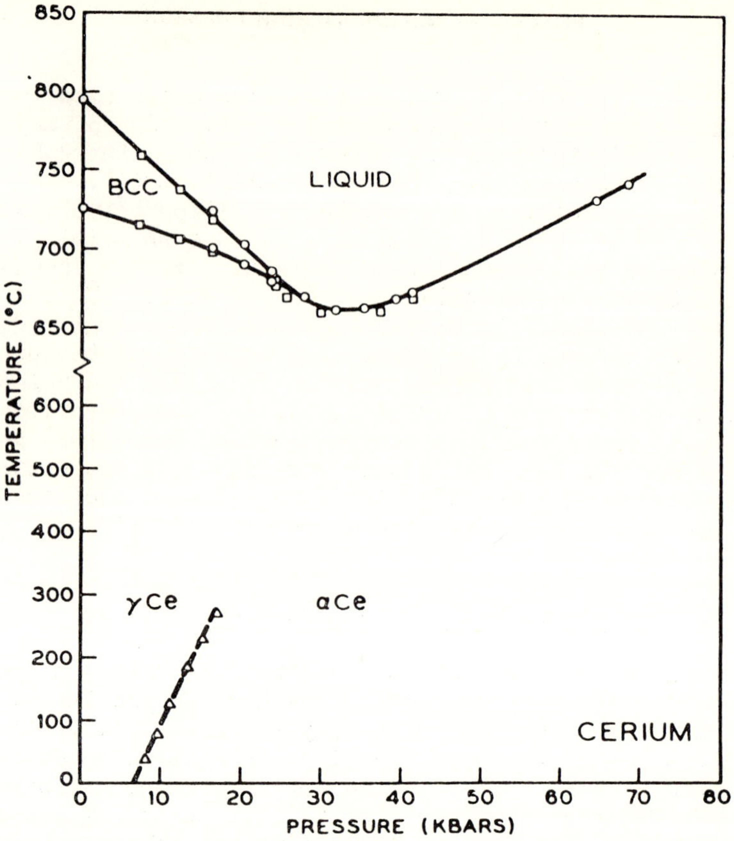

Fig. 2. Phase diagram of cerium (ref. 2).

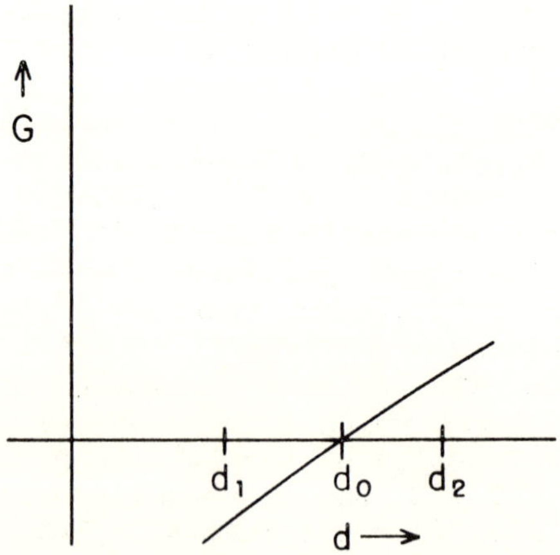

Fig. 3. Energy gap as function of lattice parameter d.

trons, as well as their energy, are entirely smooth functions of d, even at $d = d_o$, and there is then no sharp transition between the metallic and insulating states.

The situation is however quite different when electron-electron interactions are included. Consider first T = 0. Suppose d is very slightly less than d_o. Then, in the absence of interactions we have a metal which may be regarded as an insulator with a very small number of electrons and an equal number of holes. When the interaction is now included it is clear that electron-hole pairs will bind (3, 4) since their Coulomb attraction, while reduced by the dielectric constant of the medium, is not screened as in a metal and, therefore, no matter how weak, always leads to bound states. When d becomes equal to some critical value d_1 ($d_1 < d_o$), the bound pairs are sufficiently numerous to screen the attractive forces between electron-hole pairs to such an extent that these will dissolve. For $d < d_1$ we have then a "normal" metal. Now suppose that d is slightly larger than d_o. Then in the absence of interactions the valence band will be full and the conduction band empty, and there are excited states with free electrons and holes. When the interaction is now turned on, the lowest excited state is an exciton state whose excitation energy is G - B, where B is the binding energy of the excitor. Now clearly, if d is very near d_o, G is very small and hence G - B becomes negative, showing an instability of the "filled band" state with respect to exciton formation (5, 6). Under these circumstances the ground state acquires a new character in which the occupied states are combinations of states from the valence and the conduction band (6). This state of affairs persists up to a critical value d_2 ($d_2 > d_o$), beyond which the ground state becomes a normal insulator.

What is the situation at finite temperatures? We have just started to study this question and do not yet have any definite results. However, we conjecture that as function of d and T we will have the following phase diagram (see Fig. 4). This would represent a partial answer to the questions we have raised.

The conjectured phase diagram of Fig. 4 refers to materials with an even number of electrons per atom which, in the normal insulating state, can be described as having a filled band. As is well known, Mott (7) has suggested the existence of another type of insulator, which in the band picture would not have filled bands. The question of the phase diagram for such insulators seems to be very difficult and we do not enter into it here.

There is one possible point of confusion to which I would like to draw attention. As far as the electronic

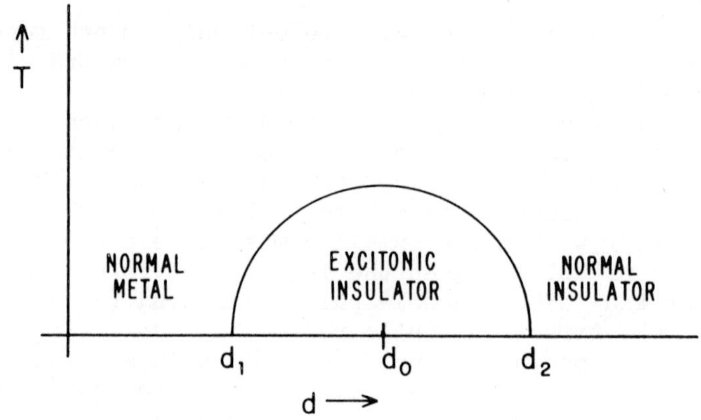

Fig. 4. Conjectured phase diagram for metal-insulator transition.

structure is concerned, one is particularly interested in non-analyticities of electronic properties when parameters characterizing the dimensions and geometry of the lattice are varied continuously. In the laboratory, of course, one usually controls the temperature and pressure. One often finds then, as a function of these parameters, discontinuous changes of the lattice dimensions and geometry which, of course, necessarily entail discontinuous changes of the electronic structure. Such first order phase transitions are of less fundamental interest. The most interesting experimental results would be observation of second order phase transitions which do not involve a discontinuous change of the lattice parameters.

References:

1. Phys. Rev. in press.
2. A. Jayaraman, Phys. Rev. 137, A179 (1965).
2a. W. Kohn, Phys. Rev. 133A, 171 (1964).
3. N. F. Mott, Phil. Mag. 6, 287 (1961).
4. L. V. Keldysh and I. V. Kolaev, paper presented at the Novosibirsk Symposium on Many Body Problems, Mar. 1965.
5. R. S. Knox, Solid State Physics, Supp. 5, p. 100.
6. J. Des Cloizeaux, J. Phys. Chem. Solids 26, 259 (1965).
7. N. F. Mott, Proc. Roy. Soc. (London) 62, 416 (1949).

THEORY OF SEMICONDUCTOR TO METAL TRANSITIONS*

David Adler and Harvey Brooks
Division of Engineering and Applied Physics
Harvard University, Cambridge, Mass.

We consider a semiconductor material in which the intrinsic energy gap depends explicitly on the number of carriers excited across the gap. We assume that the variation of the energy gap is linear, so that

$$E_G = E_{Go} - \beta n. \qquad (1)$$

For the intrinsic case it is possible to write the free energy for this semiconductor in the effective mass approximation

$$F = n\,E_{Go} - 1/2\beta n^2 - 2nkT\ln\frac{N_i}{n} - 2nkT \qquad (2)$$

where we have used Boltzmann statistics and where

$$N_i = (N_c N_v)^{1/2};\; N_c = 2\left[\frac{2\pi m_n kT}{h^2}\right]^{3/2};$$

$$N_v = 2\left[\frac{2\pi m_p kT}{h^2}\right]^{3/2}. \qquad (3)$$

Here m_n and m_p are electron and hole effective masses. Minimizing (2) with respect to the density of excited carriers n, we obtain the relation

$$n = N_i\, e^{-\frac{E_{Go}}{2kT}}\, e^{\frac{\beta n}{2kT}} \qquad (4)$$

*Supported by Advanced Research Projects Agency.

For low temperatures this transcendental equation has two roots. The one for the smaller value corresponds to a minimum of the free energy, while the upper root corresponds to a maximum and has no physical significance. The second derivative of (2) takes the form

$$F'' = \frac{2kT}{n} - \beta. \tag{5}$$

It follows that for

$$T_c = \frac{\beta n(T_c)}{2k} \tag{6}$$

there is no longer a minimum for this and higher temperatures. In consequence, there occurs a first order phase change at which n increases abruptly and the material becomes metallic. The phase change may actually occur at smaller values of n and T, depending on the free energy of the metallic phase. The relations (4) and (6) predict a ratio E_{Go}/kT_c which depends only on the parameter β, usually expressed as a multiple of E_{Go}/N, where N is the number of cells in the crystal. E_{Go}/kT_c is asymptotic to about 6 as β increases. Similar analyses may be carried out for Fermi statistics and also for the limiting case of bands narrow compared with kT and with E_{Go}. Both cases lead to larger values of E_{Go}/kT_c for a given β, at least if one uses the condition (6). The value of $n(T_c)$ always comes out less than 1/10 the number of atoms in the crystal and is often very much less. Thus the linear approximation (1) is usually quite good.

A thermodynamic argument adapted from Figielski (1) makes it possible to relate the coefficient β in (1) to other quantities. The relationship is

$$\beta = \frac{\gamma^2}{K} - \left[\frac{\partial E_G}{\partial n}\right]_{V,T} \tag{7}$$

where

$$\beta = -\left[\frac{\partial E_G}{\partial n}\right]_{p,T}, \quad \gamma = -\left[\frac{\partial E_G}{\partial p}\right]_{n,T} \tag{8}$$

and K is the compressibility. A relation analogous to (7) can be derived for any macroscopic stress component. The first term on the right of (7) depends on quantities which can be measured independently, including the pressure variation of the energy gap. It is often the dominant term.

We have studied two physical models of the origin of β. One is essentially that proposed by Slater (2) and Morin (3), in which itinerant electrons are disposed in different orbitals for different directions of spin so as to produce a net antiferromagnetic spin distribution. The resulting super-lattice potential arising from exchange produces an extra energy gap, which we identify with the gap in (1). Provided the exchange energy is sufficiently strong compared with the band widths, this model gives rise to

$$\beta = \frac{2E_{Go}}{zN} \tag{9}$$

where N is the number of cations in the crystal, and z is the number of electrons per cation.

A second mechanism considered is one in which the energy gap is produced by a distortion of crystal symmetry. As an explicit model the case of a linear chain in which alternate pairs of atoms move closer together is discussed in some detail. The energetics of the distortion is assumed and its effect on the energy gap computed. Depending on the band parameters and the percentage distortion, β may be written

$$\beta = 4(1 + \delta) \frac{E_{Go}}{N} \tag{10}$$

where β is often quite small. In most of the calculations the distortion was taken as about 1% at the maximum and the band width and gap were varied. The electrons below the gap may be thought of as occupying a bonding band with respect to the more closely spaced atoms. The electrons above the gap may thus be thought of as occupying an anti-bonding band.

In using models of the above type to explain observed transitions from the semiconducting to the metallic state, it is important to distinguish two cases. One is that of very narrow band semiconductors in which conductivity in the semiconducting phase takes place essentially by the hopping mechanism (4). At the transition the abrupt increase in the number of carriers not only destroys the symmetry-breaking component of the potential which gives rise to the energy gap, but also screens the Coulomb interactions which give rise to the localization of the carriers and thus to an activated mobility. Hence, in addition to an increase in the carrier density of from one to two orders of magnitude at the transition, there can be a very much larger discontinuous increase in mobility in the metallic phase. This mechanism is believed to account for those transition metal oxides which show jumps in conductivity of the order of 10^6 to 10^8 at the transition. In this sense the transition may be visualized as a combination of the mechanism described in the present paper with that originally envisioned by Mott (5).

The other case is that of relatively wide band semiconductors where the jump conductivity is only one or two orders of magnitude and where the mobility is not activated in the semiconducting phase. The present theory is more directly applicable by itself to such cases, the best current example of which we believe to be Ti_2O_3 with the antiferromagnetic mechanism accounting for the gap.

REFERENCES

1. T. Figielski, Phys. stat. sol. 3, 1876 (1963).
2. J. C. Slater, Phys. Rev. 82, 538 (1951).
3. F. J. Morin, Phys. Rev. Letters 3, 34 (1959).
4. R. R. Heikes and W. D. Johnston, J. Chem. Phys. 26, 582 (1957).
5. N. F. Mott, Proc. Phys. Soc. (Lond.) A62, 416 (1949).

SEMICONDUCTOR TO METAL TRANSITION IN V_2O_3*

Julius Feinleib† and William Paul

Division of Engineering and Applied Physics

Harvard University

Cambridge, Massachusetts

* This work was supported in part by the Office of Naval Research under Contract Nonr 1866(10), and by the Division of Engineering and Applied Physics, Harvard University.

† Present address: Lincoln Laboratory, Lexington, Massachusetts

I. Introduction

From the several reviews of the properties of the transition metal oxides (1), it is apparent that the one-electron band theory is inadequate to explain transport in all of these materials. Notably it has been found that the magnetic oxides of the NiO class are insulators despite the apparent occurrence of partially filled 3d bands. Some oxides of vanadium and titanium exhibit properties characteristic of semiconductors at some temperatures and properties characteristic of poor metals at others. For example, V_2O_3 shows a practically discontinuous change of resistivity from about 10^4 ohm-cm to about 10^{-3} ohm-cm near 150^oK.

There have been several proposals, some involving the properties of localized electrons (2) and some "band" electrons (3), to explain the semiconducting behavior in this and similar compounds, and also several proposals to explain the sharp semiconductor to metal transition.(4)

The initial localization at particular lattice sites may be caused by either electron-electron (5) or electron-phonon (6) interaction. Whichever is the cause, the conductivity will then follow an exponential dependence on temperature with an activation energy, as in the usual diffusion theory. The activation energy q may consist of two parts: the first is related to the production of the localizable carriers and may be zero; the second to the carrier transfer between sites. Although the activation energy is precisely defined from the conductivity according to $\sigma = \sigma_o \exp(-q/kT)$, its relation to the electronic energy spectrum is undefined because the physical nature of the activation and "hopping" are not closely examined. Also, the optical absorption, possible in principle as a carrier activation or as a charge transfer, is not simply related to the electronic energy spectrum.

In one band theory approach (3), otherwise half-filled bands are split into a full valence and an empty conduction band by an antiferromagnetic exchange interaction. The conductivity will again show a carrier activation energy, which will be a direct measure of the exchange energy. The optical absorption will be forbidden since the bands involved have the same symmetry but breakdown of the rule will give an absorption at an energy equal to that for the carrier activation. A similar explanation of the existence of a filled valence and an empty conduction band involves lattice distortions, described in the paper by Adler and Brooks (7).

Both approaches have been applied to explain a transition from semiconductor to metal as a function of volume or of temperature. In the band theory method, the solid is a semiconductor below the Neel temperature, and a metal above it, or alternatively, a semiconductor or a metal according to whether there exists a lattice distortion or not. The transition from semiconductor to metal as a function of temperature may be sharp, as has been most graphically argued in the paper by Adler and Brooks.

The second approach views the transition as one from conduction by "hopping" electrons to conduction by "flowing" or band electrons. In 1949 Mott (8) argued that, quite generally, any periodic array of atoms, and particularly one whose atoms had an odd number of electrons in their outermost shells, would behave like a perfect insulator at sufficiently large interatomic distances. On the other hand, the very existence of metals required that odd-electron atoms form metals at the normal lattice constants. Mott argued further that the transition as a function of interatomic separation would be very sharp and would occur when the electron-hole attraction was shielded by a high density of carriers. That the material be an insulator at large interatomic distances is intuitive from our knowledge of atoms and the simplest molecules, and depends obviously on electron correlation effects. Quantitative many-electron treatments by Kohn (4) and by Hubbard (4) more recently have established this part of the solution on a firm basis, and have shown that the low conductivity is not just due to a very greatly increased value of the effective mass. However, the treatment of the region of transition from semiconductor to metal, where electron correlation effects are still large, but also the tendency for electrons to spread out into bands is also large, is a good deal harder and has not been satisfactorily completed. The present position seems to be that the theory shows the transition to be sharp or gradual depending on the approximations that have been made in order to reduce the problem to one that is analytically manageable.

Since the separation of the ions appears as an important parameter in these theories, we have studied the effects of hydrostatic and uniaxial stresses on transport in both the semiconducting and metallic phases of V_2O_3 and on the transition temperature between the two. We have thereby been enabled to establish to what extent these models apply to this specific solid. In particular, we have examined the possible applicability to this situation of the statistical model presented by Adler and Brooks.

In order to set the investigation on V_2O_3 in perspective, we shall first describe the properties of this compound other than those reported in the main body of the paper. At room temperature, V_2O_3 has a corundum structure (9) with rhombohedral symmetry: there is a unique c-axis perpendicular to the basal plane which contains three identical crystal axes. Near 150°K, a phase transition occurs to monoclinic.(9) The symmetry change may be thought of as occurring by a pairing distortion of the vanadium ions along one of the three basal plane axes, and a simultaneous tilting of the c-axis. The volume increases by 3.5% in the transition to the low temperature phase. (10)

The transition is accompanied by a change in magnetic susceptibility (11), but no magnetic ordering transition has been definitely confirmed by neutron diffraction. (12) Nuclear resonance data at the transition are conflicting but appear to favor an antiferromagnetic ordering (13), although the local magnetic moment is less than 0.5 Bohr magnetons. (12) Other changes at the transition from semiconductor to metal are observed in the specific heat (14) and thermoelectric power (15). The most striking change in the resistivity is illustrated in Figure 1, along with similar changes in the resistivity of other members of the same family.

II. Transport measurements.

A. Metallic region.

A full description of the history and processing of the single crystal samples used, of the pressure and stress apparati, and of the other techniques involved, is given in a Technical Report from our laboratory (16). The general features of the temperature dependence of the resistivity have previously been reported for ceramic materials, and for single crystals. The single crystals used in this investigation appear to be at least as pure and stoichiometric as any hitherto used, and several new features of the resistivity curve have been added.

Figure 2 shows the resistivity, normalized to the 0°C value, for the two principal crystallographic directions in the temperature range from 350°K down to the metal-semiconductor transition near 150°K. The absolute values of resistivity are:

$$\rho_{11}(0^{\circ}C)\ \text{(basal plane)} = 6.3 \times 10^{-4}\ \text{ohm - cm}$$

$$\rho_{33}(0^{\circ}C)\ \text{(c-axis)} = 5.6 \times 10^{-4}\ \text{ohm - cm} \qquad (1)$$

The error in absolute value is ±10% because of the difficulty in obtaining accurate length measurements at the soldered contacts. The agreement among several samples was within 5%. The anisotropy is seen to be small and the anisotropy ratio, defined as $\rho_{33}(T)/\rho_{11}(T)$, is 0.9 at 0°C. In this temperature

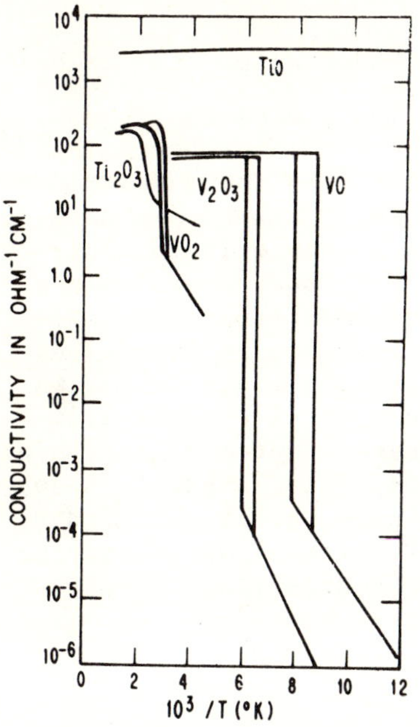

Fig. 1. Conductivity versus temperature for several transition metal oxides (after Morin, reference 1).

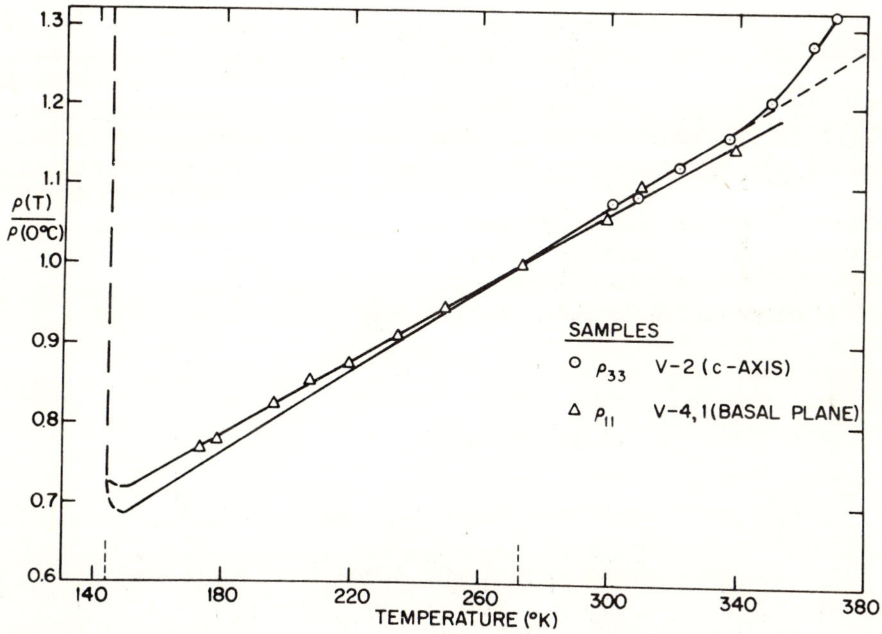

Fig. 2. Resistivity versus temperature for two directions in single crystal V_2O_3.

range the resistivity is seen to be a linear function of temperature with a positive coefficient and is calculated to be:

$$\rho(T)/\rho(0^{o}C) = [0.36 + 2.3 \times 10^{-3}\ T(^{o}K)] \tag{2}$$

The characteristic metallic resistivity shows anomalous behavior at higher temperatures, as seen in Figure 3. Both the c-axis and basal plane resistivities show a rapid increase with temperature in the region near 525°K, but remain monotonically increasing up to the highest temperatures. This result is in sharp contrast with that reported by Foex (17). In his measurements on sintered bars of V_2O_3 there is a similar rise in resistivity with temperature until a maximum is reached near 525°K. With further increase in temperature his data show a decrease in resistivity which is linear on a log ρ versus 1/T plot and yield an activation energy for conduction of 0.04 eV. The absolute resistivities of our single crystal samples are about 10^{-3} ohm-cm in this range compared to about 2×10^{-2} ohm-cm for the sintered material. We may also note that the anisotropy ratio for our crystals has increased only to a value of 1.1 at 625°K.

This high temperature anomaly in the resistivity of V_2O_3 is found to be accompanied by an anomaly in specific heat (11), a marked change in magnetic susceptibility (11), and an anomalous change in expansion coefficient (11).

The pressure effect on the room temperature resistivity for the two crystal orientations is shown in Figure 4. The decrease of resistance with pressure is similar to that observed in several other metals (18). It is seen that the anisotropy in resistivity is decreased by pressure, an effect observed by Bridgman (18) for metals such as Zn, Cd and Sn. The initial pressure coefficients are:

$$\left(-\frac{1}{\rho_{33}}\frac{\partial\rho_{33}}{\partial P}\right) = (7.6 \pm 0.2) \times 10^{-6}\ \text{bar}^{-1}$$

$$\left(-\frac{1}{\rho_{11}}\frac{\partial\rho_{11}}{\partial P}\right) = (14.0 \pm 0.2) \times 10^{-6}\ \text{bar}^{-1} \tag{3}$$

B. Transition from metal to semiconductor.

The temperature of the transition from metal to semiconductor in V_2O_3 has been variously reported to range from 168°K in sintered materials to 140°K in single crystals. The measurements on our crystals indicated a transition at atmospheric pressure to occur at 140 ± 2°K. The reverse transition on heating was less well defined because the samples usually fractured, but a best estimate would be

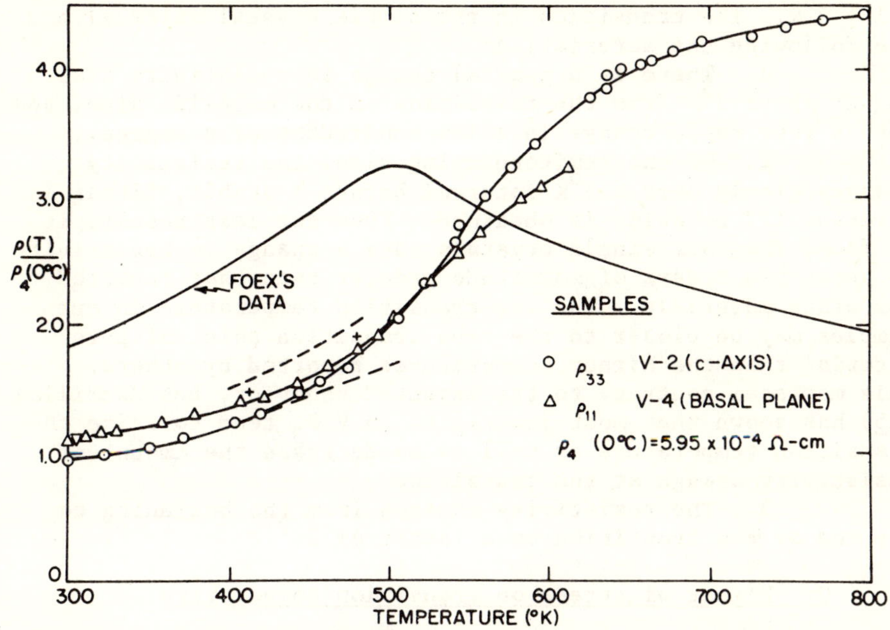

Fig. 3. High temperature resistivity of V_2O_3, showing anomalous behavior near 525°K. Data by Foex (17) on sintered bars are also included.

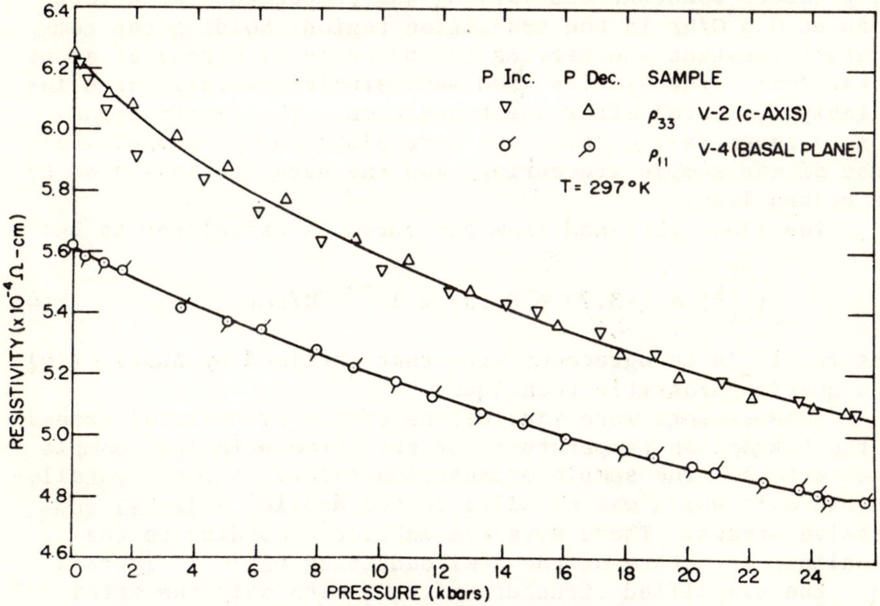

Fig. 4. Resistivity versus hydrostatic pressure for oriented samples of V_2O_3.

151 ± 3°K. The transition in the single crystal material had the following characteristics:

1. There is a gradual change in resistivity starting at about 1°K from the transition on the metallic side, and then a very rapid change in a few hundredths of a degree.

2. On the semiconducting side, the resistivity changes slowly over a 5°K interval before a stable, linear ln ρ versus 1/T behavior is obtained. From the last result, it is found that our single crystals have a change in resistivity by about two orders of magnitude greater than that reported for other materials. The low transition temperature in our samples may be closer to the true transition point of pure material than the higher temperatures reported by others. This may seem contrary to the expected behavior, but MacMillan (15) has shown that most impurities in V_2O_3 tend to raise the transition temperature as well as to decrease the amount of resistivity change at the transition.

3. The resistivity changes from the beginning to the end of the transition by a factor of 10^7.

C. Effect of stress on transition temperature.

Figure 5 summarizes the results of the effect of hydrostatic pressure on the metal to insulating transition temperature. The data were taken by two techniques: holding the pressure constant and varying the temperature at a rate of about 0.5°C/hr in the transition region; holding the temperature constant and varying the pressure at a rate of about 10 bars/min. The samples used were single crystals which invariably fractured after one transition. The return transition to the metallic state was more sluggish, presumably because of the sample fracturing, and the data are indicated by the broken line.

The slope obtained from the curve is calculated to be:

$$\left(\frac{dT_t}{dP}\right) = (-3.78 \pm 0.15) \times 10^{-3}\ {}^{o}\mathrm{K/bar} \qquad (4)$$

This result is in agreement with that obtained by Austin (19) by a quasi-hydrostatic technique.

Measurements were made of the effect of uniaxial stress on the transition temperature for the three principal sample orientations. The sample orientation refers to the crystallographic axis which was parallel to the applied uniaxial compressive stress. These axes are labeled according to the monoclinic structure of the semiconducting phase. Figure 6 shows the simplified structure of V_2O_3 with only the metal ions indicated and shows the relation between the monoclinic axes and the axes of the trigonal metallic phase (referred to

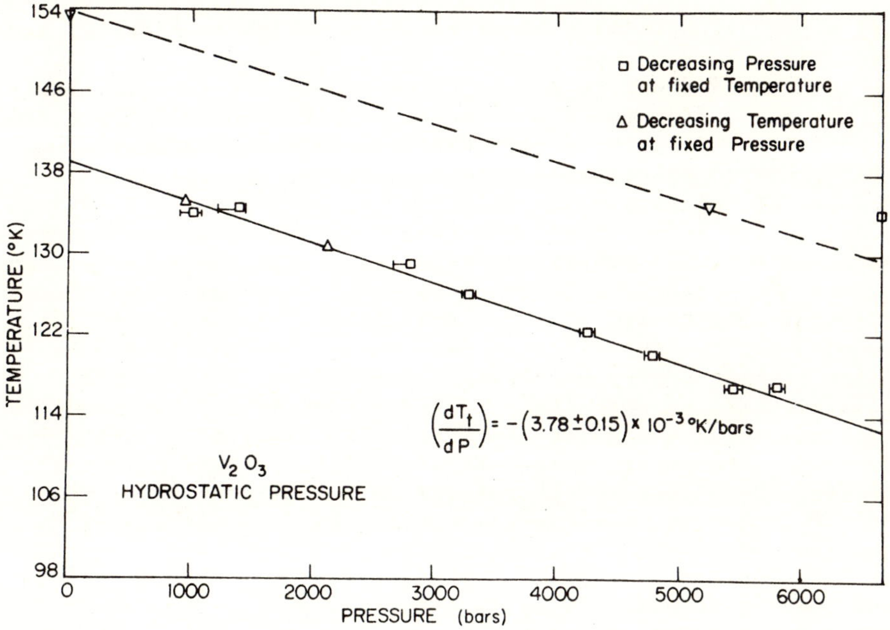

Fig. 5. Transition temperature versus hydrostatic pressure.

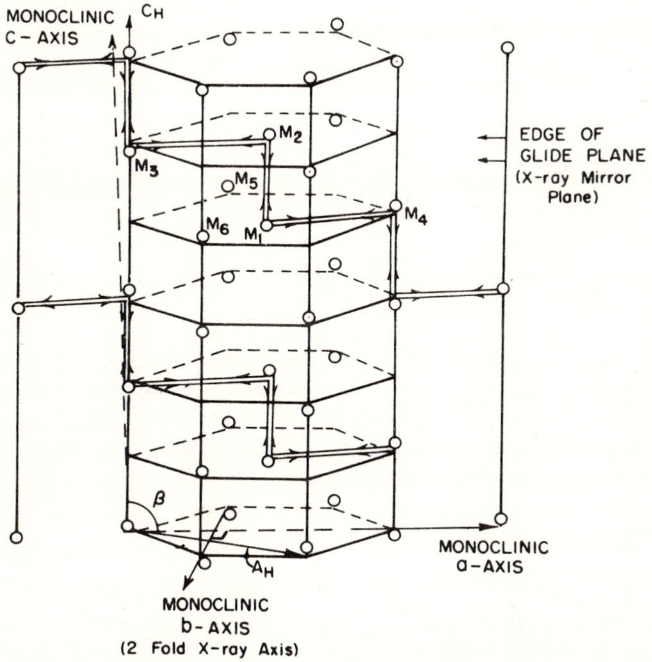

Fig. 6. Simplified structure of V_2O_3 showing only vanadium ion positions.

the hexagonal system). The results are shown in Figures 7, 8, 9.

Most of the data points recorded are for cooling transitions, i.e., from metal to insulator, since this is the first transition undergone by each sample. For those samples in which several transitions were obtainable, the different data points for the sample are indicated by the symbol near the points: the subscript is the number of the sample, and the superscript indicates the order in which the transitions occurred. Generally, it was found that the application of uniaxial compression to the a-axis samples allowed these samples to pass through several transitions without fracturing. A few b-axis samples behaved in this manner, but no c-axis samples survived more than one transition even with stress applied. A square surrounds the points which are deemed unreliable because the sample was severely fractured upon removal from the apparatus. While the heating transition from semiconductor to metal was less well defined, all the data obtained indicated that the hysteresis in the transition remained constant, independent of stress, and that the heating and cooling curves were separated by 10-12°C. The slopes of the two curves are therefore shown parallel. The dashed curve in each figure is the hydrostatic pressure data.

From the figures above, the stress coefficients for each orientation are calculated:

$$\left(\frac{dT_t}{dS}\right)_a = -(6.8 \pm 0.5) \times 10^{-3}\ ^{o}K/bar \tag{5}$$

$$\left(\frac{dT_t}{dS}\right)_b = -(4.1 \pm 0.5) \times 10^{-3}\ ^{o}K/bar \tag{6}$$

$$\left(\frac{dT_t}{dS}\right)_c = -(0.5 \pm 0.5) \times 10^{-3}\ ^{o}K/bar \tag{7}$$

To make these coefficients directly comparable to the hydrostatic coefficient, the sign of the stress is chosen to be positive for a uniaxial compression. The most striking result of these experiments is the large difference between the c-axis coefficients and the others.

D. Semiconducting region.

Accurate measurements of the transport behavior in the semiconducting region were difficult to obtain because the single crystals fractured when cooled through the

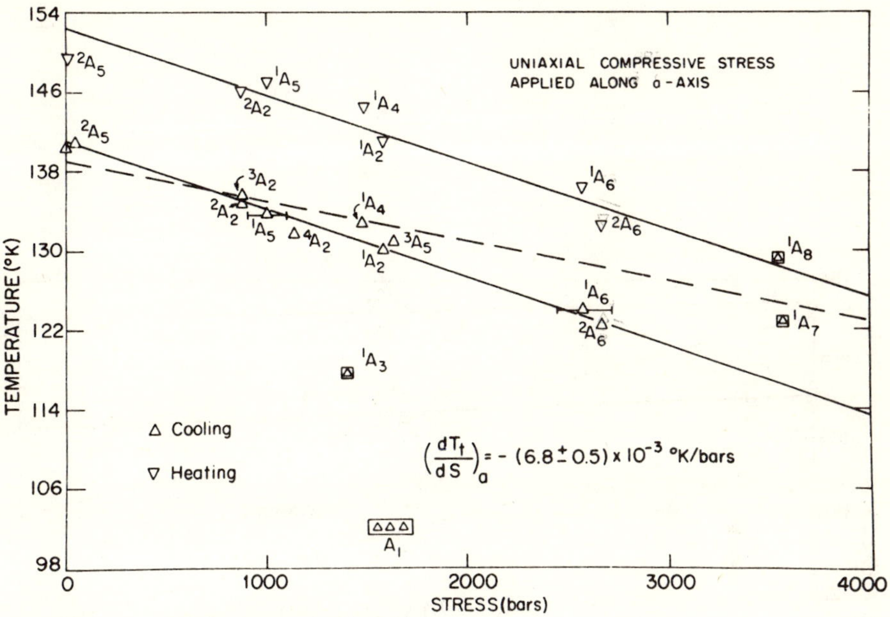

Fig. 7. Transition temperature versus uniaxial stress along the a-axis.

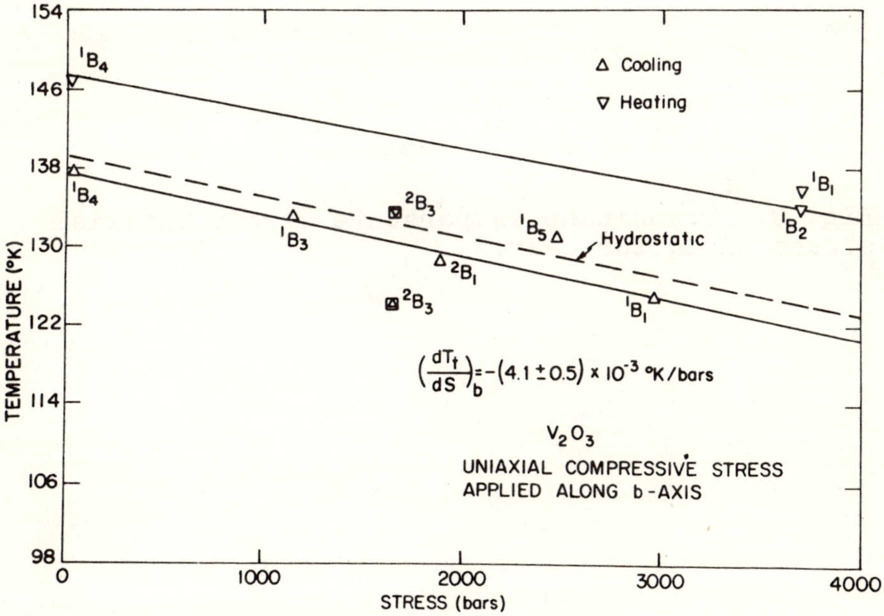

Fig. 8. Transition temperature versus uniaxial stress along the b-axis.

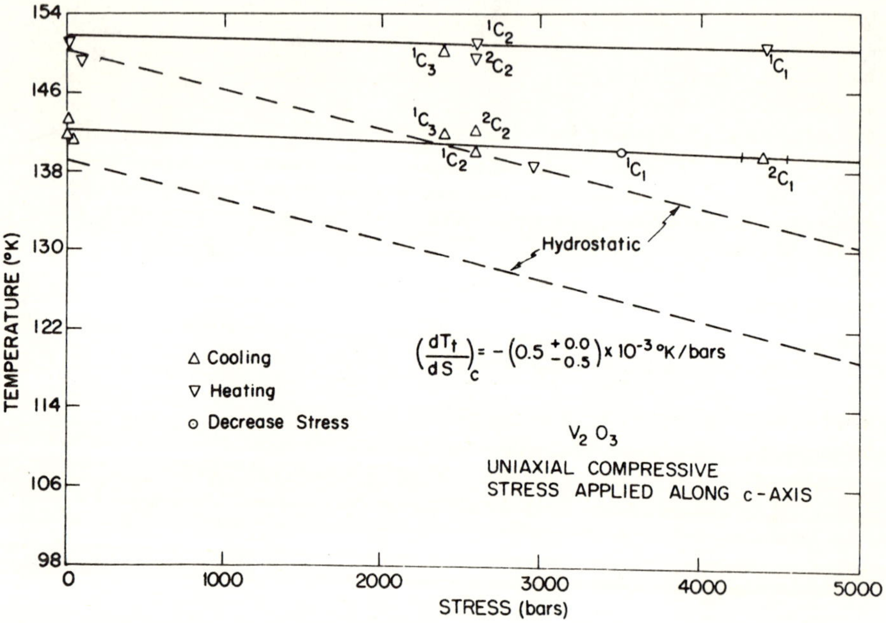

Fig. 9. Transition temperature versus uniaxial stress along the c-axis.

transition. When it was found that the application of uniaxial stress could preserve the crystal, several measurements were made of the effect of uniaxial stress on the resistivity. The effect of hydrostatic pressure was also measured by an arrangement in which the sample in the pressure vessel was stressed uniaxially by the force of a small spring. In the two terminal measuring technique, the unavoidable effects of contact resistance reduce the accuracy of the absolute values of resistivity, but several measurements in different crystals gave fairly consistent results.

Figure 10 shows two curves of resistivity versus temperature from which the activation energy for conduction is calculated. The data for sample C_3 were taken with a constant stress of 2,550 bars applied along the c-axis. This sample showed some longitudinal cracks. The data for sample B_5 were taken with a uniaxial stress of 2,480 bars applied along the b-axis and remained intact. The resistivity is measured parallel to the stress axis. The activation energy, q, is calculated from the slope of the curve and is defined by the relation:

$$\rho(T) = A\rho(T_o) \exp (q/kT) \tag{8}$$

where $\rho(T)$ is the resistivity at temperature T and A is a constant. The results obtained are:

$$\begin{aligned} q_c &= 0.12 \text{ eV} \qquad \rho_c(132^oK) = 0.5 \times 10^4 \text{ ohm-cm} \\ q_b &= 0.18 \text{ eV} \qquad \rho_b(132^oK) = 1.3 \times 10^4 \text{ ohm-cm} \end{aligned} \tag{9}$$

The constant applied stress has a very different effect on the transition temperatures of the two samples and this could also account for the difference in the resistivity and activation energy, although there are insufficient data to bear this out.

Figure 11 shows the results of several attempts to measure the effect of uniaxial stress and hydrostatic pressure on the resistivity of the single crystals in the semiconducting phase. There is a large scatter in the data, but a reasonably accurate hydrostatic pressure coefficient and stress coefficient for the b-axis were obtained. These coefficients, α, are defined by the relation:

$$\rho(S,T) = A\ \rho(0,T_o) \exp \left(\frac{q + \alpha S}{kT}\right) \tag{10}$$

where S is the stress or hydrostatic pressure. The coefficients obtained from these data are:

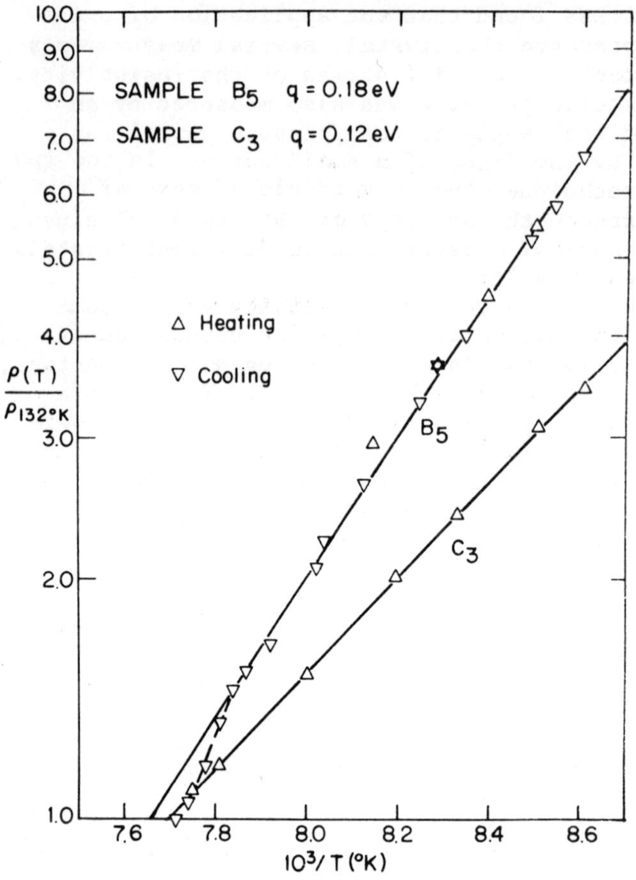

Fig. 10. Resistivity versus temperature for the semiconducting phase.

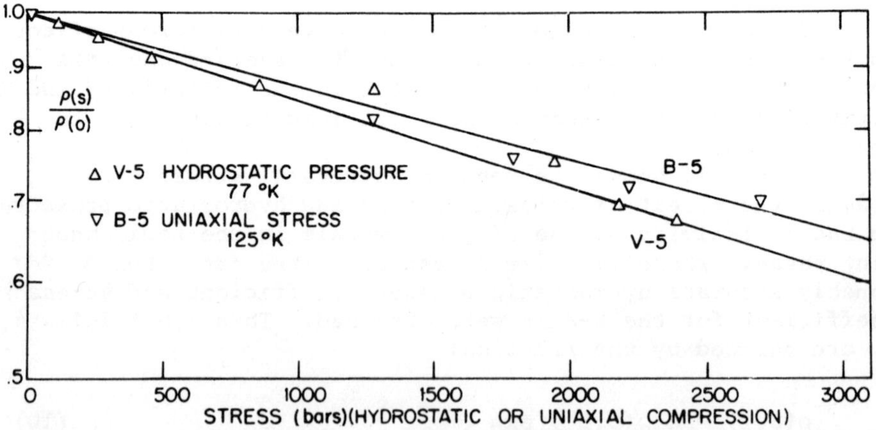

Fig. 11. Resistivity versus hydrostatic pressure and uniaxial stress for the semiconducting phase.

$$\alpha(\text{uniaxial})_b = -1.5 \times 10^{-6} \text{ eV/bar} \tag{11}$$

$$\alpha(\text{hydrostatic}) = -1.1 \times 10^{-6} \text{ eV/bar} \tag{12}$$

A similar measurement has been made by Austin (19) using a quasi-hydrostatic apparatus and the less sensitive technique of holding pressure constant and varying the temperature. His coefficient is almost a factor of five greater and is given as approximately -5×10^{-6} eV/bar.

III. Optical measurements.

In these measurements, a single crystal slice was cut to 1/2 cm^2 cross-section and then polished to a thickness of 25 microns. The crystal was kept from fracturing at the low temperature by applying a uniaxial stress through a spring while the sample was clamped between two 1/4" thick polished KBr plates. The holder then fitted in the cold finger of an optical nitrogen dewar. The holder was movable in the dewar so that in one position the light path went through the KBr plates and the sample, while in the other position, the light passed only through the KBr plates. With this arrangement, the relative optical transmission through the sample could be monitored while the sample was slowly cooled.

A Perkin-Elmer spectrometer with KBr and LiF prisms was used with a globar source and thermocouple detector. With KBr windows in the dewar and spectrometer, the system was usable to a wavelength of 20μ.

At room temperature, in the metallic phase, no transmission was observed in the range from 20 to 0.2μ. The sensitivity was limited by the large amount of light scattered around the sample through the KBr plates. The scattered light intensity was about 0.1% of the light incident on the sample.

As the sample was cooled, no transmission was obtained until the temperature was lower than the transition temperature. Figure 12 shows the results of the transmission measurement at 77°K. Because of the large amount of scattered light, shown as the dashed curve, it was difficult to calculate the absolute absorption coefficient α, as given by the relation:

$\frac{I}{I_o} = e^{-\alpha x}$ where x is 25μ for our sample. A rough estimate is that α extends from ~1000 at the transmission maximum to ~3000 where the scattered light masks the transmission. No correction was made for the reflectivity since it has not

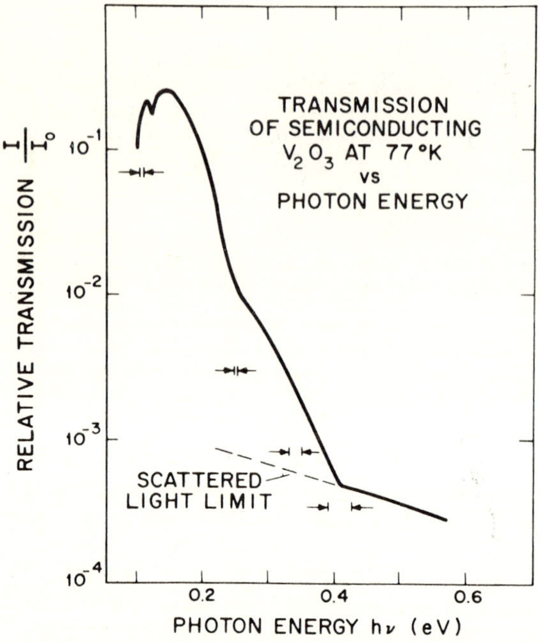

Fig. 12. Relative transmission versus photon energy for a sample 25 microns thick, at 77°K.

been measured for the semiconducting material.

As seen in Figure 12, the absorption edge is not sharp, but extends over several tenths of an electron volt. The calculated resolution of the instrument over the measured range is indicated on the figure. It is seen that this could not account for the large spread in energy. From the data on this one sample it is useless to speculate on the cause of the spreadout edge. However, from the band structure contemplated for this material, it is not surprising that the edge should be different from the wide band semiconductors. We have taken the energy gap to be about 0.1 eV from this measurement because the absorption coefficient is fairly large even at this energy. Free carrier and impurity effects could account for the high absorption coefficient at lower energies. The fine structure at low energies was reproducible in this sample, but no further measurements in other samples have been made for comparison.

IV. Discussion.

The transport behavior of V_2O_3 can be separated into three distinct temperature regions: the semiconducting range below $140^{o}K$, the metallic range between $140^{o}K$ and about $525^{o}K$ and the high temperature metallic range. The metallic behavior is most easily understood in terms of a narrow conduction band, presumably formed from the d-electron orbitals of the vanadium ions; and we can obtain some quantitative characteristics of this band from the data presented here. The semiconducting phase and the transition to a metal are less well understood. It is possible that the lack of conduction comes about from the trapping of the conduction electrons on cation sites, or in cation pair bonds, whereas a band model would suggest that an energy gap forms between a filled and an empty band of Bloch electron states. In the latter case, the activation energy for conduction refers only to the production of free carriers, whereas in the former there is also an activation energy for the mobility.

A. The metallic region.

A qualitative idea of the conduction in the metallic phase is afforded by an examination of the driving force for the transition from semiconductor to metal. The discontinuity in volume and entropy at the transition are associated with a latent heat L, given by

$$L = T_t \, \Delta S$$

where ΔS, the total change in entropy between the two phases, is composed of electronic terms, lattice terms, and spin ordering terms

$$\Delta S = \Delta S_{electronic} + \Delta S_{lattice} + \Delta S_{spin}$$

The latent heat may be calculated from the Clausius-Clapeyron relation for a first order transition

$$\left(\frac{dP}{dT}\right)_{T_t} = \frac{L}{T_t \Delta V}$$

That the transition is first order has been established by the measurements of lattice constant change at the transition of Minomura and Nagasaki (10). Their volume change, our pressure coefficient and transition temperature combine to give

$$L = 1020 \text{ cal./mole} = 0.044 \text{ eV per } V_2O_3 \text{ formula unit.}$$

Indirect arguments suggest that magnetic effects are not the main cause. From neutron diffraction measurements Abrahams has put an upper limit of 0.2 Bohr magnetons on the local moment near T_o. Although we cannot calculate the spin ordering contribution ΔS_{spin} directly, we may estimate it by comparison with the work of Bean and Rodbell on MnAs. In this material, a first order magnetic transition occurs, and these workers have calculated and measured the spin ordering entropy change to contribute about 234 cal/mole to the latent heat. The local moment in MnAs is larger than the upper limit given to that in V_2O_3 and therefore this contribution to the latent heat is probably the upper limit to that expected in V_2O_3. We may thus conclude that the ΔS_{spin} term and therefore the magnetic contribution to the transition in V_2O_3, is not as important as other factors.

The electronic entropy change may be more easily estimated. We may assume the electronic entropy in the semiconducting phase to be sensibly zero, because of the small number of carriers. The entropy of the carriers in the metallic phase may be estimated by assuming the simplest possible model of a spherical Fermi surface and a parabolic energy band. Then

$$|\Delta S_{electronic}| = \frac{z\pi^2 k^2 T_t}{2E_F}$$ where E_F is the Fermi energy measured from the bottom of the band, and z is the number of conduction electrons per molecule.

If we were to ignore the contribution of $\Delta S_{lattice}$, then

$L = T_t \, \Delta S = \frac{z\pi^2 k^2 T_t}{2E_F}$. Insertion of the values for L and T_t leads to an E_F of 0.07 eV. This value must be corrected for lattice entropy changes, which are taken to include contributions from the changes in the phonon spectrum, changes in the lattice symmetry and changes in the states of electrons other than conduction electrons. There are in the literature arguments for entropy changes of the order of k per molecule, of either sign (21). If it is indeed of this order the Fermi energy will be changed by about 0.02 eV. The exact number is not important, for it does appear that the bandwidth in the metallic conduction band is very narrow, say of the order of 0.1 eV at most. We can use this value in the formula

$$E_F = \frac{\hbar^2}{2m^*} (3\pi^2 N)^{2/3}$$

where N is the carrier concentration, to calculate an effective mass of 50 m_o, and the conductivity data quoted earlier to give a mobility and relaxation time:

$$\mu = \frac{\sigma}{Ne} = 0.2 \text{ cm}^2/\text{volt-sec.}$$

and $$\tau = \frac{\mu m_o}{e} \times \frac{m^*}{m_o} = 6 \times 10^{-15} \text{ sec.}$$

The relaxation time is of the order found in typical metals. The mobility, however, is much lower than that in metals such as copper ($\mu \sim 30$ cm^2/volt-sec). The carriers in V_2O_3 metal are therefore slower because of their large effective mass, and therefore have a smaller mean free path in the crystal than in typical metals. This is seen from the following calculation of carrier velocity at the Fermi surface U_F and the mean free path ℓ:

$$U_F = \left(\frac{2E_F}{m^*}\right)^{\frac{1}{2}} = 3 \times 10^6 \text{ cm/sec.}$$

$$\ell = \tau U_F = 2 \text{ Å.}$$

For V_2O_3, a mean free path of 2 Å is of the order of one lattice constant. In copper, the corresponding number is about 100 lattice constants. This, therefore, is the extreme narrow band limit for the concept of band states. In fact, the uncertainty in energy of the carriers with the relaxation time given is ~ 0.1 eV which is a large fraction of the bandwidth of 0.2 eV.

B. The semiconducting region.

The activation energy for conduction lies between 0.12 and 0.18 eV. Our optical measurements indicate that the band gap is about 0.1 eV so that, in an intrinsic material, the activation energy for the production of carriers would account for 0.05 eV of the total. The remaining large fraction of the activation energy is presumably related to the hopping motion of localised electrons and should lead to a mobility substantially lower than in the metallic region. That hopping may occur in this substance is not surprising in view of the narrowness of the bands even in the metallic state.

C. The semiconductor to metal transition.

We shall first recount the results we shall use from the statistical theory described by Adler and Brooks. They showed that an abrupt semiconductor to metal transition could occur where the energy gap depended on carrier excitation after the form $E_g = E_g^o - \beta n$. Two models possibly applicable to V_2O_3 were considered: one where the energy gap between narrow bands arises from antiferromagnetic order, and another where the cause is a crystal distortion of the appropriate type and magnitude. They found that

(1) the ratio of zero temperature energy gap to transition temperature, E_g^o/kT_t, lay between 5 and 8.

(2) $$\frac{d\,\ell n\, E_g^o}{dX} = \frac{d\,\ell n\, T_t}{dX}$$ where X is any external parameter.

(3) the number of carriers just before the transition was $n(T_t) = 2kT_t/\beta$ and the size of the gap at this point, $E_g(T_t) = E_g^o - 2kT_t$. Since β was shown also to be about $\frac{4E_g^o}{N}$, where N is the number of atoms per cc

$$\frac{n(T_t)}{N} = \frac{kT_t}{2E_g} = \frac{1}{10} \text{ to } \frac{1}{16} \,.$$

Thus the carrier density changed by at most a factor of 16 at the transition.

We can test the applicability of this statistical model to V_2O_3 by comparing relations (1) and (2) with experiment.

(1) From the optical measurement results, we take $E_g^o = 0.1$ eV. Using $T_o = 140^oK$, we find $E_g^o/kT_t = 8$.

(2) We equate $\frac{d\,E_g^o/2}{dX} = \frac{dq}{dX}$. The quantity q contains both the activation energy for carrier production, $E_g^o/2$, and the activation energy for hopping, E_H. Experimental evidence on NiO (16) indicates that the pressure dependence of E_H is an order of magnitude less than that measured for q in V_2O_3. Also, Adler (22) has made a rough estimate of the pressure coefficient of the polaron hopping energy, starting from Holstein's relations for "small" polarons. He finds that the predicted pressure coefficient is much smaller than those measured here for

$$q = \frac{E_g^o}{2} + E_H.$$

Thus in testing relation (2), we may safely set

$\frac{d\,E_g^o/2}{dX} = \frac{dq}{dX}$. Table 1 shows a comparison of these variations for applied hydrostatic pressure and two directions of uniaxial stress. The equality holds remarkably well.

Because relations (1) and (2) are deduced specifically from the Adler-Brooks statistical model, it is unlikely that the good agreement with experiments on V_2O_3 is fortuitous. Assuming then the validity of the model, we deduce from relation (3) that the carrier density change at the transition is just over one order of magnitude. The remaining six orders of magnitude change in conductivity experimentally measured must be caused by an increase in mobility on the metallic side. On the semiconductor side the difference between the measured $\frac{E_g^o}{2}$, i.e., 0.05 eV and q = 0.12 to 0.18 eV (equation (9)) is the activation energy for hopping, which reduces the mobility to a value far below that appropriate for unbound electrons, even of very heavy mass. We speculate that the large carrier density increase at T_t leads to greatly increased screening of the electrons from the ions, which eliminates the mobility activation energy entirely, and converts hopping electron conduction by a few electrons to conduction by N electrons per cc moving in a narrow band. We shall next use this information and the other experimental results to draw more conclusions about the conduction process in the three main regions.

D. Discussion of other models for the transition and for the conduction processes.

The results obtained permit a comparison of several other models that have been proposed to explain the semiconductor to metal transition in V_2O_3 and the main features of the conduction process.

(1) Morin had suggested the Slater model for V_2O_3 with T_t as the Néel temperature. The difficulties of this model are (a) the magnetic ordering transition is in doubt, (b) even if it occurs, the associated latent heat seems to be much less than that observed, (c) the crystal distortion is not incorporated (d) the disappearance of long range order does not mean there is no short range order which for these electrons of small mean free path can cause a band splitting.

(2) Mott, on the other hand, had argued that increased shielding would force a semiconductor to metal transition at a critical interatomic separation. Unfortunately, the lattice constant at quite low temperature in the semiconducting phase becomes smaller than it is in the high temperature metallic phase, so that this explanation will not serve without elaboration. From our earlier discussion, we suggest that the increased shielding comes not from an increase in the density of the crystal but from a band collapse (with associated lattice distortion). This is not what Mott described but the philosophy is similar.

(3) Goodenough emphasized the role of direct d electron cation-cation interaction and of critical ionic separations. He explained the transition to the semiconducting phase as a pairing of atoms in the basal plane. It is to be noted immediately from Figure 6 that the lattice distortion at T_t described in our introduction includes just such an atom pairing. It is also plausible that the effect of stress applied in the basal plane will be much greater in displacing T_t or changing E_g than stress applied along the c-axis. An extension of this model explains the high temperature anomaly as a disappearance of similar c-axis pairing. We note immediately that this envisages a large anisotropy in conductivity between 140°K and 525°K, and this is not found.

Thus it appears that none of the previous models are adequate in themselves, but also elements of most of them have been incorporated into the model suggested in this report.

E. The high temperature anomaly; spin disorder scattering.

The behavior of the resistivity at high temperatures is illustrated in Figure 3. We shall consider together the increase near 525°K, and the large residual resistivity indicated by Equation 2. Some of this large residual resistivity is probably due to the large concentration of defects inherent in crystals grown by flame fusion techniques. Another possibility is spin disorder scattering as has been discussed by deGennes and Friedel (23). Adler (22) has made a simple estimate of this effect based on the scattering relation given by Gibson (24) and assuming an exchange energy of 0.06 eV, (the value appropriate to Ti_2O_3). This temperature independent resistivity is calculated to be:

$$\rho_{so} = 1.4 \times 10^{-4} \text{ ohm-cm}$$

compared with the measured value in (2) of 2.2×10^{-4} ohm-cm.

Spin disorder scattering may also be responsible for the anomalous change in the resistivity of the metallic phase above 525°K. The shape of the resistivity curve seen in Figure 3 is very similar to the resistivity change calculated by deGennes and Friedel for a spin disorder scattering mechanism at a magnetic ordering transition. The magnitude of this resistivity increase has been calculated by Adler exactly as above except that the exchange energy assumed is that calculated from the theory discussed in (22), which gave $E_g^o = 0.4$ eV. The spin disorder resistivity was found to be

$$\rho_{so} = 12 \times 10^{-4} \text{ ohm-cm}$$

By extrapolating the linear region of the curve in Figure 3 to higher temperatures, we find that the added resisitivity above 525°K is:

$$\rho_A = 12 \times 10^{-4} \text{ ohm-cm}$$

This close agreement may certainly be accidental because of the crude band model used and the rough quantitative values used. However, the high temperature anomaly has been interpreted by other workers (25) to be caused by an antiferromagnetic to paramagnetic transition as evidenced by their magnetic susceptibility measurements. Such a transition may be a contributing part of the band rearrangement similar to the one which occurs at the semiconductor to metal transition.

V. Conclusions.

The main conclusion of this work is that the semiconductor to metal transition may be described by the statistical model of Adler and Brooks. This model explains that part of the conductivity discontinuity ascribable to the change in carrier density. The remaining, and larger, fraction of the conductivity jump corresponds to a change from carrier transport by hopping motion in the semiconducting phase to transport in a narrow band in the metallic phase.

Table 1.

$\frac{d \ln T_t}{dP} = -2.6 \times 10^{-5}\ \text{bar}^{-1}$	$\frac{d \ln E_g^o}{dP} = -2.2 \times 10^{-5}\ \text{bar}^{-1}$
$\left.\frac{d \ln T_t}{dS}\right)_b = -2.8 \times 10^{-5}\ \text{bar}^{-1}$	$\left.\frac{d \ln E_g^o}{dS}\right)_b = -3.0 \times 10^{-5}\ \text{bar}^{-1}$
$\left.\frac{d \ln T_t}{dS}\right)_c = < 0.3 \times 10^{-5}\ \text{bar}^{-1}$	——

References.

1. See, for example, Goodenough, J.B., Magnetism and the Chemical Bond, Interscience Publishers, New York (1963); or Morin, F.J., Semiconductors, Hannay (editor), Reinhold Publishers, New York (1959); Transition Metal Compounds, Schatz (editor), Gordon and Breach, New York (1964).
2. See, for example, Mott, N.F., Phil. Mag., 6, 287 (1961); Heikes, R.R. and Johnston, W.D., J. Chem. Phys. 26, 582 (1957); DeBoer, J.H. and Verwey, E.J.W., Proc. Phys. Soc., A49, 59 (1937); Yamashita, J. and Kurosawa, T., J. Phys. Soc. Japan, 15, 802, 1211 (1960).
3. See, for example, Slater, J.C., Phys. Rev. 82, 538 (1951); Morin, F.J., Phys. Rev. Letters 3, 34 (1959).
4. See Mott, loc. cit.; Kohn, W., Phys. Rev. 133, 171A (1964); Hubbard, J., Proc. Roy. Soc. A276, 238 (1963); A281, 401 (1964); Goodenough, J.B., Phys. Rev., 120, 67 (1960); J. Phys. Chem. Solids 6, 287 (1958); and references under 3.

5. See Mott, Kohn, and Hubbard, references under 3. Also, Anderson, P.W., Solid State Physics, Vol. 14, Seitz and Turnbull (editors), Academic Press, New York (1963).
6. Holstein, T., Ann. of Phys. 8, 325, 343 (1959); for several papers on the "small" polaron, see Polarons and Excitons, Kuper and Whitfield (editors), Oliver and Boyd, London (1963).
7. Adler, D. and Brooks, H., this volume.
8. Mott, N.F., Proc. Phys. Soc., A62, 416 (1949).
9. Warekois, E.P., J. App. Phys. Supp. 31, 346S (1960).
10. Minomura, S., and Nagasaki, H., J. Phys. Soc. Japan, 19, 131 (1964).
11. Foex, M.; Goldsztaub, J.; Jaffrey, R.; Lyand, R.; Wey, R. and Wucher, J.; J. des Recherches du C.N.R.S., Bellevue (Paris) 21, 237, (1952); Carr, P.H.; and Foner, S., J. App. Phys. Supp., 31, 344S (1960).
12. Abrahams, S.C., Phys. Rev. 130, 2230 (1963); Paoletti, A. and Pickart, S.J., J. Chem. Phys. 32, 308 (1960).
13. Jones, E.D., Phys. Rev., 137, A978 (1965).
14. Anderson, C.T., J. Am. Chem. Soc., 58, 564 (1936).
15. MacMillan, A.J., Lab. Ins. Res., Tech. Rept. 172, M.I.T. (1962).
16. Feinleib, J., Tech. Rept. No. HP-11, Gordon McKay Laboratory, Harvard University (1963).
17. Foex, M., reference 11.
18. For a recent review, see Paul, W., High Pressure Physics and Chemistry, Bradley (editor), p. 299, Academic Press, London (1963).
19. Austin, I.G., Phil. Mag. 7, 961 (1962).
20. Bean, C.P., and Rodbell, D.S., Phys. Rev. 126, 104 (1962).
21. Mott, N.F., reference 2; Bean, C.P. and Rodbell, D.S., reference 21.
22. Adler, D., Tech. Rept. ARPA-12, Division of Engineering and Applied Physics, Harvard University (1964).
23. deGennes, P.G., and Friedel, J., J. Phys. Chem. Solids 4, 71 (1958).
24. Gibson, J.B., J. Phys. Chem. Solids 1, 27 (1956).
25. Wucher, J., Compt. Rend. 241, 288 (1955); Teranishi, S., and Tarama, K., J. Chem. Phys. 27, 1217 (1957).

